Advanced Nanofibrous Materials Manufacture Technology Based on Electrospinning

Editors

Yanbo Liu

School of Textile Science and Engineering
Wuhan Textile University
Wuhan City
P. R. China

Ce Wang

Alan G. MacDiarmid Institute
College of Chemistry
Jilin University
Changchun
P. R. China

CRC Press
Taylor & Francis Group
Boca Raton London New York

CRC Press is an imprint of the
Taylor & Francis Group, an **informa** business
A SCIENCE PUBLISHERS BOOK

Cover credit

- The upper left image by Dr. Yanbo Liu is the schematic of single-needle electrospinning process
- The upper right image produced by Miss Sumei Ma is the illustration of meshing for five-needle electrospinning process
- The lower image is the nephogram of electric field intensity distribution in five-needle electrospinning process which was simulated by Mr. Dongyue Qi.

CRC Press
Taylor & Francis Group
6000 Broken Sound Parkway NW, Suite 300
Boca Raton, FL 33487-2742

First issued in paperback 2020

ISBN-13: 978-1-4987-8112-1 (hbk)
ISBN-13: 978-0-367-78005-0 (pbk)

Library of Congress Cataloging-in-Publication Data

Names: Liu, Yanbo, 1965- editor. | Wang, Ce (Professor), editor.
Title: Advanced nanofibrous materials manufacture technology based on electrospinning / editors, Yanbo Liu, School of Textile Science and Engineering, Wuhan Textile University, Wuhan City, P. R. China, Ce Wang, Alan G. MacDiarmid Institute, College of Chemistry, Jilin University, Changchun, P. R. China.
Description: Boca Raton, FL : Taylor & Francis Group, LLC, CRC Press is an imprint of Taylor & Francis Group, an Informa Business, [2018] | Includes bibliographical references and index.
Identifiers: LCCN 2018041060 | ISBN 9781498781121 (hardback)
Subjects: LCSH: Nanofibers--Design and construction. | Electrospinning.
Classification: LCC TA418.9.F5 A385 2018 | DDC 620.1/9--dc23
LC record available at https://lccn.loc.gov/2018041060

Visit the Taylor & Francis Web site at
http://www.taylorandfrancis.com

and the CRC Press Web site at
http://www.crcpress.com

Preface

Electrospinning is an electric driven process where the polymer solution is drawn by electric field force forming nanofibers on the receiver with or without dominant fiber alignment direction by solvent evaporation and polymer solidification. Hundreds and thousands of researchers have dedicated their energy and time in this area and a large amount of research has been carried out and published worldwide regarding principles, process control, parameter optimization, product development and applications, as well as development and manufacture of equipment. However, currently only a few books are available regarding electrospinning technology, none of which can be used as a textbook-like reference. This book is suitable for readers at different levels from the student beginners to the professionals such like engineers, researchers, professors, etc., whose research areas range from electrospinning technology, nanofibers, nanomaterials, nanofiber-based composite nonwovens/ membranes, to nanofiber yarn and fabrics, and so on.

This book comprehensively addresses the advanced nanofiber manufacture technology based on electrospinning technology, covering the category, working principle, relationships between process parameters and the structure, morphology and performance of electrospun nanofibers and nanomaterials. The electric field intensity and distribution during electrospinning are also comparatively analyzed based on finite element analysis on both the needle electrospinning and needleless electrospinning. The methods for enhanced field intensity and uniform distribution are also discussed. Furthermore, the modification techniques for improved nanomaterial strength are also covered, aiming to provide the readers effective methods for the manufacture of stronger nanofiber or nanomaterial products. This book also deals with the development and applications of electrospun products, as well as recent advances and issues in electrospinning technology.

Possibly there be some errors or limitations in the research results, analyses, and opinions in the book due to the relatively short writing time. We thank readers for your kind understanding.

We also thank our family members for their kind understanding and immense support. We are thankful for the kind help and assistance from our students, colleagues and friends.

Yanbo Liu and Ce Wang
July 12, 2018

Contents

Chapter 1

Introduction to Electrospinning Technology

Guangdi Nie,[1,2] *Xiaofeng Lu*[2,*] and *Ce Wang*[2,*]

Introduction

Since the discovery of carbon nanotubes (CNTs) by Sumio Iijima in 1991, one-dimensional (1D) nanomaterials with unique physical and chemical properties have been extensively studied in various areas, such as sensors, catalysis, biomedicine, filtration, energy storage, electronic and optical devices (Lu et al. 2009, Greiner and Wendorff 2007, Li and Xia 2004a). Up till now, numerous synthetic strategies including electrospinning, chemical vapor deposition, liquid phase, and template-assisted methods have been developed for the preparation of 1D nanostructures in the forms of fibers/wires, belts, tubes, rods, rings, and spirals (Wei et al. 2017, Cavaliere et al. 2011). Among them, electrospinning seems to be currently the simplest and highly versatile technique to produce continuous nanofibers with complex architectures from a wealth of materials such as polymers, inorganic ceramics, and composites, etc. (Lu et al. 2016). It is also possible to generate single fibers and their ordered arrays by special electrospinning set-ups. Electrospinning is not only applied in laboratories, but is also increasingly being employed in industry, which has attracted widespread attention in the field of nanotechnology over the recent years (Greiner and Wendorff 2007).

Definition for Electrospinning and Nanofibers

Generally, nanomaterials can be divided into four categories, which are, zero, one, two, and three dimensions, among which 1D nanostructures with two dimensions constrained at the nanoscale have become research hotspots due to their superior

[1] Industrial Research Institute of Nonwovens & Technical Textiles, College of Textiles and Clothing, Qingdao University, Ningxia Road 308, Qingdao, 266071, P. R. China.
[2] Alan G. MacDiarmid Institute, College of Chemistry, Jilin University, Qianjin Street 2699, Changchun, 130012, P. R. China.
* Corresponding authors: xflu@jlu.edu.cn; cwang@jlu.edu.cn

functional characteristics (Yu et al. 2015). Significantly different from the short nanorods, nanofibers possess larger aspect ratios—more than 10 (Chen et al. 2007). In a narrow sense, the diameters of nanofibers fall in the range of 1~100 nm, while broadly speaking, fibers with diameters down to 1000 nm are also referred to as nanofibers (Ding and Yu 2011).

At present, electrospinning provides a facile and efficient approach for generating uniform fiber-shaped nanostructures with solid or hollow interiors that are ultralong in length, ultrafine in diameter, and diversified in composition (Li and Xia 2004a). Normally, the basic apparatus for electrospinning in the lab is assembled with three major parts (Fig. 1.1): a high-voltage power supply (DC), a spinning nozzle, and a grounded collector. Sometimes, a syringe pump is needed to guarantee the constant and controlled feed rate of precursor through the spinneret (Lu et al. 2016). In a typical electrospinning process, the pendent droplet of polymer solution or melt is stretched and deformed into a "Taylor cone" under external electric field, and once the electrostatic force exceeds the surface tension, the charged jet flow is sprayed from the tip of the conical structure followed by the continuous elongation and unstable whipping movement, resulting in the evaporation of solvent or the solidification of melt accompanied with the formation of nanofibers on the conductive collector (Wang and Lu 2011).

Till date, a variety of fiber assemblies involving non-woven fabrics, aligned arrays, yarns, patterned mesh, convoluted fibers, random three-dimensional (3D) architectures, and so forth have been achieved by changing the collector types (Teo and Ramakrishna 2006, Ding et al. 2009). Furthermore, by mixing additional components into the precursor solution or combining with thermal treatment, electrospinning also can be extended to the construction of porous, tubular, and heterogeneous 1D inorganic nanomaterials or nanocomposites (Mai et al. 2014, Wang et al. 2015). It is well known that both the electrospinning and annealing parameters play crucial roles in determining the morphology and composition of the final nanostructures (Niu et al. 2015, Hong et al. 2015, Wali et al. 2014, Peng et al. 2015a). For the former, the precursor solution (polymer type, inorganics content, solvent volatility, solution viscosity, and conductivity), applied voltage, collection

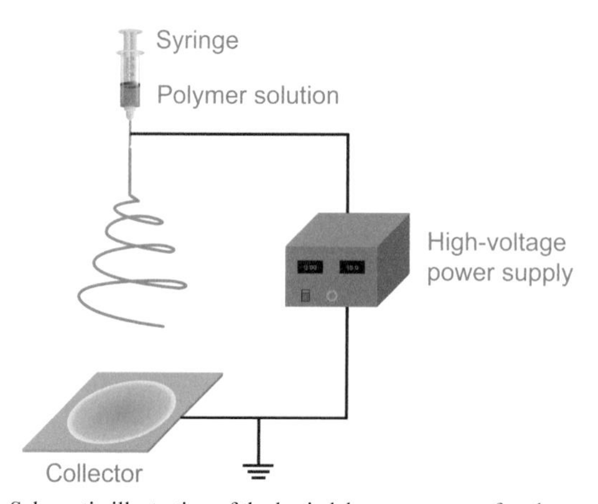

Fig. 1.1: Schematic illustration of the basic laboratory set-up for electrospinning.

distance, ambient temperature, and humidity are important factors, while the latter usually includes the heating rate, annealing temperature, atmosphere, and time.

Features of Electrospun Fibers and their Products

With the rapid development of nanoscale science and technology, the electrospun nanofibers (the general name for the 1D nanostructures fabricated by electrospinning technique) which exhibit the unique features of the nanostructured materials will find potential applications in biomedicine, filtration, protective clothing, catalysis, sensors, energy storage, electronic and optical devices (Lu et al. 2009, Greiner and Wendorff 2007, Li and Xia 2004a, Teo and Ramakrishna 2006).

Advantages

In the field of biomedicine, the electrospun nanofibers can simulate the structure and biological functions of natural extracellular matrix due to their smaller diameter compared to that of cells. Most of the tissues and organs of the human body are similar in form and structure to nanofibers, providing the possibility for the electrospun nanofibers to be used in tissue engineering and organ repair (Sill and Recum 2008, Jiang et al. 2015a). Some raw materials for electrospinning have favorable biocompatibility and biodegradability, which can serve as the built-in drug carriers that are easy to be absorbed by the human body (Hsu et al. 2014). Moreover, the excellent properties of the electrospun nanofibers, such as large specific surface area and porosity, endow them with advantages in biomedical applications (Agarwal et al. 2008).

It is universally acknowledged that the filtration efficiency of the fiber filter material will increase along with the decrease of the fiber diameter (Mei et al. 2013). In addition to the reduced diameter, the electrospun nanofibers also have the merits of small aperture, high porosity, and good homogeneity, making them a promising candidate used in gas filtration, liquid filtration, and individual protection (Choi et al. 2015, Jiang et al. 2016a).

For the nanoparticle catalysts, the obvious aggregation has an adverse impact on their utilization (Nie et al. 2013). Impressively, the electrospun fibrous substrate with flexibility not only acts as a template to disperse and recycle the active materials, but also plays an important role in improving the catalytic performance of the composites by introducing the synergistic effects of components and functions (Tong et al. 2014, Ji et al. 2017). For example, Sun et al. reported the synthesis of a highly dispersed palladium/polypyrrole/polyacrylonitrile (Pd/PPy/PAN) composite membrane via a one-pot redox polymerization process in the presence of electrospun PAN nanofibers (Sun et al. 2014). The Pd nanoparticles are uniformly encapsulated in the PPy shells on the surface of the flexible PAN cores, enabling them to exhibit good catalytic behavior toward hydrogen generation from ammonia borane. Meanwhile, the as-prepared catalyst is stable against poisoning and can be readily separated from the suspension system for reuse. Yang et al. decorated small Pd nanoparticles on the electrospun $CoFe_2O_4$ nanotubes through a simple *in situ* reduction route, which displayed enhanced peroxidase-like activity for the colorimetric detection of H_2O_2

in comparison to the individual Pd nanoparticles, thanks to the hollow structure of $CoFe_2O_4$ nanotubes, as well as the even distribution of Pd nanoparticles (Yang et al. 2016).

The large specific surface area and high porosity of the electrospun nanofibers are also beneficial to increase the interaction region between the sensing materials and the detected objects, which will remarkably enhance the property of the sensors (Jiang et al. 2016b, Zhang et al. 2017a). Liu et al. fabricated 1D SnO_2 and SnO_2/CeO_2 nanostructures with multifarious morphologies (solid nanofibers, nanotubes, nanobelts, and wire-in-tubes) by a single-spinneret electrospinning, and a subsequent annealing treatment (Liu et al. 2015). In comparison, the SnO_2/CeO_2 nanotubes revealed superior response to ethanol gas over other nanostructures because of their hollow interior, good crystallinity, and the interaction between the two metal oxides.

Furthermore, the electrospun nanofibers as both the separator and electrode materials for energy storage devices show excellent electrochemical performance owing to the following reasons: (1) Nanofibers provide direct pathways for the electron transfer as compared to the particle electrodes; (2) The accelerated ion diffusion caused by the high porosity of nanofibers is helpful in elevating the rate capability; (3) Nanofibers with large specific surface area ensure the effective contact of the electrode-electrolyte interface; (4) The empty space in nanofibers accommodates the volume changes, thus restraining the mechanical degradation during the long-term cycling tests (Wei et al. 2017, Mai et al. 2014).

It is convenient to control the fine structures of the fiber products by the electrospinning technique. Combined with the materials having low surface energy, the superhydrophobic fibrous coatings can be obtained, which are expected to be applied in the ship shell, the inner wall of pipelines, high-rise windowpanes, and other self-cleaning areas (Wang et al. 2011, Wu et al. 2012).

Disadvantages

There are some key problems now to be solved for the preparation of nanofibers using the electrospinning technique. First, it is tough to realize the precise control of the size, morphology, and composition of the electrospun nanofibers, due to the fact that the influencing factors are more complex than those for the traditional spinning process, which imposes restrictions on the improvement of the fiber material performance, hence resulting in the limited practical application scope. Second, the production efficiency of nanofibers using the electrospinning apparatus in laboratory is relatively low, so the development of electrospinning machine for mass manufacturing of nanofibers is still an urgent issue to tackle. Typically, Elmarco company in the Czech Republic supplies a series of small, medium, and large Nanospider™ electrospinning equipment for laboratory and industrial nanofiber production, laying the foundation for the further evolution of the electrospinning technique. However, the high cost of Nanospider™ product portfolio is an ineluctable challenge facing researchers. Finally, the intrinsic strength, tenacity, and other mechanical properties of the electrospun nanofibers are usually unsatisfactory, which should be enhanced for some special applications, including individual protection, self-cleaning coatings, and recyclable adsorbents.

Development of Electrospinning

To be exact, electrospinning is evolved from the electrospray ionization (ESI) process based on high voltage. The research of ESI provides a theoretical basis for the electrospinning system. To the best of our knowledge, the differences between the two electrostatic techniques mainly lie in two aspects: the working medium/ precursor solution and the morphology of the obtained samples. For the former, a Newtonian fluid with lower viscosity is used for ESI, and a non-Newtonian fluid with higher viscosity for electrospinning. For the latter, ESI is a simple method to produce monodispersed polymer micro/nanospheres, while electrospinning usually generates 1D nanostructures.

History of Electrospinning

Electrospinning was first reported by Formhals in a sequence of patents that continued from 1934 to 1944 (Formhals 1934, 1937, 1938a, b, 1939a, b, c, 1940, 1943). He invented experimental apparatus for the construction of artificial fibers using electrostatic forces, and described in detail the formation process of polymer jet flows between the electrodes, which is recognized as the beginning of electrospinning, giving us a preliminary understanding of this fiber processing technology.

During the period from the 1930's to the 1980's, the development of electrospinning was relatively slow. Researchers were mostly focused on the study of the electrospinning equipment or process, and issued a large number of patents (Fig. 1.2). In 1966, Simons exhibited an electrospinning device for producing patterned and textured non-woven fabrics by changing the arrangement mode of the collecting electrode, and discussed the effect of the solution viscosity, dielectric constant, conductivity, and volatility on the efficiency of the electrospinning process (Simons 1966). Between 1964 and 1969, Taylor proposed the theoretical underpinning of electrospinning (Taylor 1964, 1966, 1969). He mathematically simulated the shape of the conical droplet formed at the moment when the electrostatic repulsive force was balanced with the surface tension of the polymer solution. This typical

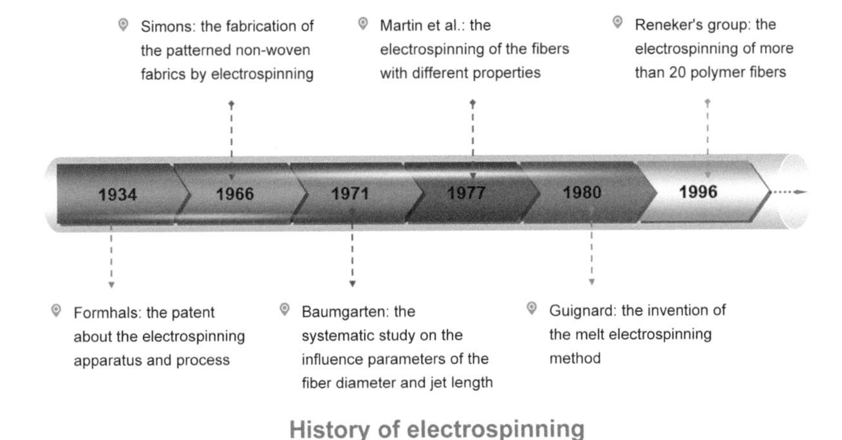

Fig. 1.2: Chronicle of historical events of electrospinning.

droplet shape was later known as the "Taylor cone", the half-angle of which was calculated to be 49.3°. In 1971, Baumgarten fabricated acrylic resin fibers with a diameter range of less than 1 μm using high-voltage electrospinning technique (Baumgarten 1971). He also investigated the influence of solution viscosity, flow rate, surrounding gas, and voltage on the fiber diameter and jet length, which was a rather in-depth and comprehensive study on the preparation of polymer fibers through electrospinning. In 1977, Martin et al. disclosed the electrospinning of multicomponent polymer solutions by using a single nozzle (Martin and Cockshott 1977, Martin et al. 1977). Besides, different polymer fibers were also simultaneously electrospun from a plurality of spinnerets. They found that the non-woven fiber mat with appropriate thickness deposited on the static or mobile collectors have a wide range of applications. In 1980, Guignard published a patent about the melt electrospinning method (Guignard 1980). Subsequently in 1981, Larrondo and Manley succeeded in synthesizing polyethylene (PE) and polypropylene (PP) fibers by electrospinning the polymer melt (Larrondo and Manley 1981). As evidenced by the experimental results, the higher the applied voltage or the melt temperature, the smaller is the diameter of fibers, and the size of the spinning nozzle seemed to be of minor importance. These discoveries created a precedent of the melt electrospinning technique.

By the 1990's, the research interest in electrospinning was reignited along with the development of nanotechnology. In 1996, Reneker's group indicated that more than 20 kinds of polymers had been electrospun from the solution or melt in their laboratory (Reneker and Chun 1996). They also explored the mechanism of the electrospinning process by experimental investigation and mathematical calculation, and then put forward the bending instability of the charged liquid jets in electrospinning (Reneker et al. 2000). As shown in Fig. 1.3, we witnessed the rapid growth of electrospinning after 2000, which has drawn great attention from worldwide academic and industrial fields.

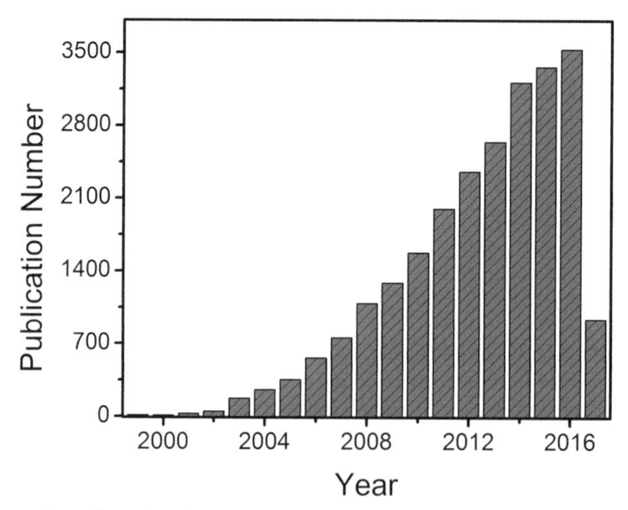

Fig. 1.3: The annual number of publications on the research topic of "electrospinning or electrospun" provided by Web of Science in Jilin University on April 10, 2017.

Current Situation of Electrospinning

Over the past two decades, fabricating nanofibers by electrospinning has been one of the most important approaches in material science, and researchers have turned their attention from the simple preparation and characterization of electrospun fibers to the understanding of the electrospinning process and mechanism. On this basis, a versatile rule of regulating the structure of electrospun fibers is generated, and the applications of the obtained fibers have also been extensively studied. The main results achieved are listed as follows.

(1) More and more polymers can be used for electrospinning. So far, over 100 types of polymer fibers have been successfully electrospun (Xu et al. 2016), such as synthetic polymers, natural macromolecules, and polyblends.
(2) The formation process and mechanism of electrospun fibers are gradually revealed, and the theoretical research has made new breakthroughs. We have some knowledge about the initial movement phase and spiral swing unsteady stage of the jet flows in the electrospinning process, where the precursor solution is first pulled out of the fluid level at the nozzle, moving in a straight line to a certain position, and then the charged jet starts with a spiral swing motion accompanied by further stretching and refining (Reneker et al. 2000, Shin et al. 2001a, b).
(3) The morphologies of the electrospun fibers tend to be multifarious. Based on the improvement of electrospinning devices, as well as the universal relationship between the electrospinning parameters and the shapes of the obtained fibers, a variety of 1D architectures have been fabricated by electrospinning, including porous fibers, flat ribbons, hollow fibers, coaxial/core-shell cables, side-by-side fibers, aligned yarns, necklace-like fibers, dendritic heterostructures, helical fibers, and so on (Chen et al. 2010, 2011a, He et al. 2015, Li and Xia 2004b, Sun et al. 2003, Liu et al. 2007, Yu et al. 2017, Jin et al. 2010, Ostermann et al. 2006, Kessick and Tepper 2004).
(4) The compositions of the electrospun fibers manifest the characteristics of diversification. Single-component or multicomponent polymer fibers, organic hybrid fibers, inorganic fibers, and organic/inorganic composite fibers have been successfully synthesized by combining electrospinning with other post-processing techniques (Lee et al. 2017, Zainab et al. 2016, Abouali et al. 2015, Chen et al. 2011b, Zheng et al. 2017).
(5) Researchers are now dedicated to broadening the application scope and enhancing the performance of the electrospun fibers which show some superiority in the fields of catalysis, sensors, biomedicine, filtration, environmental protection, and energy storage, electronic and optical devices owing to the various 1D architectures, tunable compositions, and their unique properties like large specific surface area and high porosity.
(6) Some electrospinning machines for mass production of fiber materials have already come into the public eye. The key to raise the yield of the electrospun fibers is to steadily increase the polymer jet flows at the spinneret. Therefore, the current research is mainly concentrated on the multi-jet methods with or without the spinning nozzle (Ding and Yu 2011).

Development Tendency of Electrospinning

It can be seen from the development of electrospinning that the formation process and mechanism of the electrospun polymer fibers and the optimal electrospinning conditions for different kinds of polymers have been extensively studied in the initial period. Further research work is primarily about the functionalization of polymer fibers and the fabrication of inorganic fibers and organic/inorganic composite fibers. The application of the electrospun nanofibers is one of the hot subjects at present, which is also the ultimate goal of the scientists engaged in this area. Nevertheless, there is still a long way left for the actual implementation of the commercialization of electrospun nanofibers (Wang and Lu 2011).

There exist some important problems to be settled in the preparation of nanofibers by electrospinning. First, the species of the natural macromolecules available for electrospinning are still very limited. Second, the electrospun inorganic nanofibers suffer from a serious defect of brittleness. Third, it is difficult to accurately adjust the performance of the obtained organic/inorganic composite nanofibers on account of the complex influencing factors. Finally, the lack of study on the structures and properties of electrospun nanofibers results in the application of the products only in the experimental stage.

Regarding the issues above, the authors consider that more efforts will be made in expanding the variety of spinnable polymers, developing the continuous inorganic nanofibers with high flexibility, synthesizing the high-performance and multifunctional composite nanofibers and their two- or three-dimensional (3D) assemblies, and realizing the industrial production of electrospun fibers for some time to come.

Category of Electrospinning

For a profound understanding of electrospinning, there are several categorizations according to the original fluids, equipment types, and production scales. Here we will give a specific introduction.

Category Based on the Precursors for Electrospinning

In the light of the original fluid used for electrospinning, it can be divided into two types: electrospinning from polymer solution or melt. The solution electrospinning at normal temperature is now the main research direction because of its simple device, while the melt electrospinning requires an additional heating unit at the risk of high-temperature operation, leading to relatively little attention from scholars (Lian and Meng 2017, Zhang et al. 2017b). However, for the polymers without suitable solvents, the melt electrospinning is a good choice demonstrating a commercial application prospect, which has no pollution problem caused by the solvent evaporation (Bhardwaj and Kundu 2010, Hutmacher and Dalton 2011).

Solution Electrospinning. It should be noted that the related reports usually refer to the solution rather than the melt electrospinning. Indeed, the thorough and deep investigations on the solution electrospinning have been carried out ranging from the apparatus and theory to the control parameters due to its simple and versatile manipulation in preparing ultrafine fibers from various materials. However, there inevitably exist some drawbacks for solution electrospinning. (1) The prerequisite for the given polymer used in solution electrospinning is to have a compatible solvent to generate a homogeneous solution. (2) The solvent accounting for a large proportion of the spinning solution is volatilized during the electrospinning process, thus offering a low yield of fibers. (3) The solvents available for electrospinning are generally organic ones with strong toxicity and high cost, which is difficult to recycle and will bring about the environmental pollution. All these deficiencies greatly restrict the further industrial application of solution electrospinning.

Melt Electrospinning. As an indispensable branch of electrospinning technique, melt electrospinning has received some attention since its birth in the early 1980's. Compared to the conventional solution electrospinning, melt electrospinning needs an additional heating part in order to maintain the molten state of polymers, which can be mainly classified into three groups: heat-conduction, thermal radiation, and convection devices according to the heat-transfer mechanism. There are also modular melting apparatus that combine multiple heating modes (Wang and Lu 2011). For instance, the electrical heater, circulating fluid, heated air, and laser have been employed as the heating appliances in melt electrospinning (Larrondo and Manley 1981, Brown et al. 2016, Zhang et al. 2016, Ogata et al. 2007). Moreover, the fiber solidification in melt electrospinning is contributed by the cooling of the polymer jet flow in the absence of both solvent evaporation and obvious whipping instability, different from that in solution electrospinning (Zhang et al. 2017b).

Interestingly, the electrospinning of polymer melts has distinct advantages over solution electrospinning. First, there is no environmental contamination in the melt electrospinning process for the reason that it has no demand for any solvent. Second, some polymers such as PE and PP without proper solvents at room temperature can be treated by melt electrospinning. Third, some thermoplastic polymers including polycaprolactone (PCL) and polylactic acid (PLA) that can be electrospun in the form of solutions are also appropriate for melt electrospinning, which will dramatically increase the yield of fibers with a high precursor utilization level, and avoid the existence of the residual solvent in products. Next, molding the melt electrospinning is beneficial to gain a better understanding of the electrospinning mechanism (Ding and Yu 2011, Wang and Lu 2011).

Up till now, polyolefin, polyester, polyamide, and other polymers have been successfully used in melt electrospinning to fabricate fiber materials with reduced diameters even below the micron scale (Zhang et al. 2017b, Hutmacher and Dalton 2011). Zhmayev et al. obtained the PLA mats down to 200~300 nm in fiber diameter by the melt electrospinning with a gas-assisted system installed at the end of the spinneret (Zhmayev et al. 2010). Melt electrospinning is a direct approach to deposit nontoxic fibers *in vitro*, which are similar to the structure of human extracellular matrix, and thus it has a bright future in biomedical applications. A polymer blend of poly(ethylene oxide-block-ε-caprolactone) (PEO-b-PCL) and PCL was melt and electrospun by Dalton et al. with the fibroblasts cultured *in vitro* as the collection target (Dalton et al. 2006). Cell vitality was maintained well even at the time of 72 hours after melt electrospinning.

In spite of what has been achieved in melt electrospinning, it is still in its infancy owing to the following shortcomings (Bhardwaj and Kundu 2010). (1) Melt electrospinning is often forced to be performed under a higher applied voltage than that of solution electrospinning as a result of the high viscosity and poor conductivity of the polymer melt, which is prone to cause electrostatic breakdown. (2) The absence of both solvent evaporation and obvious whipping instability is responsible for the larger diameters of the melt electrospun fibers. (3) There are also some technical problems existing in melt electrospinning that confine its further development. For example, the heating unit should be exact enough to prevent the decomposition of the polymer melt at high temperatures. It is suggested that the electrostatic interference between the heating part and the high-voltage power supply be avoided. The thermal uniformity and flow rate of the polymer melt need to be controlled and improved. In addition, the polymers available for melt electrospinning are still very limited (Ding and Yu 2011, Wang and Lu 2011). Therefore, more efforts should be devoted to the fundamental research of melt electrospinning (e.g., the establishment of spinning model, the thorough comprehension about the formation of fibers, exploring the effect of the processing parameters on the fiber morphology, etc.).

Other novel solvent-free electrospinning techniques, such as supercritical CO_2-assisted electrospinning, UV-curing electrospinning, anion-curing electrospinning, and thermo-curing electrospinning, are also perceived as high-efficiency, controllable, and environmentally friendly methods to produce micro/nanofibers. Nevertheless, the solvent-free electrospinning is subjected to the rigorous requirements for complicated set-ups, the larger diameters of the final fibers, which has not been widely concerned (Zhang et al. 2017b).

Category Based on the Types of Electrospinning Equipment

Commonly, the electrospinning equipment can be divided into three clusters, namely, single-spinneret, multi-nozzle, and needleless electrospinning, which have been discussed in detail elsewhere (Zhou et al. 2009, 2010). On the basis of the geometry of the fiber generators, there are four types of spinnerets: dot, linear, curvilinear, and areal spinnerets used in electrospinning for the construction of fibers in the laboratory or on a large scale.

Dot Jetting Spinneret. Single nozzle is the alleged dot spinneret including single tip/needle and hole. It is the simplest and frequently-used spinneret in electrospinning that is conducive to studying the basic process and mechanism of fiber formation and mastering the control parameters of fiber morphology. However, the low fiber output of the single-spinneret electrospinning slowed its industrial progress. It is reported that the design of a nozzle with multiple grooves at its inner wall (Yamashita et al. 2007), the adoption of the curved collectors with conspicuous curvature (Vaseashta 2007), and the application of a sufficiently large tangential stress provided by the transverse electric field to the cross-section of the jet (Paruchuri and Brenner 2007) can effectively increase the jet flows, thereby improving the yield of electrospun fibers. Even so, the corresponding theories of the three strategies are not yet clear, and their feasibility remains to be verified (Ding and Yu 2011, Zhou et al. 2009).

Linear Jetting Spinneret. Linear-spinneret electrospinning can realize the mass production of fibers to some extent by generating a plurality of jet flows. The linear distribution of multiple needles or holes, slit-type nozzle, and wire electrode are common examples of linear spinneret.

For the multi-needle electrospinning, the relatively simple apparatus and the easy synthesis of the mixed fibers from different precursors with desired ratios define it into one of the popular approaches to improve the yield of electrospun fibers (Teo and Ramakrishna 2006). The arrangement, number, and the occupied space of nozzles are three important factors that affect the quantity and quality of fiber membranes (Ding and Yu 2011). Although the batch manufacturing of nanofibers through the multi-spinneret electrospinning has made some progress in recent years, it is still faced with a series of difficulties such as interference between the electrospinning jets, uneven distribution of electric field, nozzle clogging, and so on (Ding et al. 2004, Theron et al. 2005), and so is the electrospinning with linear hole array on the porous tube nozzle (Varabhas et al. 2008).

The electrospinning from the free liquid surface in an open area using the linear emitting electrode of a slit-type nozzle or a metal wire is often assigned to needleless electrospinning. On the one hand, the slit-type electrospinning of polymer melts or solutions with the maximal field intensity centered at the outward edges of the slot can effectively increase the fiber production rate per unit area (Komárek and Martinová 2010, Zhang and Jin 2011), whereas, it also suffers from the inhomogeneity of the electrospun fibers. On the other hand, Forward and Rutledge demonstrated a wire electrode for electrospinning, in which the rotation of the wire spindle in the charged polyvinylpyrrolidone (PVP)-ethanol solution bath resulted in the entrainment of liquid droplets that were eventually transformed into jets (Forward and Rutledge 2012). A linear electrode spinning system of Nanospider™ G2 was also developed by Elmarco company in 2011. In a typical electrospinning process, the polymer solution placed in the storage tank was coated on the surface of the static wire electrode by the reciprocating movement of the carriage module. Then the droplets were subjected to the electrostatic traction, forming a plurality of jet flows that were continuously drawn and split. Finally, the obtained fibers were deposited on the moving substrate. It is easy to control the thickness of the fiber mat and the morphology and size of the electrospun fibers by adjusting the moving speed of the collector as well as changing the solution property and electric field intensity. Hsieh et al. incorporated the linear electrodes composed of one, two, three, or four twisted stainless steel wires into the electrospinning (Fig. 1.4a) (Hsieh et al. 2016). Their results indicated that the 3- and 4-wire electrodes with larger twisted intersections prevented the aqueous solution of polyvinyl alcohol (PVA) from generating Taylor cones, leading to the uneven diameters of the electrospun nanofibers, which were inversely proportional to the number of the stainless steel wires in the electrode. In addition to the twisted wire spinneret (Holopainen et al. 2015), the bead wire (Fig. 1.4b) was also used in

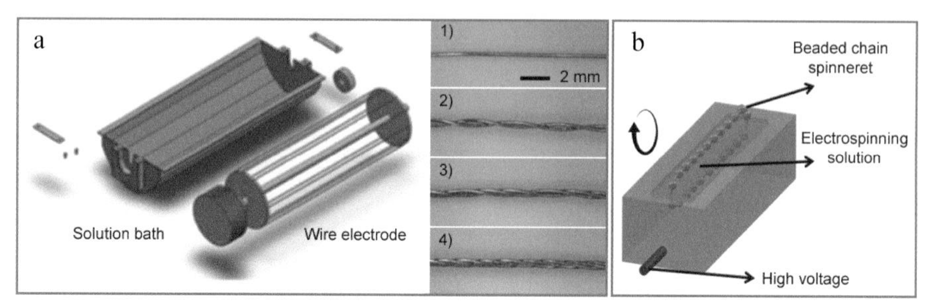

Fig. 1.4: Schematic illustrations of (a) metal wire electrode (Hsieh et al. 2016) and (b) beaded chain spinnerets (Niu and Lin 2012) used in electrospinning. Insets in (a): (1) one, (2) two, (3) three, and (4) four twisted stainless steel wires.

electrospinning, where numerous jets were ejected from the circular rotary bead-wire electrode that was wrapped with polymer solution through the solution brush (Liu et al. 2014).

Curvilinear Jetting Spinneret. Another type of needleless electrospinning set-up equipped with the curvilinear spinneret has been developed in the past decade (Niu et al. 2009, 2012, Lin et al. 2014, Wang et al. 2012, 2014a, b). Among them, a shining example is the rotating metal disk electrospinning. Niu et al. demonstrated in 2009 that a relatively low applied voltage could initiate the spinning process of PVA solution, and the liquid jets were mainly formed on the top edge of the disk nozzle (Fig. 1.5a) (Niu et al. 2009). Besides, in 2011, their group further investigated the influence of fiber generator geometry on electrospinning, and drew the conclusion that the thinner disk spinneret increased the electric field intensity, hence resulting in the finer nanofibers and higher productivity (Niu et al. 2012).

However, it still needed a more efficient spinneret with a delicate shape to achieve lower power usage and higher yield of thinner fibers. Later on, a spiral coil spinneret (Fig. 1.5b) was employed to electrospin nanofibers with smaller diameters and narrower size distribution contributed by the highly concentrated electric field on the coil surface (Lin et al. 2014, Wang et al. 2012). It was indicated that the fiber production rate increased along with the increase of coil length or coil diameter, or the decrease of spiral distance or wire diameter, but the coil dimension had a minor impact on the fiber width (Wang et al. 2012). In order to further optimize the needleless electrospinning technique and scale up the throughput of ultrafine nanofibers, Wang

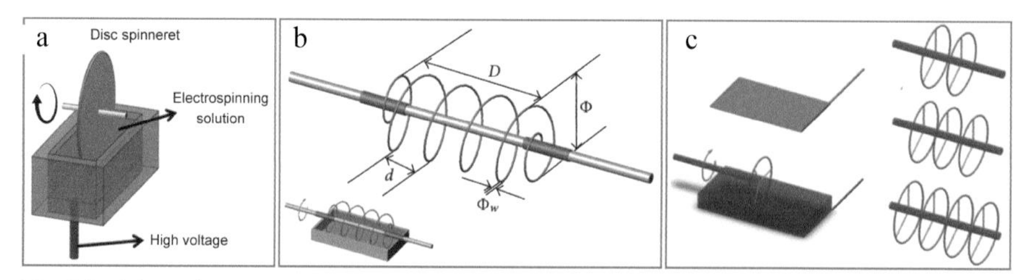

Fig. 1.5: Schematic summary of curvilinear spinnerets: (a) rotating disk (Niu and Lin 2012), (b) spiral coil (Wang et al. 2012), and (c) ring structures (Wang et al. 2014b).

and Lin et al. proposed a ring structure (Fig. 1.5c) and found that the yield of uniform nanofibers could be easily improved to the desired extent by using more rings in the spinneret (Wang et al. 2014a, b). In the electrospinning process, a plurality of jet flows was generated from the top of each ring due to the higher-intensity electric field on their surface. It is believed that the multiple-ring electrospinning holds great promise for both laboratory research and industrial manufacture of nanofibers (Wang et al. 2014b).

Areal Jetting Spinneret. As the name implies, most of the areal-spinneret electrospinning also belongs to the needleless technique. Much like the multi-needle/hole electrospinning with nonlinear aligned nozzle sets (Yang et al. 2010, Zheng et al. 2013), the porous tubular electrospinning using random hole array as spinneret (Fig. 1.6a) can scale up the production rate of nanofibers with similar average diameter to those from the single-needle electrospinning (Zhou et al. 2009, Dosunmu et al. 2006). However, it inevitably suffers from disadvantages, such as the possible clogging of the small holes, the disordered jet distribution on the tube surface, and the wider fiber size range.

In 2005, Jirsak et al. patented a unique electrospinning design with a roller as the fiber generator (Jirsak et al. 2009), which has been commercialized by Elmarco company (Nanospider™ G1). Niu et al. compared the rotating cylinder (Fig. 1.6b), disk and ball spinnerets (Fig. 1.6c) used in electrospinning (Niu et al. 2009, 2012), but all the three modes had a loose control on solution feeding, thus easily causing the larger fiber diameter (Zhou et al. 2009). It was revealed that the fibers were first drawn out from the cylinder end and then the middle surface when the applied voltage arrived at a certain level, owing to the uneven distribution of the electric field along the spinneret surface (Niu et al. 2009). With the decrease of the cylinder diameter, the fiber yield was greatly improved, but the fiber width exhibited little change (Niu et al. 2012). Besides, at the same operating conditions, the electric field intensity on the cylinder end was much lower than that on the disk edge, leading to the coarser fibers produced by the cylinder electrospinning (Niu et al. 2009). In comparison to cylinder electrode, the ball nozzle with uniformly distributed electric field on its surface generated thicker nanofibers at a lower productivity (Niu et al. 2012).

Wang et al. reported a novel needleless electrospinning system using a conical wire coil as the open spinneret that could retain viscous fluid (Fig. 1.6d) (Wang et al. 2009, Wang and Xu 2012). In the spinning process, multiple polymer jets were mainly formed on the outer surface of the cone where an electric field with higher intensity was observed. Compared to the typical needle electrospinning, this needleless technique could electrospin finer nanofibers on a larger scale of > 2.5 g/h. Lu et al. demonstrated a new electrospinning device with a rotating metal cone as the nozzle (Fig. 1.6e) (Lu et al. 2010). The electrospinning rate of this approach is about 600 g/h, which is thousands of times more than that of the conventional single-spinneret electrospinning. In addition, it overcomes the drawback of the high applied voltage required by the conical coil-typed electrospinning set-up described above.

Centrifugal electrospinning (Fig. 1.6f) emerges as another high-output method for the preparation of ultrafine nanofibers, which has aroused widespread attention

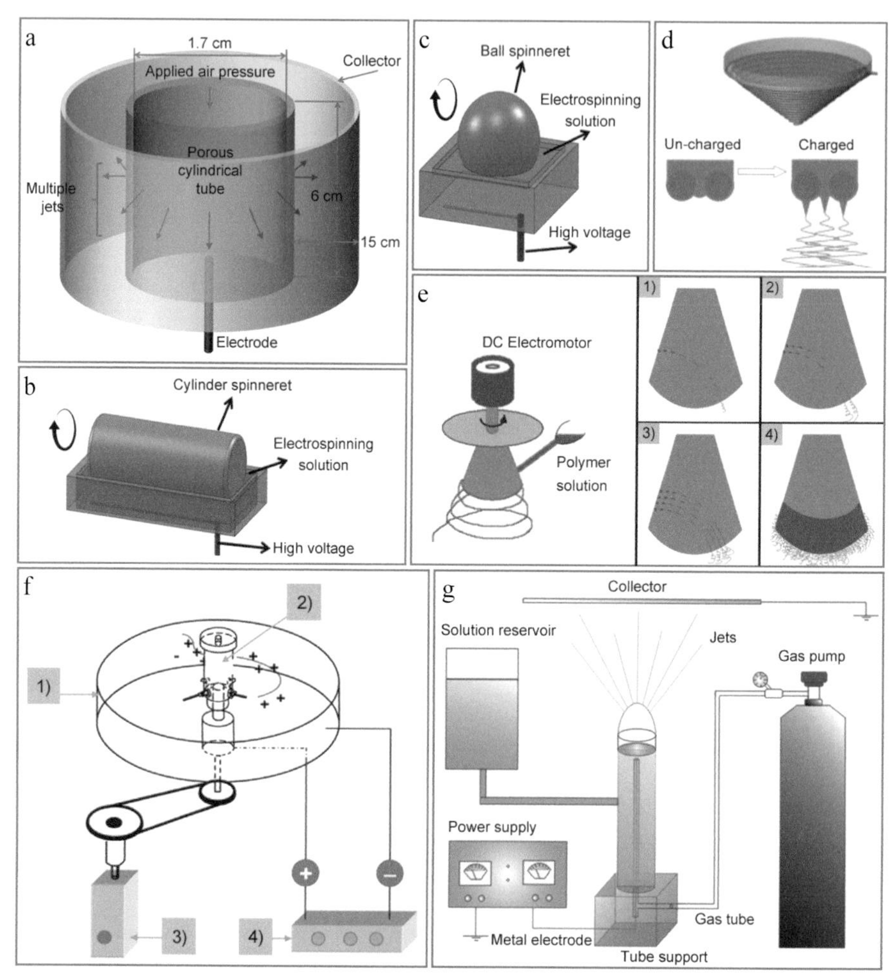

Fig. 1.6: (a–e) Schematic summary of areal spinnerets: (a) porous tube (Dosunmu et al. 2006), (b) rotating cylinder (Niu and Lin 2012), (c) ball nozzle (Niu and Lin 2012), (d) conical wire coil (Wang et al. 2009), and (e) metal cone (Lu et al. 2010). (1)–(4) Movement and transformation of substance on the cone surface. Schematic illustration of the (f) centrifugal (Kancheva et al. 2014) and (g) bubble electrospinning apparatuses (Yang et al. 2009). Insets in (f): (1) stationary cylinder collectors, (2) rotating feed unit, (3) electric motor, and (4) high-voltage power supply.

recently (Peng et al. 2017, Kancheva et al. 2014, Erickson et al. 2015). Generally, the polymer solution or melt is electrospun based on the combination of electrostatic, centrifugal, gravitational, and auxiliary forces (Peng et al. 2017). The centrifugal field guarantees the full stretching of the charged jets and significantly eliminates their bending instability in the course of movement, thereby improving the orientation and mechanical properties of the obtained nanofibers (Peng et al. 2015b). However, the centrifugal electrospinning apparatus is diversified and complicated, including high-voltage power supply, rotation feed unit, spinneret, collector, heating system, and other complementary parts (Peng et al. 2017). These various components need to be reasonably arranged so as to ensure the safe operation of the overall device.

Bubble electrospinning (Fig. 1.6g) was first presented by He et al. in 2007 (Liu and He 2007), and it has been used in fabricating a variety of polymer nanofibers such as PVA, PVP, PAN, etc. (Jiang et al. 2015b, Liu et al. 2008). Inspired by the fantastic spider spinning, the typical process of bubble electrospinning is as given below: a gas pump is introduced to generate bubbles on the liquid surface which is charged with a high voltage, and then multiple jets are initiated on the bubble surface (Chen et al. 2015, Niu and Lin 2012). It is thought that the flexibility of bubble electrospinning makes it a promising candidate for mass production of nanofibers (Yang et al. 2009).

Moreover, there are other areal-spinneret electrospinning techniques. For example, Yarin and Zussman adopted a magnetic field in electrospinning (Yarin and Zussman 2004). On the basis of the edge electrospinning from a bowl-shaped spinneret (Thoppey et al. 2011), Jiang et al. demonstrated a stepped pyramid-like nozzle for high-throughput production of quality nanofibers (Jiang et al. 2013, 2015b). No further details are given here, and please refer to the following sections.

Category Based on Production Scales

According to the production scale of nanofibers, electrospinning can be mainly divided into two classes: laboratory-scale and industrial electrospinning. There is no clear boundary between the two production lines, except that the traditional single-spinneret electrospinning with a lower productivity, typically less than 0.3 g/h (Wang et al. 2012), can only be used in the laboratory. Much effort has been made to achieve the mass production of electrospun nanofibers, which must meet the following two requirements: (1) multiple polymer jets are sprayed from the nozzle simultaneously; (2) the applied voltage is high enough to provide strong electric field around the nozzle so as to initiate the spinning (Wang et al. 2014b). Up till now, multi-needle and needleless electrospinning with increased fiber yield have been reported and used both in industry and in laboratories.

Manufacturers of Electrospinning Equipment and Products

With the continuous expansion of the application of nanofibers, the market demand for nanofiber products is gradually increasing. As a result, the industrial production technology of nanofibers has attracted wide attention from both the academic and industrial circles in the last decade. So far, about 50 companies, as well as more than 100 universities and research institutions at home and abroad engage in the study of electrospinning technique, nanofibers, and textiles. Among them, Elmarco company in the Czech Republic is very famous for the needleless electrospinning equipment called Nanospider™, which has been employed in preparing organic and inorganic nanofibers on a large scale.

The representative manufactures of electrospinning apparatus and nanofiber products in the world are summarized and listed in Table 1.1.

Table 1.1: Representative manufactures of electrospinning apparatus and nanofiber products.

Company/Country	Model series	Spinneret	Products	Service	Website
Spinbow™/Italy	Lab unit	Sliding spinneret, Max.: 4 needles	Breadth: 405 mm	Lab set-up and accessories	http://www.spinbow.it/
Kato Tech/Japan	NEU series	3 needles, linearly arranged	Fiber diameter: 50~800 nm; Breadth: 330 mm	Lab scale set-up	http://english.keskato.co.jp/products/neu.html
MECC/Japan	EDEN series, NF-1000	Needleless design, Unpublished	Breadth: 1.3 m	Lab and pilot scale equipment	http://www.mecc-nano.com
Yflow®/Spain	FibeRoller	10-nozzle injector, 8 × annular array	Breadth: 1.7 m	Lab and industrial level set-up	http://www.yflow.com/
Inovenso/Turkey	Nanospinner 416	Multi-needle matrix, up to 110 nozzles	Fiber diameter: 50~400 nm; Breadth: 1 m; Productivity: 180~5000 m^2/day, 5 kg/day	Industrial and lab scale set-up	http://www.inovenso.com
Elmarco/Czech Republic	Nanospider™, NS 8S1600U	Needle-free system, Metal wire	Fiber diameter: 80~700 nm (+/−30%); Breadth: 1.6 m	Industrial and lab scale set-up	http://www.elmarco.com/
Ucalery/China	Ispun series	Needleless design, Unpublished	Productivity: 550 m^2/day/unit	Lab and Industrial level set-up	http://www.ucalery.com
Tlwnt/China	TL-20M	Multi-needle matrix, 50~200 needles	Breadth: 200~660 mm; 1, 1.6, 2 m	Lab and Industrial level set-up	http://www.tlwnt.com

Acknowledgements

The authors wish to acknowledge grant support from the National Natural Science Foundation of China (21474043 and 51473065), and the China Postdoctoral Science Foundation (2018M630745).

References

Abouali, S., Garakani, M. A., Zhang, B., Xu, Z. L., Heidari, E. K., Huang, J. Q. et al. 2015. Electrospun carbon nanofibers with *in situ* encapsulated Co_3O_4 nanoparticles as electrodes for high-performance supercapacitors. ACS Applied Materials & Interfaces 7: 13503–13511.

Agarwal, S., Wendorff, J. H. and Greiner, A. 2008. Use of electrospinning technique for biomedical applications. Polymer 49: 5603–5621.

Baumgarten, P. K. 1971. Electrostatic spinning of acrylic microfibers. Journal of Colloid and Interface Science 36: 71–79.

Bhardwaj, N. and Kundu, S. C. 2010. Electrospinning: a fascinating fiber fabrication technique. Biotechnology Advances 28: 325–347.

Brown, T. D., Dalton, P. D. and Hutmacher, D. W. 2016. Melt electrospinning today: an opportune time for an emerging polymer process. Progress in Polymer Science 56: 116–166.

Cavaliere, S., Subianto, S., Savych, I., Jones, D. J. and Rozière, J. 2011. Electrospinning: designed architectures for energy conversion and storage devices. Energy & Environmental Science 4: 4761–4785.

Chen, H. Y., Wang, N., Di, J. C., Zhao, Y., Song, Y. L. and Jiang, L. 2010. Nanowire-in-microtube structured core/shell fibers via multifluidic coaxial electrospinning. Langmuir 26: 11291–11296.

Chen, H. Y., Di, J. C., Wang, N., Dong, H., Wu, J., Zhao, Y. et al. 2011a. Fabrication of hierarchically porous inorganic nanofibers by a general microemulsion electrospinning approach. Small 7: 1779–1783.

Chen, J. Y., Wiley, B. J. and Xia, Y. N. 2007. One-dimensional nanostructures of metals: large-scale synthesis and some potential applications. Langmuir 23: 4120–4129.

Chen, R. X., Wan, Y. Q., Si, N., He, J. H., Ko, F. and Wang, S. Q. 2015. Bubble rupture in bubble electrospinning. Thermal Science 19: 1141–1149.

Chen, X., Unruh, K. M., Ni, C. Y., Ali, B., Sun, Z. C., Lu, Q. et al. 2011b. Fabrication, formation mechanism, and magnetic properties of metal oxide nanotubes via electrospinning and thermal treatment. The Journal of Physical Chemistry C 115: 373–378.

Choi, J., Yang, B. J., Bae, G. N. and Jung, J. H. 2015. Herbal extract incorporated nanofiber fabricated by an electrospinning technique and its application to antimicrobial air filtration. ACS Applied Materials & Interfaces 7: 25313–25320.

Dalton, P. D., Klinkhammer, K., Salber, J., Klee, D. and Möller, M. 2006. Direct *in vitro* electrospinning with polymer melts. Biomacromolecules 7: 686–690.

Ding, B., Kimura, E., Sato, T., Fujita, S. and Shiratori, S. 2004. Fabrication of blend biodegradable nanofibrous nonwoven mats via multi-jet electrospinning. Polymer 45: 1895–1902.

Ding, B. and Yu, J. Y. 2011. Electrospinning and Nanofibers. China Textile & Apparel Press, Beijing, China.

Ding, Z. W., Salim, A. and Ziaie, B. 2009. Selective nanofiber deposition through field-enhanced electrospinning. Langmuir 25: 9648–9652.

Dosunmu, O. O., Chase, G. G., Kataphinan, W. and Reneker, D. H. 2006. Electrospinning of polymer nanofibres from multiple jets on a porous tubular surface. Nanotechnology 17: 1123–1127.

Erickson, A. E., Edmondson, D., Chang, F. C., Wood, D., Gong, A., Levengood, S. L. et al. 2015. High-throughput and high-yield fabrication of uniaxially-aligned chitosan-based nanofibers by centrifugal electrospinning. Carbohydrate Polymers 134: 467–474.

Formhals, A. 1934. Process and apparatus for preparing artificial threads. U.S. Patent # 1,975,504.

Formhals, A. 1937. Production of artificial fibers. U.S. Patent # 2,077,373.

Formhals, A. 1938a. Artificial fiber construction. U.S. Patent # 2,109,333.

Formhals, A. 1938b. Method and apparatus for the production of fibers. U.S. Patent # 2,123,992.

Formhals, A. 1939a. Method and apparatus for spinning. U.S. Patent # 2,160,962.

Formhals, A. 1939b. Method and apparatus for the production of fibers. U.S. Patent # 2,158,416.

Formhals, A. 1939c. Method of producing artificial fibers. U.S. Patent # 2,158,415.

Formhals, A. 1940. Artificial thread and method of producing same. U.S. Patent # 2,187,306.

Formhals, A. 1943. Production of artificial fibers from fibers forming liquids. U.S. Patent # 2,323,025.

Forward, K. M. and Rutledge, G. C. 2012. Free surface electrospinning from a wire electrode. Chemical Engineering Journal 183: 492–503.

Greiner, A. and Wendorff, J. H. 2007. Electrospinning: a fascinating method for the preparation of ultrathin fibers. Angewandte Chemie International Edition 46: 5670–5703.

Guignard, C. 1980. Process for the manufacture of a plurality of filaments. U.S. Patent # 4,230,650.

He, Z. Y., Liu, Q., Hou, H. L., Gao, F. M., Tang, B. and Yang, W. Y. 2015. Tailored electrospinning of WO_3 nanobelts as efficient ultraviolet photodetectors with photo-dark current ratios up to 1000. ACS Applied Materials & Interfaces 7: 10878–10885.

Holopainen, J., Penttinen, T., Santala, E. and Ritala, M. 2015. Needleless electrospinning with twisted wire spinneret. Nanotechnology 26: 025301.

Hong, Y. J., Yoon, J. W., Lee, J. H. and Kang, Y. C. 2015. A new concept for obtaining SnO_2 fiber-in-tube nanostructures with superior electrochemical properties. Chemistry-A European Journal 21: 371–376.

Hsieh, C. T., Lou, C. W., Pan, Y. J., Huang, C. L., Lin, J. H., Lin, Z. I. et al. 2016. Fabrication of poly(vinyl alcohol) nanofibers by wire electrode-incorporated electrospinning. Fibers and Polymers 17: 1217–1226.

Hsu, Y. H., Chen, D. W. C., Tai, C. D., Chou, Y. C., Liu, S. J., Ueng, S. W. N. et al. 2014. Biodegradable drug-eluting nanofiber-enveloped implants for sustained release of high bactericidal concentrations of vancomycin and ceftazidime: *in vitro* and *in vivo* studies. International Journal of Nanomedicine 9: 4347–4355.

Hutmacher, D. W. and Dalton, P. D. 2011. Melt electrospinning. Chemistry-An Asian Journal 6: 44–56.

Ji, D. X., Peng, S. J., Lu, J., Li, L. L., Yang, S. Y., Yang, G. R. et al. 2017. Design and synthesis of porous channel-rich carbon nanofibers for self-standing oxygen reduction reaction and hydrogen evolution reaction bifunctional catalysts in alkaline medium. Journal of Materials Chemistry A 5: 7507–7515.

Jiang, G. J., Zhang, S. and Qin, X. H. 2013. High throughput of quality nanofibers via one stepped pyramid-shaped spinneret. Materials Letters 106: 56–58.

Jiang, G. J., Zhang, S., Wang, Y. T. and Qin, X. H. 2015b. An improved free surface electrospinning with micro-bubble solution system for massive production of nanofibers. Materials Letters 144: 22–25.

Jiang, S. H., Hou, H. Q., Agarwal, S. and Greiner, A. 2016a. Polyimide nanofibers by "Green" electrospinning via aqueous solution for filtration applications. ACS Sustainable Chemistry & Engineering 4: 4797–4804.

Jiang, T., Carbonea, E. J., Lo, K. W. H. and Laurencin, C. T. 2015a. Electrospinning of polymer nanofibers for tissue regeneration. Progress in Polymer Science 46: 1–24.

Jiang, Z. Q., Zhao, R., Sun, B. L., Nie, G.D., Ji, H., Lei, J. Y. et al. 2016b. Highly sensitive acetone sensor based on Eu-doped SnO_2 electrospun nanofibers. Ceramics International 42: 15881–15888.

Jin, Y., Yang, D. Y., Kang, D. Y. and Jiang, X. Y. 2010. Fabrication of necklace-like structures via electrospinning. Langmuir 26: 1186–1190.

Jirsak, O., Sanetrnik, F., Lukas, D., Kotek, V., Martinova, L. and Chaloupek, J. 2009. Method of nanofibres production from a polymer solution using electrostatic spinning and a device for carrying out the method. U.S. Patent # 7,585,437 B2.

Kancheva, M., Toncheva, A., Manolova, N. and Rashkov, I. 2014. Advanced centrifugal electrospinning setup. Materials Letters 136: 150–152.

Kessick, R. and Tepper, G. 2004. Microscale polymeric helical structures produced by electrospinning. Applied Physics Letters 84: 4807–4809.

Komárek, M. and Martinová, L. 2010. Design and evaluation of melt-electrospinning electrodes. 2nd Nanocon International Conference, Olomouc, Czech Republic, EU 72–77.

Larrondo, L. and Manley, R. S. J. 1981. Electrostatic fiber spinning from polymer melts. I. experimental observations on fiber formation and properties. Journal of Polymer Science: Polymer Physics Edition 19: 909–920.

Lee, K. S., Eom, K. H., Lim, J. H., Ryu, H., Kim, S., Lee, D. K. et al. 2017. Aqueous boron removal by using electrospun poly(vinyl alcohol) (PVA) mats: a combined study of IR/Raman spectroscopy and computational chemistry. The Journal of Physical Chemistry A 121: 2253–2258.

Li, D. and Xia, Y. N. 2004a. Electrospinning of nanofibers: reinventing the wheel? Advanced Materials 16: 1151–1170.

Li, D. and Xia, Y. N. 2004b. Direct fabrication of composite and ceramic hollow nanofibers by electrospinning. Nano Letters 4: 933–938.

Lian, H. and Meng, Z. X. 2017. Melt electrospinning vs. solution electrospinning: a comparative study of drug-loaded poly (ε-caprolactone) fibres. Materials Science and Engineering: C 74: 117–123.

Lin, T., Wang, X. G. Wang, X. and Niu, H. T. 2014. Electrostatic spinning assembly. U.S. Patent # 8,747,093 B2.

Liu, S. L., Huang, Y. Y., Zhang, H. D., Sun, B., Zhang, J. C. and Long, Y. Z. 2014. Needleless electrospinning for large scale production of ultrathin polymer fibres. Materials Research Innovations 18: S4-833–S4-837.

Liu, Y. and He, J. H. 2007. Bubble electrospinning for mass production of nanofibers. International Journal of Nonlinear Sciences and Numerical Simulation 8: 393–396.

Liu, Y., He, J. H. and Yu, J. Y. 2008. Bubble-electrospinning: a novel method for making nanofibers. Journal of Physics: Conference Series 96: 012001.

Liu, Y. S., Yang, P., Li, J., Matras-Postolek, K., Yue, Y. L. and Huang, B. B. 2015. Morphology adjustment of SnO_2 and SnO_2/CeO_2 one dimensional nanostructures towards applications in gas sensing and CO oxidation. RSC Advances 5: 98500–98507.

Liu, Z. Y., Sun, D. D., Guo, P. and Leckie, J. O. 2007. An efficient bicomponent TiO_2/SnO_2 nanofiber photocatalyst fabricated by electrospinning with a side-by-side dual spinneret method. Nano Letters 7: 1081–1085.

Lu, B., Wang, Y. J., Liu, Y. X., Duan, H. G., Zhou, J. Y., Zhang, Z. X. et al. 2010. Superhigh-throughput needleless electrospinning using a rotary cone as spinneret. Small 6: 1612–1616.

Lu, X. F., Wang, C. and Wei, Y. 2009. One-dimensional composite nanomaterials: synthesis by electrospinning and their applications. Small 5: 2349–2370.

Lu, X. F., Wang, C., Favier, F. and Pinna, N. 2016. Electrospun nanomaterials for supercapacitor electrodes: designed architectures and electrochemical performance. Advanced Energy Materials 1601301.

Mai, L. Q., Tian, X. C., Xu, X., Chang, L. and Xu, L. 2014. Nanowire electrodes for electrochemical energy storage devices. Chemical Reviews 114: 11828–11862.

Martin, G. E. and Cockshott, I. D. 1977. Fibrillar product of electrostatically spun organic material. U.S. Patent # 4,043,331.

Martin, G. E., Cockshott, I. D. and Fildes, F. J. T. 1977. Fibrillar lining for prosthetic device. U.S. Patent # 4,044,404.

Mei, Y., Wang, Z. M. and Li, X. S. 2013. Improving filtration performance of electrospun nanofiber mats by a bimodal method. Journal of Applied Polymer Science 128: 1089–1094.

Nie, G. D., Zhang, L., Lu, X. F., Bian, X. J., Sun, W. N. and Wang, C. 2013. A one-pot and *in situ* synthesis of CuS-graphene nanosheet composites with enhanced peroxidase-like catalytic activity. Dalton Transactions 42: 14006–14013.

Niu, C. J., Meng, J. S., Wang, X. P., Han, C. H., Yan, M. Y., Zhao, K. N. et al. 2015. General synthesis of complex nanotubes by gradient electrospinning and controlled pyrolysis. Nature Communications 6: 7402.

Niu, H. T., Lin, T. and Wang, X. G. 2009. Needleless electrospinning. I. a comparison of cylinder and disk nozzles. Journal of Applied Polymer Science 114: 3524–3530.

Niu, H. T. and Lin, T. 2012. Fiber generators in needleless electrospinning. Journal of Nanomaterials 2012: 725950.

Niu, H. T., Wang, X. G. and Lin, T. 2012. Needleless electrospinning: influences of fibre generator geometry. The Journal of the Textile Institute 103: 787–794.

Ogata, N., Yamaguchi, S., Shimada, N., Lu, G., Iwata, T., Nakane, K. et al. 2007. Poly(lactide) nanofibers produced by a melt-electrospinning system with a laser melting device. Journal of Applied Polymer Science 104: 1640–1645.

Ostermann, R., Li, D., Yin, Y. D., McCann, J. T. and Xia, Y. N. 2006. V_2O_5 nanorods on TiO_2 nanofibers: a new class of hierarchical nanostructures enabled by electrospinning and calcination. Nano Letters 6: 1297–1302.

Paruchuri, S. and Brenner, M. P. 2007. Splitting of a liquid jet. Physical Review Letters 98: 134502.

Peng, S. J., Li, L. L., Hu, Y. X., Srinivasan, M., Cheng, F. Y., Chen, J. et al. 2015a. Fabrication of spinel one-dimensional architectures by single-spinneret electrospinning for energy storage applications. ACS Nano 9: 1945–1954.

Peng, H., Zhang, J. N., Li, X. H., Song, Q. S. and Liu, Y. 2015b. Modes of centrifugal electrospinning. Engineering Plastics Application 43: 138–142.

Peng, H., Liu, Y. and Ramakrishna, S. 2017. Recent development of centrifugal electrospinning. Journal of Applied Polymer Science 134: 44578.

Reneker, D. H. and Chun, I. 1996. Nanometre diameter fibres of polymer, produced by electrospinning. Nanotechnology 7: 216–223.

Reneker, D. H., Yarin, A. L., Fong, H. and Koombhongse, S. 2000. Bending instability of electrically charged liquid jets of polymer solutions in electrospinning. Journal of Applied Physics 87: 4531–4547.

Shin, Y. M., Hohman, M. M., Brenner, M. P. and Rutledge, G. C. 2001a. Electrospinning: A whipping fluid jet generates submicron polymer fibers. Applied Physics Letters 78: 1149–1151.

Shin, Y. M., Hohman, M. M., Brenner, M. P. and Rutledge, G. C. 2001b. Experimental characterization of electrospinning: the electrically forced jet and instabilities. Polymer 42: 9955–9967.

Sill, T. J. and Recum, H. A. V. 2008. Electrospinning: applications in drug delivery and tissue engineering. Biomaterials 29: 1989–2006.

Simons, H. L. 1966. Process and apparatus for producing patterned non-woven fabrics. U.S. Patent # 3,280,229.

Sun, W. N., Lu, X. F., Tong, Y., Lei, J. Y., Nie, G. D. and Wang, C. 2014. A one-pot synthesis of a highly dispersed palladium/polypyrrole/polyacrylonitrile nanofiber membrane and its recyclable catalysis in hydrogen generation from ammonia borane. Journal of Materials Chemistry A 2: 6740–6746.

Sun, Z. C., Zussman, E., Yarin, A. L., Wendorff, J. H. and Greiner, A. 2003. Compound core-shell polymer nanofibers by co-electrospinning. Advanced Materials 15: 1929–1932.

Taylor, G. 1964. Disintegration of water drops in an electric field. Proceedings of the Royal Society of London. Series A, Mathematical and Physical Sciences 280: 383–397.

Taylor, G. 1966. The force exerted by an electric field on a long cylindrical conductor. Proceedings of the Royal Society of London. Series A, Mathematical and Physical Sciences 291: 145–158.

Taylor, G. 1969. Electrically driven jets. Proceedings of the Royal Society of London. Series A, Mathematical and Physical Sciences 313: 453–475.

Teo, W. E. and Ramakrishna, S. 2006. A review on electrospinning design and nanofibre assemblies. Nanotechnology 17: R89–R106.

Theron, S. A., Yarin, A. L., Zussman, E. and Kroll, E. 2005. Multiple jets in electrospinning: experiment and modeling. Polymer 46: 2889–2899.

Thoppey, N. M., Bochinski, J. R., Clarke, L. I. and Gorga, R. E. 2011. Edge electrospinning for high throughput production of quality nanofibers. Nanotechnology 22: 345301.

Tong, Y., Lu, X. F., Sun, W. N., Nie, G. D., Yang, L. and Wang, C. 2014. Electrospun polyacrylonitrile nanofibers supported Ag/Pd nanoparticles for hydrogen generation from the hydrolysis of ammonia borane. Journal of Power Sources 261: 221–226.

Varabhas, J. S., Chase, G. G. and Reneker, D. H. 2008. Electrospun nanofibers from a porous hollow tube. Polymer 49: 4226–4229.

Vaseashta, A. 2007. Controlled formation of multiple Taylor cones in electrospinning process. Applied Physics Letters 90: 093115.

Wali, Q., Fakharuddin, A., Ahmed, I., Rahim, M. H. A., Ismail, J. and Jose, R. 2014. Multiporous nanofibers of SnO_2 by electrospinning for high efficiency dye-sensitized solar cells. Journal of Materials Chemistry A 2: 17427–17434.

Wang, C. and Lu, X. F. 2011. Functional organic nanomaterials—High-voltage electrospinning technology and nanofibers. Science Press, Beijing, China.

Wang, H. G., Yuan, S., Ma, D. L., Zhang, X. B. and Yan, J. M. 2015. Electrospun materials for lithium and sodium rechargeable batteries: from structure evolution to electrochemical performance. Energy & Environmental Science 8: 1660–1681.

Wang, X., Niu, H. T., Lin, T. and Wang, X. G. 2009. Needleless electrospinning of nanofibers with a conical wire coil. Polymer Engineering & Science 49: 1582–1586.

Wang, X. and Xu, W. L. 2012. Effect of experimental parameters on needleless electrospinning from a conical wire coil. Journal of Applied Polymer Science 123: 3703–3709.

Wang, X., Niu, H. T., Wang, X. G. and Lin, T. 2012. Needleless electrospinning of uniform nanofibers using spiral coil spinnerets. Journal of Nanomaterials 2012: 785920.

Wang, X., Wang, X. G. and Lin, T. 2014a. 3D electric field analysis of needleless electrospinning from a ring coil. Journal of Industrial Textiles 44: 463–476.

Wang, X., Lin, T. and Wang, X. G. 2014b. Scaling up the production rate of nanofibers by needleless electrospinning from multiple ring. Fibers and Polymers 15: 961–965.

Wang, X. F., Ding, B., Yu, J. Y. and Wang, M. R. 2011. Engineering biomimetic superhydrophobic surfaces of electrospun nanomaterials. Nano Today 6: 510–530.

Wei, Q. L., Xiong, F. Y., Tan, S. S., Huang, L., Lan, E. H., Dunn, B. et al. 2017. Porous one-dimensional nanomaterials: design, fabrication and applications in electrochemical energy storage. Advanced Materials 1602300.

Wu, J., Wang, N., Wang, L., Dong, H., Zhao, Y. and Jiang, L. 2012. Electrospun porous structure fibrous film with high oil adsorption capacity. ACS Applied Materials & Interfaces 4: 3207–3212.

Xu, M. J., Wang, M. X., Xu, H., Xue, H. G. and Pang, H. 2016. Electrospun-technology-derived high-performance electrochemical energy storage devices. Chemistry-An Asian Journal 11: 2967–2995.

Yamashita, Y., Tanaka, A. and Ko, F. 2007. Characteristics of elastomeric nanofiber membranes produced by electrospinning. Journal of Textile Engineering 53: 137–142.

Yang, R. R., He, J. H., Xu, L. and Yu, J. Y. 2009. Bubble-electrospinning for fabricating nanofibers. Polymer 50: 5846–5850.

Yang, Y., Jia, Z. D., Li, Q., Hou, L., Liu, J., Wang, L. M. et al. 2010. A shield ring enhanced equilateral hexagon distributed multi-needle electrospinning spinneret. IEEE Transactions on Dielectrics and Electrical Insulation 17: 1592–1601.

Yang, Z. Z., Zhang, Z., Jiang, Y. Z., Chi, M. Q., Nie, G. D., Lu, X. F. et al. 2016. Palladium nanoparticles modified electrospun $CoFe_2O_4$ nanotubes with enhanced peroxidase-like activity for colorimetric detection of hydrogen peroxide. RSC Advances 6: 33636–33642.

Yarin, A. L. and Zussman, E. 2004. Upward needleless electrospinning of multiple nanofibers. Polymer 45: 2977–2980.

Yu, H. Q., Li, Y., Song, Y., Wu, Y. B., Lan, X. J., Liu, S. M. et al. 2017. Ultralong well-aligned TiO_2. Ln^{3+} (Ln = Eu, Sm, or Er) fibres prepared by modified electrospinning and their temperature-dependent luminescence. Scientific Reports 7: 44099.

Yu, Z. N., Tetard, L., Zhai, L. and Thomas, J. 2015. Supercapacitor electrode materials: nanostructures from 0 to 3 dimensions. Energy & Environmental Science 8: 702–730.

Zainab, G., Wang, X. F., Yu, J. Y., Zhai, Y. Y., Babar, A. A., Xiao, K. et al. 2016. Electrospun polyacrylonitrile/polyurethane composite nanofibrous separator with electrochemical performance for high power lithium ion batteries. Materials Chemistry and Physics 182: 308–314.

Zhang, B., Yan, X., He, H. W., Yu, M., Ning, X. and Long, Y. Z. 2017b. Solvent-free electrospinning: opportunities and challenges. Polymer Chemistry 8: 333–352.

Zhang, L. and Jin, C. X. 2011. High-voltage electrospinning apparatus with electron shuttle. CN Patent # 102268747 A.

Zhang, L. H., Duan, X. P., Yan, X., Yu, M., Ning, X., Zhao, Y. et al. 2016. Recent advances in melt electrospinning. RSC Advances 6: 53400–53414.

Zhang, M. F., Zhao, X. N., Zhang, G. H., Wei, G. and Su, Z. Q. 2017a. Electrospinning design of functional nanostructures for biosensor applications. Journal of Materials Chemistry B 5: 1699–1711.

Zheng, T., Yue, Z. L., Wallace, G. G., Du, Y., Martins, P., Lanceros-Mendez, S. et al. 2017. Local probing of magnetoelectric properties of PVDF/Fe_3O_4 electrospun nanofibers by piezoresponse force microscopy. Nanotechnology 28: 065707.

Zheng, Y. S., Liu, X. K. and Zeng, Y. C. 2013. Electrospun nanofibers from a multihole spinneret with uniform electric field. Journal of Applied Polymer Science 130: 3221–3228.

Zhmayev, E., Cho, D. and Joo, Y. L. 2010. Nanofibers from gas-assisted polymer melt electrospinning. Polymer 51: 4140–4144.

Zhou, F. L., Gong, R. H. and Porat, I. 2009. Mass production of nanofibre assemblies by electrostatic spinning. Polymer International 58: 331–342.

Zhou, F. L., Gong, R. H. and Porat, I. 2010. Needle and needleless electrospinning for nanofibers. Journal of Applied Polymer Science 115: 2591–2598.

Chapter 2

Theories and Principles behind Electrospinning

*Yichun Ding, Wenhui Xu, Tao Xu, Zhengtao Zhu** and *Hao Fong**

Introduction

The fast developing technology of electrospinning is a unique way to produce polymer, ceramic, carbon/graphite, composite, and hierarchically structured fibers with diameters typically being hundreds of nanometers (commonly known as electrospun nanofibers) (Bognitzki et al. 2001, Deitzel et al. 2001, Ding et al. 2003, Ding et al. 2016a, Doshi et al. 1995, Dzenis 2004, Fong et al. 2001, Huang et al. 2003, Jaeger et al. 1998, Kim et al. 2003, Lai et al. 2014, Li et al. 2003, Nan et al. 2017, Reneker et al. 1996, Zhang et al. 2014a, Zhou et al. 2009). Unlike conventional fiber spinning techniques (i.e., melt spinning, wet spinning, dry spinning, etc.), which are capable of producing fibers with diameters down to the micrometer range (ca., 5–15 μm), the electrospinning technique is capable of producing fibers with diameters down to the nanometer range (ca., 10–1000 nm). In electrospinning, the electrostatic force is utilized to drive the process and to produce the fibers. Polymer nanofibers are electrospun directly from polymer solutions or melts. Ceramic nanofibers can be made by electrospinning the sol-gels of their alkoxide precursors followed by high temperature pyrolysis. Carbon/graphite nanofibers can be made through carbonization/graphitization of their polymer nanofiber precursors.

Electrospun nanofibers possess many extraordinary properties, including small diameters and the concomitant large specific surface areas, ordered molecular orientation, and the resulting superior directional strength, high aspect ratio (i.e., greater than 1000:1) with rarely microscope identifiable fiber ends, etc. The nanofibers

Program of Biomedical Engineering, South Dakota School of Mines and Technology, 501 East Saint Joseph Street, Rapid City, South Dakota 57701, USA.
* Corresponding authors: Zhengtao.Zhu@sdsmt.edu; Hao.Fong@sdsmt.edu

can be prepared with various morphologies (e.g., cylinder-shaped, beaded, wrinkled, foamed, and ribbon-shaped). Different nanoscale fillers (e.g., layered silicates and carbon nanotubes) can be incorporated into nanofibers with the filler particles highly aligned along the fiber axes (Fong et al. 2002, Ge et al. 2004). In addition, the non-woven fabrics made of electrospun nanofibers offer unique capabilities to control the pore sizes among nanofibers. Moreover, the nanofibers can also serve as templates for the preparation of various nanotubes (Hou et al. 2002, Hou et al. 2004, Li et al. 2004, Loscertales et al. 2004). Unlike nanorods, nanowires, nanotubes, and nano-whiskers, which are prepared mostly by synthetic, bottom-up methods, electrospun nanofibers are produced through a top-down nano-manufacturing process, which results in low-cost electrospun nanofibers that are also relatively easy to align, assemble, and process into applications; furthermore, electrospun nanofibers are often continuous and do not require further expensive purification. As a result, the electrospun nanofibers have been of significant scientific, military, and commercial interests, including composite, filtration, protective clothing, catalysis, agriculture, biomedical applications (e.g., tissue engineering and drug delivery), electronic applications (e.g., sensors and detectors), and energy-related applications (e.g., solar cells, batteries, fuel cells, and supercapacitors) (Bergshoef et al. 1999, Ding et al. 2016b, Fong 2004, Guo et al. 2017, Kim et al. 1999a, 1999b, Li et al. 2002, Ma et al. 2017, Peng et al. 2014, Presser et al. 2011, Tsai et al. 2002, Wang et al. 2014, Wang et al. 2016, Wang et al. 2015, Wang et al. 2017a, Wang et al. 2017b, Xu et al. 2015, Yao et al. 2017).

Electrospinning Process

The electrospinning process involves a complex combination of fluid mechanics, polymer science, and electrostatics. Many factors such as spin dope characteristics (e.g., viscosity, dielectric constant, and surface tension coefficient), process variables (e.g., applied voltage, distance between spinneret and collector, and spin dope flow rate), and environmental conditions (e.g., humidity, temperature, and solvent vapor pressure) can significantly affect the process, and further influence the morphological structures and properties of nanofibers. Till date, the comprehensive understanding of electrospinning is still under further investigation, and the processing control has not yet been fully accomplished.

Electrostatic aerosol spraying of small molecules can be considered to be the forerunner of electrospinning, and this has been studied for a long time. It has been adopted for applications such as spraying of paints in the painting industry, spraying of pesticides and insecticides in agriculture, spraying of vitamins and other additives in the food industry, as well as spraying of coatings in the semiconductor industry. More than a century ago, Lord Raleigh studied the instabilities that occurred with electrically charged liquid droplets. His research revealed that when the electrostatic force overcame the surface tension, a liquid jet could be created (Rayleigh 1882). In 1952, Vonnegut and Neubauer (Vonnegut et al. 1952) produced uniform streams of highly charged droplets with diameters in the micron range by applying electrostatic potential of 5–10 kilovolts to liquids flowing from capillary tubes. Their experiment proved that mono-dispersed aerosols with particle sizes of a micron or less could be formed from the pendent droplet at the end of a pipette, and that the size/diameter of the droplet was sensitive to the applied potential. Unlike electrostatic aerosol spraying that applies to small molecules, electrospinning applies to macromolecules or sol-gels. It is believed that the topological makeup and the physical entanglement of macromolecular chains (and the related viscoelasticity of spin dope) can prevent the capillary breakup of the electrospinning jet and result in the formation of nanofibers.

In general, the electrospinning process can be considered to comprise of three stages: (1) the initiation of an electrospinning jet/filament and the extension of the jet along a straight trajectory; (2) the growth of the bending instability and the further extreme elongation of the jet which allows the jet to become very long and thin while following a looping and spiraling path; (3) the solidification of the jet (either through solvent evaporation or through melt cooling), leading to the formation of nanofibers. Figure 2.1 schematically depicts the electrospinning process. A spin dope is placed in a container (e.g., a regular glass pipette) and direct current (DC) high voltage, usually in the range of 5–40 kilovolts, is applied to the spin dope through an electrode (e.g., a metal wire). An electrically grounded collector is placed at a certain distance (known as gap distance) away from the spinneret. The gap distance is usually in the range of several centimeters to approximately one meter. When the electrostatic field reaches a critical value and the electrostatic force overcomes the surface tension and the viscoelastic force, a jet/filament ejects and travels straight for a certain distance (known as jet length). The jet then starts to bend, forming helical loops. This phenomenon is termed as "bending (or whipping) instability" (Hohman et al. 2001, Reneker et al. 2008, Reneker et al. 2000, Shin et al. 2001, Theron et al.

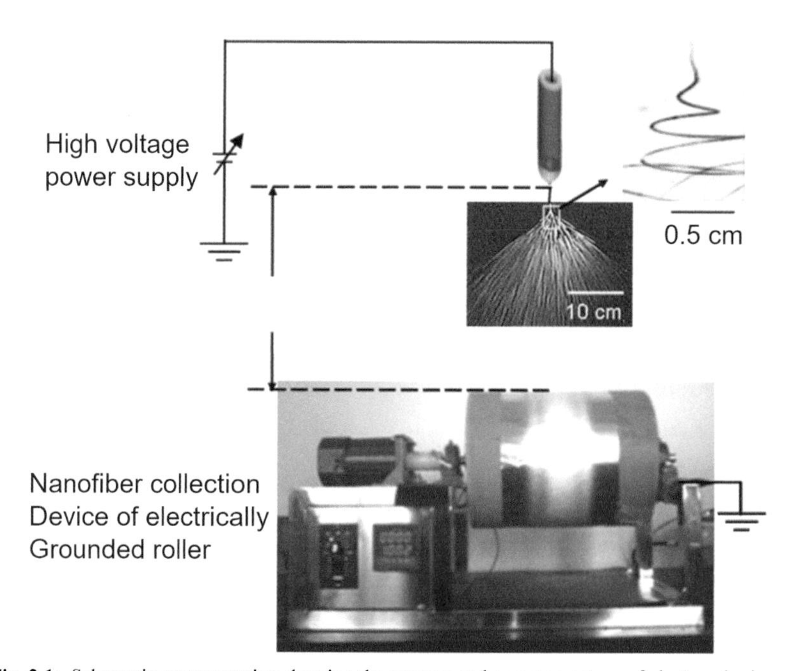

Fig. 2.1: Schematic representation showing the process and common set-up of electrospinning.

2001). Typically, the bending instability causes the diameter of an electrospinning jet to reduce by more than 100 times and the length of the jet to elongate by more than 10,000 times in a very short time period of 50 milliseconds or less. As a result, the elongational flow rate during bending instability is extremely large (up to 1,000,000 s^{-1}). This huge elongational flow rate allows the close alignment along the nanofiber axis of macromolecules as well as nanofillers (e.g., layered silicates and carbon nanotubes, if they are present). In electrospinning (especially during bending instability), the traveling jets solidify very rapidly; and the resulting nanofibers can be collected either as non-woven fabrics or as yarns.

Before high-speed digital cameras were available, visual observations, and home video images with the speed of 30 frames per second (Fig. 2.1), of electrospinning jet/filament were interpreted as evidence that electrospinning was a process that split/splayed the primary jet into many smaller jets. The splitting/splaying jets were believed to have emerged from the region at the end of the straight segment (known as conical envelope). Such visual observations were misleading. After a high-speed digital camera (with the speed of 2,000 per second or higher) was employed, it became obvious that the conical envelope consisted of one looping, spiraling, continuous, and gradually thinner jet, as shown in the inset of Fig. 2.1. The lines observed in visual observations and in home video images were due to the fast movements of bright specular reflection spots on the bending jet. It is necessary to note that, although the bending instability is considered to be the primary reason for the formation of nanofibers, splitting/splaying has also been seen sometimes; and it is believed that the splitting/splaying process provides a viable alternative mechanism for the formation of nanofibers (Reneker et al. 2000).

Jet Initiation

Jet initiation is a complex and intriguing process. While jet initiation is generally considered less important in the electrospinning process than the maintenance of a stable jet, it is an essential step that deserves separate attention. In a simple electrospinning set-up, a pendent droplet of spin dope is supported by surface tension at the tip of a spinneret. When an electrostatic field (typically between 0.1 and 5.0 kV/cm) is applied between the spinneret and the grounded collector, the motion of ions charges the surface of spin dope droplet. As soon as the electrostatic force overcomes the surface tension and the viscoelastic force (which is associated with the viscoelasticity of spin dope), a liquid jet ejects from a conical protrusion (commonly known as the Taylor Cone) that forms on the surface of pendant droplet. The jet is electrically charged and it carries away the charges generated by the electrostatic field. Increasing the field strength usually results in (1) shorter initiation time, (2) higher charge density carried by the jet, (3) faster flow rate, and (4) sometimes, unstable and/or multiple jets.

Figure 2.2 shows a representative sequence of spontaneous jet initiation taken with a digital camera using the speed of 500 frames per second. A 3.0 wt.% polyethylene oxide (PEO) aqueous solution was used as the model spin dope. The solution flowed through an orifice with the diameter of 300 μm at the bottom of a metal spoon. The electrostatic field along the axis of the jet was 0.5 kV/cm. The

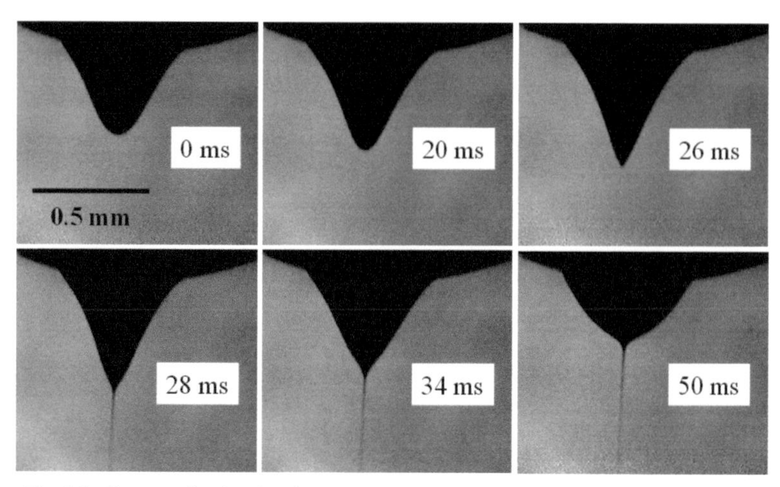

Fig. 2.2: Photographs showing the representative sequence of spontaneous jet initiation.

photographs in Fig. 2.2 were taken sequentially as the high potential was applied to the stable pendent droplet. During the spontaneous jet initiation, the conical protrusion became shaper and sharper due to the movement and accumulation of field-induced charges, which further resulted in the local electrostatic field at the tip of the conical protrusion becoming stronger and stronger. When the electrostatic force was high enough to overcome the surface tension and viscoelastic forces, an electrospinning jet was ejected. After the jet ejection, the conical protrusion relaxed to a rounded and steady shape in a short time period (ca., 20 ms). It is noteworthy that the electrospinning jet can also be initiated at a much lower potential by mechanically pulling a jet out of the pendent droplet, since the potential required for initiation is typically two or more times higher than that required for maintaining the jet flow. It has also been found that a well-maintained shape of the pendent droplet is crucial to keep the electrospinning process steady.

The typical diameters of electrospinning jets/filaments near the conical protrusion are in the range of 20 to 100 mm. The diameter decreases with the jet traveling away from the pendent droplet, which implies the jet is stretched during traveling. A higher electrostatic field and/or a lower solution surface tension coefficient often favor the formation of a thicker jet. Addition of an electrolyte (e.g., sodium chloride) into spin dope, with other parameters unchanged, often reduces the jet diameter. Increasing the concentration (i.e., viscosity) of the spin dope does not always increase the jet diameter. The largest jet diameter occurs when the solution viscosity is in a medium range; both higher and lower viscosity favors a thinner jet. The effect of the molecular weight of the solute/polymer on the jet diameter is complex and is not yet well understood.

Bending Instability and Extreme Elongation of the Jet

After initiation, the path of the electrospinning jet is straight for a distance typically from ~ 0.1 to ~ 10 cm, depending upon the spin dope properties and processing conditions. Thereafter, an electrically driven bending instability starts to grow, triggered by the perturbation of the lateral position and lateral velocity of the jet. During bending instability, the repelling forces between the charges carried with the jet cause every segment of the jet to elongate continuously along a changing path until the jet solidifies. Since the jet can be stretched up to 100,000 times longer during electrospinning, the geometrically simple idea is that the jet would elongate in a straight line along its axis, leading to an implausibly high velocity (up to 100,000 m/s) at the thin, leading end of such a straight jet. Instead, the jet bends and develops a series of lateral movements that grow into spiraling loops. Each of these loops grows larger in diameter as the jet grows longer and becomes thinner, as shown in the inset of Fig. 2.1.

The electrically-driven bending instability generally occurs in three steps: (1) the jet, traveling in a straight path, suddenly develops an array of bends; (2) as the segments of the jet in each bend elongate, the linear array of bends becomes a series of spiraling loops with increasing diameters; and (3) as the perimeter of each loop increases, the cross-sectional diameter of the jet becomes smaller, followed by the development of smaller scale bends along the loops. After the bending instability, the axis of a particular segment of the jet may lie in any direction. The continuous elongation of each segment is most strongly influenced by the repulsion between the charges carried by adjacent segments of the jet. The external electrostatic field, by acting on the charged jet, causes the entire jet to drift towards the grounded collector.

The high-speed photographs in Fig. 2.3 illustrate these three steps. A straight electrospinning jet entered the region (top-left in the images) where the electrically-driven bending instability produced an array of helical bends. While the jet ran continuously, it shifted through a series of similar but changing paths. Most of the loops moved downward at a velocity of about 1 m/s, but some loops with larger diameters remained in the field of view for a longer time. For example, the straight thin segment that ran horizontally across the left image in Fig. 2.3 was part of such a loop that remained in view for over 15 milliseconds. This segment was smooth until, in a time interval of one millisecond, the bends shown in the right image of Fig. 2.3

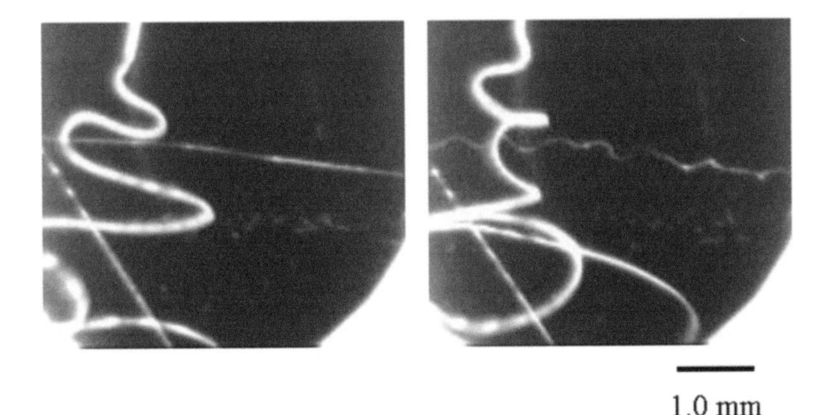

Fig. 2.3: High-speed photographs showing the bending instability occurs in three steps.

developed. During this 15 millisecond period, many bends and loops formed and moved downward through the field of view. The diameter of every segment became smaller, and the length of every segment increased.

It is often possible to follow the evolution of the shape of segments, such as those shown in the image of Fig. 2.3, back to the time at which they entered the upper left corner of the image, when stepping backwards through time sequences in the image files created by the high speed digital camera. It is more difficult to follow the evolution of the jet into the smaller scale of bending because the images become fainter, and are soon ambiguous as the jet becomes thinner. The elongation and the resultant thinning of the jet continue as long as the charges on the jet supply enough force to overcome the surface tension and viscoelastic forces. The viscosity of the jet increases as the solvent evaporates, and eventually the elongation stops.

The "area reduction ratio", which is defined as the ratio of the cross-sectional area of the upper end of a segment to the cross-sectional area at the lower end of the same segment, is equal to the draw ratio if the volume of material in the segment is conserved. Consider a jet, with a 6 wt.% concentration of a polymer in a volatile solvent. The jet reduces from a diameter of 50 μm (right after jet initiation) to a dry nanofiber with a relatively large diameter of 0.5 μm. Thus the cross-sectional area reduction ratio is 10,000. The drying process accounts for a factor of 16, and the elongation of initial straight part of the jet contributes an additional factor of 5. The remaining area reduction ratio (i.e., 125) occurs during the bending instability. Nanofibers with diameters of 0.05 μm or less have often been observed in experiments. The corresponding area reduction ratio is then 12,500 or higher. While 12,500 is a large area reduction ratio, it occurs when many segments of the jet are drawn with expanding loops in different directions at the same time. If the jet were drawn in a straight line to a ratio of 12,500, the required velocity at the nanofiber end of the jet would be much faster than the speed of sound in air. The actual bending and looping of the electrospinning jet achieve very high elongation without such an unreasonably high jet/fiber-traveling velocity.

The elongational flow rate can be simply estimated by using the area reduction ratio and the time of the jet travel. The traveling time that a typical segment of the jet is in flight ($\delta t = 0.2$ s) is estimated using the gap distance (20 cm) divided by the downward velocity of the jet (1 m/s). The elongated flow rate is defined as ($\delta\zeta/\delta t\ \zeta$), where ζ is the initial segment length and $\delta\zeta$ is the final segment length. The draw ratio $\delta\zeta/\zeta$ has the same value of the area reduction ratio, which is around 125 to 12,500, as shown above. Therefore, the elongational flow rates are 625 and 62,500 s^{-1}, respectively, for the two cases described above. This estimation is very conservative, and the elongational flow rate during the bending instability can be much higher (i.e., up to 1,000,000 s^{-1}). Based on previous research, the transformation from a random coil to an elongated macromolecule occurred when the elongational flow rate, multiplied by the conformational relaxation time of the molecule, is greater than 0.5 (De Gennes 1974, Smith et al. 1998). Assuming that the conservatively estimated relaxation time of spin dope is 0.01 seconds, the macromolecules are expected to be closely aligned along the fiber axis during electrospinning (particularly during bending instability). It is necessary to note, however, if the fibers are not dry enough after the elongation stops, the relaxation and the continuation of solvent evaporation can disrupt the ordered structures and significantly weaken the nanofibers. This may happen at the late period of jet/fiber traveling or after the fibers are collected.

Jet Solidification and Formation of the Nanofibers

In electrospinning, the traveling jet solidifies very quickly and the solidified jet turns into nanofiber. Details about the jet solidification remain a mystery. Given an example of 6 wt.% PEO aqueous solution in an isolated system, and the evaporation heat of water of 539 cal/g, the temperature of the other half has to decrease by 539°C to evaporate half of the water in the solution. During electrospinning, the 6 wt.% PEO aqueous solution becomes almost completely dry PEO nanofibers in a period of less than one second. Where the evaporation heat comes from and how the heat can be transferred so quickly are interesting questions, even when the electrospinning jets are small, the surface area is large, and the evaporation heat of water in the charged jet under the electrostatic field may be less. It is possible that the evaporation heat comes from two sources: (1) the environment and (2) the energy stored in the electrospinning jet.

During the stretching of a charged jet, the jet remains continuous without capillary breakup. The longer the solidification time is, the more the jet can be lengthened. The solidification time is related to many factors, such as solvent volatility, solvent vapor pressure in the environment, volumetric charge density carried by the jet, and the strength of the applied electrostatic field. It is crucial to systematically and quantitatively study the jet solidification through designing an electrospinning apparatus with controllability over processing variables (e.g., electrostatic field, gap distance, and flow rate) and environmental conditions (e.g., temperature, vapor pressure, humidity).

Since detailed understanding of bending instability and jet solidification has not been accomplished, fully controlling the morphological structures and properties of the electrospun nanofibers remains a technological bottleneck; for example, there is still no effective way to control the diameters of nanofibers. Although it was reported that the smallest diameters occurred during the combination of excessive charge density carried by the jet, low flow rate, and with low solution concentration, these conditions are either difficult to adjust in experiment or limited to a narrow window (Deitzel et al. 2001, Fong et al. 2001). An analytical model which simplified the fiber diameter as a function of surface tension, flow rate, and electric current, predicted the existence of minimum diameters for electrospun nanofibers (Fridrikh et al. 2003). Such a conclusion is questionable, because the solvent evaporation and the resulting viscoelasticity variations of the jet were not taken into account. Studies of the variations in diameters of PEO nanofibers have shown that, as long as dry and smooth (i.e., no beads and/or beaded fibers) PEO nanofibers are formed, the diameters do not change significantly with the strength of the electric field (no matter the field strength is changed by adjusting the applied voltage or by adjusting the gap distance). The diameters also do not change appreciably with the solution concentration (Fong et al. 2001).

In general, the formation of nanofibers is a delicate and complicated balance of three major forces involved in the electrospinning process: the electrical force, surface tension, and the viscoelastic force. Among these three forces, the electrical force always favors the formation of the product with the highest surface areas. Surface tension always favors the formation of the product with the smallest surface areas. Viscoelastic force is a force which varies significantly with the evaporation of solvent and is the main reason which prevents the break up of the electrospinning jet/filament into droplets/beads. When the electrical force is dominant, viscoelastic force works against the electrical force. When surface tension is dominant, viscoelastic force works against surface tension. Theoretically, the smallest nanofibers are capable of being formed under two conditions: (1) when the excess charge density carried by the electrospinning jet is high, and (2) when the time period is long enough and the viscoelastic force is high enough to prevent the capillary breakup of the jet, but low enough to allow the electrical force to effectively stretch the jet. Both conditions have not yet been systematically investigated. For condition (1), it has been revealed that the addition of soluble electrolytes to spin dope (e.g., addition of sodium chloride to PEO aqueous solution) can significantly increase the excess charge density carried by the jet and cause the formation of smaller diameter nanofibers (Fong et al. 2001). This method, however, also creates negative effects, such as (a) a smaller flow rate and the resulting decrease in nanofiber productivity, and (b) the contamination of the prepared nanofibers by the electrolytes. The removal of the electrolytes without sacrificing the properties of nanofibers may be difficult. For condition (2), further understanding of jet solidification is required.

Electrospinning Parameter Variations and the Resulting Polymer Nanofibers

Electrospinning of polymer nanofibers can be traced back to 1934 (Formhals 1934), when Formhals invented a process for making polymer fibers by using only electrostatic force. Formhals obtained a series of patents on his electrospinning inventions (Formhals 1934, 1937, 1939a, 1939b, 1940a, 1940b, 1944). Later, Gladding (Gladding 1939) and Simons (Simons 1966) improved the electrospinning apparatus and made the process more stable. In the 1970's, the DuPont Company continued to investigate the process of electrospinning, and Baumgarten published photographs with other data describing the continuous formation of acrylic polymer fibers having diameters less than one micrometer (Baumgarten 1971). The common feature of these early researches was that the spin dopes were polymer solutions. Moreover, polymer fibers could also be electrospun from their melts. Manley and Larrondo (Larrondo et al. 1981a, 1981b, 1981c) reported in 1981 that polyethylene and polypropylene fibers could be electrospun into fibers from their melts. Mostly due to the limited knowledge and the resulting lack of control over the process, electrospinning was not adopted as a commercial process for the manufacture of polymer fibers. Since the 1990's, rapid developments in nanotechnology have led to an increased attention on electrospinning and the polymer nanofibers created by this process. Numerous research groups all over the world have contributed to the understanding of the electrospinning process, the characterization of the electrospun nanofibers, and the identification of potential applications for the manufactured nanofibers.

In principle, polymer nanofibers can be electrospun from either solutions or melts. The electrospinning from melts has, however, encountered many technical difficulties. These include: (1) the difficulty in getting the excess charge density carried by the electrospinning jet high enough to make fibers with diameters in nanometer range, (2) the difficulty to control the cooling rate of the electrospinning jet: the cooling rate is usually too fast to allow sufficient elongation of the jet, (3) the use of the spinning process has to be in a vacuum or in an inert atmosphere to avoid polymer degradation. In consequence, not many polymer nanofibers have been successfully prepared through melt electrospinning. The fibers reportedly prepared by melt electrospinning usually have a diameter much larger than one micrometer (Larrondo et al. 1981a, 1981b, 1981c, Reneker et al. 2003), thus it is arguable whether these products can be termed as "nanofibers". Nevertheless, the melting electrospinning process does not involve solvents, so no solvent recycling issues are involved. This may lead to lower costs and less environmental concerns. As a result, melt electrospinning remains to be found of interest to industry.

The polymer fibers prepared by solution electrospinning usually have diameters ranging from a few nanometers to one micron (i.e., they are "nanofibers"). Currently, many synthetic and natural polymers have been successfully electrospun into nanofibers, including (1) thermoplastic homopolymers, including vinyl polymers, acrylic polymers, polyamides, polyesters, polyethers, and polycarbonates, (2) thermoplastic copolymers, including vinyl-co-vinyl polymers, acrylic-coacrylic copolymers, and vinyl-co-acrylic polymers, (3) elastomeric polymers, including triblock copolymer elastomers, polyurethane elastomers, and ethylene-properlene-diene-elastomers, (4) high performance polymers, such as polyimides and aromatic polyamides, (5) liquid crystalline polymers, such as poly(p-phenylene terephthalamide) and polyaramid, (6) textile polymers, such as polyethylene terephthalate and polyacrylonitrile, (7) electrically conductive polymers, such as polyaniline, as well as (8) biocompatible polymers like polycarprolactone, polylactide, and polyglycolide, and (9) natural polymers, including proteins, polysaccharides, and nucleic acids.

As long as a polymer is soluble and able to form a solution with high enough concentration, it is possible that the polymer can be electrospun into nanofibers. If the molecular weight of the polymer is low, a high solution concentration is usually required to form continuous nanofibers. As an example, the required minimum concentration for electrospinning nylon 6 (with the average molecular weight of ~ 15,000 g/mol) hexafluoroisopropanol solution into nanofibers is ~ 1.5 wt.%; while for the nylon 6 with the average molecular weight of 10,000 g/mol, the minimum concentration is ~ 3.5 wt.%. When the concentration of polymer solution is lower than the minimum requirement, beaded nanofibers and/or beads (instead of continuous nanofibers) will be formed (Fong et al. 1999a). The effects of molecular weight on the formation of polymer nanofibers are not only limited to the minimum concentration requirement. If a polymer has too high a molecular weight, then the solution electrospinning is very difficult and usually will not be stable enough, or the continuous nanofibers cannot be prepared reliably. For example, PEO is a well-studied polymer in electrospinning. When the average molecular weight of PEO exceeds 4 million, the electrospinning process becomes unstable and no dry/continuous PEO nanofiber can be prepared.

In solution electrospinning, the solvent plays an important role. If a solvent is too non-polar (e.g., cyclohexane and toluene), other polar solvents may be required to form a mixture solvent in order to successfully perform electrospinning and prepare continuous nanofibers. For example, the solution of polystyrene in pure toluene cannot be electrospun because the jet cannot be formed, but the addition of a few drops of *N*,*N*-dimethylformamide (DMF) into the solution can successfully generate the jet, resulting in the formation of continuous polystyrene fibers. This suggests that, in the electrospinning process, it is likely that the excess charges carried by the jet/fiber are associated with the solvent molecules and impurities (i.e., trace amount of electrolytes) instead of the polymer molecules.

Fiber Diameter and Electrospinning Parameters

Studies of nanofiber diameter variations on processing parameters have been carried out, showing that when other parameters are held constant, the diameters of PEO nanofibers would not change appreciably with electric field. Similar results have been observed when the electric field is changed by either changing the voltage or by changing the distance between the pendent drop and the collector; moreover, diameters of PEO nanofibers would not vary distinguishably with the change of solution viscosity. However, when NaCl is added into the solution, the PEO nanofibers would be much thinner (Fig. 2.4a); additionally, partial substitution of water with ethanol would make the diameter of PEO nanofibers significantly larger (Fig. 2.4b).

The current resulted from collection of charged electrospinning PEO jet/fiber is increased with the square of applied voltage, as shown in Fig. 2.5a. The current is proportional both to the mass flow rate of the spin dope and to the excess charge density carried by the jet, each of which is proportional to applied voltage, as shown in Figs. 2.5b and 2.5c.

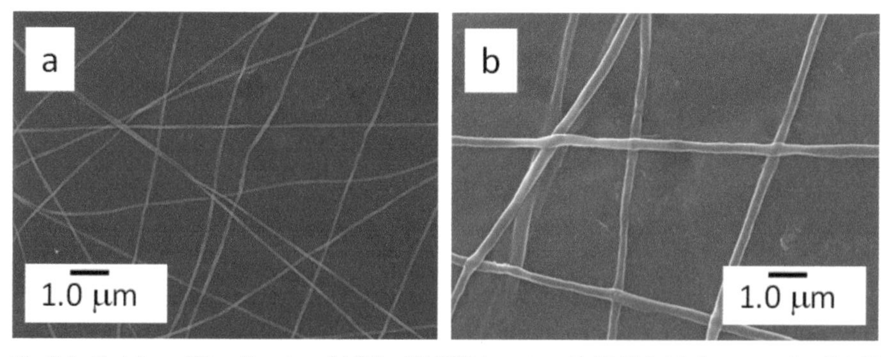

Fig. 2.4: Variation of fiber diameters. (a) 3.0 wt.% PEO in water with NaCl (with the mass ratio of NaCl to PEO being 0.5). (b) 3.0 wt.% PEO in water/ethanol mixture (with the mass ratio of ethanol to water being 2/3).

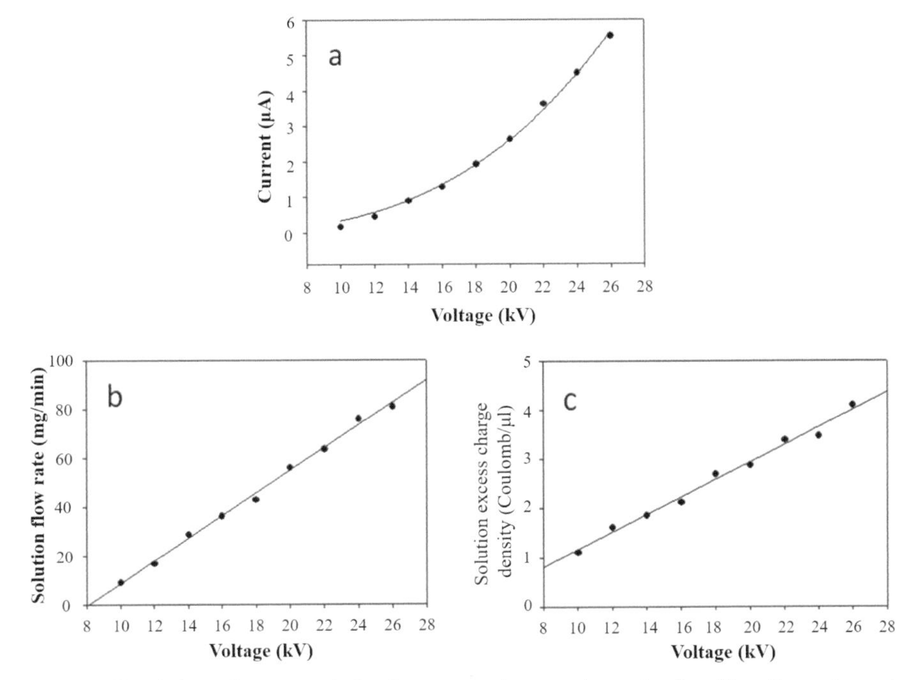

Fig. 2.5: Correlations of current, solution flow rate, and excess charge density with voltage, observed from a solution of 2.44 wt.% PEO in water. The distance from the spinneret to the collector was set at 21 cm.

The length of the straight segment before the onset of bending increases with the electric field and solution viscosity, and decreases with the surface tension coefficient. The higher excess charge density would lead to the shorter straight segment (Fig. 2.6).

The flow rate of the solution was calculated from the mass of collected nanofibers, the concentration of solution, and the collection time. A typical jet has a flow rate in the range from a few milligrams per minute to tens of milligrams per minute. As shown in Fig. 2.7, the flow rate of solution increases with the electric field, while it decreases with the viscosity of solution. Using the same concentration of PEO solution, but changing the solvent from water to a mixture of water and ethanol, would cause the flow rate of solution to increase; on the other hand, the addition of NaCl into PEO aqueous solution would decrease the flow rate.

Morphological Structures of Polymer Nanofibers

The morphological structures of electrospun polymer nanofibers vary. There are many reasons for different morphological structures. The most important ones are that the electrospinning process is driven by electrostatic instability, and that the solvent evaporation plays an important role in the formation of polymer nanofibers. Polymer nanofibers are predominately cylindrically shaped, as shown in Fig. 2.8a, an SEM image of electrospun nylon 6 nanofibers. The formation of cylindrically shaped polymer nanofibers is easy to understand, and cylindrically shaped nanofibers can

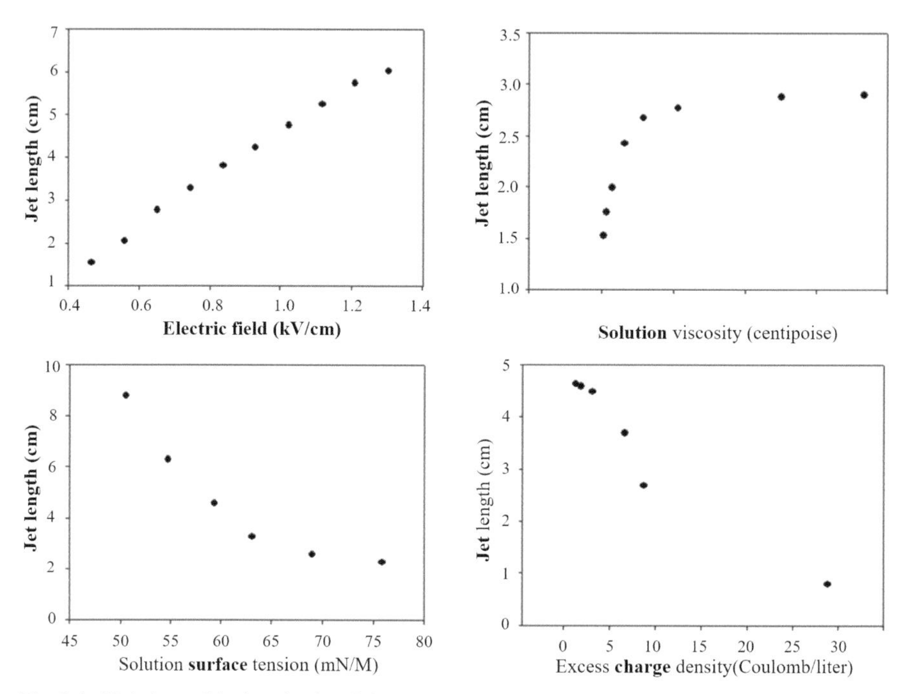

Fig. 2.6: Variations of the length of straight segment (i.e., jet length) on electric field, solution viscosity, solution surface tension coefficient, and excess charge density.

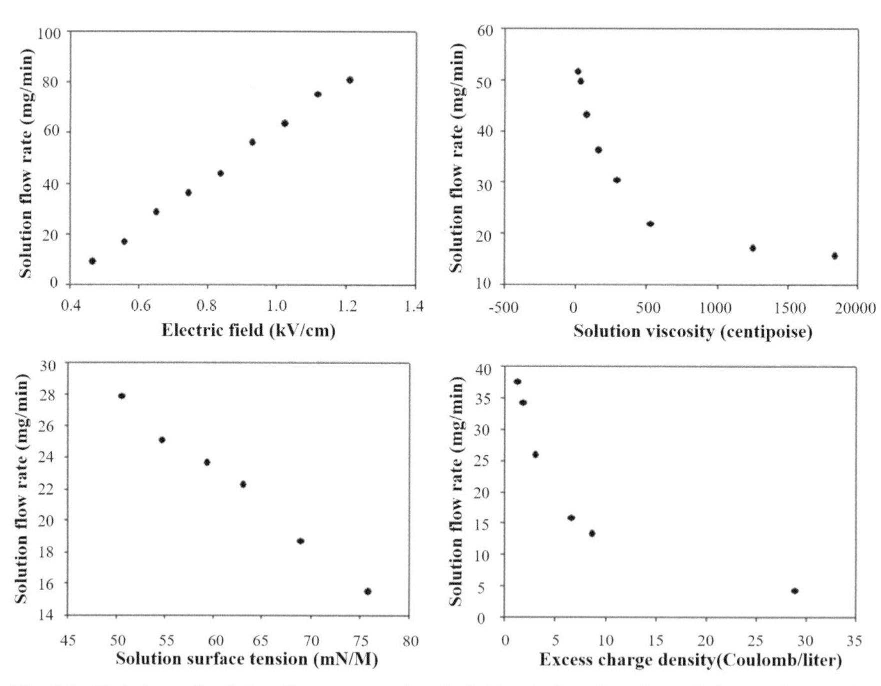

Fig. 2.7: Variations of solution flow rate on electric field, solution viscosity, solution surface tension coefficient, and excess charge density (carried by the electrospinning jet).

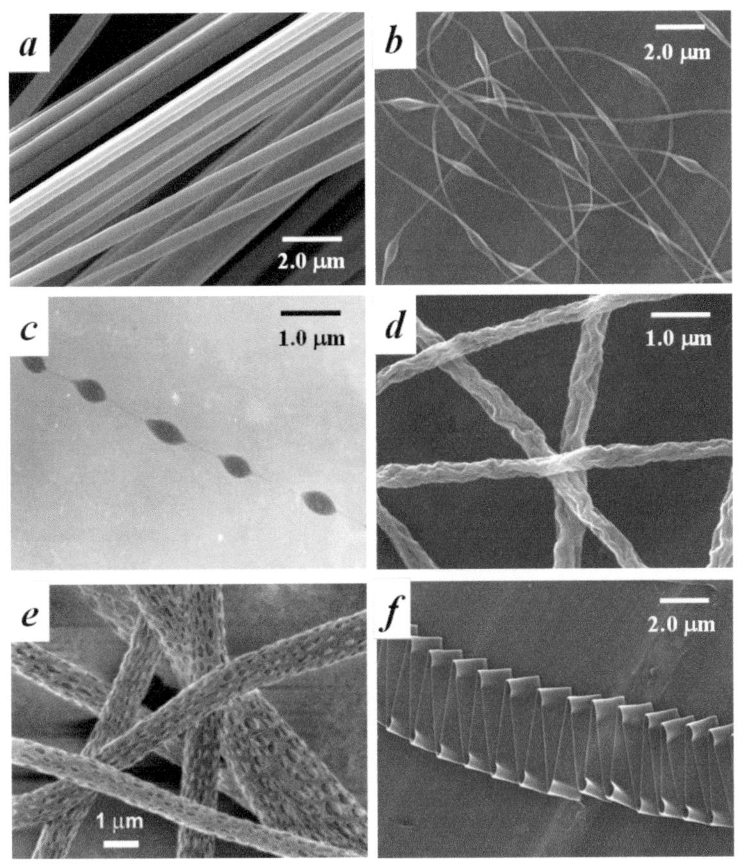

Fig. 2.8: Morphological structures of polymer nanofibers: (a) cylindrically shaped (Fong et al. 2002) (Copyright 2002, reproduced with permission from Elsevier B.V.), (b) beaded-shaped (Fong et al. 1999a) (Copyright 1999, reproduced with permission from Elsevier B.V.), (c) beaded-shaped with ultra-thin nanofibers between beads (Fong et al. 1999b) (Copyright 1999, reproduced with permission from John Wiley & Sons, Inc.), (d) wrinkled (Fong et al. Unpublished Data), (e) foamed (Bognitzki et al. 2001) (Copyright 2001, reproduced with permission from John Wiley & Sons, Inc.), and (f) ribbon-shaped (Fong et al. Unpublished Data).

be electrospun from a wide range of polymers. The morphology of beaded-shaped polymer nanofibers is also commonly observable. The reason for such morphology is due to the capillary breakup of the electrospinning jet by surface tension (Fong et al. 1999a). This is similar to when an electrically driven jet of a small molecule solution forms droplets/beads, which is the case during electrospraying. Unlike small molecule solutions, the pattern of the capillary breakup for polymer solutions is different. Instead of breaking completely, the jets between the droplets form nanofibers; the contraction of the radius of the jet, driven by surface tension, causes the remaining solution to form beads (Fig. 2.8b). As the viscosity of the solution is increased, the beads are larger, the average distance between beads is longer, the average fiber diameter is larger, and the shape of the beads changes from spherical to spindle-like. As the excess charge density increases, the beads become smaller and more spindle-like; and the diameters of fibers become smaller. Decreasing the surface tension by adding alcohol to the solution reduces the size and number of beads. Neutralization of the excess charge carried by the jet using ions generated from an air corona discharger results in the formation of a large number of beads. The reason is that, after the bending instability, the tension along the axis of the fiber, which resists the capillary breakup (i.e., the formation of beads), depends on the repelling force generated by the excess charges carried on the jet. Jaeger and coworkers (Jaeger et al. 1996) studied the surface of beads on PEO nanofibers using atomic force microscopy (AFM). Their results indicated that the beads possessed a highly ordered surface at the molecular level. Sometimes the fibers between the beads can be very thin and the beads can be very small. The TEM image (Fig. 2.8c) of a beaded nanofiber, which was made by electrospinning of styrene-butadiene-styrene tri-block copolymer (SBS), showed that the fibers between the beads had diameters as small as 3 nm (Fong et al. 1999b). Since the diameter of a macromolecule is approximately 0.5 nm, there must be fewer than 50 macromolecular chains in a typical cross-section of such a thin fiber. It is expected that many polymers may form such very thin beaded nanofibers, while the very thin SBS beaded nanofibers were actually observed using TEM because the sample was stained with OsO_4.

Unlike the morphologies of cylindrically shaped and beaded-shaped nanofibers, the morphologies of wrinkled, foamed, and ribbon-shaped nanofibers (as shown in Figs. 2.8d, 2.8e, and 2.8f, respectively) are less commonly observed. Figure 2.8d depicts the wrinkled poly(octoxycarbonyl phenylene) nanofibers, which were electrospun using a solution that contained 30 wt.% poly(octoxycarbonyl phenylene) in a mixture solvent (i.e., 75 wt.% chloroform and 25 wt.% DMF). Although it is suspected that the low molecular weight of polymer and the unusually high solution concentration may be attributed to the formation of such wrinkled nanofibers, the detailed formation mechanism is not clear. Figure 2.8e shows the foamed poly-L-lactide (PLLA) nanofibers, which were electrospun using a solution that contained 5 wt.% PLLA in dichloromethane (Bognitzki et al. 2001). Dichloromethane is a very volatile solvent with the boiling point of 40°C and the vapor pressure of 475 mbar at 20°C. Such a high solvent volatility may be accountable for the formation of PLLA nanofibers with a regular pore structure. The average pore size was in the order of 100 nm in width and 250 nm in length, with the long axis of the pore being oriented along the fiber axis. Such orientation may be the result of uni-axial extension of the jet during electrospinning. The replacement of dichloromethane by solvents with lower vapor pressure (e.g., chloroform) reduced the tendency towards pore formation significantly. Similarly, foamed nanofibers may also be electrospun using other polymers (e.g., polycarbonate and polyvinylcarbazole) with dichloromethane being the solvent. Among all types of uncommon morphologies, the most mysterious and interesting one may be the ribbon-shaped nanofibers (as shown in Fig. 2.8f), which were electrospun using a solution that contained 15 wt.% elastin (a synthetic protein) in a mixture solvent (i.e., 50 wt.% water and 50 wt.% ethanol). The ribbon-shaped elastin nanofibers had very uniform width of ~ 2 μm and thickness of ~ 200 nm. The regular folding structure was believed to be formed during the landing of the ribbon-shaped nanofibers on the collector of aluminum foil. Although Reneker and coworkers proposed an explanation that the formation of ribbon-shaped nanofibers resulted from a thin skin formed upon rapid evaporation of the solvent (Koombhongse et al. 2001), the explanation is debatable because the same solvent mixture (i.e., 50 wt.% water and 50 wt.% ethanol) has not been found to result in the formation of ribbon-shaped nanofibers for other water-soluble polymers (e.g., PEO, polyvinyl alcohol, and polypyrrolidone). The ribbon-shaped nanofibers were also found when electrospinning solutions containing polymers with the capability to form strong inter-molecular hydrogen bonds (e.g., nylon 6/hexafluoroisopropanol solution) (Fong et al. 2002). It is therefore speculated that the formation of ribbon-shaped nanofibers might be somehow related to the nature of polymer, including the formation of hydrogen bonding and other intermolecular interactions. Compared to the sizes of β-pleated sheet found in proteins, however, the size of such ribbon-shaped nanofibers is about three orders of magnitudes larger, which makes the above explanation questionable. The accurate and detailed description of formation mechanisms is still dependent upon further investigations.

Properties of Polymer Nanofibers

Compared to other types/forms of polymeric materials, the most distinctive physical property of electrospun polymer nanofibers is that the macromolecular chains are aligned along the fiber axis, while the overall nanofiber structures are relatively amorphous. Almost all kinds of polymer nanofibers show strong birefringence, even those made of amorphous polymers, including non-tactic polystyrene, non-tactic polymethyl methacrylate, and the triblock copolymer of poly (styrene-co-butadiene-co-styrene). The extremely large elongational rate involved in the electrospinning process (especially during the bending instability) is responsible for the macromolecular alignment and the resulting birefringence of nanofibers. The macromolecular alignment can also be confirmed by polarized FT-IR spectra. As shown in Fig. 2.9a (Liu et al. 2007a), the polarized FT-IR spectra of the aligned nylon 6 nanofibers (similar to those in Fig. 2.8a) show differing absorption intensities at 1550 cm^{-1}, which is associated with the in-plane stretching of N–H bond (Vasanthan et al. 2001). From top to bottom, the 5 curves represent the polarized FT-IR spectra collected at the incident angles (i.e., the angle between the vibrational direction of polarized infrared beam and the axis of aligned nanofibers) of 0° (i.e., parallel), 30°, 45°, 60°, and 90° (i.e., perpendicular), respectively. According to the results of previous research (Vasanthan et al. 2001), the in-plane stretching of N–H bond is sensitive to the incident angle; while the absorption of C=O (at 1644 cm^{-1}) is not. Hence, the sequential decrease in the infrared absorption while the incident angle changes from parallel to perpendicular indicates the nylon 6 macromolecules are aligned along the nanofiber axes. Interestingly, although macromolecules are aligned in the nanofibers, they are less likely to fit into a crystal lattice to form polymer crystallites. During the electrospinning process (especially during the bending instability), the macromolecular chains not only are stretched, but also quickly lose mobility due to the extraordinarily fast solidification rate. Figure 2.9b shows the DSC traces of electrospun nylon 6 nanofibers and nylon 6 solution precipitant at the heating rate of 10°C/min (Liu et al. 2007a). The solution precipitant was prepared by using the same electrospinning solution for making nanofibers, which

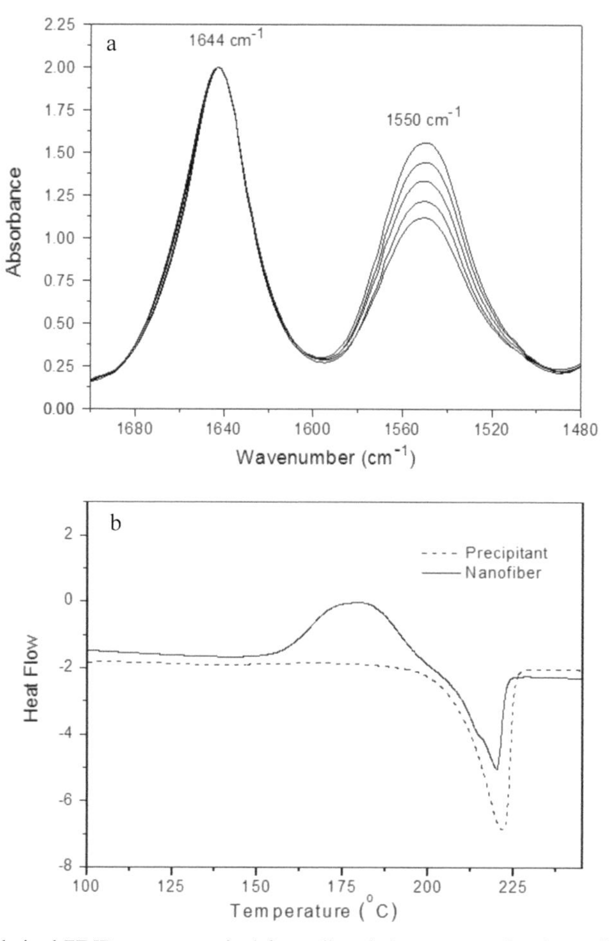

Fig. 2.9: (a) Polarized FT-IR spectra acquired from aligned electrospun nylon 6 nanofibers (Liu et al. 2007a), (b) DSC traces of electrospun nylon 6 nanofibers (solid line) and nylon 6 solution precipitant (dashed line) at the heating rate of 10°C/min (Copyright 2007, reproduced with permission from American Chemical Society) (Liu et al. 2007a).

was 5 wt.% nylon 6 in 1,1,1,3,3,3-hexafluoroisopropanol, and the precipitation was conducted in diethyl ether with vigorous stirring. The DSC traces show that there is a crystallization peak for nanofibers (note that such peak may also be partially due to the tension releasing), while no crystallization peak can be identified for the solution precipitate. This clearly indicated that, even compared to the solution precipitate, the electrospun nanofibers were more amorphous.

Such properties of polymer nanofibers (i.e., molecularly oriented but not crystallized) have also been suggested by X-ray diffraction patterns (XRD). In Fig. 2.10 (Fong et al. 2002), image "a" represents the XRD of a nylon 6 solution cast film, and image "b" represents the XRD of aligned electrospun nylon 6 nanofibers. The two diffraction rings in the solution cast film arise from the (200) and (002), (202) planes of the α-crystalline form of nylon 6. Crystallization in the solution cast

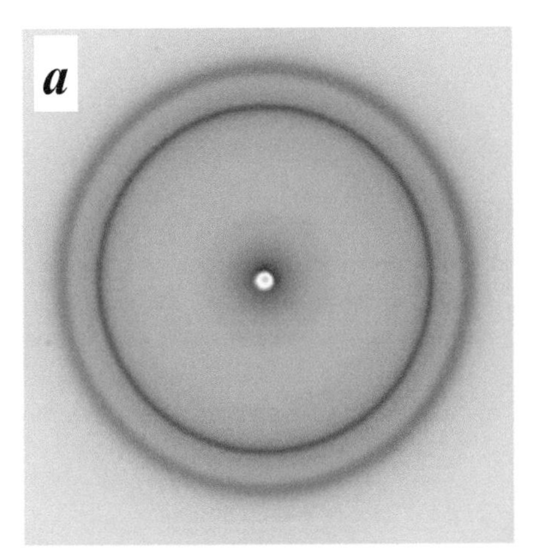

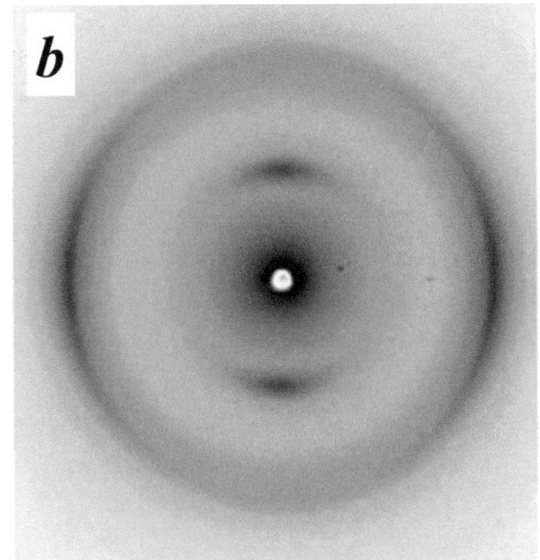

Fig. 2.10: X-ray diffraction patterns of nylon 6: (a) solution cast film, and (b) aligned electrospun nanofibers with vertical fiber axes (Copyright 2002, reproduced with permission from Elsevier B.V.) (Fong et al. 2002).

sample does not exhibit any preferential orientation (i.e., uniform diffraction rings). In contrast, for the aligned electrospun nanofibers, the crystal structure of nylon 6 adopts the meta-stable γ-form (i.e., equatorial reflection from (001) and (200), meridian reflection from (020)), with the layer normal of the crystallites parallel to the fiber axis. This is consistent with previous results that the γ form is preferred in both as spun and spun drawn fibers of nylon 6 (Murthy et al. 1985). That the γ crystal form is preferred in electrospun nylon 6 nanofibers has been indicated by DSC results as well. In Fig. 2.9b, an exothermal peak associated with re-crystallization appears at 175°C, followed by an endothermal peak corresponding to the melting transition temperature at approximately 220°C, which is associated with the γ crystal form. The shoulder at 216°C is attributed to the α crystal form.

The unique macromolecular-oriented but relatively un-crystallized structure of the electrospun polymer nanofibers has profound influences on their mechanical properties as well as other properties. Based on the highly ordered molecular orientation, one could predict that the polymer nanofibers are mechanically strong. One could also predict the polymer nanofibers are mechanically weak because the structure is quite amorphous. Since the measurements of the mechanical properties of single nanofiber encounter technical difficulties, there still lacks sufficient experimental evidence to support whether the polymer nanofibers are mechanically strong. There have been some research efforts to characterize the overall mechanical properties of nanofibers, either through direct measurement of nanofiber non-woven fabrics, or through impregnating the fabrics into a polymer matrix and measuring the mechanical properties of the composites (Fong 2004). Neither method can precisely characterize the properties of nanofibers because many other factors could significantly vary the results. These factors include the extent of fiber-entanglement in the fabric as well as the interfacial properties between the fiber and the matrix. Detailed characterizations depend upon, and merit, further investigation.

Different Types of Electrospinning

Nowadays, the electrospinning technique has become a widely adopted method, which has not only been employed in academic research but also in industrial production (Persano et al. 2013, Yu et al. 2017). Furthermore, various types of electrospinning devices have been designed and developed to prepare nanofibers with different morphologies/properties and/or to resolve/mitigate the problem of low productivity. For example, multi-needle electrospinning has been developed to improve the productivity and to prepare the hybrid nanofiber mats/membranes (Chen et al. 2015, Xi et al. 2014a, Xi et al. 2014b). Additionally, co-axial electrospinning has been developed for making novel structural nanofibers such as core-shell, hollow, and multi-channel nanofibers (Chen et al. 2010, He et al. 2013, Li et al. 2004, Sun et al. 2003, Zhao et al. 2007); and needleless electrospinning and external force-assisted electrospinning (e.g., gas-assisted electrospinning and centrifugal electrospinning) have been designed for large-scale production of nanofibers (Edmondson et al. 2012, Liu et al. 2007b, Lu et al. 2010, Yu et al. 2017, Zhmayev et al. 2010). Although the fundamental principles of these electrospinning techniques are similar to that of single-needle electrospinning (i.e., nanofibers are produced under an applied electric field), the electrospinning processes are more complicated due to the complex electric field distribution; as a result, these electrospinning techniques have both advantages and limitations, which are described in the following sections.

Single-Needle Electrospinning

The initial and probably the most common electrospinning process is based on the single needle equipment; as described above, the process appears to be simple while actually being quite complex; and it is subjective to comprehensive influences of processing parameters, spin dope characteristics, and environmental conditions. Single-needle (Fig. 2.11a) electrospinning is usually capable of making nanofibers from only one type of spin dope, which could be a solution consisting of one solute in one solvent, a liquid consisting of multiple solutes in mixture solvents, as well as a colloid suspension or an emulsion. Thus, the resulting nanofibers could be the nanofibers with one component (e.g., polymer, carbon, and ceramic nanofibers), hybrid/composite nanofibers, as well as hierarchical structured nanofibers and core-shell nanofibers. Typically, the hourly spin dope consumption rate of single-needle electrospinning is in the range of several hundreds of micrograms to a few grams. Evidently, the nanofiber productivity is very low; hence, the single-need electrospinning is not suitable for large-scale production of nanofibers. On the other hand, the set-ups are easy to make; the technique has thus been widely adopted in research laboratories for making nanofibers.

The electrospinning jet/filament is subjected to bending instability which makes the traveling path completely chaotic; as a result, the nanofibers are usually collected as randomly overlaid membrane; however, numerous research endeavors have been devoted to align/pattern the electrospun nanofibers. Unidirectional aligned nanofibers have higher mechanical strength and may have some unique optical and electrical properties, and thus they can be used for the fabrication of composites

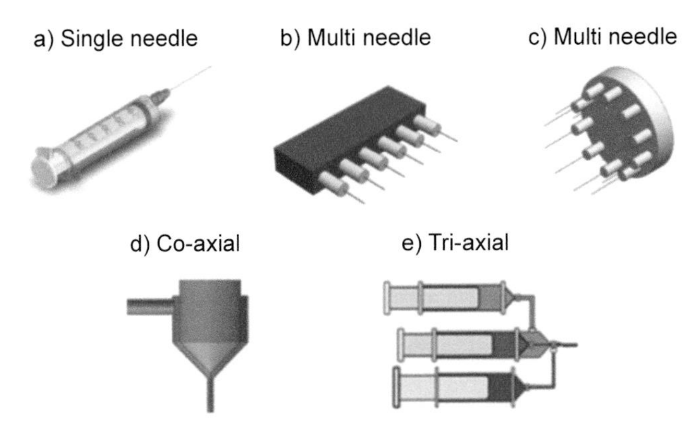

Fig. 2.11: Possible geometries/configurations of needle electrospinning spinnerets (Copyright 2013, reproduced with permission from John Wiley & Sons, Inc.) (Persano et al. 2013).

and the production of electronic devices. Based on the single-needle electrospinning, various types of set-ups have been developed for collecting aligned nanofibers; for example, by using a metal frame, two parallel conducting stripes, and a rotating disk as collectors, and by employing magnetic field assisted collection (Ding et al. 2016a). Near-field electrospinning that is operated with reduced spinning voltage and decreased collection distance is a recently developed technique for effectively controlling the deposition of nanofibers. As described before, in the common electrospinning process, the spin jet/filament travels straight in the initial stage, and then undergoes bending instability because of the repulsive forces (He et al. 2017). However, in the near-field electrospinning process, the bending instability can be significantly restricted due to the shorter collection distance and reduced voltage; consequently, the nanofibers can be deposited in a controllable manner to form different patterns. Therefore, the near-field electrospinning provides an interesting approach for achieving the position-controlled deposition of individual nanofibers for applications such as nanogenerators, wearable sensors, nanodevices, microelectromechanical systems, and tissue engineering (He et al. 2017).

Multi-Needle Electrospinning

Compared to that of single-needle electrospinning, the productivity of multi-needle (multi-spinneret) electrospinning is substantially higher; additionally, this electrospinning technique is also capable of making hybrid nanofiber mats/membranes when two or more spin dopes are simultaneously used. Till date, many spin dopes and different needle/spinneret configurations have been investigated for multi-needle electrospinning. For the configuration in multi-needle electrospinning system, the number and arrangement of needles should be carefully considered. For example, the needle arrays can be arranged in linear (Fig. 2.11b), circular (Fig. 2.11c), or other configurations (Persano et al. 2013).

As the multi-needle electrospinning system/equipment has many needles/spinnerets, the electric field distribution among the needles is uneven and even unstable; and such a situation has profound impacts on the resulting nanofibers and nanofiber mats/membranes. One major issue in the multi-needle electrospinning process is jet deviation, which is caused by the uneven distribution of electric field, particularly if the distance among needles is very close; and the jet deviation may cause problems such as dripping of the spin dope and clogging of some needles (Liu et al. 2013b, Xie et al. 2012). In general, the variations of local electric field at the needles can substantially affect the electrospinning process and lead to significant discrepancies on the morphological structures of the produced nanofibers. Moreover, the nanofiber collection is another issue, because the strong charge repulsion among adjacent electrospinning jets makes nanofibers fly in a chaotic pattern; thus they are difficult to be collected (Zheng et al. 2015). Hence, for continuous, stable, and large-scale production of nanofiber mats/membranes with desired morphological/structural properties (e.g., uniform thickness and appropriate mass per unit area), the multi-needle electrospinning system/equipment has to be judiciously designed.

Multi-Nozzle Electrospinning

Multi-nozzle (multi-hole) electrospinning set-up with several nozzles on a single needle enables the production of nanofibers with interesting morphological structures, such as core-shell and hollow nanofibers. The configuration of nozzles on the needle can be parallel or concentric (e.g., co-axial and tri-axial), as shown in Figs. 2.11d–e (Persano et al. 2013). The simplest multi-nozzle electrospinning set-up has a spinneret equipped with two nozzles, and the set-up can be used to make core-shell nanofibers when the two nozzles are co-axially arranged (Fig. 2.12a) (Sun et al. 2003); when the two nozzles are arranged side by side, it can result in the formation of nanofibers with Janus structures (Fig. 2.12b) (Peng et al. 2016, Starr et al. 2013). Furthermore, if the two spin dopes have significantly different properties (e.g., one made of a polymer with rigid macromolecular chains and the other one made of another polymer with flexible macromolecular chains), this technique can also result in the formation of helical nanofibers (i.e., nano-springs, as shown in Fig. 2.12c) due to different viscoelastic contractions during the process of jet solidification (Chen et al. 2009, Jiang et al. 2014).

Hollow and multi-channel nanofibers are other types of structures that can be achieved by multi-nozzle electrospinning. It is interesting to note that the inside component of core-shell nanofibers (made by the coaxial electrospinning) can be removed upon thermal treatment and/or solvent extraction if it is degradable and/or dissolvable, leading to the formation of hollow nanofibers (Fig. 2.12d) (Li et al. 2004). For example, Chen and coworkers prepared the nanofibers with nanowire-in-microtube structure (Fig. 2.12e) by tri-axial electrospinning followed by removal of the intermediate layer component (Chen et al. 2010). In another example, Zhao and coworkers reported the preparation of multi-channel microtubes by multi-fluidic compound-jet electrospinning (Fig. 2.12f), and the resulting microtubes had different numbers of channels by using spinnerets with different numbers of nozzles (Zhao et al. 2007). Therefore, the multi-nozzle electrospinning is a very effective and versatile technique for making the above-mentioned hierarchically structured nanofibers; however, the practical applications of such nanofibers are limited, because these

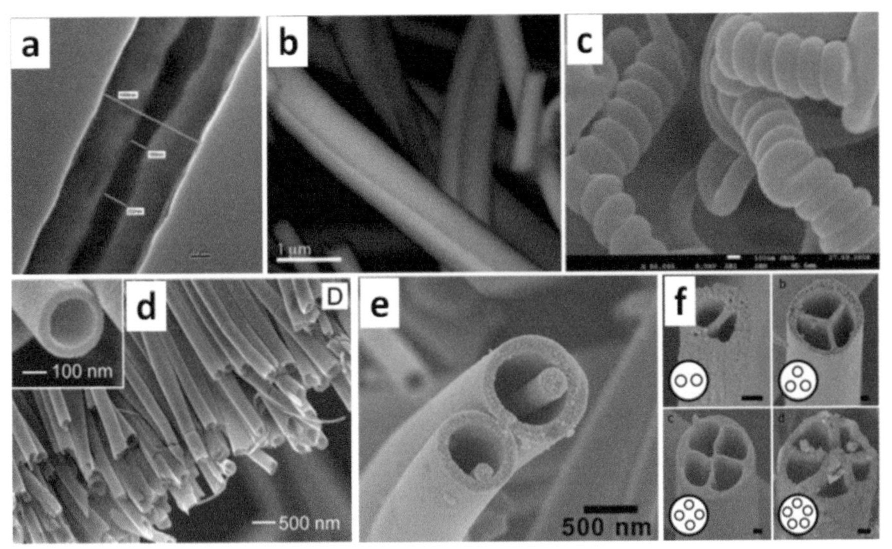

Fig. 2.12: Different morphological structures of electrospun nanofibers prepared by multi-nozzle electrospinning: (a) PDT/PEO core-shell nanofiber (Copyright 2003, reproduced with permission from John Wiley & Sons, Inc.) (Sun et al. 2003), (b) Janus-type bi-component nanofibers (Copyright 2013, reproduced with permission from Royal Society of Chemistry) (Starr et al. 2013), (c) Helical polymer nanofibers (i.e., nano-springs, Copyright 2009, reproduced with permission from John Wiley & Sons, Inc.) (Chen et al. 2009), (d) anatase TiO_2 hollow nanofibers (Copyright 2004, reproduced with permission from American Chemical Society) (Li et al. 2004), (e) Nanofibers with nanowire-in-microtube structure (Copyright 2010, reproduced with permission from American Chemical Society) (Chen et al. 2010), (f) TiO_2 multi-channel nanotubes (Copyright 2007, reproduced with permission from American Chemical Society) (Zhao et al. 2007).

electrospinning processes are highly unstable and the resulting nanofibrous materials are not very morphologically/structurally uniform. Usually and practically, the multi-nozzle electrospinning (e.g., the coaxial electrospinning) is rather difficult to control; thus the large-scale preparation of nanofibers with desired morphological structures remains a technological challenge.

Needleless Electrospinning

The major challenge of electrospinning technique is the low production rate of nanofibers, which significantly limits the commercial applications. The multi-needle electrospinning can considerably improve the productivity, but it is still far away from industrial-scale production. The needleless electrospinning has emerged as an effective technique for further up-scaling the production of nanofibers (Niu et al. 2011). In recent years, various types of needleless electrospinning set-ups have been developed; for example, needleless spinnerets of stationary wire and rotating cylinder have been made for large-scale production of electrospun nanofibers (Yu et al. 2017).

The general strategy of various needleless electrospinning set-ups is to make a "spinneret" that can concurrently produce a large amount of jets from open surface of spin dope under the applied high voltage. Such spinnerets can be divided into two types of rotating and stationary set-ups (Yu et al. 2017). As shown in Fig. 2.13, there are different devices of rotating spinnerets, such as disk, wire, ball, cylinder, cone, spiral coil, beaded chain, and others (Yu et al. 2017). During the process of needleless electrospinning, the spinneret is partially immersed in spin dope; and due to the rotation, a thin layer of spin dope can be automatically loaded on the surface of spinneret. The rotation can also create conical spikes on the surface of solution. When a high voltage is applied, these conical spikes are drawn to form the "Taylor cones", and jets/filaments are then ejected from the spikes under sufficient electric force to result in the formation of nanofibers.

Compared to needle electrospinning, needleless electrospinning has some advantages, which include that the nanofiber productivity can be substantially improved and that the needle clogging problem can be avoided. One the other hand, morphological properties of the produced nanofibers are often difficult to precisely control. This is because the electric field distribution during needleless electrospinning is more complicated than that during needle electrospinning; additionally, the shape/geometry of spinneret and the properties (e.g., rheological properties) of spin dope also have significant effects (Niu et al. 2012). Another serious issue of needleless electrospinning is that spin dope droplets can often be generated, and the resulting beads/particles (particularly if there is some leftover solvent) can distinguishably ruin the collected nanofiber mat/membrane. Additionally, even though the needleless electrospinning is an effective method for large-scale production of nanofibers, the designs of set-up/equipment are still upon further improvements to meet the practical demands of industrial production.

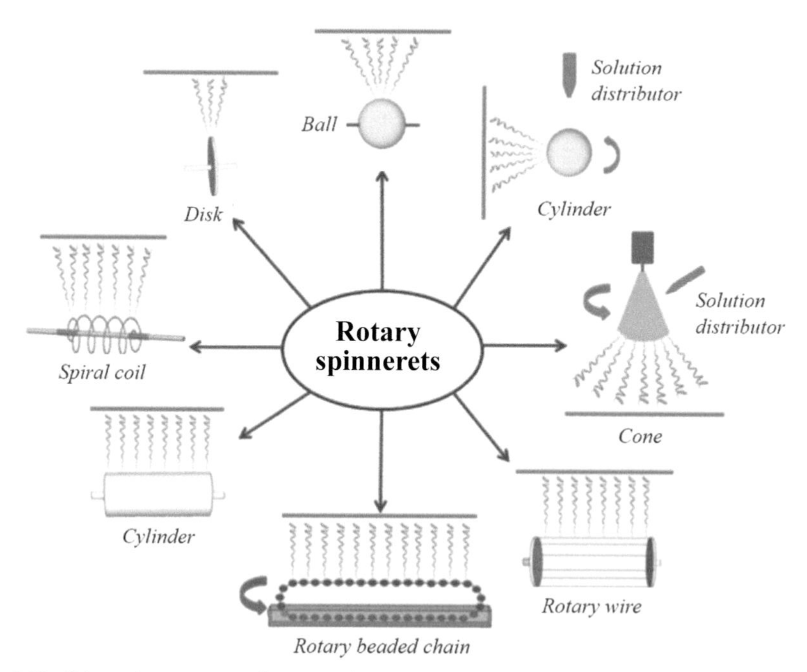

Fig. 2.13: Schematic summary of rotary spinnerets for needleless electrospinning (Copyright 2017, reproduced with permission from John Wiley & Sons, Inc.) (Yu et al. 2017).

Bubble Electrospinning

For the afore-mentioned needleless electrospinning that utilizes stationary spinnerets, such as wires and flat metals, external forces are typically required to feed spin dopes. Stable feed of spin dope is a critical issue in maintaining a continuous and high-quality production. Recently, another interesting type of electrospinning which has been termed as "bubble electrospinning" (Liu et al. 2007b, Yang et al. 2009), has been developed, which uses gas bubbles to initiate the electrospinning jets (i.e., feed the spin dope). Numerous bubbles can be generated simultaneously and continuously in bubble electrospinning process; hence, the productivity of nanofibers can be substantially improved.

The mechanism of bubble electrospinning is similar to that of conventional electrospinning. Specifically, by placing the nozzle at the bottom, bubbles are generated on the surface by compressed air; when a high voltage is applied, charges are induced to the bubble surface. The coupling of surface charge and external electric field creates a tangential stress, resulting in the deformation of a small bubble into a protuberance induced upward directed reentrant jet. Once the electric field exceeds the critical value and the surface tension is overcome, a plurality of temporary jets/filaments are generated and then ejected to form nanofibers (Liu et al. 2008, Liu et al. 2007b). The critical voltage to overcome the surface tension depends on the size of bubble and the pressure of inlet air, while the size of a bubble typically depends on the viscosity of solution. In practical applications, temperature can be adopted to adjust the bubble sizes (He et al. 2010).

Compared to other electrospinning processes, a key advantage of the bubble electrospinning process is that smaller nanofibers can be produced without the need for nozzles; and the applied voltages can be lower. Moreover, a large number of protruded bubbles can be easily produced, which can be simultaneously electrospun into nanofibers. Additionally, whether a spin dope can be electrospun into nanofibers does not strongly depend on its viscosity; thus a major issue/problem in the traditional needle electrospinning can be overcome. The average diameter of nanofibers can be relatively easy to adjust, and the thinnest nanofibers can have the diameters as small as 50 nm (He et al. 2010). Nevertheless, a major concern/problem of bubble electrospinning is that the nanofiber diameter is highly affected by the size of bubbles; whereas it is difficult to control the generation of uniform bubbles with similar sizes, thus the obtained nanofibers are often non-uniform in diameter.

External Force Assisted Electrospinning (Electroblowing and Centrifugal Electrospinning)

Air assisted electrospinning (i.e., electroblowing) and centrifugal electrospinning are more recently developed external force assisted electrospinning techniques that combine the electrospinning process with conventional spinning techniques (i.e., air-blowing and centrifugal spinning) (Edmondson et al. 2012, Erickson et al. 2015, Hsiao et al. 2012, Kancheva et al. 2014, Wang et al. 2005, Zhmayev et al. 2010). In the air-assisted electrospinning, nanofibers are produced by two simultaneously applied forces of electrical force and air-blowing shear force. Due to the assistance of air-blowing force, this method is especially useful for spin dopes with high viscosities (e.g., polymer solution with high concentration and polymer melt), because the air-blowing force could help to overcome the surface tension (Zhmayev et al. 2010). In the centrifugal electrospinning, nanofibers are produced upon the external assistance of centrifugal force. In the conventional centrifugal spinning, high speed is required; and the diameters of the produced fibers are in micrometers (Zhang et al. 2014b). After applying a high voltage, the rotating speed can be reduced from thousands of rounds per minute (rpm) to hundreds of rpm; while the diameters of the resulting fiber can be decreased to several hundred nanometers. Furthermore, owing to the centrifugal force, the centrifugal electrospinning can be operated at a substantially lower voltage. It is also reported that the nanofibers produced by centrifugal electrospinning possess good alignment (Erickson et al. 2015, Liu et al. 2013a).

The air-assisted electrospinning and centrifugal electrospinning may provide possible approaches to large-scale nanofiber production, since both techniques have higher productivities than the needle electrospinning. Nevertheless, they also have limitations/concerns including (1) the set-ups are more complex and expensive, and (2) both techniques are difficult to be utilized for continuous production. Additionally, the morphological structures of the resulting fibers and fiber mats/membranes are hard to control because of various issues, including the intrinsic properties of spin dope and the operational conditions such as air blowing velocity (or rotating speed) and applied voltage. For example, in centrifugal electrospinning, the driving force parameters, device parameters, and materials parameters, as well as other operating parameters (e.g., the perturbation frequency, density, environmental temperature, and humidity) have complicated influences on properties/characteristics of the produced fibers (Peng et al. 2017). Moreover, the spinneret designs also need further improvements; and the running systems working with a high voltage, high rotating speed, and/or high heating temperature may result in serious safety concerns.

Concluding Remarks

In recent decades, electrospinning has become a widely adopted technique to prepare nanofibers and/or nanofibrous materials for a variety of applications. Various nanofibers (e.g., polymer, ceramic, carbon/graphite, composite, and hierarchically structured nanofibers) with different morphologies have been prepared and studied. Several types of electrospinning have also been investigated to meet the requirements/demands of academic research and industrial production. Although the electrospinning process appears quite simple, it is actually a complex combination of fluid mechanics, polymer science, and electrostatics. Till date, the fundamental understanding of electrospinning process is still under further investigation; even though many designs of spinnerets have been studied for large-scale production of nanofibers, the productivity and stability are still not able to meet the requirements/demands of many industrial applications. Herein, fundamental theories and principles on the electrospinning process are introduced; the preparation and properties of polymer nanofibers are then specifically discussed. Additionally, the principles, advantages, and limitations of several types of electrospinning (e.g., needle electrospinning, needleless electrospinning, and external force assisted electrospinning) are also described.

Acknowledgements

The authors would like to acknowledge the funding supports from the EPSCoR program of (US) National Science Foundation (under the award number of IIA-1335423) and from the State of South Dakota (under the award number of UP1500172). The authors would also like to acknowledge the Biomedical Engineering Program at the South Dakota School of Mines and Technology.

References

Baumgarten, P. K. 1971. Electrostatic spinning of acrylic microfibers. Journal of Colloid and Interface Science 36(1): 71–79.

Bergshoef, M. M. and Vancso, G. J. 1999. Transparent nanocomposites with ultrathin, electrospun nylon-4,6 fiber reinforcement. Advanced Materials 11(16): 1362–1365.

Bognitzki, M., Czado, W., Frese, T., Schaper, A., Hellwig, M., Steinhart, M. et al. 2001. Nanostructured fibers via electrospinning. Advanced Materials 13(1): 70–72.

Chen, H., Wang, N., Di, J., Zhao, Y., Song, Y. and Jiang, L. 2010. Nanowire-in-microtube structured core/shell fibers via multifluidic coaxial electrospinning. Langmuir 26(13): 11291–11296.

Chen, S., Hou, H., Hu, P., Wendorff, J. H., Greiner, A. and Agarwal, S. 2009. Effect of different bicomponent electrospinning techniques on the formation of polymeric nanosprings. Macromolecular Materials and Engineering 294(11): 781–786.

Chen, W., Liu, Y., Ma, Y. and Yang, W. 2015. Improved performance of lithium ion battery separator enabled by co-electrospinnig polyimide/poly(vinylidene fluoride-co-hexafluoropropylene) and the incorporation of TiO_2-(2-hydroxyethyl methacrylate). Journal of Power Sources 273: 1127–1135.

De Gennes, P. G. 1974. Coil-stretch transition of dilute flexible polymers under ultrahigh velocity gradients. The Journal of Chemical Physics 60(12): 5030–5042.

Deitzel, J. M., Kleinmeyer, J., Harris, D. and Beck Tan, N. C. 2001. The effect of processing variables on the morphology of electrospun nanofibers and textiles. Polymer 42(1): 261–272.

Ding, B., Kim, H., Kim, C., Khil, M. and Park, S. 2003. Morphology and crystalline phase study of electrospun TiO_2–SiO_2 nanofibres. Nanotechnology 14(5): 532–537.

Ding, Y., Hou, H., Zhao, Y., Zhu, Z. and Fong, H. 2016a. Electrospun polyimide nanofibers and their applications. Progress in Polymer Science 61: 67–103.

Ding, Y., Yang, J., Tolle, C. R. and Zhu, Z. 2016b. A highly stretchable strain sensor based on electrospun carbon nanofibers for human motion monitoring. RSC Advances 6(82): 79114–79120.

Doshi, J. and Reneker, D. H. 1995. Electrospinning process and applications of electrospun fibers. Journal of Electrostatics 35(2): 151–160.

Dzenis, Y. 2004. Spinning continuous fibers for nanotechnology. Science 304(5679): 1917–1919.

Edmondson, D., Cooper, A., Jana, S., Wood, D. and Zhang, M. 2012. Centrifugal electrospinning of highly aligned polymer nanofibers over a large area. Journal of Materials Chemistry 22(35): 18646–18652.

Erickson, A. E., Edmondson, D., Chang, F. -C., Wood, D., Gong, A., Levengood, S. L. et al. 2015. High-throughput and high-yield fabrication of uniaxially-aligned chitosan-based nanofibers by centrifugal electrospinning. Carbohydrate Polymers 134: 467–474.

Fong, H. and Reneker, D. H. Unpublished Data.

Fong, H., Chun, I. and Reneker, D. H. 1999a. Beaded nanofibers formed during electrospinning. Polymer 40(16): 4585–4592.

Fong, H. and Reneker, D. H. 1999b. Elastomeric nanofibers of styrene-butadiene-styrene triblock copolymer. Journal of Polymer Science Part B Polymer Physics 37(24): 3488–3493.

Fong, H. and Reneker, D. H. 2001. Electrospinning and formation of nanofibers. *In*: D. R. Salem (ed.). Structure Formation in Polymeric Fibers. Carl Hanser Verlag, Cincinnati, OH, USA.

Fong, H., Liu, W., Wang, C. -S. and Vaia, R. A. 2002. Generation of electrospun fibers of nylon 6 and nylon 6-montmorillonite nanocomposite. Polymer 43(3): 775–780.

Fong, H. 2004. Electrospun nylon 6 nanofiber reinforced BIS-GMA/TEGDMA dental restorative composite resins. Polymer 45(7): 2427–2432.

Formhals, A. 1934. Process and apparatus for preparing artificial threads. U.S. Patents, US1975504.
Formhals, A. 1937. Production of artificial fibers. U.S. Patents, US2077373.
Formhals, A. 1939a. Method and apparatus for spinning. U.S. Patents, US2160962.
Formhals, A. 1939b. Method and apparatus for the production of artificial fibers. U.S. Patents, US2158416.
Formhals, A. 1940a. Artificial thread and method of producing same. U.S. Patents, US2187306.
Formhals, A. 1940b. Production of artificial fibers from fiber forming liquids. U.S. Patents, US 2323025.
Formhals, A. 1944. Method and apparatus for spinning. U.S. Patents, US 2349950.
Fridrikh, S. V., Yu, J. H., Brenner, M. P. and Rutledge, G. C. 2003. Controlling the fiber diameter during electrospinning. Physical Review Letters 90(14): 144502.
Ge, J. J., Hou, H., Li, Q., Graham, M. J., Greiner, A., Reneker, D. H. et al. 2004. Assembly of well-aligned multiwalled carbon nanotubes in confined polyacrylonitrile environments: Electrospun composite nanofiber sheets. Journal of the American Chemical Society 126(48): 15754–15761.
Gladding, E. K. 1939. Apparatus for the production of filaments, threads, and the like. U.S. Patents, US 2168027.
Guo, Z., Tang, G., Zhou, Y., Shuwu, L., Hou, H., Chen, Z. et al. 2017. Fabrication of sustained-release CA-PU coaxial electrospun fiber membranes for plant grafting application. Carbohydrate Polymers 169: 198–205.
He, G., Wang, X., Xi, M., Zheng, F., Zhu, Z. and Fong, H. 2013. Fabrication and evaluation of dye-sensitized solar cells with photoanodes based on electrospun TiO_2 nanotubes. Materials Letters 106: 115–118.
He, J. H., Liu, Y. and Xu, L. 2010. Apparatus for preparing electrospun nanofibres: a comparative review. Materials Science and Technology 26(11): 1275–1287.
He, X., Zheng, J., Yu, G., You, M., Yu, M., Ning, X. et al. 2017. Near-field electrospinning: progress and applications. The Journal of Physical Chemistry C 121(16): 8663–8678.
Hohman, M. M., Shin, M., Rutledge, G. and Brenner, M. P. 2001. Electrospinning and electrically forced jets. I. Stability theory. Physics of Fluids 13(8): 2201–2220.
Hou, H., Jun, Z., Reuning, A., Schaper, A., Wendorff, J. H. and Greiner, A. 2002. Poly(p-xylylene) nanotubes by coating and removal of ultrathin polymer template fibers. Macromolecules 35(7): 2429–2431.
Hou, H. and Reneker, D. H. 2004. Carbon nanotubes on carbon nanofibers: A novel structure based on electrospun polymer nanofibers. Advanced Materials 16(1): 69–73.
Hsiao, H. -Y., Huang, C. -M., Liu, Y. -Y., Kuo, Y. -C. and Chen, H. 2012. Effect of air blowing on the morphology and nanofiber properties of blowing-assisted electrospun polycarbonates. Journal of Applied Polymer Science 124(6): 4904–4914.
Huang, Z. -M., Zhang, Y. Z., Kotaki, M. and Ramakrishna, S. 2003. A review on polymer nanofibers by electrospinning and their applications in nanocomposites. Composites Science and Technology 63(15): 2223–2253.
Jaeger, R., Schönherr, H. and Vancso, G. J. 1996. Chain packing in electro-spun poly(ethylene oxide) visualized by atomic force microscopy. Macromolecules 29(23): 7634–7636.
Jaeger, R., Bergshoef, M. M., Batlle, C. M. I., Schönherr, H. and Julius Vancso, G. 1998. Electrospinning of ultra-thin polymer fibers. Macromolecular Symposia 127(1): 141–150.
Jiang, S., Duan, G., Zussman, E., Greiner, A. and Agarwal, S. 2014. Highly flexible and tough concentric triaxial polystyrene fibers. ACS Applied Materials & Interfaces 6(8): 5918–5923.
Kancheva, M., Toncheva, A., Manolova, N. and Rashkov, I. 2014. Advanced centrifugal electrospinning setup. Materials Letters 136: 150–152.
Kim, C. and Yang, K. S. 2003. Electrochemical properties of carbon nanofiber web as an electrode for supercapacitor prepared by electrospinning. Applied Physics Letters 83(6): 1216–1218.
Kim, J. -S. and Reneker, D. H. 1999a. Mechanical properties of composites using ultrafine electrospun fibers. Polymer Composites 20(1): 124–131.
Kim, J. -S. and Reneker, D. H. 1999b. Polybenzimidazole nanofiber produced by electrospinning. Polymer Engineering & Science 39(5): 849–854.
Koombhongse, S., Liu, W. and Reneker, D. H. 2001. Flat polymer ribbons and other shapes by electrospinning. Journal of Polymer Science Part B: Polymer Physics 39(21): 2598–2606.
Lai, C., Zhou, Z., Zhang, L., Wang, X., Zhou, Q., Zhao, Y. et al. 2014. Free-standing and mechanically flexible mats consisting of electrospun carbon nanofibers made from a natural product of alkali

lignin as binder-free electrodes for high-performance supercapacitors. Journal of Power Sources 247: 134–141.

Larrondo, L. and St. John Manley, R. 1981a. Electrostatic fiber spinning from polymer melts. I. Experimental observations on fiber formation and properties. Journal of Polymer Science: Polymer Physics Edition 19(6): 909–920.

Larrondo, L. and St. John Manley, R. 1981b. Electrostatic fiber spinning from polymer melts. II. Examination of the flow field in an electrically driven jet. Journal of Polymer Science: Polymer Physics Edition 19(6): 921–932.

Larrondo, L. and St. John Manley, R. 1981c. Electrostatic fiber spinning from polymer melts. III. Electrostatic deformation of a pendant drop of polymer melt. Journal of Polymer Science: Polymer Physics Edition 19(6): 933–940.

Li, D. and Xia, Y. 2003. Fabrication of titania nanofibers by electrospinning. Nano Letters 3(4): 555–560.

Li, D. and Xia, Y. 2004. Direct fabrication of composite and ceramic hollow nanofibers by electrospinning. Nano Letters 4(5): 933–938.

Li, W. -J., Laurencin, C. T., Caterson, E. J., Tuan, R. S. and Ko, F. K. 2002. Electrospun nanofibrous structure: A novel scaffold for tissue engineering. Journal of Biomedical Materials Research 60(4): 613–621.

Liu, S. -L., Long, Y. -Z., Zhang, Z. -H., Zhang, H. -D., Sun, B., Zhang, J. -C. et al. 2013a. Assembly of oriented ultrafine polymer fibers by centrifugal electrospinning. J. Nanomaterials 2514103(2514103): 1–9.

Liu, Y., Cui, L., Guan, F., Gao, Y., Hedin, N. E., Zhu, L. et al. 2007a. Crystalline morphology and polymorphic phase transitions in electrospun nylon-6 nanofibers. Macromolecules 40(17): 6283–6290.

Liu, Y. and He, J. H. 2007b. Bubble electrospinning for mass production of nanofibers. International Journal of Nonlinear Sciences and Numerical Simulation 8(3): 393–396.

Liu, Y., He, J. -H., Xu, L. and Yu, J. -Y. 2008. The principle of bubble electrospinning and its experimental verification. Journal of Polymer Engineering 28(1-2): 55–65.

Liu, Y. and Guo, L. 2013b. Homogeneous field intensity control during multi-needle electrospinning via finite element analysis and simulation. Journal of Nanoscience and Nanotechnology 13(2): 843–847.

Loscertales, I. G., Barrero, A., Márquez, M., Spretz, R., Velarde-Ortiz, R. and Larsen, G. 2004. Electrically forced coaxial nanojets for one-step hollow nanofiber design. Journal of the American Chemical Society 126(17): 5376–5377.

Lu, B., Wang, Y., Liu, Y., Duan, H., Zhou, J., Zhang, Z. et al. 2010. Superhigh-throughput needleless electrospinning using a rotary cone as spinneret. Small 6(15): 1612–1616.

Ma, X., Kolla, P., Yang, R., Wang, Z., Zhao, Y., Smirnova, A. L. et al. 2017. Electrospun polyacrylonitrile nanofibrous membranes with varied fiber diameters and different membrane porosities as lithium-ion battery separators. Electrochimica Acta 236: 417–423.

Murthy, N. S., Aharoni, S. M. and Szollosi, A. B. 1985. Stability of the γ form and the development of the α form in nylon 6. Journal of Polymer Science: Polymer Physics Edition 23(12): 2549–2565.

Nan, W., Zhao, Y., Ding, Y., Shende, A. R., Fong, H. and Shende, R. V. 2017. Mechanically flexible electrospun carbon nanofiber mats derived from biochar and polyacrylonitrile. Materials Letters 205: 206–210.

Niu, H., Wang, X. and Lin, T. 2011. Needleless electrospinning: developments and performances. *In*: Nanofibers-Production, Properties and Functional Applications: InTech.

Niu, H., Wang, X. and Lin, T. 2012. Needleless electrospinning: influences of fibre generator geometry. The Journal of the Textile Institute 103(7): 787–794.

Peng, H., Liu, Y. and Ramakrishna, S. 2017. Recent development of centrifugal electrospinning. Journal of Applied Polymer Science 134(10): n/a–n/a.

Peng, L., Jiang, S., Seuß, M., Fery, A., Lang, G., Scheibel, T. et al. 2016. Two-in-one composite fibers with side-by-side arrangement of silk fibroin and poly(l-lactide) by electrospinning. Macromolecular Materials and Engineering 301(1): 48–55.

Peng, X., Ye, W., Ding, Y., Jiang, S., Hanif, M., Liao, X. et al. 2014. Facile synthesis, characterization and application of highly active palladium nano-network structures supported on electrospun carbon nanofibers. RSC Advances 4(80): 42732–42736.

Persano, L., Camposeo, A., Tekmen, C. and Pisignano, D. 2013. Industrial upscaling of electrospinning and applications of polymer nanofibers: A review. Macromolecular Materials and Engineering 298(5): 504–520.

Presser, V., Zhang, L., Niu, J. J., McDonough, J., Perez, C., Fong, H. et al. 2011. Flexible nano-felts of carbide-derived carbon with ultra-high power handling capability. Advanced Energy Materials 1(3): 423–430.

Rayleigh, L. 1882. XX. On the equilibrium of liquid conducting masses charged with electricity. Philosophical Magazine 14(87): 184–186. https://www.tandfonline.com/doi/abs/10.1080/14786448208628425?journalCode=tphm16.

Reneker, D. H. and Chun, I. 1996. Nanometre diameter fibres of polymer, produced by electrospinning. Nanotechnology 7(3): 216–223.

Reneker, D. H., Yarin, A. L., Fong, H. and Koombhongse, S. 2000. Bending instability of electrically charged liquid jets of polymer solutions in electrospinning. Journal of Applied Physics 87(9): 4531–4547.

Reneker, D. H., Hou, H., Rangkupan, R. and Lennhoff, J. 2003. Electrospinning polymer nanofibers in a vacuum. Polymer Preprints 44(2): 68–69.

Reneker, D. H. and Yarin, A. L. 2008. Electrospinning jets and polymer nanofibers. Polymer 49(10): 2387–2425.

Shin, Y. M., Hohman, M. M., Brenner, M. P. and Rutledge, G. C. 2001. Electrospinning: a whipping fluid jet generates submicron polymer fibers. Applied Physics Letters 78(8): 1149–1151.

Simons, H. L. 1966. Process and apparatus for producing patterned non-woven fabrics. U.S. Patents, US 3280229.

Smith, D. E. and Chu, S. 1998. Response of flexible polymers to a sudden elongational flow. Science 281(5381): 1335–1340.

Starr, J. D. and Andrew, J. S. 2013. Janus-type bi-phasic functional nanofibers. Chemical Communications 49(39): 4151–4153.

Sun, Z., Zussman, E., Yarin, A. L., Wendorff, J. H. and Greiner, A. 2003. Compound core–shell polymer nanofibers by co-electrospinning. Advanced Materials 15(22): 1929–1932.

Theron, A., Zussman, E. and Yarin, A. L. 2001. Electrostatic field-assisted alignment of electrospun nanofibres. Nanotechnology 12(3): 384–390.

Tsai, P. P., Schreuder-Gibson, H. and Gibson, P. 2002. Different electrostatic methods for making electret filters. Journal of Electrostatics 54(3): 333–341.

Vasanthan, N. and Salem, D. R. 2001. FTIR spectroscopic characterization of structural changes in polyamide-6 fibers during annealing and drawing. Journal of Polymer Science Part B: Polymer Physics 39(5): 536–547.

Vonnegut, B. and Neubauer, R. L. 1952. Production of monodisperse liquid particles by electrical atomization. Journal of Colloid Science 7(6): 616–622.

Wang, X., Um, I. C., Fang, D., Okamoto, A., Hsiao, B. S. and Chu, B. 2005. Formation of water-resistant hyaluronic acid nanofibers by blowing-assisted electro-spinning and non-toxic post treatments. Polymer 46(13): 4853–4867.

Wang, X., Xi, M., Fong, H. and Zhu, Z. 2014. Flexible, transferable, and thermal-durable dye-sensitized solar cell photoanode consisting of TiO_2 nanoparticles and electrospun TiO_2/SiO_2 nanofibers. ACS Applied Materials & Interfaces 6(18): 15925–15932.

Wang, X., Xi, M., Zheng, F., Ding, B., Fong, H. and Zhu, Z. 2015. Reduction of crack formation in TiO_2 mesoporous films prepared from binder-free nanoparticle pastes via incorporation of electrospun SiO_2 or TiO_2 nanofibers for dye-sensitized solar cells. Nano Energy 12: 794–800.

Wang, X., Xi, M., Wang, X., Fong, H. and Zhu, Z. 2016. Flexible composite felt of electrospun TiO_2 and SiO_2 nanofibers infused with TiO_2 nanoparticles for lithium ion battery anode. Electrochimica Acta 190: 811–816.

Wang, Z., Crandall, C., Sahadevan, R., Menkhaus, T. J. and Fong, H. 2017a. Microfiltration performance of electrospun nanofiber membranes with varied fiber diameters and different membrane porosities and thicknesses. Polymer 114: 64–72.

Wang, Z., Sahadevan, R., Yeh, C. -N., Menkhaus, T. J., Huang, J. and Fong, H. 2017b. Hot-pressed polymer nanofiber supported graphene membrane for high-performance nanofiltration. Nanotechnology 28(31): 31LT02.

Xi, M., Wang, X., Zhao, Y., Feng, Q., Zheng, F., Zhu, Z. et al. 2014a. Mechanically flexible hybrid mat consisting of TiO_2 and SiO_2 nanofibers electrospun via dual spinnerets for photo-detector. Materials Letters 120: 219–223.

Xi, M., Wang, X., Zhao, Y., Zhu, Z. and Fong, H. 2014b. Electrospun ZnO/SiO_2 hybrid nanofibrous mat for flexible ultraviolet sensor. Applied Physics Letters 104(13): 133102.

Xie, S. and Zeng, Y. 2012. Effects of electric field on multineedle electrospinning: Experiment and simulation study. Industrial & Engineering Chemistry Research 51(14): 5336–5345.

Xu, T., Miszuk, J. M., Zhao, Y., Sun, H. and Fong, H. 2015. Electrospun polycaprolactone 3D nanofibrous scaffold with interconnected and hierarchically structured pores for bone tissue engineering. Advanced Healthcare Materials 4(15): 2238–2246.

Yang, R., He, J., Xu, L. and Yu, J. 2009. Bubble-electrospinning for fabricating nanofibers. Polymer 50(24): 5846–5850.

Yao, Q., Cosme, J. G. L., Xu, T., Miszuk, J. M., Picciani, P. H. S., Fong, H. et al. 2017. Three dimensional electrospun PCL/PLA blend nanofibrous scaffolds with significantly improved stem cells osteogenic differentiation and cranial bone formation. Biomaterials 115: 115–127.

Yu, M., Dong, R. -H., Yan, X., Yu, G. -F., You, M. -H., Ning, X. et al. 2017. Recent advances in needleless electrospinning of ultrathin fibers: From academia to industrial production. Macromolecular Materials and Engineering 302(7): 201700002.

Zhang, L., Aboagye, A., Kelkar, A., Lai, C. and Fong, H. 2014a. A review: carbon nanofibers from electrospun polyacrylonitrile and their applications. Journal of Materials Science 49(2): 463–480.

Zhang, X. and Lu, Y. 2014b. Centrifugal spinning: An alternative approach to fabricate nanofibers at high speed and low cost. Polymer Reviews 54(4): 677–701.

Zhao, Y., Cao, X. and Jiang, L. 2007. Bio-mimic multichannel microtubes by a facile method. Journal of the American Chemical Society 129(4): 764–765.

Zheng, Y., Gong, R. H. and Zeng, Y. 2015. Multijet motion and deviation in electrospinning. RSC Advances 5(60): 48533–48540.

Zhmayev, E., Cho, D. and Joo, Y. L. 2010. Nanofibers from gas-assisted polymer melt electrospinning. Polymer 51(18): 4140–4144.

Zhou, Z., Lai, C., Zhang, L., Qian, Y., Hou, H., Reneker, D. H. et al. 2009. Development of carbon nanofibers from aligned electrospun polyacrylonitrile nanofiber bundles and characterization of their microstructural, electrical, and mechanical properties. Polymer 50(13): 2999–3006.

Chapter 3

Raw Materials and Solution Preparation for Electrospinning

Kai Pan, Yangxiu Liu* and *Qiutong Wang*

Introduction

Electrospinning has got great development in recent years. In order to meet different needs, such as the research and applications, more and more materials have been developed for the technology, and solvents at the same time. However, no matter what the properties of the resulting materials, polymers are the indispensable parts of electrospinning process. We can directly produce polymer or composite fibers by electrospinning, and can obtain the ceramics or carbon materials as well by post processing. Therefore, deep understanding and grasp over the nature of the raw materials and the classification will be helpful to us. In this chapter, we will start our discussion from polymers, and aim to bring some new information about electrospinning. Besides, the preparation including solvent and solution properties will be mentioned.

Organic Polymers and Solvents for Solution Electrospinning

As we mentioned above, polymers are indispensable in electrospinning. Therefore, we will start with the introduction of the classification of polymeric materials in electrospinning, and their matched solvent systems. With the development of polymer species and the electrospinning, more and more polymers and their solvent systems have been reported, and can be used in electrospinning processes. However, exploring new polymer varieties and their applications was still the goal of electrospinning researchers. Especially in recent years, with the rapid development of electrospinning, combined with the major problems and needs of global science

College of Materials Science and Engineering, Beijing University of Chemical Technology, 15 North Third Ring Road, Chaoyang District, Beijing, 100029, P.R. China.
* Corresponding author: pankai@mail.buct.edu.cn

and technology, the development of natural polymer materials has become a hot spot even in the field of electrospinning. Herein, the natural polymers and their solvent systems in electrospinning will be gradually introduced at first.

Natural Polymers and their Solvents

With the decrease of world oil resources and the rising of crude oil prices, the development of the traditional synthetic polymer industry has been restricted. At the same time, synthetic polymer materials are difficult to degrade, and the environmental pollution is becoming more and more serious. Renewable natural polymers come from natural, biological, plant, and microbial resources are inexhaustible resources. What's more, these materials are easily decomposed into water, carbon dioxide, and inorganic small molecules by natural microorganisms, which means they are environmentally friendly materials. In particular, natural polymers with various functional groups can be modified into new materials by chemical or physical methods, and various functional materials can also be prepared through the emerging nanotechnology system, so they will probably become the main alternative material to synthetic plastic chemical products. For example, in recent years, the research and application of cellulose nanocrystals gradually increased. This natural polymer has broad application prospects in the fields of materials, energy, environmental, and personal health care for its excellent performance (Raquez et al. 2012, Hossain et al. 2011).

In addition, silk, spider dragline, and other natural polymer materials with special high strength performance also attracted researchers' attention. They are likely to be used in military, medical, health, and other fields (Ding et al. 2010, Shengyuan et al. 2011, Li et al. 2009, Mele 2016, Jiang et al. 2016, Yang et al. 2011). There are a lot of natural polymers which can be processed by electrospinning technique in nature. Compared to synthetic polymers, the biggest advantage of natural polymers are their good biocompatibility and biodegradability. In combination with the advantages of natural polymers and electrospinning technology, the resulting products could be widely used in biological tissue engineering and medical applications. Electrospun nanofibers have large specific surface area, high porosity, super light, super thin thickness, and even similar fiber structures to the extracellular matrix. As we know, there are many three-dimensional network structures in the connective tissue of organisms, so it is possible to design and construct human tissue engineering scaffolds using electrospun nanofibers. However, most natural polymers are polyelectrolyte that are easily ionized in the solvent and get charged. It may increase the instability in electrospinning process. Besides, most proteins, polysaccharides, and nucleic acids maintain a regular three-dimensional structure in common solutions, which means they are difficult to dissolve evenly in the solvent that makes the electrospinning more difficult. The interaction of dissolved protein fiber's polypeptide chain structure has a significant impact on mechanical properties of the obtained fiber. Especially cellulose and chitin, the interaction of polysaccharide chain and the effect of crystallization are the main factors that influence the mechanical properties of fiber. In electrospinning, the interaction between natural polymer chains and solvent in spinning solution is an important research topic.

Cellulose. As a kind of non-toxic and biodegradable natural polysaccharide, cellulose has a large number of hydroxyl groups on the macromolecular chains, and forms strong hydrogen bonds between cellulose molecules. Figure 3.1 gives the molecular formula of cellulose and a plant rich in cellulose—bamboo. As many organic and inorganic solvents are difficult to dissolve cellulose in, so, the cellulose superfine fiber is usually obtained by electrospinning the derivative of cellulose and then hydrolysis. For example, previous researchers use acetone/water mixed solution or acetone/DMF mixture as solvent to dissolve cellulose acetate for electrospinning, then soak the superfine fiber in KOH or NaOH solution with ethanol, which do not change the fiber structure after hydrolysis, and the obtained product is usually a mixture of cellulose and its derivatives. In addition, the cellulose can be dissolved by N-methyl morpholine oxide/water (NMMO/H_2O) system (Fink et al. 2001, Kulpinski 2005, Dong et al. 2002, Kim et al. 2006). A new hydrogen bond is formed between cellulose and NMMO solvent as the NMMO enters the amorphous and crystalline regions of the cellulose; the distance between the macromolecules of cellulose increases, and eventually cellulose dissolves in NMMO to form a homogeneous solution. This method which increases the solubility of cellulose in solvents and reduces the viscosity of the system is beneficial to electrospinning. By using lithium chloride/dimethylacetamid solvent system, higher cellulose concentration can be obtained, and the solution system is stable (Kim et al. 2005, Dupont 2003, Röder et al. 2001, Xu et al. 2008). However, the solvent is highly corrosive and difficult to separate, purify, and recycle. In contrast, NMMO is non corrosive and has a recovery of more than 90%, so NMMO is known as an environmentally friendly cellulose solvent.

In summary, the preparation of cellulose solution is the key to electrospinning. It requires solvents with strong polarity, but these solvents have high boiling point and are difficult to volatilize, which makes it difficult for electrospun fibers. The problem still needs to be solved by improving solvent system and optimizing spinning equipment. Electrospinning of cellulose product has good thermal stability, chemical stability, and biodegradability. It can be widely used in affinity membranes, biosensors, chemical sensors, protective clothing, and reinforced nanomaterials.

Collagen. Collagen is the most widely found protein in animals which exists in the connective tissue of the animal's skin, bone, and tendon. The amino acid sequence of collagen mainly consists of three repeating units of glycine, proline,

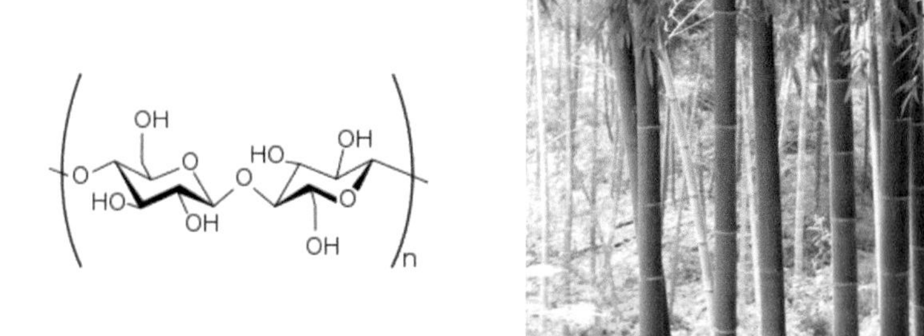

Fig. 3.1: Molecular formula of cellulose and a photo of bamboo which is rich in cellulose.

and hydroxyproline, and is composed of three polypeptide chains which form three helix structures, as shown in Fig. 3.2 (Meng et al. 2013, Gobeaux et al. 2010). In general solvents, the solubility of collagen is limited. By enzyme hydrolyzing, it can be dissolved in acid solution and form soluble collagen. However, only 3%(wt) collagen can be dissolved without damaging the three helix structure (Gobeaux et al. 2010). In addition, the evaporation rate of acid solvent is slow, and it has strong binding force with collagen. In electrospinning, when the acid is used as a solvent, the collagen solution always drops into the receiving device in the form of droplets and fails to prepare a continuous fiber (Burck et al. 2013). Trifluoroethanol and hexafluoroisopropanol (HFIP) with high volatilization rate are good solvents for polypeptide biopolymer, but these solvents cause collagen to become gelatin (Zeugolis et al. 2008, Kim et al. 2008, Chakrapani et al. 2012). In order to avoid the problems caused by the use of fluoro solvents, concentrated salt/ethanol solvent systems and ionic liquids were developed as solvents for collagen, and good results have been achieved (Dong et al. 2009, Kew et al. 2011).

One of the main research subjects of electrospun collagen is to develop an environmentally friendly and non-toxic solvent system without damaging the structure of collagen itself. Secondly, the collagen denaturation and the modification of fiber after spinning are also the problems that need to be solved urgently. Electrospun collagen may be used in wound healing, hemostasis, and tissue engineering.

Chitin and Chitosan. There are many chitins in the shells of invertebrates and fungi, as well as in protozoa and some green algae. Chitosan is the product after deacetylation of chitin. Their molecular formula are shown in the left and right sides of Fig. 3.3, respectively. Due to the presence of large amounts of free ammonia in its molecular structure, the solubility of chitosan has greatly improved. Chitosan has good biocompatibility, biodegradability, anti-coagulation, and wound healing functions. In conventional spinning, chitin is soluble in the mixture of trichloroacetic

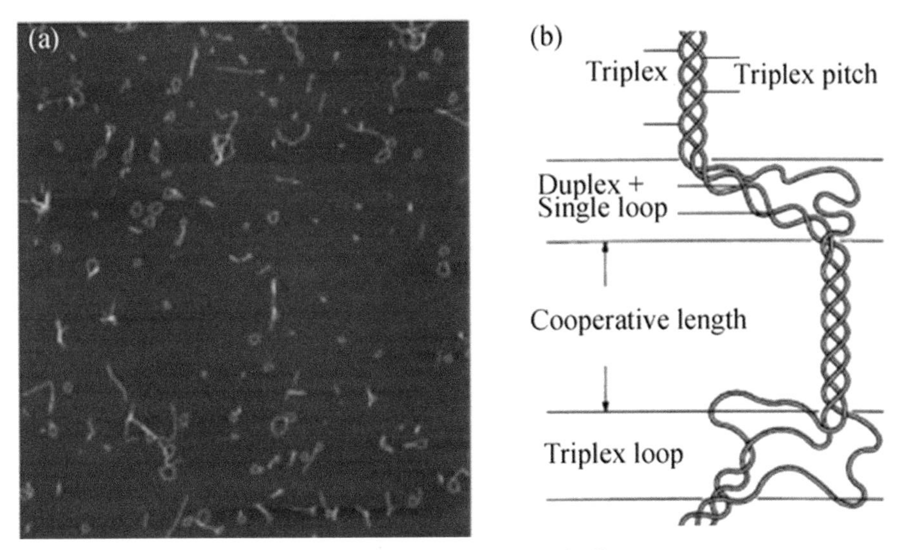

Fig. 3.2: The three-helix structure of collagen.

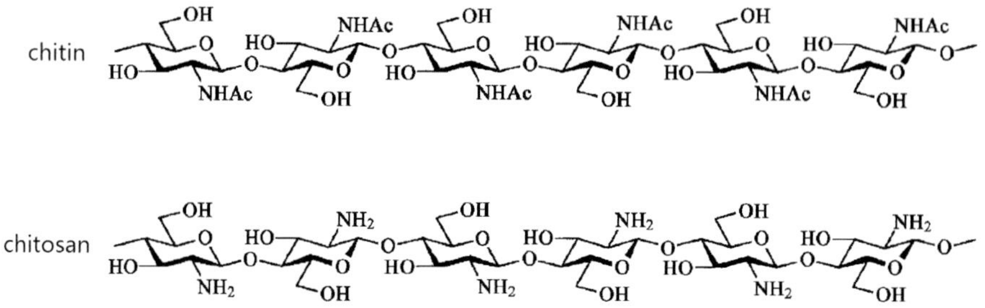

Fig. 3.3: The molecular formula of chitin and chitosan.

acid and dichloromethane, and is also soluble in dimethylformamide containing lithium chloride with acetone, methanol, or isopropyl alcohol as coagulating bath. Chitin is also soluble in concentrated strong acids, such as concentrated hydrochloric acid, sulfuric acid, and phosphoric acid. The regenerated chitin, which is soluble in phosphoric acid and then precipitated, is soluble in formic acid. Due to the presence of amino groups in chitosan, it becomes a polyelectrolyte when chitosan is protonated, causing electrospinning of pure chitosan fibers to be very difficult (Min et al. 2004). Electrospun chitosan nanofibers were successfully obtained with trifluoroacetic acid (TFA) as solvent. That is because the salt formed by the amino groups of TFA with chitosan effectively reduces the interaction between chitosan molecules, and makes spinning easier (Ohkawa et al. 2004). In addition to TFA, another effective solvent is concentrated acetic acid. When the concentration of acetic acid reaches 90%, uniform and smooth fibers can be obtained via electrospinning (Geng et al. 2005).

Chitin and its derivatives, chitosan, are the second major biological resources after plant fiber, and they have excellent biocompatibility. Due to its non-toxic, non irritating, antibacterial, anticoagulant, and biodegradable properties, it has been widely used in the biomedical field. However, the single chitin or chitosan electrospun products have the disadvantages of poor mechanical properties, low yield, and poor reproducibility. Further improvements should be made by mixing with other polymer matrices, spinning, optimizing spinning parameters, and equipment.

Silk Fibroin. Silk fibroin is a filamentous protein that forms silk fibers. The fiber, gel, film which were prepared using silk fibroin have good biocompatibility, oxygen permeability, water vapor permeability, and biodegradable advantages, so they can be used in cosmetics, biological, and medical materials. In electrospinning, silk fibroin is usually dissolved in hexafluoroisopropanol, hexafluoroacetone, formic acid, or other organic solvents to make spinning solution (Jeong et al. 2007). However, these solvents are toxic, and the residual in the material will inevitably affect the application of this material in the field of biomedicine. Therefore, electrospun silk fibroin without organic solvent is more suitable for the application of biomaterials in the field of biomedicine. Electrospinning with water as a solution faces the following problems: low concentration of fibroin aqueous solution, low viscosity, and slow evaporation. The silk fibroin solution was concentrated and then blended with salt before electrospinning, and achieved good results.

In recent years, some reports have been made on the preparation of superfine fibers by electrospinning with biomacromolecular such as spider dragline and silkworm silk. Figure 3.4 shows photos of silkworm, silk, and spider dragline. Spider dragline is made up of a variety of amino acids. Inside the spider dragline silk is a compact silk protein fiber. The mechanical strength of spider silk fiber is close to carbon fiber, and high-strength synthetic fibers like Aramid and Kelve, while its toughness is obviously superior to the above several kinds of fibers. However, a single component spider silk can be prepared by electrospinning only when hexafluoroisopropanol (HFIP) is used as solvent. Moreover, the product has poor stability and is easy to shrink. Therefore, spider silk is usually blended with polymers of excellent mechanical properties, then electrospun to prepare nanofibers with reinforcing and toughening functions. It can also be blended with excellent biocompatible polymers to prepare biological tissue regeneration support materials, as the solvent system of these composites is similar to that of a single component; it is not repeated here.

According to a new report by the National Academy of Sciences, a US lab is studying bullet-proof clothing (Fig. 3.5) with spider silk as a reinforcing material. Such new bullet-proof materials are said to be lighter and more resilient, and are expected to be deployed in the US Army if successful. What's more, researchers in Italy and Britain fed spiders with corresponding aqueous dispersions, and spider filaments of composite graphene or carbon nanotubes were obtained. The tensile strength and toughness of these spider fibers are much higher than those of conventional spider silk, and their strength is comparable to that of the strongest carbon fibers (Lepore et al. 2017).

Gelatin. Gelatin is a water-soluble polymer and a kind of denatured collagen, and has similar composition and properties as collagen. The conventional method of gelatin spinning is to squeeze the aqueous solution of gelatin into the saturated solution of sodium sulfate. After wet spinning, the gelatin fibers are obtained by stretching, post-treatment and drying. The solution was prepared by dissolving gelatin with trifluoroethanol. It was found that the solution could be electrospun when the concentration is between 5% and 12.5%. Researchers have also dissolved the gelatin in deionized water or formic acid water to electrospin. However, the pure gelatin fibers prepared by electrospinning have the defects of fragility, easily deforming, and poor moisture resistance. In order to realize the functional application of gelatin, industrial production mainly blends gelatin with other materials and then electrospinning. For example, An et al. (An et al. 2010) blended gelatin and lactide, then electrospun, and investigated the effects of formic acid concentration, gelatin concentration, and voltage on the composite nanofibers. Li et al. (Li et al. 2006) mixed polyaniline with gelatin, and prepared the gelatin composite fiber scaffold by electrospinning. The results showed that the material not only has excellent mechanical properties, but also has good biocompatibility and structural compatibility.

At present, electrospun gelatin fiber has many disadvantages, such as poor spinning continuity, low spinning yield, poor moisture resistance, volatility, and easy to break. It needs to be improved by adding polymer matrix, product crosslinking, and improving the spinning equipment and process parameters.

Sodium Alginate (SA). As shown in Fig. 3.6, sodium alginate is a polysaccharide substance extracted from algae. Alginate fiber has a great moisture gel performance,

Fig. 3.4: Silkworm silk and spider dragline.

Fig. 3.5: Bullet-proof cloth.

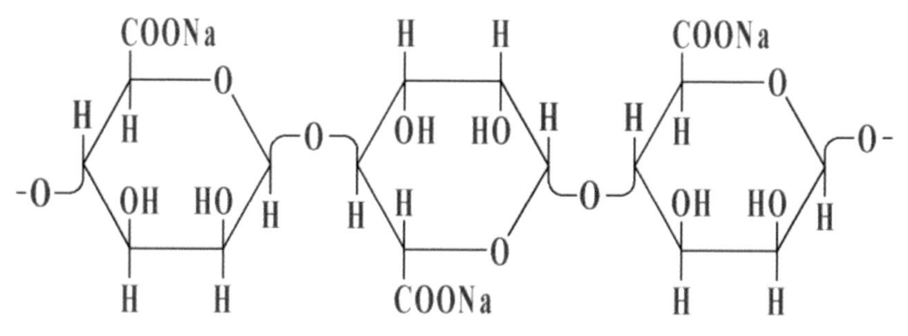

Fig. 3.6: Sodium alginate.

high oxygen permeability, excellent biocompatibility, biodegradation ability, and high ion adsorption property. When the alginate fiber is used on the wound contact layer, it interacts with the wound to produce sodium alginate and calcium alginate gel. The gel is hydrophilic and allows oxygen to pass through, while bacteria cannot. Compared to traditional dressings, alginate dressing has high moisture absorption, good hemostatic properties, good biocompatibility, and can promote wound healing. After the wound is healed, it can be removed without pain. However, it is very difficult to prepare pure SA nanofibers by electrospinning. It is because the SA molecular chain is rigid; the molecular chain is always closely overlapped to form chain entanglement together effectively. This leads to the polymer chain's lack of flexibility in the spinning solution, and is not enough to stabilize the electrojet. At present, the only achievable single component electrospinning in the SA electrospinning case is by adding co-solvent glycerin to an aqueous solution of sodium alginate (Nie et al. 2009). Rheological studies showed that the addition of glycerol enhanced the flexibility and entanglement of SA chains in the solution, and thus improved the spinnability of SA solutions. The addition of glycerol to the SA solution also increases the viscosity of the SA solution and reduces the surface tension and conductivity of the SA solution, which makes electrospinning easier.

Lu et al. (Lu et al. 2006) successfully prepared SA/PEO composite nanofibers by the electrospinning method (Fig. 3.7). The effects of solution viscosity, conductivity, and surface tension on the composite nanofibers morphology, mechanical properties,

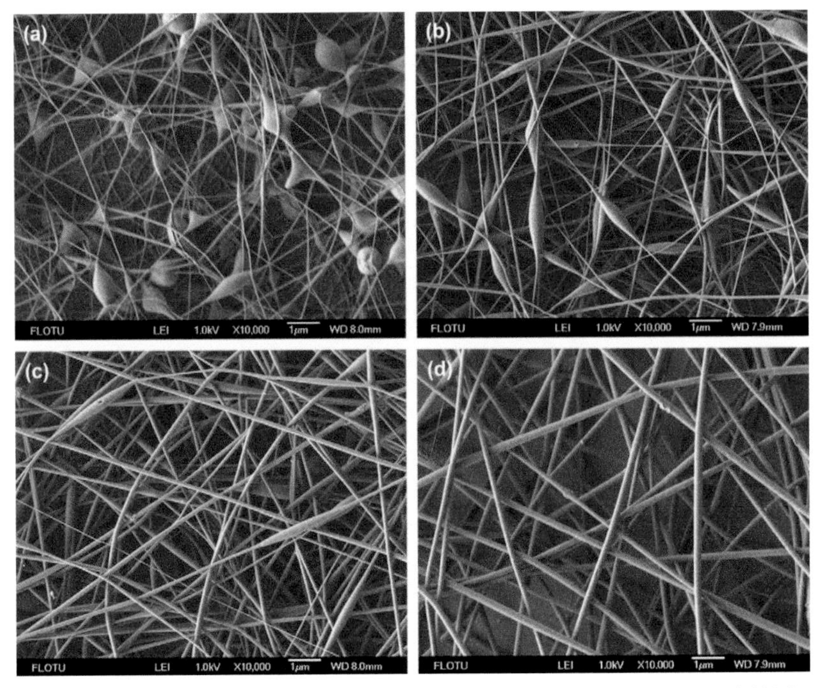

Fig. 3.7: SEM of electrospun nanofibers with different solution concentrations (Lu et al. 2006).

and water resistance were investigated. Results showed that with solution concentration varied from 1% to 3%, the electrospun nanofiber became more and more smooth and uniform. The water resistance of the composite nanofiber membrane can be improved by subsequent crosslinking. Safi et al. (Safi et al. 2007) prepared SA/PVA composite nanofibers by electrospinning. The results showed that this nanofiber membrane could accelerate wound healing better.

Hyaluronic Acid (HA). It can be seen in Fig. 3.8 that hyaluronic acid is a linear molecule mucopolysaccharide formed by glucuronic acid and N-acetyl glucosamine disaccharide units which are alternately connected with good biodegradability, biocompatibility, high viscoelasticity, and specific binding to specific receptors on the cell surface. Hyaluronic acid with its unique molecular structure and physicochemical properties shows many physiological functions *in vivo*, such as lubricating joints, adjusting the permeability of blood vessel wall, regulatory proteins, and promoting wound healing. It plays a special role in water conservation, which can improve skin nutrition and metabolism. Therefore, hyaluronic acid has great potential in cosmetics, medicine, ophthalmology, and plastic surgery.

The temperature of previous preparation of pure hyaluronic acid nanofiber is higher than 37°C to destroy the intramolecular hydrogen bond in hyaluronic acid and reduce intermolecular chain entanglement. Li et al. electrospun HA at 40 ± 3°C in ethanol solution and obtained hyaluronic acid nanofibers by a series of post-treatment (Li et al. 2006). Latest research shows that by choosing a solvent that destroys the single helix structure of hyaluronic acid, this purpose can be realized. Through using deionized water, formic acid, and dimethylformamide as solvents, researchers successfully complete the electrospinning process at room temperature.

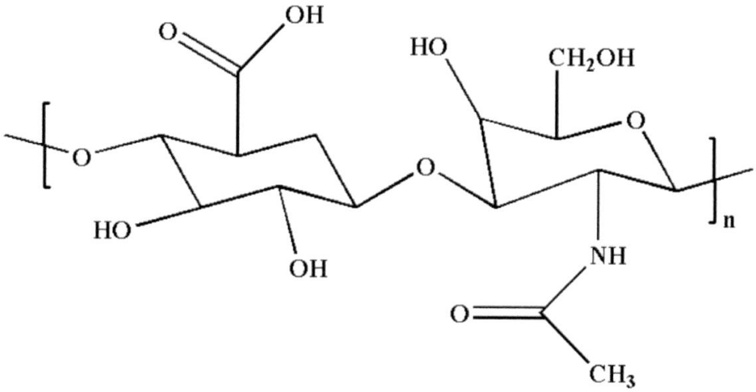

Fig. 3.8: Hyaluronic acid.

Synthetic Polymers and their Solvents

In recent years, many polymers have been used in electrospinning technology, including synthetic flexible polymers, such as polyester, nylon, polyvinyl alcohol produced by conventional techniques, and elastomer polymers such as polyurethane, block copolymer of butadiene and styrene (SBS), and even rigid polymers in liquid crystalline state such as polyparaphenylene terephthalamide. Electrospinning can be used to obtain twisted polymer fiber that has a wide diameter distribution (usually a few microns to a few tens of nanometers). Compared to the natural polymer, the nonwoven fabrics of synthetic polymers obtained by electrospinning widely used for it can be easily modified according to the specific requirements and the chemical composition, or surface morphology of the products can be optimized. Recent studies have shown that the nanofiber prepared by electrospinning can be used in lithium battery separator, cell scaffold material, adsorption, and separation membrane, which present a better performance than traditional materials.

In principle, all soluble and/or fusible polymers can be processed into fibers by electrospinning, provided that material properties (such as solubility, glass transition temperature, melting point, molecular weight, molecular weight distribution, entanglement density, solvent properties including volatility, dielectric constant, and others, as well as solution/melt properties including viscosity, surface tension coefficient, and other), processing parameters (such as applied voltage, feed rate and consumption rate, electrode separation, and geometry), and environmental conditions (such as temperature, relative humidity, solvent vapor pressure) are correctly adjusted. In this respect, empirical knowledge is crucial; theoretical models can increasingly be applied to predict the dimensions and structures of the fibers produced.

It is not amenable to make a general recommendation for particular concentrations, as well as viscosities, electrical conductivities, and surface tensions, because the ideal values of these parameters vary considerably with the polymer-solvent system. The polarity of the electrodes and the use of alternating or direct current often do not seem to have much influence.

There are many ways to classify synthetic polymers. For example, according to the method they are polymerized, different solvent, or decomposition methods, polymers are divided into several different types. So we won't go into much detail here. Instead, a summary of synthetic polymers and their solvents and applications commonly used in electrospinning will be presented in tabular form. They are roughly divided into water-soluble, organic-soluble, biodegradable polymers, and other types (Tables 3.1–3.3).

Table 3.1: Water-soluble polymers.

Polymer	Solution	Researchers
polyacrylic acid (PAA)	Water	(Baştürk et al. 2012)
polyvinylpyrrolidone (PVP)	Water, ethanol, dimethylformamide	(Jiang et al. 2012)
polyvinyl alcohol (PVA)	Water	(He et al. 2009)
polyethylene oxide (PEO)	Water, chloroform, isopropyl alcohol, acetone, ethanol	(Aliabadi et al. 2013)

Table 3.2: Organic-soluble polymers.

Polymer	Solution	Researchers
nylon 6, nylon 6, 6, nylon 4, 6 (PA)	formic acid	(Li et al. 2013, Kang et al. 2011)
Polyurethane (PU)	dimethylformamide	(Barakat et al. 2009, Pedicini and Farris 2003)
Polycarbonate (PC)	dimethylformamide, dichloromethane, chloroform, tetrahydrofuran	(Moon et al. 2008, Im et al. 2008)
polyacrylonitrile (PAN)	dimethylformamide	(Im et al. 2008)
polyimide (PI)	dimethylacetamide, N-methyl-2-pyrrolidone	(Shen et al. 2015)
polymethyl methacrylate (PMMA)	tetrahydrofuran, acetone, chloroform	(Ding et al. 2009)
polystyrene (PS)	dimethylformamide, tetrahydrofuran	(Mazinani et al. 2009)
polyvinyl chloride (PVC)	dimethylformamide, tetrahydrofuran	(Chiscan et al. 2012)
polybenzimidazole (PBI)	dimethylformamide	(von Graberg et al. 2008)
Polyethyleneterephth-alate	trifluoroacetic acid	(Ma et al. 2005)
polyvinyl butyral (PVB)	ethanol	(Lin et al. 2016)
polyvinylidene fluoride (PVDF)	acetone:dimethylacetamide = 7:3	(Kim et al. 2004)

Table 3.3: Biodegradable polymers.

Polymer	Solution	Researchers
polylactic acid (PLA)	N,N-dimethylformamide	(Shao et al. 2011)
polyglycolic acid (PGA)	Hexafluoroisopropanol	(You et al. 2006)
polycaprolactone (PCL)	chloroform:methanol = 3:1, toluene:methanol = 1:1, dichloromethane:methanol = 3:1	(Cipitria et al. 2011)

Organic Polymers for Melt Electrospinning

The electrospinning of molten polymers avoids the use of solvents and is, therefore, attractive from the perspective of productivity and environmental considerations. However, the method is limited by the fact that fibers with diameters of less than 500 nm and with a narrow diameter distribution cannot yet be fabricated.

An alternative might be electrospinning from the melt, whereas for all but a few exceptions (polymers with low-melting temperatures), this method leads to average fiber diameters that are considerably larger than 1 μm and to a broad distribution of diameters, as a result of the high melt-viscosities of the polymers. Due to their high viscosities, the electrospinning of polymer melts requires high electric fields. Under normal atmosphere, such high electric fields lead to the danger of electric shock. Variation of the atmospheric composition and even variation of the humidity can have a significant impact on the electrospinning process. High vacuum conditions could allow electrospinning at higher voltages.

In 2004, Jason and Frank (2005) summed up the theoretical models of electrospinning of polymer melts and made a comprehensive study of melt electrospinning. Then in 2008, Reneker et al. (2007) made a special elaboration on the development of melt electrospinning as one of the future directions of electrospinning. In general, melt electrospinning is particularly suitable for polymers that have no suitable solvents at room temperature, or are highly toxic and difficult to recycle in solvent systems. To date, polymers that have been successfully melt electrospun mainly include polyolefins, polyesters, and polyamides. The following Table 3.4 gives a brief overview of some representative polymers.

Table 3.4: Polymers for melt electrospinning.

Polymers	**Temperature/ °C**	**Voltage/ kV**	**Distance/ cm**	**Diameter/ μm**	**Researchers**
polyurethane (PU)	180–240	20–30	1–3	≈ 5.8	(Sanders et al. 2005)
polypropylene	200–370	30	2	≈ 3.5	(Lyons et al. 2004, Rangkupan 2003, Kong et al. 2009)
low density polyethylene (LDPE)	240	60	15	≈ 1.3	(Deng et al. 2009)
polylactic acid (PLA)	180–220	30	10	0.5–3	(Zhou et al. 2006, Hunley et al. 2008)
polyester	laser heating	18–20	2.5	1–3	(Ogata et al. 2007)

Inorganic Materials Preparation by Electrospinning

Inorganic materials refer to materials made of inorganic substances alone or in combination with other substances. As an important industrial fiber material, inorganic fibers have established a stable position in the field of materials. The refinement, ordering, and flexibility of fibers are not only the inevitable trend of scientific and technological development, but also the actual demand of high-tech industry. It has been found that inorganic fibers, including metal oxide fibers, metal fibers, carbon fibers, metal sulfides, and nitride fibers, could be prepared by electrospinning nanofibers as templates, which has greatly expanded the application area of one-dimensional materials by electrospinning technology, such as catalysis (Nalbandian et al. 2015), sensing (Kathiravan et al. 2017), filtration (Yu et al. 2013), energy (Thavasi et al. 2008), aerospace (Hung et al. 2010), biomedicine (RadhakrishnanSridhar et al. 2013), and so on. Compared to polymer nanofibers, the preparation of inorganic nanofibers is more complex, whether from material selection or process design. However, with the development of science and technology, the requirements of materials are more demanding. Inorganic materials with better heat resistance and more function seem to be more popular. In recent years, in the field of electrospinning, the studies of inorganic nanofibers are also increasingly hot, with a tendency to exceed the polymer nanofibers, and are thus the focus of this chapter. In view of the rapid development of electrospinning technology, the preparation process, technology, and methods of traditional organic/inorganic, composite nanofibers have been introduced many times and will not be repeated here. The following we will prepare the basic process of electrospinning inorganic nanofibers and introduce them step by step. We will gradually classify and introduce the inorganic nanofibers from the basic process of electrospinning.

Preparation progress of electrospun inorganic fibers (Fig. 3.9, Table 3.5) consists of the following three steps: preparation of the spinnable precursor solution, electrospinning the obtained solutions, and the thermal treatment, and the preparation details and process parameters vary with the kinds of inorganic fiber.

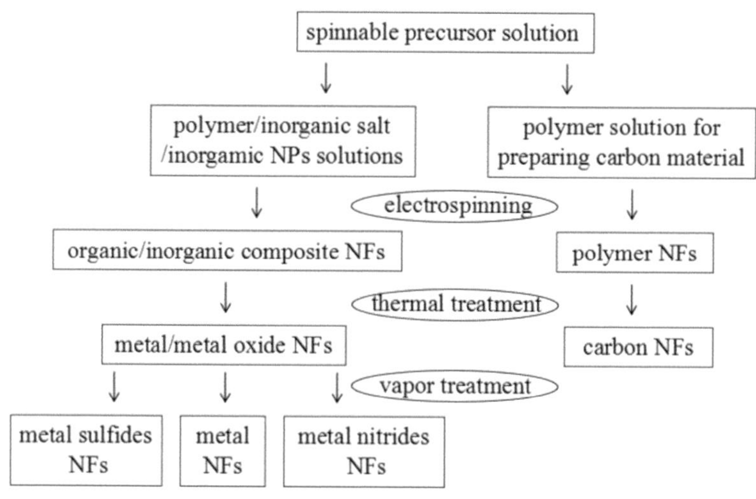

Fig. 3.9: Fabrication process of inorganic nanofibers by electrospinning.

Carbon Nanofibers

Carbon fiber, which refers to carbon fibers with carbon content of more than 95% after high temperature carbonization, is a sort of high performance fiber. As its name implies, it has not only the excellent properties of carbon materials, but also the flexibility and machinability of textile fibers, which makes it become the new generation of reinforcing fiber. Carbon fiber has the characteristics of high strength, high modulus, low density, and it is resistant to high temperature and corrosion, with good electrical conductivity and thermal conductivity (Jong and Geus 2000). Carbon nanofibers prepared by electrospinning technology have unique one-dimensional micro-nanostructures and extremely high length-diameter ratio. These excellent features make carbon fiber have a very wide range of applications, such as civil construction, aerospace, automotive, sports recreation supplies, energy, and health care (Liu et al. 2010, Ma et al. 2016).

There are many types of carbon fibers that can be divided into electrospinning and gas phase growth according to the manufacturing process and raw materials. The theoretical tensile strength of carbon fibers exceeds 180 GPa, but the tensile strength of the most powerful industrial carbon fiber (Toray T1000) is only 7 GPa (Liu et al. 2009). This is because carbon fibers may produce many defects in the

Table 3.5: Electrospinning of inorganic nanofibers.

Inorganic nanofibers	**Precursors**	**Polymer templates**	**Additives**	**Volatile solvents**	**Reference**
Al_2O_3	$AlAc_3$	PVP	acetone	ethanol	(Azad 2006)
ZnO	$ZnAc_2$	PAA		H_2O; ethanol	(Zhang et al. 2009)
SiO_2	TEOS		HCl	H_2O; ethanol	(Choi et al. 2003)
SnO_2	$SnCl_2$	PVA		H_2O; IPA; propyl alcohol	(Yang et al. 2008)
Fe_2O_3	$Fe(NO_3)_3$	PVA		H_2O	(Zheng et al. 2009)
NiO	$NiAc_2$		citric acid	H_2O	(Xu 2006)
Co_3O_4	$CoAc_2$		citric acid	H_2O	(Gu et al. 2008)
CeO_2	$Ce(NO_3)_3$	PVA		H_2O	(Yang et al. 2005)
GeO_2	Germanium	PVAc	propionic acid	acetone-H_2O-IPA	(Viswanathamurthi et al. 2004)
Mn_3O_4	$MnAc_2$	PMMA		trichloromethane; DMF	(Fan and Whittingham 2007)
MoO_3	$(NH_4)_6Mo_7O_{24}$	PVA		H_2O; ethanol	(Li et al. 2006)
V_2O_5	$OV(OCH(CH_3)_2)_3$	PMMA		trichloromethane; DMF	(Ban et al. 2009)
WO_3	Tungsten isopropoxide	PVAc		propyl alcohol; DMF	(Wang et al. 2006)
ZrO_2	Zirconium acetylacetonate	PVP	acetic acid	ethanol	(Formo et al. 2009)
ZnS	$ZnAc_2$; H_2S	PVP		ethanol	(Lin et al. 2010)
$BaTiO_3$	Barium titanium ethylhexano-isopropoxide	PVP	acetylacetone	IPA	(Mccann et al. 2006)
$LiCoO_2$	LiAc; $CoAc_2$		citric acid	H_2O	(Gu et al. 2005)
$MgTiO_3$	$Mg(OC_2H_5)_2$; TIPT	PVAc		DMF; ethanediol; methyl ether	(Dharmaraj et al. 2004)
Co-doped ZnO	$ZnAc_2$; $CoAc_2$	PVP		H_2O; ethanol	(Yang et al. 2007)
Na-doped ZnO	$ZnAc_2$; NaCl	PVA	Triton X-100	H_2O	(Zhang et al. 2010)

process of preparation, including the following three categories: (1) cracks, nicks, punctures, diametrical bulges, and other surface defects; (2) bulk defects such as holes, large cavities, refractory inclusions, cracks and flaws induced by internal stress, entanglements, and disordered structures; (3) sheath-core structure and other non-uniform structures (Fig. 3.10). However, the preparation of carbon fibers by electrospinning technology can greatly reduce such defects and enable a substantial reduction in fiber diameter to produce high-performance carbon fibers with nano-size effects. In recent years, Hao Fong's team (Reneker et al. 2000, Chen et al. 2001, Liu et al. 2009, Hedin et al. 2011, Zhao et al. 2012) has worked on the theory

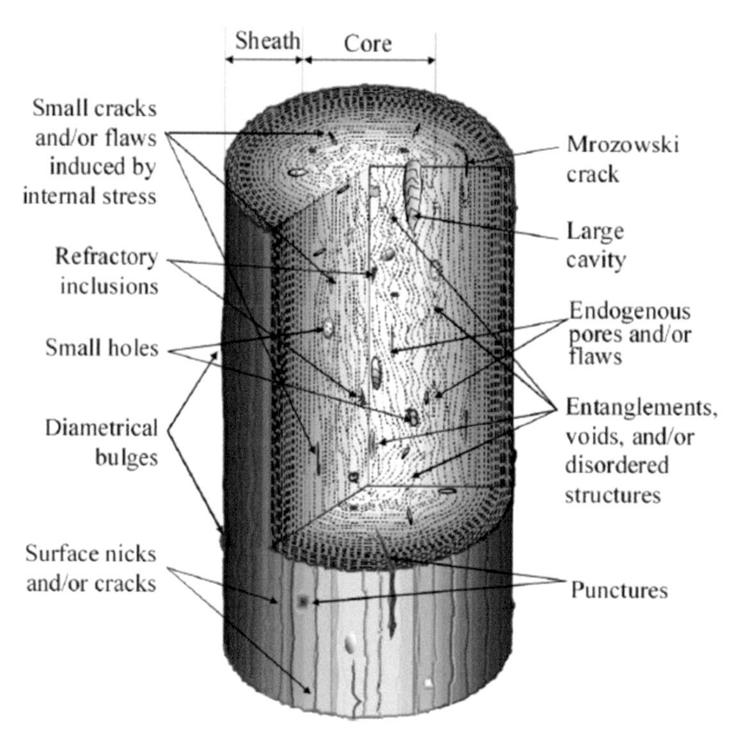

Fig. 3.10: Schematic representation of carbon fibers showing numerous types of structural imperfections.

and application of electrospun carbon nanofibers and obtained a series of results. They fabricated micro-nano-sized carbon fibers by electrospinning technology, and combined them with surface modification, chemical deposition, physical doping, and other methods to continuously improve the mechanical, electrical, physical, and chemical properties of carbon nanofibers, expanding the application prospects of carbon nanofibers.

For the carbon nanofibers fabricated by electrospinning, only the polymer components are needed in the preparation of the spinning solution without adding inorganic components, and depending on the type of polymer components, carbon fibers can be divided into polyacrylonitrile (PAN) based carbon fibers, phenolic resin based carbon fibers, and graphite fibers. The faster the heating rate, the better the crystallinity of the carbon fiber. With the increase of the carbonization temperature, the elastic modulus of the fiber increases, and the tensile strength increases first and then decreases.

The degree of graphitization of carbon fibers is different depending on the temperature at which carbon fibers are carbonized. In general carbon fiber under N_2 or H_2 atmosphere, the carbonization temperature is 800–1500°C, but when the carbonization temperature rises to 2000–3000°C, the fiber will be graphite fiber, and this process should be carried out under Ar atmosphere. This means that, with the increase of carbonization temperature, the graphitization degree of carbon fiber is getting better and better. Graphite fiber with stable chemical properties is a good conductor of heat and electricity, and it has many unique physicochemical properties such as excellent heat resistance, corrosion resistance and lubricity, which depends on the crystal structure and electronic configuration of graphite fiber materials (Fu 2010). The carbon content of the ordinary carbon fiber is 99%, while the carbon content of the graphite fiber is more than 98–99%. If the carbonization temperature is 700–1000ºC, when the water vapor or CO_2 is injected during the nitrogen atmosphere, the carbon fibers are activated. These carbon fibers are called activated carbon fibers (Wang et al. 2017).

Phenolic resin base carbon materials have higher carbon yield and porosity, compared to polyacrylonitrile, adhesion, asphalt, etc., and have an advantage in the preparation of activated carbon fiber (Cai et al. 2016). As early as 1966, the United States Carborundum Company began to develop the phenolic resin base carbon fiber for the aerospace industry. The traditional phenolic fiber preparation includes the melting spinning method of linear phenolic resin and the solution spinning method of thermosetting phenolic resin, but the curing time is too long and the fiber performance is not good. Therefore, in recent years, researchers continue to develop spinning technology, in which the electrospinning is the most representative. Electrospinning can effectively improve the fiber flexibility on the macro level, enhance its mechanical properties, and get phenolic resin based activated carbon fibers with a higher specific surface area. However, due to the low molecular weight and poor spinnability of the phenolic resin, it is necessary to add spinning aids to improve the viscosity of the spinning solution (Imaizumi et al. 2009).

In addition, sometimes in order to enhance the crystallinity of carbon fibers, inorganic components are added into the precursor solution, followed by electrospinning technology to fabricate one-dimensional nanofibers, and finally one-dimensional carbon nanofibers can be obtained after heat treatment.

Metal Oxide Nanofibers

In recent years, metal oxide nanomaterials with outstanding performance have been more and more favored, and they can be widely used in sensor, photoelectricity, catalysis, and lithium, etc. The preparation of metal oxide nanofibers by electrospinning has also become a hot topic. For early instance, Xia et al. prepared anatase crystalline TiO_2 hollow fibers by electrospinning (Fig. 3.11) that had high specific surface area and therefore exhibited excellent catalytic properties (Li and Xia 2004a). Interiorly, Chen et al. prepared flexible Al_2O_3 nanofibers with great mechanical property by using electrospinning technology (Fig. 3.12), and these fibers can be used as high-temperature insulation materials, high temperature catalyst supports, engineering strengthening materials, and automotive exhaust gas treatment agents (Wang et al. 2014). In combination with effective electrospinning technology, the preparation, research, and application of one-dimensional metal oxide nanofibers are becoming more and more popular.

To obtain metal oxide nanofibers by electrospinning, the spinning solution must contain metal oxide nanoparticles or inorganic precursors such as metal alkoxides and metal salts. However, the inorganic compound solution lacks viscoelasticity, so only electrospraying can occur in high voltage electrostatic field instead of

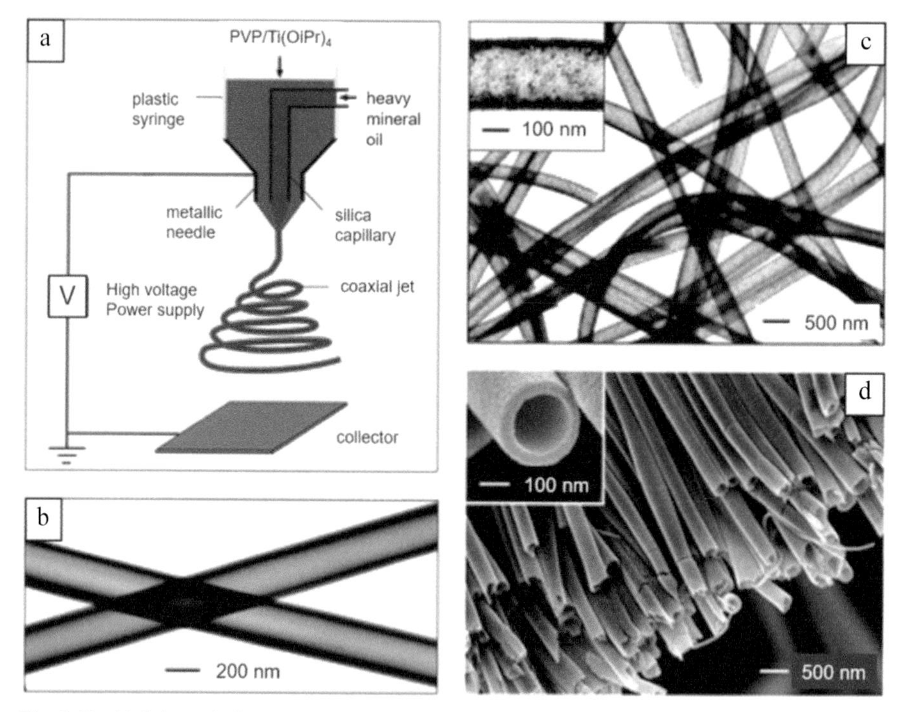

Fig. 3.11: (a) Schematic illustration of the set-up for electrospinning core-sheath nanofibers; (b) TEM image of two as-spun hollow fibers after the oily cores had been extracted with octane; (c, d) TEM and SEM images of TiO_2 hollow fibers.

Fig. 3.12: Optical images (a) and SEM images (b) of the flexible Al_2O_3 nanofiber membranes that obtained at 700°C and after kept for eight months.

electrospinning. Therefore, the preparation of inorganic nanofibers is based on the polymers' spinning ability. To inhibit the hydrolysis and gelation of metal alkoxides and metal salts, ethanol, acetic acid, and other substances are added into the spinning solution generally. In a word, the spinning solution for preparing metal oxide nanofibers should contain inorganic salts or inorganic nanoparticles, polymer templates, additives and volatile solvents such as DMF, DMSO, ethanol, and dichloromethane.

Unlike polymer precursors of carbon fibers, the precursors of metal oxide nanofibers are inorganic/polymer composite nanofibers. When calcining this inorganic/polymer composite nanofibers prepared by electrospinning in the muffle furnace, the inorganic matter will grow along the nanofiber polymer template while the polymer template is gradually decomposing. Finally, the one-dimensional metal oxide fibers are obtained. In the calcination process, the decomposition of the polymer leads to a smaller fiber diameter, and the metal oxide single crystal or polycrystalline nanoparticles is deposited so that the surface of the inorganic nanofibers is roughened. By adjusting the heat treatment temperature and time, the diameter, crystal phase composition, and fiber surface roughness of metal oxide nanofiber can be controlled. In view of the increasingly metal oxide nanofiber varieties, we organize the metal oxide nanofibers having been reported into the following table, including precursors, polymer templates, additives and volatile solvents, as the details of the experiment will not be repeated.

Metal Nanofibers

As everyone knows, metal materials have good optical properties, magnetic properties, superconducting properties, corrosion resistance, and catalytic properties, playing an extremely important role in industry, national defense, science, and technology. In recent years, the preparation of metal nanofibers with controllable size is still a challenge. In addition to metal vapor deposition, electrochemical method, and template synthesis, electrospinning is also used in the preparation of metal nanofibers. Similar to metal oxide nanofibers, synthesizing metal nanofibers needs to prepare spinnable solutions with corresponding metal alkoxides and polymer templates. After that, precursor nanofibers are prepared by electrospinning and then reprocessed to obtain metal nanofibers.

For some inorganic salts, the metal nanofibers such as Au and Pt can be directly obtained by carbonizing precursor nanofibers (Pol et al. 2008, Shui and Li 2009), but because of their rapid growth, it is difficult for polymer template to inhibit the growth rate, so the nanofibers obtained are not uniform. In order to improve this situation, the precursor fibers can be treated by heat treatment to obtain metal oxide nanofibers and then reduced to metal nanofibers. Bignitzki and coworkers fabricated $Cu(NO_3)_2$/PVB precursor nanofibers and thermal oxidized them in the atmosphere to produce CuO, and finally Cu nanofibers were fabricated by reducing in H_2 atmosphere (Bognitzki et al. 2006). Ji prepared precursor composite nanofibers with PVA as template and $Ni(NO_3)_2$ as metal salt precursor, then heated these fibers at 460°C in the atmosphere for thermal oxidation, and the Ni nanofiber with 135 nm diameter were finally prepared after H_2 reduction (Yi et al. 2013). One-dimensional metal sulfides and nitride nanofibers can be prepared by the same method.

In conclusion, compared to traditional methods such as cutting, drawing, melt spinning, and thermal decomposition method, electrospinning can be used to prepare micro-nano-sized metal fibers with good morphology and controllable diameter.

However, because of its complex operations, reports about electrospun metal nanofibers are not common.

Composite Nanomaterials Preparation by Electrospinning

With the development of electrospinning technology, disadvantages of mechanical properties and functional design of single component nanofibers are becoming more and more obvious. The demand for nanofibers with two or more components is increasing day by day, so composite nanofibers came into being.

Nanocomposite fiber is a kind of important functional fiber material which emerges with the development of science and technology. It is composed of two or more phases, and at least one phase is compounded in nanometer range. These phases can be organic, inorganic, or both, usually with a continuous phase called a matrix and another as a dispersed phase called an enhancement. Due to its excellent comprehensive performance, especially its performance design, it has been widely used in aerospace, national defense, transportation, sports, and other fields. In addition, based on the pursuit of functionality, the research preparing composite nanofibers with electrospinning method is relatively mature, and the relevant review is more (Sawicka and Gouma 2006, Sahay et al. 2012, Li and Xia 2004b, Lu et al. 2009). So, we will mainly introduce the classification of composite nanofibers, and won't make in-depth discussion of the details of each work.

Polymer/Polymer

There are two or more polymers existing in this system, so that the fibers can contain the advantages of multiple materials at the same time, such as magnetic, electrical conductivity, thermostability, mechanical properties, corrosion resistance, etc., and even create synergistic effects between different materials to prepare high performance composite fiber materials. Here, we only discuss blended polymer nanofibers.

In order to improve the performance of polymers and broaden its application, two or even more kinds of polymers can be blended to prepare multicomponent composite nanofibers with excellent properties and realize the functional composite of different organic polymers according to the principle of complementary composite material performance. In recent years, the research on electrospinning of natural polymers has been developing rapidly. Composite nanofiber materials made of natural polymer with good biocompatibility, biological activity, and biodegradability, have been widely given attention in tissue engineering, wound dressings, and surgery sutures. Delgadillo et al. successfully prepared hybrid fibrous mats by electrospinning mixed solutions of PET and chitosan, which could be a new kind of material for useful biomaterial applications, functionalized clothing, and affinity-based filtration applications (Lopes-Da-Silva et al. 2009). Chong et al. prepared composite nanofiber scaffolds by electrospinning the mixed solution of gelatin and PCL (Fig. 3.13), and studied their effects on repairing wounds and building skin. The results showed that the mouse fibroblasts were able to adhere normally to the surface of composite

Fig. 3.13: FESEM micrographs of PCL/gelatin nanofibrous scaffold.

nanofiber scaffolds, and the addition of gelatin significantly reduced the healing time of wound, which was expected to produce suitable artificial dermis (Chong et al. 2007). The electrospun nanofiber scaffolds have unique microstructure, good mechanical properties, and can bite the structure of extracellular matrix, therefore electrospinning technology is increasingly expected to be the ideal technique for preparing tissue engineering scaffolds. In addition to natural polymer materials, many synthetic polymers are also electrospun to fabricate composite nanofibers. Jiang et al. created a series of micro/nanostructured PNIPAAm/PS composite films with controllable wettability and morphology by simply controlling the concentration of polymer. When the temperature is changed from 20°C to 50°C, the wettability of the film surface can be switched between superhydrophilicity and superhydrophobicity (Wang et al. 2008). Park et al. fabricated electrospun biodegradable nanofiber meshes composed of PCL and PEO to control protein release, and the meshes can be potentially applied for the release of angiogenic growth factors to the tissue defect site (Kim et al. 2007). Zhang et al. prepared PEDOT:PSS/PVP nanofibers via electrospinning for CO gas sensing, and the result showed that this nanofiber membrane was sensitive to low concentration (5–50 ppm) CO gas due to its nano-size effect (Zhang et al. 2016).

The composite organic materials can compensate for the lack of chemical or structural incompatibility of a single polymer with natural polymers. What's more, the mechanical properties, biological activity, and degradation behavior of the fibers can be regulated by adjusting the proportions between components, in order to meet the demand of various biological materials.

Polymer/Inorganic

Polymer/inorganic composite nanofibers are composites obtained by the combination of organic polymers and nanoparticles. For example, the PVP/α-Fe_2O_3 composite nanofibers, PAN/Ag nanofibers and PVDF/graphene composite nanofibers are nanocomposites formed by sheet material or inorganic particles dispersed in polymer matrix in nanoscale. In polymer/inorganic composite nanofibers, a considerable fraction of nanomaterials are functional, such as the ZnO in PVA/ZnO nanofibers has good photocatalytic activity; PLA/Ag composite nanofibers have antibacterial properties and biodegradability; polymer nanocomposites with carbon nanotubes exhibit high mechanical properties, electrical properties, and unique thermal, optical, field emission, absorption and other excellent properties. Electrospun nanomaterials can also provide orientation, and fiber materials with different orientation degree can be prepared according to the stress and function characteristics. Moreover, the polymer/inorganic composite nanofibers can create unique functions and synergistic effects through the structure design.

Mandal et al. developed a novel flexible and ultrasensitive nanopressure sensor based on Ce^{3+} doped electrospun PVDF/graphene NFs, which offered exceptional piezoelectric characteristics, to enable ultrahigh sensitivity for measuring pressure, even at extremely small values (2 Pa) (Garain et al. 2016). Barakat et al. doped the PSF electrospun nanofibers by amorphous SiO_2 nanoparticles and GO individually to prepare effective PSF-based nanofiber membrane for petroleum oil fractions/water (Fig. 3.14), and the final result showed that SiO_2-doped PSF nanofiber membranes had higher mechanical strength and daily flux that was 100 (Obaid et al. 2015), 115, and 187 m^3/m^2 for gasoline, kerosene, and hexane, respectively (Khan et al. 2015). Khan et al. reported the fabrication and characterization of electrospun PAN and PMMA nanofiber separators embedded with graphene nanoflakes, and test results revealed that the physical properties, such as wettability, dielectric constant, ionic conductivity, and thermal conductivity values of the nanocomposite separators were significantly increased as a function of graphene concentrations (Khan et al. 2015). Chen et al. reported the design and fabrication of flexible and photorewritable PVP/a-WO_3 hybrid membranes through electrospinning, on which images with high resolution can be photoprinted and heat-erased for over 40 cycles (Wei et al. 2016). This photochromic membrane has great potential as a robust rewritable system to replace regular fabric prints for sustainable developments. Electrospun polymer/noble metal hybrid nanofibers have developed rapidly as surface-enhanced Raman scattering (SERS) active substrates over the last few years. Cheng et al. showed a general and facile approach for the preparation of PAN/noble metal/SiO_2 nanofibrous mats with PEF activity for the first time by combining electrospinning and controlled silica coatings, which had advantages including good selectivity, remarkable sensitivity, and recyclability (Zhang et al. 2015).

Due to the strong interaction between the inorganic nanomaterial, it's easy to agglomerate and affect the final properties of composites, but electrospinning technology is an effective way to avoid agglomeration and prepare polymer/inorganic composite nanofibers. At present, there are three main methods for preparing polymer/inorganic composite nanofibers: one is to have prepared nanoparticles directly dispersed in polymer solution, then electrospin the solution to obtain composite nanofibers, called hybrid method; two is to prepare nanoparticles *in situ* in polymer solution, and then obtain the composite fibers by electrospinning, called *in situ* composite method; the third one is to mix the precursor of the corresponding mixed nanoparticles in polymer solution and the corresponding polymer composite

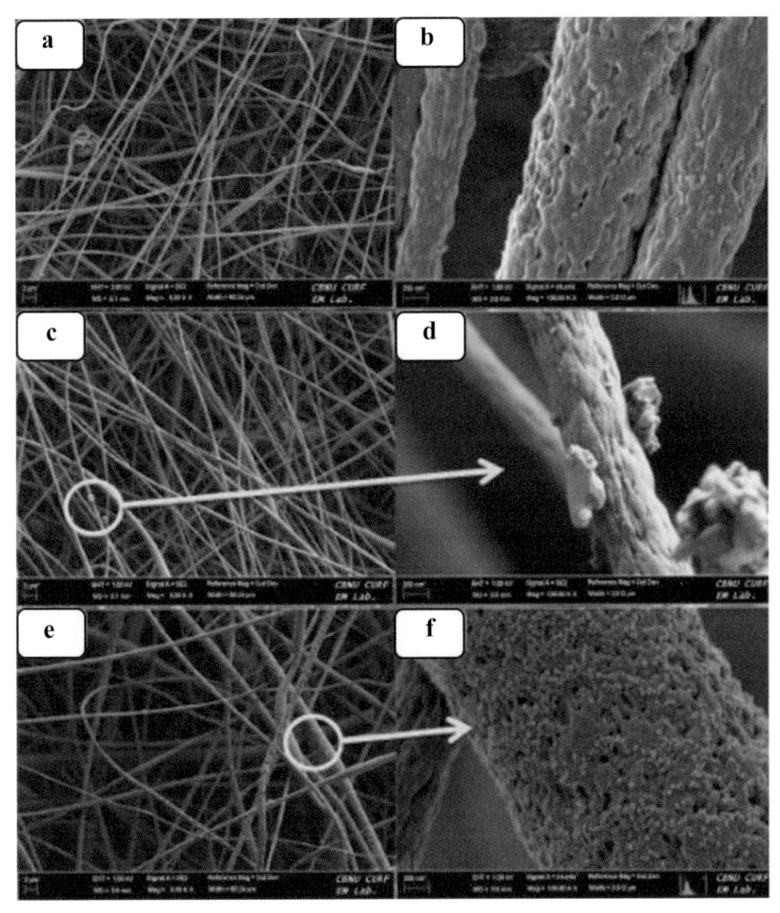

Fig. 3.14: FE-SEM images of the original PSF (a, b), PSF-GO (c, d), and PSF-SiO_2 (e, f) electrospun nanofibers.

fiber containing the precursor is obtained by electrospinning, then use UV reduction method to get the polymer/inorganic composite nanofibers, called after processing method.

Inorganic/Inorganic

Inorganic composite fiber is composed of two or more inorganic substances. Except for maintaining the nature of their own, they can complement each other to enhance functionality to meet the various requirements, which has aroused widespread concern in recent years. It is mainly divided into metal-oxide composite nanofibers and oxide-oxide composite nanofibers, which are usually prepared by two-step method. Firstly, add inorganic nanoparticles or precursors into spinnable precursor solution and fabricate 1D polymer/inorganic nanocomposite nanofibers by electrospinning. Secondly, obtain inorganic composite nanofibers by calcination.

Ding et al. constructed hierarchical TiNFNPs with robust flexibility and porosity by the combination of sol–gel electrospinning and *in situ* polymerization that possessed better photodegradation performance towards MB compared to a commercial catalyst (P25) and could be directly taken out from solution facilely without using any tedious or time-consuming sedimentation process (Zhang et al. 2015). In addition, they also took PBZ as a new kind of carbon source and introduced $SnCl_2$ to fabricate highly porous carbon nanofibrous membranes with intriguing mechanical elasticity and functionality via combining the multicomponent electrospinning and *in situ* polymerization (Fig. 3.15). The as-prepared SnO_2/CNF membranes exhibited robust elasticity and durability, which could rapidly recover to their original designed shape after serious deformations (Ge et al. 2016). Wei et al. synthesized ZnO/SnO_2 composite nanofibers by a simple electrospinning approach with subsequent calcination at 700°C using PAN as the polymer precursor, and the nanofibers showed excellent lithium storage properties in terms of cycling stability and rate capability (Wei et al. 2016). Kang et al. prepared novel

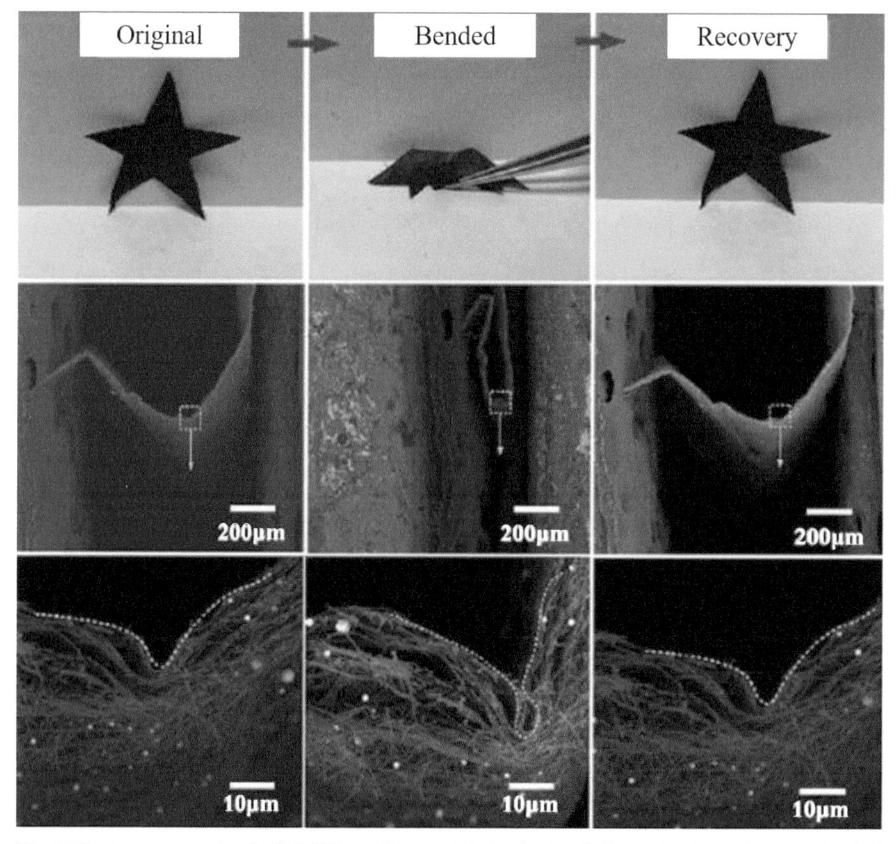

Fig. 3.15: A representative SnO_2/CNF membrane with the designed shape showing robust mechanical durability, and the *in situ* SEM observations of the bent membrane and recovery focus on a small piece (< 2 mm).

freestanding graphene oxide (GO)-embedded porous carbon nanofiber (PCNF) webs by electrospinning from polyacrylonitrile and GO (Luo et al. 2016), followed by carbonization and steam activation, and the GO-PCNF electrode exhibited excellent deionization performance (Bai et al. 2014). Yang et al. fabricated a novel and highly efficient visible-light-driven photocatalyst with robust stability that was made up of thoroughly mesoporous $TiO_2/WO_3/g\text{-}C_3N_4$ ternary hybrid nanofibers, through a foaming-assisted electrospinning process followed by a solution dipping process (Hou et al. 2016).

It can be seen that the best advantage of electrospinning method is that it is easy to prepare one-dimensional composite nanofibers with uniform diameter. Due to the combination of the advantages and properties of various components in nanofibers, the composite nanofibers are widely applied in photocatalytic, photochemistry, supercapacitor, and a series of fields.

Concluding Remarks

In this chapter, the electrospun nanofibers are classified and the raw materials for each kind of nanofibers are introduced, including the polymeric materials in electrospinning and their matched solvent systems. In addition, inorganic nanofibers such as metal oxide nanofibers, metal nanofibers, carbon nanofibers, metal sulfides nanofibers, and metal nitrides nanofibers are also introduced about the basic process of electrospinning and post-treatment. Additionally, the specific preparation processes of composite nanofibers, such as polymer/polymer nanofibers, polymer/inorganic nanofibers and inorganic/inorganic nanofibers are involved in this chapter, as well to help the readers to comprehensively understand the way the composite nanofibers are prepared and their excellent comprehensive performances.

References

Aliabadi, M., Irani, M., Ismaeili, J., Piri, H. and Parnian, M. J. 2013. Electrospun nanofiber membrane of PEO/Chitosan for the adsorption of nickel, cadmium, lead and copper ions from aqueous solution. Chemical Engineering Journal 220: 237–43.

An, K., Liu, H., Guo, S., Kumar, D. N. and Wang, Q. 2010. Preparation of fish gelatin and fish gelatin/poly(L-lactide) nanofibers by electrospinning. International Journal of Biological Macromolecules 47: 380–8.

Azad, A. M. 2006. Fabrication of transparent alumina (Al_2O_3) nanofibers by electrospinning. Materials Science & Engineering A 435: 468–73.

Bai, Y., Huang, Z. H., Yu, X. L. and Kang, F. 2014. Graphene oxide-embedded porous carbon nanofiber webs by electrospinning for capacitive deionization. Colloids & Surfaces A Physicochemical & Engineering Aspects 444: 153–8.

Ban, C., Chernova, N. A. and Whittingham, M. S. 2009. Electrospun nano-vanadium pentoxide cathode. Electrochemistry Communications 11: 522–5.

Barakat, N. A. M., Kanjwal, M. A., Sheikh, F. A. and Kim, H. Y. 2009. Spider-net within the N6, PVA and PU electrospun nanofiber mats using salt addition: Novel strategy in the electrospinning process. Polymer 50: 4389–96.

Baştürk, E. D. S., Danış, Ö. and Kahraman, M. V. 2012. Covalent immobilization of α-amylase onto thermally crosslinked electrospun PVA/PAA nanofibrous hybrid membranes. Journal of Applied Polymer Science 127: 349–55.

Bognitzki, M., Becker, M., Graeser, M., Massa, W., Wendorff, J. H., Schaper, A. et al. 2006. Preparation of sub-micrometer copper fibers via electrospinning. Advanced Materials 18: 2384–6.
Burck, J., Heissler, S., Geckle, U., Ardakani, M. F., Schneider, R., Ulrich, A. S. et al. 2013. Resemblance of electrospun collagen nanofibers to their native structure. Langmuir 29: 1562–72.
Cai, J., Fu, S., Zhang, Y., Liu, G., Qiu, J. and Zhou, A. 2016. Synthesis and capacitance performance of phenolic resin-based highly ordered mesoporous carbon materials. Engineering Plastics Application.
Chakrapani, V. Y., Gnanamani, A., Giridev, V. R., Madhusoothanan, M. and Sekaran, G. 2012. Electrospinning of type I collagen and PCL nanofibers using acetic acid. Journal of Applied Polymer Science 125: 3221–7.
Chen, Z., Foster, M. D., Zhou, W., Fong, H., Reneker, D. H. 2001. Structure of Poly(ferrocenyldimethylsilane) in electrospun nanofibers. Macromolecules 34: 6156–8.
Chiscan, O., Dumitru, I., Postolache, P., Tura, V. and Stancu, A. 2012. Electrospun PVC/Fe_3O_4 composite nanofibers for microwave absorption applications. Materials Letters 68: 251–4.
Choi, S. S., Lee, S. G., Im, S. S., Kim, S. H. and Yong, L. J. 2003. Silica nanofibers from electrospinning/sol-gel process. Journal of Materials Science Letters 22: 891–3.
Chong, E. J., Phan, T. T., Lim, I. J., Zhang, Y. Z., Bay, B. H., Ramakrishna, S. et al. 2007. Evaluation of electrospun PCL/gelatin nanofibrous scaffold for wound healing and layered dermal reconstitution. Acta Biomaterialia 3: 321–30.
Cipitria, A., Skelton, A., Dargaville, T. R., Dalton, P. D. and Hutmacher, D. W. 2011. Design, fabrication and characterization of PCL electrospun scaffolds—a review. Journal of Materials Chemistry 21: 9419.
Deng, R., Liu, Y., Ding, Y., Xie, P., Luo, L. and Yang, W. 2009. Melt electrospinning of low-density polyethylene having a low-melt flow index. Journal of Applied Polymer Science 114: 166–75.
Dharmaraj, N., Park, H. C., Lee, B. M., Viswanathamurthi, P., Kim, H. Y. and Lee, D. R. 2004. Preparation and morphology of magnesium titanate nanofibres via electrospinning. Inorganic Chemistry Communications 7: 431–3.
Ding, B., Wang, M., Wang, X., Yu, J. and Sun, G. 2010. Electrospun nanomaterials for ultrasensitive sensors. Materials Today 13: 16–27.
Ding, Y., Zhang, P., Long, Z., Jiang, Y., Xu, F. and Di, W. 2009. The ionic conductivity and mechanical property of electrospun P(VdF-HFP)/PMMA membranes for lithium ion batteries. Journal of Membrane Science 329: 56–9.
Dong, B., Arnoult, O., Smith, M. E. and Wnek, G. E. 2009. Electrospinning of collagen nanofiber scaffolds from benign solvents. Macromol. Rapid Commun. 30: 539–42.
Dong, B. K. L. Y., Lee, W. S., Jo, S. M. and Kim, B. C. 2002. Double crystallization behavior in dry-jet wet spinning of cellulose/N-methylmorpholine-N-oxide hydrate solutions. European Polymer Journal 38: 109–19.
Dupont, A. -L. 2003. Cellulose in lithium chloride/N,N-dimethylacetamide, optimisation of a dissolution method using paper substrates and stability of the solutions. Polymer 44: 4117–26.
Fan, Q. and Whittingham, M. S. 2007. Electrospun manganese oxide nanofibers as anodes for lithium-ion batteries. Electrochemical and Solid-State Letters 10: A48–A51.
Fink, H. P. W. P., Purz, H. J. and Ganster, J. 2001. Structure formation of regenerated cellulose materials from NMMO-solutions. Progress in Polymer Science 26: 1473–524.
Formo, E., Camargo, P. H. C., Lim, B., Jiang, M. and Xia, Y. 2009. Functionalization of ZrO_2 nanofibers with Pt nanostructures: The effect of surface roughness on nucleation mechanism and morphology control. Chemical Physics Letters 476: 56–61.
Garain, S., Jana, S., Sinha, T. K. and Mandal, D. 2016. Design of *in situ* poled Ce(3+)-doped electrospun PVDF/graphene composite nanofibers for fabrication of nanopressure sensor and ultrasensitive acoustic nanogenerator. Acs Applied Materials & Interfaces 8: 4532.
Ge, J., Qu, Y., Cao, L., Wang, F., Dou, L., Yu, J. et al. 2016. Polybenzoxazine-based highly porous carbon nanofibrous membranes hybridized by tin oxide nanoclusters: durable mechanical elasticity and capacitive performance. Journal of Materials Chemistry A 4: 7795–804.
Geng, X., Kwon, O. H. and Jang, J. 2005. Electrospinning of chitosan dissolved in concentrated acetic acid solution. Biomaterials 26: 5427–32.
Gobeaux, F., Belamie, E., Mosser, G., Davidson, P. and Asnacios, S. 2010. Power law rheology and strain-induced yielding in acidic solutions of type I-collagen. Soft Matter 6: 3769.

Gu, Y., Dairong Chen, A. and Jiao, X. 2005. Synthesis and electrochemical properties of nanostructured $LiCoO_2$ fibers as cathode materials for lithium-ion batteries. Journal of Physical Chemistry B 109: 17901–6.

Gu, Y., Jian, F. and Wang, X. 2008. Synthesis and characterization of nanostructured Co_3O_4 fibers used as anode materials for lithium ion batteries. Thin Solid Films 517: 652–5.

He, D. H. B., Yao, Q. F., Wang, K. and Yu, S. H. 2009. Large-scale synthesis of flexible free-standing SERS substrates with high sensitivity: electrospun PVA nanofibers embedded with controlled alignment of silver nanoparticles. Acs Nano. 3: 3993–4002.

Hedin, N., Sobolev, V., Zhang, L., Zhu, Z. and Hao, F. 2011. Electrical properties of electrospun carbon nanofibers. Journal of Materials Science 46: 6453–6.

Hossain, K. M. Z., Ahmed, I., Parsons, A. J., Scotchford, C. A., Walker, G. S., Thielemans, W. et al. 2011. Physico-chemical and mechanical properties of nanocomposites prepared using cellulose nanowhiskers and poly(lactic acid). Journal of Materials Science 47: 2675–86.

Hou, H., Gao, F., Wang, L., Shang, M., Yang, Z., Zheng, J. et al. 2016. Superior thoroughly mesoporous ternary hybrid photocatalysts of $TiO_2/WO_3/g\text{-}C_3N_4$ nanofibers for visible-light-driven hydrogen evolution. Journal of Materials Chemistry A 4: 6276–81.

Hung, N. T., Tuong, N. M. and Rakov, E. G. 2010. Acid functionalization of carbon nanofibers. Inorganic Materials 46: 1077–83.

Hunley, M. T., Karikari, A. S., McKee, M. G., Mather, B. D., Layman, J. M., Fornof, A. R. et al. 2008. Taking advantage of tailored electrostatics and complementary hydrogen bonding in the design of nanostructures for biomedical applications. Macromolecular Symposia 270: 1–7.

Im, J. S., Kim, M. I. and Lee, Y. -S. 2008. Preparation of PAN-based electrospun nanofiber webs containing TiO_2 for photocatalytic degradation. Materials Letters 62: 3652–5.

Imaizumi, S., Matsumoto, H., Suzuki, K., Minagawa, M., Kimura, M. and Tanioka, A. 2009. Phenolic resin-based carbon thin fibers prepared by electrospinning: Additive effects of poly(vinyl butyral) and electrolytes. Polymer Journal 41: 1124–8.

Jason, L. and Frank, K. 2005. Feature article: Melt electrospinning of polymers: A Review. Polymer News 30: 170–178.

Jeong, L., Lee, K. Y. and Park, W. H. 2007. Effect of solvent on the characteristics of electrospun regenerated silk fibroin nanofibers. Key Engineering Materials 342-343: 813–6.

Jiang, S., Hou, H., Agarwal, S. and Greiner, A. 2016. Polyimide nanofibers by "green" electrospinning via aqueous solution for filtration applications. ACS Sustainable Chemistry & Engineering 4: 4797–804.

Jiang, Y. N., Mo, H. Y. and Yu, D. G. 2012. Electrospun drug-loaded core-sheath PVP/zein nanofibers for biphasic drug release. International Journal of Biological Macromolecules 438: 232–9.

Jong, K. D. and Geus, J. 2000. Carbon nanofibers: Catalytic synthesis and applications. Catalysis Reviews 42: 481–510.

Kang, H. -K., Shin, H. -K., Jeun, J. -P., Kim, H. -B. and Kang, P. -H. 2011. Fabrication and characterization of electrospun polyamide 66 fibers crosslinked by gamma irradiation. Macromolecular Research 19: 364–9.

Kathiravan, D., Huang, B. R. and Saravanan, A. 2017. Self-assembled hierarchical interfaces of ZnO nanotubes/graphene heterostructures for efficient room temperature hydrogen sensors. ACS Applied Materials & Interfaces 9: 12064–72.

Kew, S. J., Gwynne, J. H., Enea, D., Abu-Rub, M., Pandit, A., Zeugolis, D. et al. 2011. Regeneration and repair of tendon and ligament tissue using collagen fibre biomaterials. Acta Biomater. 7: 3237–47.

Khan, W. S., Asmatulu, R., Rodriguez, V. and Ceylan, M. 2015. Enhancing thermal and ionic conductivities of electrospun PAN and PMMA nanofibers by graphene nanoflake additions for battery-separator applications. International Journal of Energy Research 38: 2044–51.

Kim, C. -W., Frey, M. W., Marquez, M. and Joo, Y. L. 2005. Preparation of submicron-scale, electrospun cellulose fibers via direct dissolution. Journal of Polymer Science Part B: Polymer Physics 43: 1673–83.

Kim, C. -W., Kim, D. -S., Kang, S. -Y., Marquez, M. and Joo, Y. L. 2006. Structural studies of electrospun cellulose nanofibers. Polymer 47: 5097–107.

Kim, J. R., Choi, S. W., Jo, S. M., Lee, W. S. and Kim, B. C. 2004. Electrospun PVdF-based fibrous polymer electrolytes for lithium ion polymer batteries. Electrochimica Acta. 50: 69–75.

Kim, T. G., Lee, D. S. and Park, T. G. 2007. Controlled protein release from electrospun biodegradable fiber mesh composed of poly(epsilon-caprolactone) and poly(ethylene oxide). International Journal of Pharmaceutics 338: 276.

Kim, T. G., Chung, H. J. and Park, T. G. 2008. Macroporous and nanofibrous hyaluronic acid/collagen hybrid scaffold fabricated by concurrent electrospinning and deposition/leaching of salt particles. Acta Biomaterialia 4: 1611–9.

Kong, C. S., Jo, K. J., Jo, N. K. and Kim, H. S. 2009. Effects of the spin line temperature profile and melt index of poly(propylene) on melt-electrospinning. Polymer Engineering & Science 49: 391–6.

Kulpinski, P. 2005. Cellulose nanofibers prepared by the N-methylmorpholine-N-oxide method. Journal of Applied Polymer Science 98: 1855–9.

Lepore, E., Bosia, F., Bonaccorso, F., Bruna, M., Taioli, S., Garberoglio, G. et al. 2017. Spider silk reinforced by graphene or carbon nanotubes. 2d Materials 4.

Li, B., Jiang, B., Boyce, B. M. and Lindsey, B. A. 2009. Multilayer polypeptide nanoscale coatings incorporating IL-12 for the prevention of biomedical device-associated infections. Biomaterials 30: 2552–8.

Li, B., Yuan, H. and Zhang, Y. 2013. Transparent PMMA-based nanocomposite using electrospun graphene-incorporated PA-6 nanofibers as the reinforcement. Composites Science and Technology 89: 134–41.

Li, D. and Xia, Y. 2004a. Direct fabrication of composite and ceramic hollow nanofibers by electrospinning. Nano Letters 4: 933–8.

Li, D. and Xia, Y. 2004b. Electrospinning of nanofibers: Reinventing the wheel? Advanced Materials 16: 1151–70.

Li, J., He, A., Han, C. C., Fang, D., Hsiao, B. S. and Chu, B. 2006. Electrospinning of hyaluronic acid (HA) and HA/Gelatin blends. Macromolecular Rapid Communications 27: 114–20.

Li, M., Guo, Y., Wei, Y., MacDiarmid, A. G. and Lelkes, P. I. 2006. Electrospinning polyaniline-contained gelatin nanofibers for tissue engineering applications. Biomaterials 27: 2705–15.

Li, S., Shao, C., Liu, Y., Tang, S. and Mu, R. 2006. Nanofibers and nanoplatelets of MoO_3 via an electrospinning technique. Journal of Physics & Chemistry of Solids 67: 1869–72.

Lin, D., Wu, H., Zhang, R. and Pan, W. 2010. Preparation of ZnS nanofibers via electrospinning. Journal of the American Ceramic Society 90: 3664–6.

Lin, P. -Y., Wu, Z. -S., Juang, Y. -D., Fu, Y. -S. and Guo, T. -F. 2016. Microwave-assisted electrospun PVB/CdS composite fibers and their photocatalytic activity under visible light. Microelectronic Engineering 149: 73–7.

Liu, J., Yue, Z. and Fong, H. 2009. Continuous nanoscale carbon fibers with superior mechanical strength. Small 5: 536–42.

Liu, J., Zhou, P., Zhang, L., Ma, Z., Liang, J. and Hao, F. 2009. Thermo-chemical reactions occurring during the oxidative stabilization of electrospun polyacrylonitrile precursor nanofibers and the resulting structural conversions. Carbon 47: 1087–95.

Liu, J., Tian, Y., Chen, Y., Liang, J., Zhang, L. and Hao, F. 2010. A surface treatment technique of electrochemical oxidation to simultaneously improve the interfacial bonding strength and the tensile strength of PAN-based carbon fibers. Materials Chemistry & Physics 122: 548–55.

Lopes-Da-Silva, J. A., Veleirinho, B. and Delgadillo, I. 2009. Preparation and characterization of electrospun mats made of PET/chitosan hybrid nanofibers. Journal of Nanoscience & Nanotechnology 9: 3798.

Lu, J. -W., Zhu, Y. -L., Guo, Z. -X., Hu, P. and Yu, J. 2006. Electrospinning of sodium alginate with poly(ethylene oxide). Polymer 47: 8026–31.

Lu, X., Wang, C. and Wei, Y. 2009. One-dimensional composite nanomaterials: synthesis by electrospinning and their applications. Small 5: 2349–70.

Luo, L., Xu, W., Xia, Z., Fei, Y., Zhu, J., Chen, C. et al. 2016. Electrospun ZnO–SnO_2 composite nanofibers with enhanced electrochemical performance as lithium-ion anodes. Ceramics International 42: 10826–32.

Lyons, J., Li, C. and Ko, F. 2004. Melt-electrospinning part I: processing parameters and geometric properties. Polymer 45: 7597–603.

Ma, S., Liu, J., Qu, M., Wang, X., Huang, R. and Liang, J. 2016. Effects of carbonization tension on the structural and tensile properties of continuous bundles of highly aligned electrospun carbon nanofibers. Materials Letters 183: 369–73.

Ma, Z., Kotaki, M., Yong, T., He, W. and Ramakrishna, S. 2005. Surface engineering of electrospun polyethylene terephthalate (PET) nanofibers towards development of a new material for blood vessel engineering. Biomaterials 26: 2527–36.

Mazinani, S., Ajji, A. and Dubois, C. 2009. Morphology, structure and properties of conductive PS/CNT nanocomposite electrospun mat. Polymer 50: 3329–42.

Mccann, J. T., Chen, J. I. L., Li, D., Ye, Z. G. and Xia, Y. 2006. Electrospinning of polycrystalline barium titanate nanofibers with controllable morphology and alignment. Chemical Physics Letters 424: 162–6.

Mele, E. 2016. Electrospinning of natural polymers for advanced wound care: towards responsive and adaptive dressings. Journal of Materials Chemistry B 4: 4801–12.

Meng, Z., Zheng, X., Tang, K., Liu, J. and Qin, S. 2013. Dissolution of natural polymers in ionic liquids: A review. E-Polymers 12: 317–45.

Min, B. -M., Lee, S. W., Lim, J. N., You, Y., Lee, T. S., Kang, P. H. et al. 2004. Chitin and chitosan nanofibers: electrospinning of chitin and deacetylation of chitin nanofibers. Polymer 45: 7137–42.

Moon, S. and Farris, R. J. 2008. The morphology, mechanical properties, and flammability of aligned electrospun polycarbonate (PC) nanofibers. Polymer Engineering & Science 48: 1848–54.

Nalbandian, M. J., Zhang, M., Sanchez, J., Choa, Y. -H., Cwiertny, D. M. and Myung, N. V. 2015. Synthesis and optimization of $BiVO_4$ and co-catalyzed $BiVO_4$ nanofibers for visible light-activated photocatalytic degradation of aquatic micropollutants. Journal of Molecular Catalysis A: Chemical 404-405: 18–26.

Nie, H., He, A., Wu, W., Zheng, J., Xu, S., Li, J. et al. 2009. Effect of poly(ethylene oxide) with different molecular weights on the electrospinnability of sodium alginate. Polymer 50: 4926–34.

Obaid, M., Tolba, G. M. K., Motlak, M., Fadali, O. A., Khalil, K. A., Almajid, A. A. et al. 2015. Effective polysulfone-amorphous SiO_2 NPs electrospun nanofiber membrane for high flux oil/water separation. Chemical Engineering Journal 279: 631–8.

Ogata, N., Shimada, N., Yamaguchi, S., Nakane, K. and Ogihara, T. 2007. Melt-electrospinning of poly(ethylene terephthalate) and polyalirate. Journal of Applied Polymer Science 105: 1127–32.

Ohkawa, K., Cha, D., Kim, H., Nishida, A. and Yamamoto, H. 2004. Electrospinning of chitosan. Macromolecular Rapid Communications 25: 1600–5.

Pedicini, A. and Farris, R. J. 2003. Mechanical behavior of electrospun polyurethane. Polymer 44: 6857–62.

Pol, V. G., Koren, E. and Zaban, A. 2008. Fabrication of continuous conducting gold wires by electrospinning. Chemistry of Materials 20: 3055–62.

RadhakrishnanSridhar, SubramanianSundarrajan, JayaramaReddyVenugopal, RajeswariRavichandran and SeeramRamakrishna. 2013. Electrospun inorganic and polymer composite nanofibers for biomedical applications. Journal of Biomaterials Science Polymer Edition 24: 365–85.

Rangkupan, R. R. D. 2003. Electrospinning process of molten polypropylene in vacuum. Journal of Metals 12: 81–7.

Raquez, J. M., Murena, Y., Goffin, A. L., Habibi, Y., Ruelle, B., DeBuyl, F. et al. 2012. Surface-modification of cellulose nanowhiskers and their use as nanoreinforcers into polylactide: A sustainably-integrated approach. Composites Science and Technology 72: 544–9.

Reneker, D. H., Yarin, A. L., Fong, H. and Koombhongse, S. 2000. Bending instability of electrically charged liquid jets of polymer solutions in electrospinning. Journal of Applied Physics 87: 4531–47.

Reneker, D. H., Yarin, A. H., Zussman, E. and Xu, H. 2007. Electrospinning of nanofibers from polymer solutions and melts. Advances in Applied Mechanics 41: 43–195.

Röder, T. M. B., Schelosky, N. and Glatter, O. 2001. Solutions of cellulose in N, N-dimethylacetamide/lithium chloride studied by light scattering methods. Polymer 42: 6765–73.

Safi, S., Morshed, M., Hosseini Ravandi, S. A. and Ghiaci, M. 2007. Study of electrospinning of sodium alginate, blended solutions of sodium alginate/poly(vinyl alcohol) and sodium alginate/poly(ethylene oxide). Journal of Applied Polymer Science 104: 3245–55.

Sahay, R., Kumar, P. S., Sridhar, R., Sundaramurthy, J., Venugopal, J., Mhaisalkar, S. G. et al. 2012. Electrospun composite nanofibers and their multifaceted applications. Journal of Materials Chemistry 22: 12953.

Sanders, J. E., Lamont, S. E., Karchin, A., Golledge, S. L. and Ratner, B. D. 2005. Fibro-porous meshes made from polyurethane micro-fibers: effects of surface charge on tissue response. Biomaterials 26: 813–8.

Sawicka, K. M. and Gouma, P. 2006. Electrospun composite nanofibers for functional applications. Journal of Nanoparticle Research 8: 769–81.

Shao, S., Zhou, S., Li, L., Li, J., Luo, C., Wang, J. et al. 2011. Osteoblast function on electrically conductive electrospun PLA/MWCNTs nanofibers. Biomaterials 32: 2821–33.

Shen, Y., Chen, L., Jiang, S., Ding, Y., Xu, W. and Hou, H. 2015. Electrospun nanofiber reinforced all-organic PVDF/PI tough composites and their dielectric permittivity. Materials Letters 160: 515–7.

Shengyuan, Y., Peining, Z., Nair, A. S. and Ramakrishna, S. 2011. Rice grain-shaped TiO_2 mesostructures—synthesis, characterization and applications in dye-sensitized solar cells and photocatalysis. Journal of Materials Chemistry 21: 6541.

Shui, J. and Li, J. C. 2009. Platinum nanowires produced by electrospinning. Nano Letters 9: 1307.

Thavasi, V., Singh, G. and Ramakrishna, S. 2008. Electrospun nanofibers in energy and environmental applications. Energy & Environmental Science 1: 205–21.

Viswanathamurthi, P., Bhattarai, N., Kim, H. Y., Khil, M. S., Lee, D. R. and Suh, E. K. 2004. GeO_2 fibers: preparation, morphology and photoluminescence property. Journal of Chemical Physics 121: 441–5.

von Graberg, T., Thomas, A., Greiner, A., Antonietti, M. and Weber, J. 2008. Electrospun silica-polybenzimidazole nanocomposite fibers. Macromolecular Materials and Engineering 293: 815–9.

Wang, G., Ji, Y., Huang, X., Yang, X., Gouma, P. I. and Dudley, M. 2006. Fabrication and characterization of polycrystalline WO_3 nanofibers and their application for ammonia sensing. Journal of Physical Chemistry B 110: 23777.

Wang, K., Song, Y., Yan, R., Zhao, N., Tian, X., Li, X. et al. 2017. High capacitive performance of hollow activated carbon fibers derived from willow catkins. Applied Surface Science 394: 569–77.

Wang, N., Zhao, Y. and Jiang, L. 2008. Low-cost, thermoresponsive wettability of surfaces: Poly(N-isopropylacrylamide)/Polystyrene composite films prepared by electrospinning. Macromolecular Rapid Communications 29: 485–9.

Wang, Y., Li, W., Xia, Y., Jiao, X. and Chen, D. 2014. Electrospun flexible self-standing g-alumina fibrous membranes and their potential as high-efficiency fine particulate filtration media. Journal of Materials Chemistry A 2: 15124–31.

Wei, J., Jiao, X., Wang, T. and Chen, D. 2016. Electrospun photochromic hybrid membranes for flexible rewritable media. ACS Appl. Mater. Interfaces 8: 29713–20.

Xu, S. 2006. Determination of phthalates in water samples using polyaniline-based solid-phase microextraction coupled with gas chromatography. Journal of Chromatography A 1135: 101–8.

Xu, S., Zhang, J., He, A., Li, J., Zhang, H. and Han, C. C. 2008. Electrospinning of native cellulose from nonvolatile solvent system. Polymer 49: 2911–7.

Yang, A., Tao, X., Pang, G. K. H. and Siu, K. G. G. 2008. Preparation of porous tin oxide nanobelts using the electrospinning technique. Journal of the American Ceramic Society 91: 257–62.

Yang, M., Xie, T., Peng, L., Zhao, Y. and Wang, D. 2007. Fabrication and photoelectric oxygen sensing characteristics of electrospun Co doped ZnO nanofibres. Applied Physics A Materials Science & Processing 89: 427–30.

Yang, P., Zhan, S., Huang, Z., Zhai, J., Wang, D., Xin, Y. et al. 2011. The fabrication of PPV/C60 composite nanofibers with highly optoelectric response by optimization solvents and electrospinning technology. Materials Letters 65: 537–9.

Yang, X., Mu, R., Shao, C., Liu, Y. and Guan, H. 2005. Nanofibers of CeO_2 via an electrospinning technique. Thin Solid Films 478: 228–31.

Yi, J., Zhang, X., Zhu, Y., Li, B., Wang, Y., Zhang, J. et al. 2013. Nickel nanofibers synthesized by the electrospinning method. Materials Research Bulletin 48: 2426–9.

You, Y., Youk, J. H., Lee, S. W., Min, B. -M., Lee, S. J. and Park, W. H. 2006. Preparation of porous ultrafine PGA fibers via selective dissolution of electrospun PGA/PLA blend fibers. Materials Letters 60: 757–60.

Yu, Q., Mao, Y. and Peng, X. 2013. Separation membranes constructed from inorganic nanofibers by filtration technique. Chemical Record 13: 14–27.

Zeugolis, D. I., Khew, S. T., Yew, E. S., Ekaputra, A. K., Tong, Y. W., Yung, L. Y. et al. 2008. Electrospinning of pure collagen nano-fibres—just an expensive way to make gelatin? Biomaterials 29: 2293–305.

Zhang, H. D., Yan, X., Zhang, Z. H., Yu, G. F., Han, W. P., Zhang, J. C. et al. 2016. Electrospun PEDOT:PSS/PVP nanofibers for CO gas sensing with quartz crystal microbalance technique. International Journal of Polymer Science (2016-4-27) 2016: 1–6.

Zhang, H., Li, Z., Wang, W., Wang, C. and Liu, L. 2010. Na+-doped zinc oxide nanofiber membrane for high speed humidity sensor. Journal of the American Ceramic Society 93: 142–6.

Zhang, H., Cao, M., Wu, W., Xu, H., Cheng, S. and Fan, L. J. 2015. Polyacrylonitrile/noble metal/SiO_2 nanofibers as substrates for the amplified detection of picomolar amounts of metal ions through plasmon-enhanced fluorescence. Nanoscale 7: 1374–82.

Zhang, R., Wang, X., Song, J., Si, Y., Zhuang, X., Yu, J. et al. 2015. *In situ* synthesis of flexible hierarchical TiO_2 nanofibrous membranes with enhanced photocatalytic activity. Journal of Materials Chemistry A 3: 22136–44.

Zhang, Z., Li, X., Wang, C., Wei, L., Liu, Y. and Shao, C. 2009. ZnO hollow nanofibers: Fabrication from facile single capillary electrospinning and applications in gas sensors. Journal of Physical Chemistry C 113: 19397–403.

Zhao, Y., Wang, X., Lai, C., He, G., Zhang, L., Fong, H. et al. 2012. Electrospun carbon nanofibrous mats surface-decorated with Pd nanoparticles via the supercritical CO_2 method for sensing of H2. Rsc Advances 2: 10195–9.

Zheng, W., Li, Z., Zhang, H., Wang, W., Wang, Y. and Wang, C. 2009. Electrospinning route for α-Fe_2O_3 ceramic nanofibers and their gas sensing properties. Materials Research Bulletin 44: 1432–6.

Zhou, H., Green, T. B. and Joo, Y. L. 2006. The thermal effects on electrospinning of polylactic acid melts. Polymer 47: 7497–505.

Chapter 4

Guiding Parameters for Electrospinning Process

Jing Yan,[1] *Weimin Kang*[2,*] and *Bowen Cheng*[2,*]

Introduction

Electrospinning is a convenient and versatile method to fabricate different kinds of one-dimensional nanostructures such as nanofibers, nanotubes, and nanobelts from a wide array of polymers. Fundamental studies on various aspects of the electrospinning process have identified important parameters that affect the fiber size and morphology (Kanafchian et al. 2011a, 2011b, Mottaghitalab and Haghi 2011, Ziabari et al. 2008). In order to gain a better understanding, general factors including the nature of polymer-solution system, processing parameters, and environmental parameters are considered and introduced in this chapter. It is interesting that these parameters do not work independently, but interact with each other, having a synergic or retarding effect on the fiber morphology.

Nature of Polymer-solution System

Polymer Molecular Weight

The molecular weight is the predominant factor that determines the diameter of the collected fibers. The relationship between polymer molecular weight (M_w) and the types of structures produced by electrospinning technique has been widely investigated. The M_w of the polymer may have a significant effect on the rheological properties, electrical conductivity, dielectric strength, and surface tension of the

[1] School of Textiles, Tianjin Polytechnic University, 399 Binshuixi Avenue, Xiqing District, Tianjin, 300387, China.
[2] School of Textiles, Tianjin Polytechnic University, State Key Laboratory of Separation Membranes and Membrane Processes (Tianjin Polytechnic University), 399 Binshuixi Avenue, Xiqing District, Tianjin, 300387, China.
* Corresponding authors: kweimin@126.com; bowen15@tjpu.edu.cn

spinning solutions (Koski et al. 2004), and further influence the electrospun fiber structure. Zeng et al. (Zeng et al. 2005) prepared the Poly(vinyl alcohol) (PVA) nanofibers by electrospinning of PVA/water solutions with different molecular weights of PVA: $M_w = 195{,}000$, $M_w = 125{,}000$, and $M_w = 100{,}000$, and investigated the impact of the PVA molecular weight on the fiber shape. The solution viscosity was increased and the surface tension decreased slightly by increasing the PVA molecular weight, and the electrical conductivity was unaffected. As a result, cylindrical PVA fibers with diameters ranging from 350–700 nm were obtained from PVA with higher M_w. PVA with lower M_w gave thin fibers (50–150 nm), but with numerous beads. PVA of low M_w resulted in round and much larger beads (Zeng et al. 2013). Similarly, Koski et al. (Koski et al. 2004) investigated the effects of M_w on the fiber structure of electrospun PVA with molecular weights ranging from 9,000 to 186,000 g/mol. At a molecular weight of 9,000–13,000 g/mol, the fibrous structure was not completely stabilized and a bead-on-string structure was obtained. The fibers in between the beads also have a circular cross-section, with a diameter typically between 250 nm and one µm (Fig. 4.1a). As the molecular weight increased to 13,000–23,000 g/mol, a fibrous structure was stabilized (Fig. 4.1b). The fibers still exhibit a circular cross-section with a diameter between 500 nm and 1.25 µm. Extensive coiling and looping can also be observed in the electrospun polymer. Flat fibers are observed at a molecular weight of 31,000–50,000 g/mol (Fig. 4.1c).

The formation of the fiber structure has a great relation to the solution concentration C. When $[\eta]C > 4$ (semi-dilute entangled regime), where $[\eta]$ is the intrinsic viscosity, the polymer chains in the solution begin to entangle with each other and the solution viscosity increases significantly (Hong et al. 2001). Tacx et al. (Tacx et al. 2000) have obtained the Mark-Houwink relationship for PVA in water as shown in Eq. 4.1:

$$[\eta] = 6.51 \times 10^{-4} M_w^{0.628} \tag{4.1}$$

Using this equation, $[\eta]C$ can be calculated for various conditions. It was observed that a fibrous structure could not be stabilized for $[\eta]C < 4$. The calculated values of $[\eta]C$ for stable fibrous structures were typically between five and twelve, indicating that a minimum degree of chain entanglement is needed for producing fiber structures. At each M_w, there is a minimum concentration needed to stabilize the fibrous structure and maximum concentration where the solution cannot be

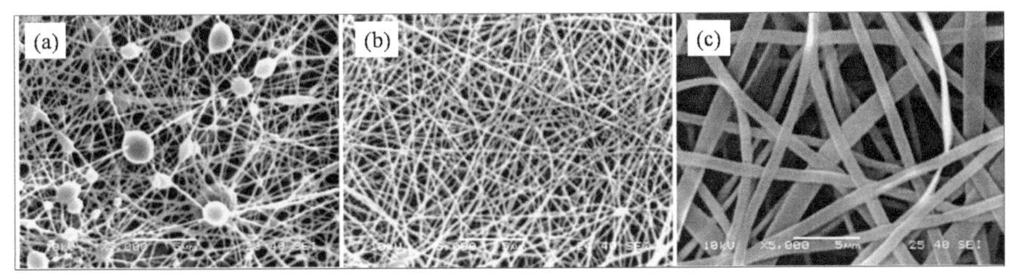

Fig. 4.1: Photographs showing the typical structure in the electrospun polymer for various PVA molecular weights: (a) 9,000–10,000 g/mol; (b) 13,000–23,000 g/mol; (c) 31,000–50,000 g/mol (solution concentration: 25 wt%).

electrospun (Koski et al. 2004). As M_w increases, the density of chain entanglements increases. Gupta and Wilkes (Gupta and Wilkes 2003) and Zeng et al. (Zeng et al. 2003) have investigated the effect of molecular weight on continuous fiber formation at a given polymer concentration. Increased chain entanglements and longer relaxation times, a consequence of increased polymer concentration, were thought to be responsible for fiber formation, and a study of the role of chain entanglements in solution during electrospinning has been done by Shenoy et al. (Shenoy et al. 2005), and the expression that relates entanglement density in solution to the molecular properties is given in Eq. 4.2:

$$(n_e)_{soln} = \frac{\varnothing_p M_w}{M_e} \tag{4.2}$$

Here, the $(n_e)_{soln}$ is the solution entanglement number (or density), $Ø_p$ is volume fraction of polymer in solution, M_w is average molecular weight of polymer, and M_e is entanglement molecular weight of polymer. The only required parameter is the entanglement molecular weight of the undiluted polymer (M_e). In general, M_e is readily available for a large number of polymers. Alternatively, in the absence of experimental values, M_e can also be theoretically estimated by employing the entanglement constraint model. Then, based on the structure requirements (fibers/beads/mixture), the predictions facilitate the proper choice of polymer concentration/molecular weight space.

Furthermore, the distribution of PVA fibers also changes as the molecular weight increases according to the research by Tao (Tao 2003). As M_w increases, a broader distribution of PVA fibers may be obtained, so the molecular weight of the polymer has a significant role in establishing the electrospun fiber structure. At a constant concentration, the structure changes from beaded fibers, to complete fibers, and to flat ribbons as the molecular weight is increased. Ojha et al. also studied the different molecular weight of nylon-6 on the morphology of electrospun fibers (Ojha et al. 2010). Assessment of the SEM images of nylon-6 with three molecular weights showed the effect of concentration on the nanofiber morphology; the morphology changes from beads, to beads on strings, to well-formed nanofibers, with a small concentration of beads as the molecular weight increases.

Concentration and Viscosity of Solution

Under certain conditions, the concentration of the solution is a decisive factor affecting the entanglement of the molecular chains in the solution. According to the concentration of polymer solution and different forms of the molecular chain, polymer solutions can be easily divided into polymer dilute solution, sub-concentrated solution, and concentrated solution, as shown in Fig. 4.2. The molecular chains are separated from each other and distributed uniformly in dilute solution, and when the solution concentration increases, the molecular chains will interpenetrate and overlap. The essential difference between a dilute solution and a concentrated solution is whether a single macromolecular strand is present in isolation (Gupta et al. 2005).

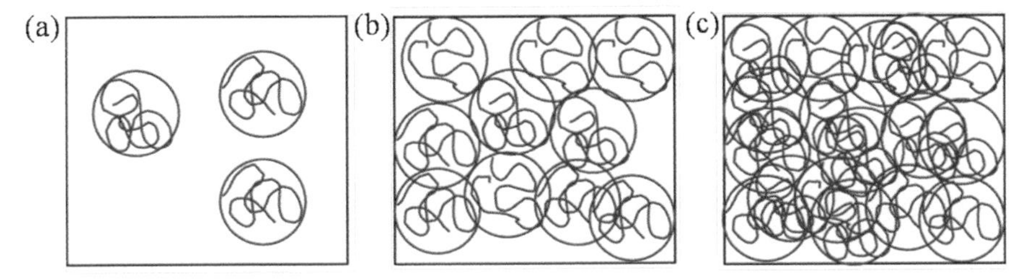

Fig. 4.2: Physical representation of the three solution regimes: (a) dilute; (b) semidilute unentangled; (c) semidilute entangled.

As the polymer solution concentration increases, the viscosity of the solution changes significantly, and the viscosity of the solution is closely related to the relative molecular weight of the polymer. According to Newton's viscosity law, the viscosity of a liquid can be defined as the ratio of shear stress to shear rate, when a fluid is laminar. The viscosity of solvent and polymer solution are named as η_0 and η, respectively, and the calculation formula of relative viscosity η_r and specific viscosity η_{sp} are defined in Eqs. 4.3 and 4.4:

$$\eta_r = \frac{\eta}{\eta_0} \tag{4.3}$$

$$\eta_{sp} = \eta_r - 1 \tag{4.4}$$

The intrinsic viscosity $[\eta]$ of the polymer solution is defined by Eq. 4.5 (Flory 1954):

$$[\eta] = [\frac{\eta}{C}]_{C \to 0} \tag{4.5}$$

The relationship between $[\eta]$ and M_w can be expressed by Eq. 4.6:

$$[\eta] = \mathrm{K} M_w^{\alpha} \tag{4.6}$$

Here, the K and a are constant, and the range of the a is 0.5 to 0.8. Therefore, the contact concentration can be estimated by Eq. 4.7:

$$C^* = \frac{1}{[\eta]} \tag{4.7}$$

Here, $[\eta]C$ is Berry number (B_e) which reflects the degree of entanglement of polymer molecular chains in solution. When the value of the B_e is over one, the molecular chains will overlap (Shenoy et al. 2005).

As early as 1971, Baumgarten (Baumgarten 1971) reported the effect of the polymer concentration and viscosity on electrospinning processes. When the polymer concentration improved from 7% to 20%, the viscosity increased from 0.17 Pa·s to 21.5 Pa·s correspondingly, and the diameter of the fiber increased with increasing solution concentration. Gupta et al. (Gupta et al. 2005) found that the experimental values of concentration were consistent with the theoretical values by study the electrospinning process of the Methyl methacrylate with different M_w. Frank (Gogotsi 2006) studied the relationship of concentration and M_w of the polymer and fibers diameter, and found that the diameter of the fibers increased with increasing solution concentration. For the polymers with high molecular weight, the fiber diameter increased with concentration faster than polymers with lower M_w. At the same time, the relationship of the fiber diameter and the value of the B_e was discussed. When the $B_e < 1$, the solution was very dilute, it could not form nanofibers by electrospinning because the molecular chains could not contact each other in the solution; when the $1 < B_e < 3$, the fiber diameter increased slowly with increasing value of the B_e, and the range of diameter was 100 to 500 nm; when the $3 < B_e < 4$, the fiber diameter increased quickly with increasing value of the B_e, and the range of diameter was 1,700 to 2,800 nm because the degree of molecular chain entanglement increased, and the viscosity of the solution increased.

A lot of researches showed that in the case of low concentration and viscosity of the electrospinning solution, polymer beads can be obtained. When the solution concentration and viscosity were higher than a certain critical value, the solution jet was stretched by the electric field force and had a longer relaxation time due to the increased degree of the entanglement between the molecular chains. The axial orientation of the entangled molecular chains inhibited the jet effectively in part of the molecular chain fracture and got a continuous electrospun fiber. Based on the general rules of the concentration and viscosity of polymer solution, the morphology of electrospun fibers can be controlled by adjusting the properties of the solution (Mckee and Long 2006).

Surface Tension

The molecules in fluid usually follow Brownian motion, and there is a mutual attraction between the molecules making up the fluid. Within a certain range, the smaller the distance between molecules, the greater the gravitation. Gravity between similar molecules is called cohesion, which makes the liquid interface molecules close to each other. The automatic contraction of the liquid surface, which acts on the liquid surface and makes the liquid surface contract to the smallest area of the force, is called surface tension. Electrospinning solution generally consists of a polymer and a solvent, which belongs to the binary system and is different from the one-component fluid. Its surface tension is not only related to the temperature and pressure, but also to the solution composition (Ding and Yu 2011).

For polymer solutions, when the solvent concentration is low, solvent molecules tend to agglomerate into spheres due to surface tension. At higher concentrations, the viscosity of the solution increases, indicating stronger interaction between the polymer chains and the solvent. And the solvent molecules tend to separate the entangled molecular chains and reduce their tendency to aggregate and shrink (Ramakrishna et al. 2005). During electrospinning, the polymeric solution is electrically charged and a conical droplet is formed at the needle tip. As the electric forces overcome the surface tension of the solution, a polymeric jet is generated from the surface of the droplet and travels towards the collector. The surface tension of the solution not only affects the formation mode of the tip cone of the Taylor cone, but also affects the jet movement and splitting under the high-voltage electric field, which ultimately affects the morphology of the electrospun fibers. For electrospinning, the electrostatic repulsion of the charged polymer solution or melt surface must be greater than the surface tension of the solution to allow the spinning process to be smooth. Owing to the Rayleigh instability in the axial direction, the surface tension tends to allow the solution to transform into spherical droplets to form beaded fibers; and the electric field force acting on the jet surface tends to increase the jet area to make the jet thinner, but not easy to form bead fiber. In this process, the viscoelastic force of the polymer solution also inhibits the rapid transition of the jets, thereby supporting the formation of fibers with a smooth surface (Li and Xia 2010). Increasing the concentration of the solution can reduce the surface tension of the solution to some extent, which facilitates the formation of continuous uniform fibers.

Doshi and Reneker (Doshi and Reneker 2002) electrospun polyethylene oxide (PEO) with different concentrations to control the surface tension. They found that the electric field strength at the apex of the cone was determined by the following equation:

$$\mathrm{E} = \sqrt{\frac{4\gamma}{\varepsilon_0 R}} \tag{4.8}$$

Here, E is the electric field strength, γ is the surface tension of polymer solution, ε_0 is the permittivity of the free space, and R is the radius of curvature of the rounded off cone apex. According this equation, in the case of a certain solution concentration, the surface morphology of the electrospun fibers can be controlled by adjusting the surface tension of the solution. Fong et al. (Fong et al. 1999) studied PEO in different proportions of water/ethanol mixture and found that the viscosity of the solution increased from 402 cP to 1,179 cP, and the surface tension decreased from 75.8 mN/m to 50.5 mN/m as the ethanol content increased. Both of these changes are conducive to the formation of bead-free continuous fibers. Lee et al. (Lee et al. 2003) studied the fiber morphology of electrospun polycaprolactone affected by the mixed solvent CH_2Cl_2/dimethylformamide (DMF) composition ratio. It was found that the surface tension of the solution decreased with the increase of DMF content, meanwhile, the diameter of fiber also decreased sharply and the distribution became uneven. Jarusuwannapoom et al. (Jarusuwannapoom et al. 2005) electrospun polystyrene (PS) with different solutions and found that the formation of beads could be the co-work of viscoelastic and surface tension (Fig. 4.3). If the charged jets do not "dry" fast before being collected on a grounded target, some parts of the "damp" fibers could retract as a result of the reduction in the Coulombic force due to partial charge neutralization. Based on the qualitative observation of the results obtained, the important factors determining the electrospinnability of the as-prepared PS solutions are high dipole moment of the solvent and high conductivity of both the solvent and the resulting solutions, high boiling point of the solvent, and not-so-high values of the viscosity and the surface tension of the resulting solutions. Fridrikh et

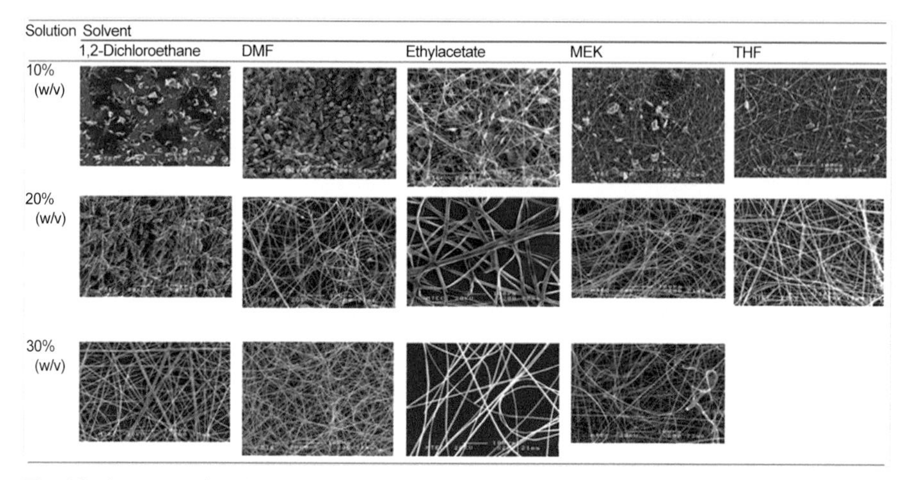

Fig. 4.3: Some scanning electron micrographs (at 200×) of as-spun polystyrene fibers from solutions of polystyrene in 1,2-dichloroethane, dimethylformamide (DMF), ethylacetate, methylethylketone (MEK), and tetrahydrofuran (THF). (Remark: the applied potential and the collection distance were 20 kV and 10 cm and the scale bar in each micrograph is for 100 μm.)

al. (Fridrikh et al. 2003) have presented a model of a charged fluid jet in an electric field under conditions applicable to the whipping instability as shown in Eq. 4.9.

$$h_t = \left(\gamma \varepsilon \frac{Q^2}{I^2} \frac{2}{\pi(2\ln \chi - 3)} \right)^{1/3} \tag{4.9}$$

The model predicts a terminal jet diameter, which is a consequence of balance between normal stress due to surface tension and surface charge repulsion, and can be determined by the flow rate, the electric current, and the surface tension of the fluid. The 2/3 scaling with the inverse volume charge density $(I/Q)^{-1}$ predicted by the model is confirmed experimentally using several concentrations of polycaprolactone, providing convincing evidence for the correctness of the model. Further support for the model is provided by the reasonable prediction of the dry fiber diameter for polycaprolactone, PEO, and polyacrylonitrile.

Conductivity of Solution

Conductivity of solution is mainly determined by the polymer type, used solvents, and the components (Bhardwaj and Kundu 2010). The availability of ionizable salts is the key factor greatly influencing the spinning jet and the morphology of electrospun nanofibers (Demir et al. 2002, Li et al. 2016, Wang et al. 2011). Usually, the conductivity of solution is proportional to the salt amount in the solution, which can be explained as the increased amount of the free ions in the solution (Barakat et al. 2009). Since the ionic charge of molecules have a direct relation with the electrical conductivity of the solution, the addition of a few salt molecules results in a higher charge density on the surface of the ejected jet during electrospinning, leading to an increase in the net electric charge carried by the jet. As the charge carried by the jet increases, a greater stretching and elongation of the jet takes place during electrospinning, thereby causing a reduction in the bead formation, which is also related to enhanced molecular alignment during the fiber formation caused by salt-induced electrical effects (Sui et al. 2011). However, highly conductive solutions will be extremely unstable in the presence of strong electric fields, leading to significant decrease in the fiber diameter and dramatic bending instabilities, and thus a broad diameter distribution (Narasimhan 2016). Matabola and Moutloali (Matabola and Moutloali 2013) studied the effect of NaCl loading on the morphology of polyvinylidene fluoride (PVDF) nanofibers, and found that the addition of the salts results in significant reduction of the bead density and improved uniformity in fiber morphology. Furthermore, lower fiber diameter was also observed as the salt concentration was increased. Arayanarakul et al. (Arayanarakul et al. 2006) did a detailed study on the effects of inorganic salts (NaCl, LiCl, KCl, $MgCl_2$, and $CaCl_2$) on PEO fibers. The authors observed that with increase of salt concentrations, there was a monotonic decrease in both surface tension and viscosity, and an increase in conductivity, attributed to the electrospinnability of the polymer. Importantly, increasing the conductivity of the solution could enhance the instability of the Taylor cone, which would greatly improve the probability of formation of microsized droplets, and thus dense nano-nets (Wang et al. 2011). Wang et al. (Wang et al. 2012) reported the effect of salts types and salt content on the morphology of polyamide-66 (PA-66) nano-fiber/net membranes. They found that the PA-66/$BaCl_2$ (0.27 mol/L) nano-fiber/net membranes possess the optimal nano-nets with a high coverage rate (over 95%) and a layer-by-layer packing structure. They also discussed the effect of NaCl content on the morphology of Poly(acrylic acid) (PAA) nano-fiber/net membranes, and found that the fibers containing 0.1 wt% NaCl showed a dramatically decreased average diameter, which corresponds to the sharply increased conductivity (Wang et al. 2011). Higher NaCl concentration (from 0.1 to 0.2 wt%) led to a fiber sticking morphology, and an increase in the diameter of the spun fibers attributed to the localized charge effects on the surface of the fibers (Patel et al. 2007), as well as changed chain conformation of PAA with increase in the ionic strength of the solution (Kim et al. 2005). In addition, for a polymer solution with a high concentration of salts, the conductivity has a different effect on the fiber diameter. The influences of different salts on the morphology and diameter of the PAN nanofibers have been explored by Qin et al. (Qin et al. 2007). The authors observed that the diameter of nanofibers electrospun by solutions with different salts size down as follows: LiCl $>$ $NaNO_3$ $>$ $CaCl_2$ $>$ NaCl, because the increase of electric conductivity of polymer solution led to the increase of surface charge of spinning jet, which made the spinning more fluent, increased jet quantity of solution, and increased diameter of fibers.

In Cheng's research group, a novel polyvinylidenefluoride (PVDF) tree-like nanofiber can be fabricated via one-step electrospinning by adding certain amount of salts, such as LiCl, $AlCl_3$, $CaCl_2$, tetrabutyl ammonium bromide (TBAB), tetraethyl ammonium chloride (TEAC), tetrabutyl ammonium chloride (TBAC), etc., into PVDF solution as shown in Fig. 4.4 (Li et al. 2016). The schematic diagrams illustrating the electrospinning process, possible mechanism for the formation of tree-like nanofibers, and typical FE-SEM image of PVDF/TBAC tree-like nanofibers are shown in Fig. 4.5. The addition of salt can increase the electrical conductivity of

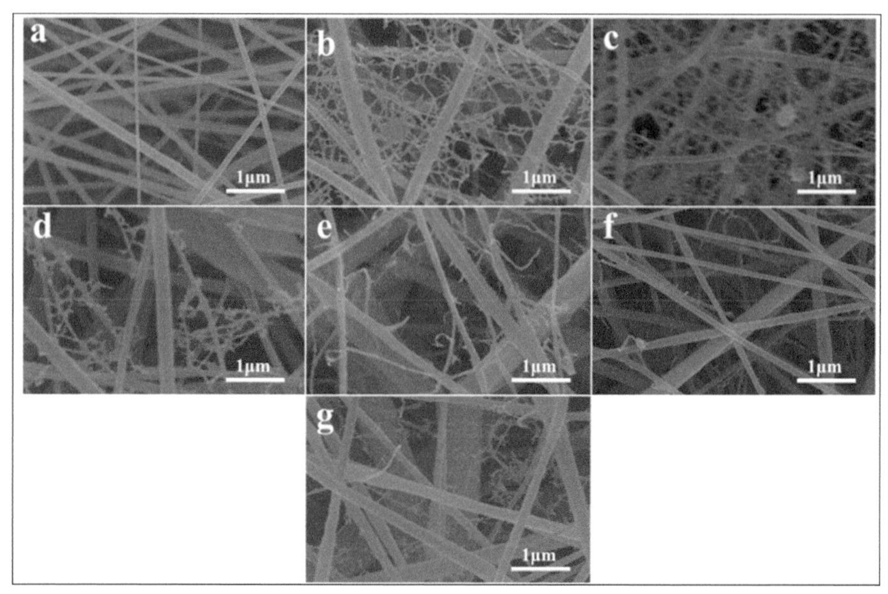

Fig. 4.4: FE-SEM images of PVDF nanofiber membranes with different types of salt: (a) no salt (pure PVDF), (b) 0.10 mol/L TBAC, (c) 0.10 mol/L TBAB, (d) 0.10 mol/L TEAC, (e) 0.10 mol/L LiCl, (f) 0.05 mol/L $AlCl_3$, (g) 0.05 mol/L $CaCl_2$. Spinning parameter: applied voltage of 30 kV, tip to collector distance of 15 cm, extrusion rate of 1 mL/hr.

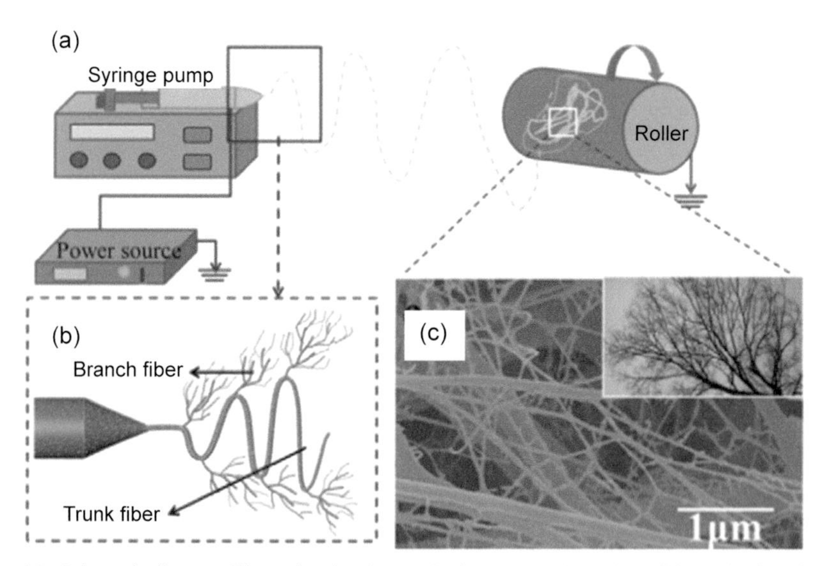

Fig. 4.5: Schematic diagrams illustrating the electrospinning process (a) and possible mechanism for the formation of tree-like nanofibers (b), (c) Typical FE-SEM image of PVDF/TBAC tree-like nanofibers and the inset is the optical image of tree branches.

the polymer solution, and has a great influence on the morphology of electrospun nanofibers. PVDF nanofibers formed from the solution without salt exhibit a common structure and have a uniform diameter of 180 nm. The formed tree-like nanofibers comprise trunk fibers with 100–500 nm diameter and branch fibers with 5–100 nm diameter due to the splitting of jets corresponding to the sharply increased conductivity caused by the incorporation of organic salts. The tree-like nanofibers have improved crystallinity, mechanical strength, significantly reduced pore size, and enhanced specific surface area, exhibiting a potential application in separation, tissue engineering, sensors, catalysis, etc. (Hekmati et al. 2013, Moon et al. 2017, Shuakat and Lin 2015). Additionally, poly(vinylidene fluoride)-graft-poly(acrylic acid) (PVDF-g-PAA) (Li et al. 2016), cellulose acetate (Thoppey et al. 2010) poly-m-phenylene isophthal amide (PMIA) (Thoppey et al. 2012), polyurethane (Liu et al. 2014), nylon 6 (Liu et al. 2011), and other polymers can also be electrospun to tree-like nanofibers by adding certain amount of above salts.

Addition of Inorganic Components

Electrospinning technique has been widely applied in the preparation of composite nanofibers, and is considered to be the most effective way to prepare composite nanofibers. The materials prepared by electrospinning organic solution with inorganic components are usually called inorganic/organic composite nanofibers, in which the polymer is the continuous phase and inorganic material is the dispersed phase. However, because of the strong interaction between inorganic nanomaterials, they are easy to reunite spontaneously, which will further affect the final service life of composite materials. Therefore, the way to get polymer composites with good dispersion and fully exploiting their function of inorganic nanomaterials without destroying the physical properties of polymers is an urgent problem in the field of inorganic/organic nanomaterials. Inorganic oxides, metal sulfides, metal materials, inorganic salts, carbon materials, etc. can be added to the polymer by mainly following four methods: (1) directly dispersing the inorganic materials in spinning solution; (2) *in situ* preparation of inorganic nano materials; (3) mixing precursors corresponding to inorganic nanomaterials in the polymer solution; (4) electrospinning composite polymer precursors with the following post-process. Chen et al. (Chen et al. 2009) prepared the ZnO nanotubes/PAA nanofiber membrane by thermally converting the Zinc ions associated with vinyl-COO^-(vi-COO^-) groups to ZnO species, which can be applicable to other inorganic materials. Kanehata et al. (Kanehata et al. 2007) prepared SiO_2/PVA composite fiber material through electrospinning technique by directly dispersing SiO_2 in polymer solution. With the increase in particle diameter, the viscosity of the solution decreases. The research shows that the addition of SiO_2 nanoparticles makes the surface of composites become very rough and develop grain-shaped protuberances. Ostermann et al. (Ostermann et al. 2006) prepared a variety of oxide/polymer composites with optical and magnetic properties using TiO_2 and $NiFe_2O_4$ as nanoparticles, and polyvinylpyrrolidone as matrix by electrospinning technique. Wang et al. (Wang et al. 2010) prepared TiO_2/polypyrrole coaxial fibers by *in situ* polymerization of TiO_2 on the electrospun fiber surface. The coaxial fiber has both high conductivity of polypyrrole and surface photoelectric properties of TiO_2, which are expected to be applied in the field of photoelectric conversion. Lu et al. (Lu et al. 2010) prepared polymer composite fibers containing metal sulfide semiconductor nanoparticles (PbS, CdS, ZnS, and Ag_2S) by the combination of electrospinning and gas-solid heterogeneous reactions. In this composite fiber, part of the sulfide lattice is replaced by another metal, which helps to improve the properties of the sulfide. Yang et al. (Yang et al. 2003) prepared Ag doped membrane by electrospinning, and found that the structure and properties of Ag nanoparticles can remain stable under high voltage electric field, and the addition of Ag nanoparticles increases the diameter and conductivity of composite fibers to a certain extent. A freestanding membrane composed of a nanofiber network of a graphene-polymer nanocomposite is fabricated by electrospinning, and applied as an optical element in fiber lasers. The functionalization of graphene with conjugated organic molecules provides a handle for improving mechanical and thermal properties, as well as tuning the optical properties. A small loading (0.0765 wt%) of functionalized graphene enhances the total optical absorption of poly(vinyl acetate) (PVAc) by ten times. The results show that electrospun graphene nanocomposites are promising candidates as practical and efficient photonic materials for the generation of ultrashort pulses in fiber lasers (Bao et al. 2010). Yan et al. (Yan et al. 2017) prepared polyacrylonitrile/polyhedral oligomericsilsesquioxane (PAN/POSS)-derived carbon nanofibers (CNFs) in terms of POSS content (0–60 wt%) for their potential utilization as freestanding supercapacitor electrodes. After stabilization and carbonization of electrospun PAN/POSS nanofibers, the POSS residues could be removed by hydrofluoric acid (HF) treatment to produce porous structure. Characterization of PAN/POSS-based CNFs revealed that the addition of POSS in precursor solutions could effectively produce porous structures on the surface of CNFs, increase the specific surface area, and finally improve the electrochemical performance. In Song's research, magnetic Fe_3O_4-POSS particles with Si-OH were prepared by hydrosilylation reaction between the Fe_3O_4-iH and POSS with hydroxyl and vinyl groups. PAN/Fe_3O_4-POSS nanofibers mats were subsequently fabricated by the electrospinning technique. The stability of the surface potential was remarkably improved and the surface potential retention reached 50% for PAN/Fe_3O_4-POSS mats with 1 wt% Fe_3O_4-POSS. Compared to pure PAN, the charge retention of PAN/Fe_3O_4-POSS was increased, the collection efficiency increased, and the filter resistance decreased when the PAN nanofibers with Fe_3O_4-POSS were used as electrets filter media (Wang et al. 2012).

Processing Parameters

Applied Voltage

The applied voltage is one of the most crucial parameters in the electrospinning process. The high voltage will induce the necessary charges on the solution, and the external electric field will initiate the electrospinning process when the electrostatic force in the solution overcomes the surface tension of the solution. Generally, a minimum voltage of 6 kV, either positive or negative, is able to cause the solution drop at the tip of the needle to distort into the shape of a Taylor Cone during jet initiation (Bhardwaj and Kundu 2010, Dong et al. 2011, Ju et al. 2016). Higher voltages may result in more charges, which will accelerate the jet, and a greater amount of solution will come out from the needle tip. A lot of study concerning the influence of applied voltage on the electrospinning was published. However, there is no common agreement on the effect of applied voltage on the diameter of electrospun fibers (Demir et al. 2002, Gu et al. 2005, Heikkil et al. 2010, Ju et al. 2017, Liu et al. 2011, Mo et al. 2004). Gu et al. (Gu et al. 2005) reported that fiber diameter was not significantly changed by altering the voltage applied. Mo et al. (Mo et al. 2004) and Heikkil et al. (Heikkil et al. 2010) concluded that the diameter of the nanofiber tends to decrease with an increase in the voltage applied. In most cases, a higher voltage may cause greater stretching of the solution due to the greater columbic forces in the jet, as well as a stronger electric field. The influence level may depend on the concentration of polymer solution. If the concentration and viscosity of polymer are low, the molecular chain will break and polymer beads will be formed due to the weak tangling force of the molecular chain and high surface tension of the solution during the orientation process. Meanwhile, these effects may lead to a reduction in fiber diameter because of the intense stretching of the fibers between polymer beads at high voltage. Deitzel et al. (Deitzel et al. 2001) have studied the effect of applied voltage on the fiber morphology by electrospinning of PEO/water solution with concentration of 7 wt%. They found that the jet would eject steadily from the end of Taylor Cone surface, and fibers with uniform diameter were obtained when the voltage was set at 5.5 kV. When the voltage increased to 7.5 kV, the jet directly ejected from the inside of the nozzle and there were many beads formed in the membrane due to the instability of the jet. Fong et al. (Fong et al. 1999) also reported that the PEO fiber morphology changed from a defect-free fiber at an initiating voltage of 5 kV to a highly beaded structure at a voltage of 18 kV, and the fiber diameter decreased as well. If concentration and viscosity of the polymer are high, the intense entanglement of molecular chains as well as the strong viscous stress of the solution will somewhat inhibit the effect of applied electric field, leading to an inconspicuous effect on fiber morphology. According to the analysis of jet stretching, the increase of applied voltage may cause more charges on jet surface, which may intensify perturbation of the jet flow and slightly reduce the fiber diameter. Ding et al. (Ding et al. 2002) have demonstrated that the fiber diameter will decrease from 240 nm to 220 nm, along with the voltage increasing from 7 kV to 19 kV by electrospinning the PVA/water solution with concentration of 11 wt%. When increasing polymer concentration during electrospinning, the applied voltage did not have a significant influence on the diameter of PVA nanofibers, and it only influenced the structure of nanofiber mat and caused an increase in the number of nanofibers. Zhang et al. (Zhang et al. 2005) noticed that there was a slight increase in the diameter distribution of PVA nanofibers with an increase of the applied voltage. The range of the diameter distribution changed from 150–300 nm to 100–500 nm, leading to an enlargement of the average diameter. Therefore, the effect of applied voltage on fiber morphology may differ for different polymer solution systems with different voltage range.

Receiving Distance

The receiving distance (between spinning tip and collector) has a significant effect on the jet path and traveling time before resting on the collector. If the voltage is kept constant and other conditions are similar, the electric field strength will be inversely proportional to the distance, and deposition time, evaporation rate, and whipping or instability interval will change accordingly, which easily affect the structure and morphology of electrospun fibers because of their dependence on these parameters (Chang et al. 2005, Subbiah et al. 2010). In a typical electrospinning set-up, the receiving distance ranges from 8 cm to 20 cm, which generally allows sufficient flight time for the solvent to vaporize so that a dry fiber strand is deposited. If the receiving distance is so short that the solvent is inadequately vaporized, fused fibers may be formed. However, there are also some researches on near-field electrospinning (Chang et al. 2014, Shao et al. 2015, Youn et al. 2016). At such close proximity, the spinning tip is generally much smaller than conventional electrospinning, so that the initial jet radius is also reduced. The smaller volume or feed rate allows sufficient vaporization of the solvent such that relatively dry fiber can be collected. It is easily understood that if the receiving distance is too short, the fiber will not have enough time to solidify before reaching the collector, whereas if the receiving distance is too long, bead fiber can be obtained. It is well known that one important physical aspect of the electrospun fiber is the dryness from the solvent, so the optimum receiving distance is usually recommended. Ghelich et al. (Ghelich et al. 2014) prepared PVA/nickel oxide-gadolinium doped ceria fibers by electrospinning, and it could be observed that fibers fused when receiving distance was 8 cm, and distinct individual fibers were collected when receiving distance was 10 cm. The diameters of the fibers collected at both receiving distances were similar. At 15 cm, fused fibers were again observed and the diameter increased significantly. The increase in fiber diameter was attributed to reduced electrostatic field strength which leads to less stretching of the fibers. As there was less stretching, the greater diameter increased the amount of solvent trapped within the fibers. The trapped solvents continued to diffuse out after the fiber had deposited, and this caused fusion of the fibers. In addition, there were two different effects of different receiving distances on fiber diameters (Bosworth and Downes 2012, Doustgani 2015). On the one hand, fiber diameter decreased with the increasing receiving distance. Longer receiving distance provides more time not only to evaporate the solvent in the electric field, but also to stretch the jet, thereby encouraging formation of thinner nanofibers. On the other hand, increasing the receiving distance means that the electric field strength ($E = V/d$) will decrease, resulting in less acceleration, which leads to fibers with larger diameters. The balance between these two effects will determine the final fiber diameter. Hence, increasing receiving distance may increase (Baker et al. 2006, Jarusuwannapoom et al. 2005), decrease (Wang and Kumar 2010), or may not change (Chang et al. 2005, Yuan et al. 2004) the fiber diameter.

In many applications, it is preferred that the fiber deposition is uniform over a specific area. This will reduce the formation of distinct fiber layers, which will reduce the tensile property of a membrane due to breakage at weaker interlayer interface. Uniformity of the fiber layer would also give the composite structure a more consistent performance and property. To facilitate uniform distribution of the fibers, a larger deposition area is preferred, and this has been demonstrated by increasing the receiving distance (Dong et al. 2006). In addition, it is worth mentioning that the interplay between the applied voltage and the receiving distance affects the solution jet. The optimal applied voltage is also dependent on the receiving distance, as both parameters are related in the determination of the electric field strength. A typical electrified jet will follow a short stable path before acquiring a whipping profile; thus selecting an appropriate receiving distance is of utmost importance (Chen et al. 2017). To sum up, the influence of the receiving distance on the structure and morphology of electrospun fibers is different because it is closely related to many other items, such as the property of spinning solution, applied voltage, ambient temperature and relative humidity, etc.

Feed Rate

The feed rate represents the amount of polymer solution to be electrospun per time, which to a certain extent reflects the efficiency of spinning. The feed rate of the polymer solution is always controlled by the syringe pump. Although feed rate is not an independent spinning parameter which is frequently related to the concentration of polymer solution, it is one of the important electrospinning parameters that control the stability of the spinning process, the diameter, and morphology of fibers. The feed rate of the solution played a decisive role in fiber formation, because there had to be a certain amount of polymer solution suspended at the tip of the nozzle. A lower rate is more desirable during the electrospinning process because the lower feed rate will allow the solvent to have more time to evaporate, and the fibers will have more time to stretch, which will favor the formation of more uniform and thinner nanofibers. However, it is not feasible to form a Taylor cone when the feed rate is too low. An increase of the feed rate enlarged the solution amount spraying from the nozzle and reduced the electric force per unit volume. Consequently, the fiber became thicker when the feed rate of the solution was improved. This increase of the fiber diameter is attributed to the shorter time taken for the jets to move from spinneret to collector, and hence, the smaller solvent evaporation rate, which leaves thicker fibers. Meanwhile, the higher the feed rate of the solution supplied in the unit time, the more fibers are obtained. If the injection rate is exorbitant, fibers are difficult to form by the electrospinning method. Henriques et al. (Henriques et al. 2009) assembled a new electrospinning apparatus and used PEO as a model polymer to perform a systematic study on the influence of processing parameters, including feed rate on the morphology and diameter of electrospun nanofibers. The results indicated that fiber diameter increased linearly with solution feed rate. With an increase of the solution feed rate, more polymer was pushed through the needle, and fiber diameter also increased. According to Cork's research (Cork et al. 2017), poly(trimethylene carbonate)-co-(l-lactide) copolymers were synthesized and processed into 3D fibrous scaffolds using solution electrospinning. A range of fiber diameters (0.5–5.9 mm) and pore sizes (3.5–19.8 mm) were achieved simply by adjusting the voltage, collector distance, and feed rate. When comparing fiber mats produced at the same collector distance and applied voltage using different feed rates, it is evident that larger fiber diameters and larger pores are obtained when increasing the feed rate. Yalcinkaya (Yalcinkaya 2015) also studied the effect of feed rate on the morphology of electrospun nanofibers, and found that the increasing feed rate provided more polymer solution to spin. When the feed rate was increased to a higher value (1.0 mL/hr), it was observed that the stable jet case range was broader than using a lower feed rate. Biber et al. (Biber et al. 2010) investigated the effect of the feed rate of the syringe on the distribution of diameters of nanofibers, with the other parameters remaining constant. The feed rate of the syringe is set to 3 μL/min, 5 μL/min, and 10 μL/min, while Nylon 6/formic acid solution with 15% (w/v) concentration is electrospun under the conditions of 20 kV electrical potential and 10 cm receive distance. The mean diameter diminishes while increasing the flow rate of the solution through the needle tip. Li et al. (Li et al. 2014) used poly(vinyl pyrrolidone) as target polymer to study the effect of feed rate on the fiber diameter. Given the other processing parameters at a certain value, the fiber diameter increased monotonously with increasing feed rate. The feed rate of the solution played a decisive role in fiber formation. It was not feasible to form a Taylor cone when the feed rate was too low. An increase in the feed rate enlarged the solution amount spraying from the nozzle and reduced the electric force per unit volume. Consequently, the fiber became thicker when the feed rate of the solution was improved. At a high solution concentration, increasing the feed rate had a visible impact on increasing the fiber diameter, whereas at a lower concentration, the fiber diameter was not remarkably changed when the feed rate was varied. According to Al-Qadhi's study (Al-Qadhi et al. 2015), the feed rate was one of the important electrospinning parameters that control the stability of Taylor cone. The average diameter of the fibers was found to increase with an increase in the feed rate attributed to the shorter time taken for the jet to move from spinneret to collector, and hence, the smaller solvent evaporation rate, which leaves thicker fibers. In addition, higher feed rate means greater amount of polymer chain entanglements, and hence thicker fibers. Halloysite nanotubes (HNT) reinforced polylactic acid (PLA) nanocomposite fibers were produced by Touny using an electrospinning approach to investigate various factors, such as type of solvent, solution concentration, and feed rate on the effect of the electrospinning process (Al-Qadhi et al. 2015). Three different feed rates of 1.5, 2.5, and 5 mL/hr were used in his study. The average diameter of the fibers spun with a rate of 1.5 mL/hr was 230 nm, and those spun with a rate of 2.5 mL/hr had a diameter of 280 nm. However, when the rate was increased to 5 mL/hr, the electrospun fibers were produced with the diameter of 2 μm. Generally, for a given voltage, as the feed rate increases, there is a corresponding increase in the fiber diameter.

Electrospinning Type/Principle/Spinneret

The most widely used electrospinning set-up usually has one spinneret. Hekmati et al. (Hekmati et al. 2013) studied the effect of the needle length on the morphological properties of polyamide-6 electrospun nanowebs. Statistical analysis of the obtained results revealed that the increase of needle length significantly increased the average nanofibers' diameter. Inversely, the diameter of the nanoweb collection zone reduced when the needle length increased. Taking advantage of the well-designed spinneret, core-shell structured fibers can be produced. Ou et al. (Ou et al. 2011) prepared novel epitaxial-like packed, super aligned, mono-layered hollow fibrous membranes by co-axial electrospinning technology, as shown in Fig. 4.6a. Poly(L-lactic acid) (PLLA) solution was used as a sheath solution dope and aqueous solution of polyethylene glycol (PEG) was used as a core solution dope. During the electrospinning process,

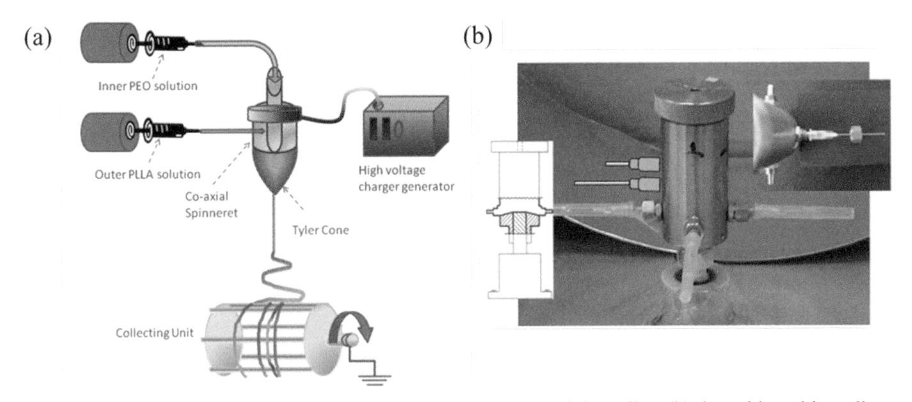

Fig. 4.6: Electrospinning set-up with spinning needles: (a) co-axial needles; (b) tips with multi needles.

dope concentrations and feed rate ratios were adjusted, separately, to evaluate the formation of resulting membranes. With lower shell solution concentration (from 8 to 15 wt%), the collected, flattened film showed arrangement of lower order. A stood-up film was clearly seen when concentration increased to 17 wt% and higher. As the collecting time increased, these fibers piled up and eventually, stood up as a thin film from the surface of the collector. Increasing flow rate ratio (FRR) also resulted in similar outcomes and verifying the cause of this phenomenon. After washing with water, scanning electron microscopy (SEM) revealed sheets of mono-layered, micron-sized, hollow fiber arrays, which were well aligned and tightly packed, just like the epitaxial growth of some semiconducting materials. To increase the productivity, double- or multi-needles are adopted. Yu et al. (Yu et al. 2014) designed a multi nozzle electrospinning spinneret with assistant sheath gas to release the multi jets injection, by which the production rate of nanofiber can be promoted. The sheath gas around the nozzle decreased the surface charge density and provided an excess stretching force to increase the motion speed of charged jet. The diameter of the charged jet was also decreased by the sheath gas further. On the other hand, the sheath gas also reduced the mutual disturbance among charged jets, by which stable multi jets injection can be gained. The stretching force increased along with air pressure of sheath gas. Then, both the nanofiber diameter and the critical voltage required for jet injection decreased with the increase of air pressure. With the increase of sheath gas pressure, the uniformity of electrospinning nanofiber can be also improved. Kancheva et al. (Kancheva et al. 2014) proposed an advanced centrifugal electrospinning set-up using a pivotal feed unit and stationary collectors of large diameters and different designs: cylindrical collectors or collectors consisting of circularly arranged metal strips (Fig. 4.6b). The use of a collector composed of strips enables the production of aligned fibers. The simultaneous use of four nozzles results in enhancement of the production rate and shortens the time for fabrication of a denser mat with a large surface area and enhanced exploitation properties.

Needleless electrospinning set-ups have attracted much attention over the past decades as an effective approach to enhance the production rate of electrospinning due to the merit of no clogging. Bubble-electrospinning was invented in 2007 by Liu and He (Liu and He 2007), using polymeric bubbles to produce multiple charged jets by applying a high voltage on the bubbles' surfaces to overcome the surface tension (Fig. 4.7a). Then they discovered that the average diameter of fibers increased with the increase of the applied voltage in bubble electrospinning, which is quite different from that in the traditional electrospinning process under similar conditions. The number of beaded fibers decreased with increasing applied voltage. Additionally, the crystallinities of polyvinylpyrrolidone (PVP) ultrafine fibers obtained in this process were higher than that of PVP powders. The production rate of bubble electrospinning was higher than that of the traditional electrospinning (Liu et al. 2011). Jiang et al. (Jiang et al. 2013) developed an electrospinning set-up using one stepped pyramid-shaped copper spinneret, which can produce finer nanofibers, and has more than 100 times productivity (4 g/hr) than the traditional needle based electrospinning. The scheme of the novel needleless electrospinning system is shown in Fig. 4.7b, which contained a high-voltage direct-current power supply, a stepped pyramid spinneret, a Teflon solution reservoir, a peristaltic pump, and a grounded collector. A stepped pyramid spinneret was utilized as the electrospinning generator. Wang

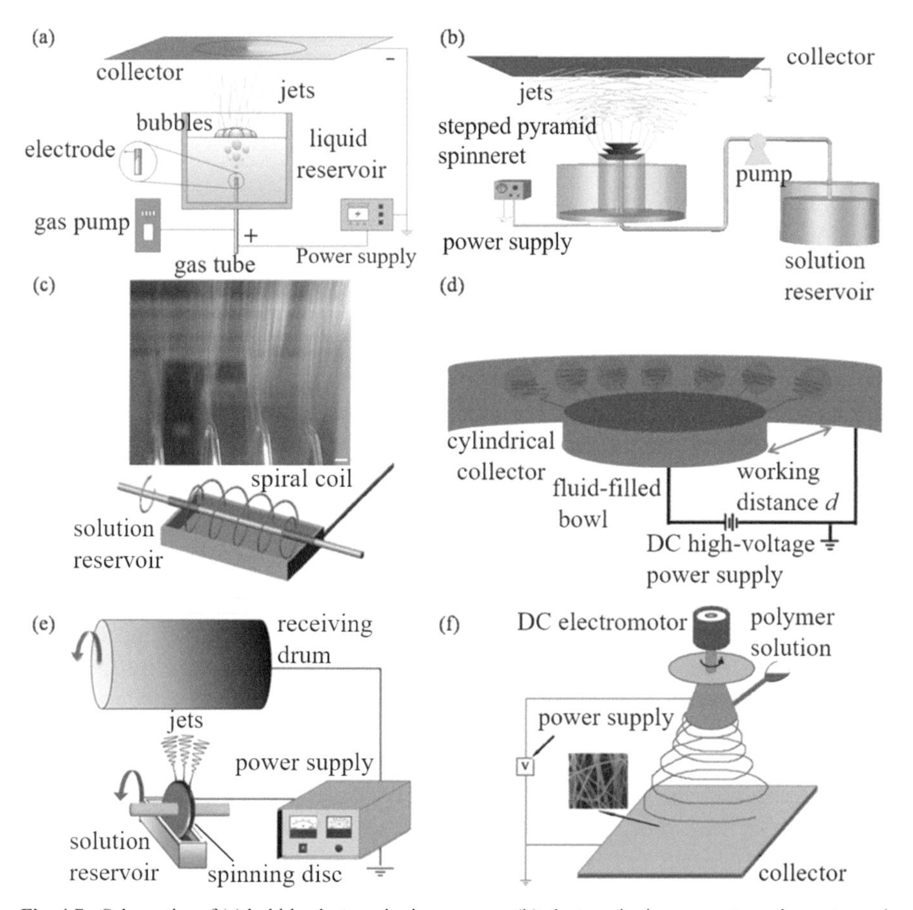

Fig. 4.7: Schematics of (a) bubble electrospinning process; (b) electrospinning apparatus using a stepped pyramid spinneret; (c) spiral coil and photos of spiral coil spinning processes; (d) bowl electrospinning set-up; (e) disk-electrospinning set-up; (f) electrospinning set-up that used a rotating cone as the spinneret.

et al. (Wang et al. 2012) used an electrospinning set-up with a rotating spiral wire coil as a spinneret to prepared Polyvinyl alcohol nanofibers (Fig. 4.7c). The coil dimension had a considerable influence on the nanofiber production rate, but a minor effect on the fiber diameter. The fiber production rate increased with the increased coil length or coil diameter, or the reduced spiral distance or wire diameter. Higher applied voltage or shorter collecting distance also improved the fiber production rate, but had little influence on the fiber diameter. Compared to the conventional needle electrospinning, the coil electrospinning produced finer fibers with a narrower diameter distribution. Thoppey et al. (Thoppey et al. 2010) demonstrated an easily-implemented, edge-plate geometry for electrospinning, and produced high quality nanofibers from unconfined polymer fluids, and found that the electric field gradient, not just the electric field amplitude, was a critical parameter for successful self-initiated jetting. Considering a single spinning site, the edge-plate configuration resulted in the same or a higher fabrication rate as traditional needle electrospinning, while producing nanofibers similar in quality (diameter, diameter distribution, and collected mat porosity). Then they designed Bowl-based edge electrospinning, which provides a useful scale-up approach of single-needle electrospinning to produce high quality nanofibers, simply and with demonstrated scaling of 40× in a single batch (Fig. 4.7d), providing a scheme to test fluid-electric field interactions and refine and develop models to predict the number of fluid instabilities and the flow through each cone jet (Thoppey et al. 2012). Li et al. (Li et al. 2016) used disk-electrospinning to produce poly(-caprolactone)/gelatin (PCL/GT) scaffolds of different structures (Fig. 4.7e), namely the nanoscale structure constructed by nanofibers, and the multiscale structure consisting of nanofibers and microfibers. It was found that, due to the inhomogeneity of PCL/GT solution, disk-electrospun PCL-GT scaffold presented multiscale structure with larger pores than that of the acid assisted one (PCL-GT-A). Scanning electron microscopy images indicated the PCL-GT scaffold was constructed by nanofibers and microfibers. Mouse fibroblasts and rat bone marrow stromal cells both showed higher proliferation rates on multiscale scaffold than nanoscale scaffolds. It was proposed that the nanofibers bridged between the microfibers enhanced cell adhesion and spreading, while the large pores on the three-dimensional (3D) PCL-GT scaffold provided more effective space for cells to proliferate and migrate. However, the uniform nanofibers and densely packed structure in PCL-GT-A scaffold limited the cells on the surface. Lu et al. (Lu et al. 2010) demonstrated a new super high-throughput electrospinning technique with a large metal rotating cone as the spinneret (Fig. 4.7f). The typical production rate by this method is 10 g/min, which is 1000 times more than that of the traditional electrospinning method. Nanofibers could be obtained as highly aligned arrays with lengths up to several centimeters by modifying the collector value of voltage affecting the diameter of the electrospinning fibers, while the electrospinning rate had not-obvious effects on the diameter.

Receiver Morphology/Specification

The electrospinning set-up usually have disk or cylinder receiver, and the fabricated nanofibers are randomly collected on the receiver. In recent years, the orderly arrangement of controlled spinning has attracted lots of attention because electrospun membranes with ordered nanofibers can realize some distinctive properties. The simplest method to make ordered nanofibers is to increase the rotating speed of the receiver to a certain degree, and the nanofibers can be aligned along the rotating direction (Jiang et al. 2013). For example, Chew et al. (Chew et al. 2005) used the cylindrical drum receiving device to get the ordered fibers, but the order of fibers was not high. Also, the speed of the drum is proportional to the order of the order of the drum arranged around the drum, but it is easy to break the fiber when the drum speed is too fast. Some researches use an additional electric field to control the electrostatic spinning jets, so that the fiber distribution on the receiving board is well controlled. In addition, a quite effective method to obtain the fiber arranged in the orientation is to use special receiving devices such as parallel electrodes, rollers, flywheels, etc. The following are the commonly used electrostatic spinning receivers to obtain aligned fibers.

Flat-panel Receiver. The parallel electrode method can be used to collect the single axially arranged electrospun fibers. The key to the success of this method is the use of a collector consisting of two pieces of electrically conductive substrates separated by a gap whose width could be varied from hundreds of micrometers to several centimeters. As shown in Fig. 4.8a, the collector contained two pieces of conductive silicon stripes separated by a gap. As driven by electrostatic interactions, the charged nanofibers are stretched to span across the gap and thus to become uniaxially aligned arrays over large areas. As the nanofibers were suspended over the gap, they could

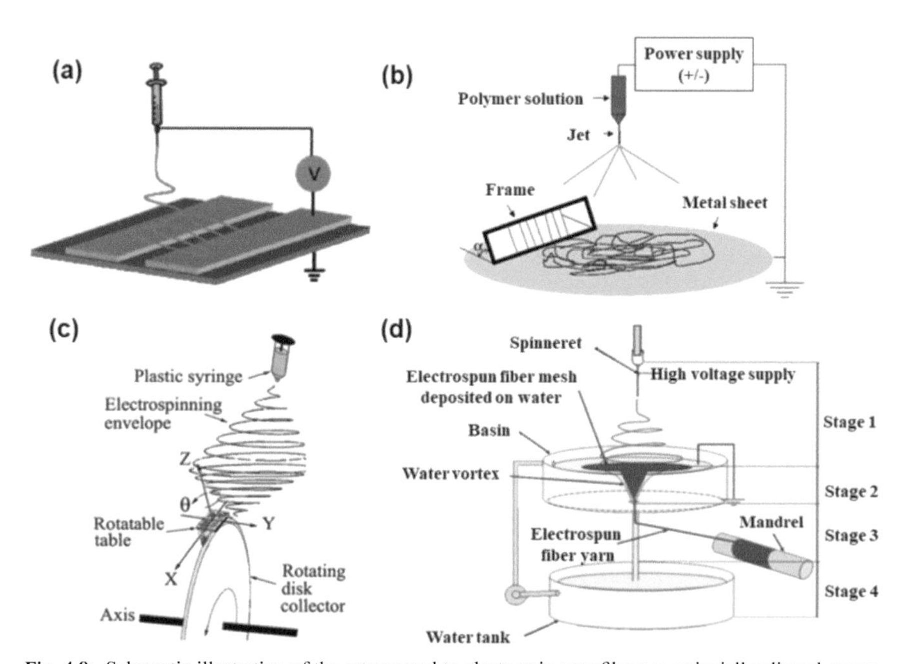

Fig. 4.8: Schematic illustration of the set-up used to electrospin nanofibers as uniaxially aligned arrays. (a) Flat-panel receiver; (b) Frame collector; (c) Rotating disk receiver; (d) Coagulation bath receiver: 1. high voltage electrostatic needle; 2. coagulation pool; 3. fiber net; 4. water swirl; 5. electrostatic spinning nanofiber yarn; 6. drum; 7. water tank.

be conveniently transferred onto the surfaces of other substrates for subsequent treatments and various applications (Li et al. 2010).

Frame Collector. In order to obtain an individual nanofiber for the purpose of experimental characterizations, Huang et al. (Huang et al. 2003) recently developed an approach to align fibers by simply placing a rectangular frame structure under the spinning jet (Fig. 4.8b). It is found that a regular arrangement of fiber net can be obtained by using aluminum and wooden triangular frame receiving device. The parallel arrangement of fiber obtained by aluminum receiving device is higher than that of wood. More investigation is under way to understand the alignment characteristics in terms of varying the shape and size of frame rods, the distance between the frame rods, and the inclination angle of a single frame.

Rotating Disk Receiver. Zussman et al. (Zussman et al. 2003) has designed a rotating disk receiver, which is a disk that rotates around the Y axis. The disk, which is made of aluminum (diameter 5,200 mm), has a tapered edge with a half angle of 26.6 to create a stronger converging electrostatic field an electrostatic lens. An electric potential difference of approximately 8 kV was created between the surface of the liquid drop and the rotating disk collector. During the spinning process, the disk was rotated at a constant speed as it collected the arriving nanofibers on its sharp edge (Fig. 4.8c). As the electrospun nanofiber reached the edge of the wheel, it was wound around the wheel. In addition to facilitating the collection of parallel rows column fiber with diameter of 10–180 nm, and lengths up to several centimeters, the disk also changes the fiber collection direction prior to the formation of the fiber network from time to time around the Z axis rotation angle, which can improve the regularity of fiber arrangement.

Coagulation Bath Receiver. Smit et al. (Smit et al. 2005) used the method of aqueous phase deposition to prepare continuous electrospun nanofibers. The forming principle of nanofibers is: from the solution jet spinneret out of the formation of fiber under the effect of the field, and then fell into the water and sediment into the water, and then through the drawing and winding to rotate at a certain speed on the roller, if the roller speed control right can obtain single fiber. Teo et al. (Teo et al. 2007) further improved the receiving device of solidification pool (Fig. 4.8d). The water in the solidification pool was formed by the pump to form eddy current, and the fibers precipitated in the solidification pool were brought out by the vortex at the bottom of the pool center. The nanofiber yarn is formed by automatic cluster.

In addition to the nanofibers, ropes or yarns fabricated from electrospinning have attracted lots of attention. Chang and Shen (Chang and Shen 2010) fabricated double helical microropes of polyvinylpyrrolidone (PVP) with diameters of less than 10 mm and lengths of up to 5 cm using a electrospinning set-up with a negatively charged rotating collector tip as shown in Fig. 4.9a, so that the two jets from two positively charged spinnerets were induced to two bundles that met at the rotating collector tip, leading to the formation of microropes. The pitch of microropes could be monitored by simply adjusting the distance between the two spinnerets. Shuakat and Lin (Shuakat and Lin 2015) reported a nanofibre yarn electrospinning technique which combines both needle and needleless electrospinning. A rotating intermediate ring collector was employed to directly collect freshly electrospun nanofibres

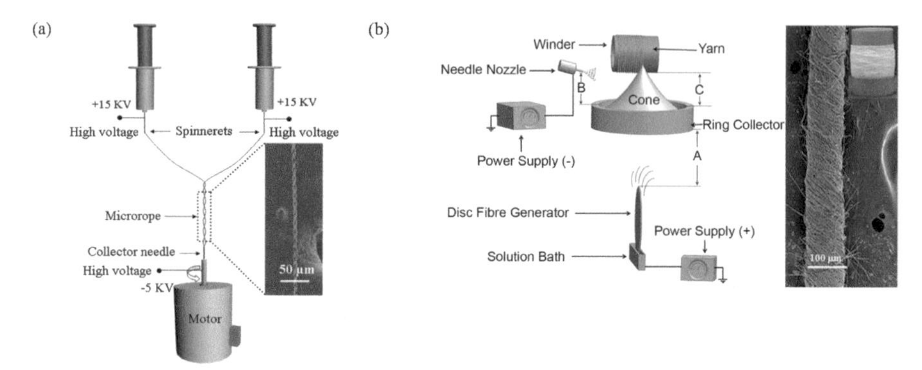

Fig. 4.9: Schematic illustration of (a) electrospinning set-up for fabricating microropes; (b) electrospinning set-up for fabricating yarns and SEM image of an electrospun nanofiber yarn (inset shows nanofiber yarn collected on a spool) (Reproduced from Ref. (Shuakat and Lin 2015) with permission from the Royal Society of Chemistry).

into a fibrous cone, which was further drawn and twisted into a nanofibre yarn (Fig. 4.9b). This novel system was able to produce high tenacity yarn (tensile strength 128.9 MPa and max strain 222.1%) at a production rate of 240 m/hr, with a twist level up to 4,700 twists per meter.

Environmental Parameters

Atmosphere Temperature

Temperature and relative humidity (RH) are two important atmosphere parameters that strongly affect the electrospinning process. Electrospinning solution is generally carried out at room temperature. The change in temperature causes two main effects on changing the average diameter (Vrieze et al. 2009): firstly, increasing the atmosphere temperature of electrospinning accelerates the flow of molecular chains in the jet and increases the conductivity of the solution; secondly, it reduces the viscosity and surface tension of the solution so that some polymer solutions cannot be electrospun. For example, due to the gel and electrolyte properties of proteins and ionic polysaccharides, their aqueous solutions are viscous at room temperature and difficult to spin. By increasing the atmosphere temperature, the viscosity of the solution can be reduced and the spinning process proceeds smoothly. In addition, increasing the ambient temperature of electrospinning can also accelerate the rate of volatilization in the jet, allowing the jet to solidify rapidly and weakening the stretching of the jet, resulting in an increase in fiber diameter. Mit-uppatham et al. (Mit-uppatham et al. 2004) have investigated the effect of temperature on the electrospinning of polyamide-6 fibers ranging from 25 to 60°C, and found that with increase in temperature, there is a yield of fibers with decreased fiber diameter owing to the decrease in the viscosity of the polymer solutions at increased temperatures (Bhardwaj and Kundu 2010). Su et al. (Su et al. 2011) report a new method to fabricate hafnium oxide (HfO_2) nanobelts by electrospinning a sol-gel solution with the implementation of heating and subsequent calcination treatment. They investigated the temperature dependent products and concluded that the heating temperature of spinning ambient was crucial to the formation of HfO_2 nanobelts. By tuning the temperature, the morphological transformation of HfO_2 from nanowires to nanobelts achieved due to the rapid evaporation of solvent played an important role in the formation process of HfO_2 nanobelts. It is found that nanobelts can only be obtained with temperature higher than 50°C and they are in the high quality monoclinic phase. In the research of Liu et al. (Liu et al. 2013), poly(vinyl alcohol) solutions with different temperatures (20, 35, 50, and 65°C) were employed to fabricate nanofibers in a crater-like electrospinning process. The influence of solution temperature on the electrospinnability of solutions and the quality of prepared nanofibers were assessed. The results showed that the solution temperature exerted an appreciable influence on the viscosity of polymer solution, which in turn influenced the process of prediction of nanofibers and their quality. With increase in the solution temperature, the critical applied voltage and air pressure decreased. An ideal temperature is about 50°C to produce nanofibers in these experiments. However, a higher temperature caused a rapid evaporation of solvent in the solution, which caused aggravation.

Humidity

The ambient humidity can directly affect the properties of the media surrounding the jet, especially its compatibility with the solvent in the jet. If the moisture and solvent are compatible, increasing the humidity of the environment will inhibit the removal of solvent in the jet and slow the rate of jet curing. Conversely, it can accelerate the solvent volatilization and the jet cure speed. At very low humidity, a volatile solvent may dry rapidly as the evaporation of the solvent is faster. Sometimes the evaporation rate is so fast compared to the removal of the solvent from the tip of the needle that the electrospinning process may only be carried out for a few minutes before the needle tip is clogged (Baumgarten 1971). Tripatanasuwan et al. (Tripatanasuwan et al. 2007) studied the average fiber diameters of PEO fibers at different relative humidity measured from SEM images. The charged jet solidified at the largest diameter when electrospun at low humidity, since water in the jet evaporated rapidly. The average diameter of PEO nanofibers gradually decreased from around 253 nm when electrospun at 5.1% relative humidity to around 144 nm when electrospun at 48.7% relative humidity. It is important to note that humidity also plays an important role in controlling the surface morphology. In Pai's study (Pai et al. 2010), the solidification rate of PS fibers electrospun from DMF is faster at high relative humidity because the water absorbed from the air into the jet acts as a nonsolvent for PS. SEM images of as-spun fibers electrospun from 30 wt% PS/DMF solutions under relative humidity ranging from 11 to 43% are shown in Fig. 4.10. The fibers electrospun at greater 24% relative humidity have smooth surfaces. Below 24% relative humidity, the smooth surface is replaced by a wrinkled surface, and the fiber diameter tends to be smaller. Below 15% relative humidity, solidification is delayed to longer time and the jet undergoes further thinning, and eventually capillary instability sets in, resulting in the beads-on-string fiber morphology; both beads and strings exhibit a wrinkled or collapsed surface morphology. The fiber diameter for beads-on-string structures is hard to estimate, particularly for the fiber obtained at 15% relative humidity, near

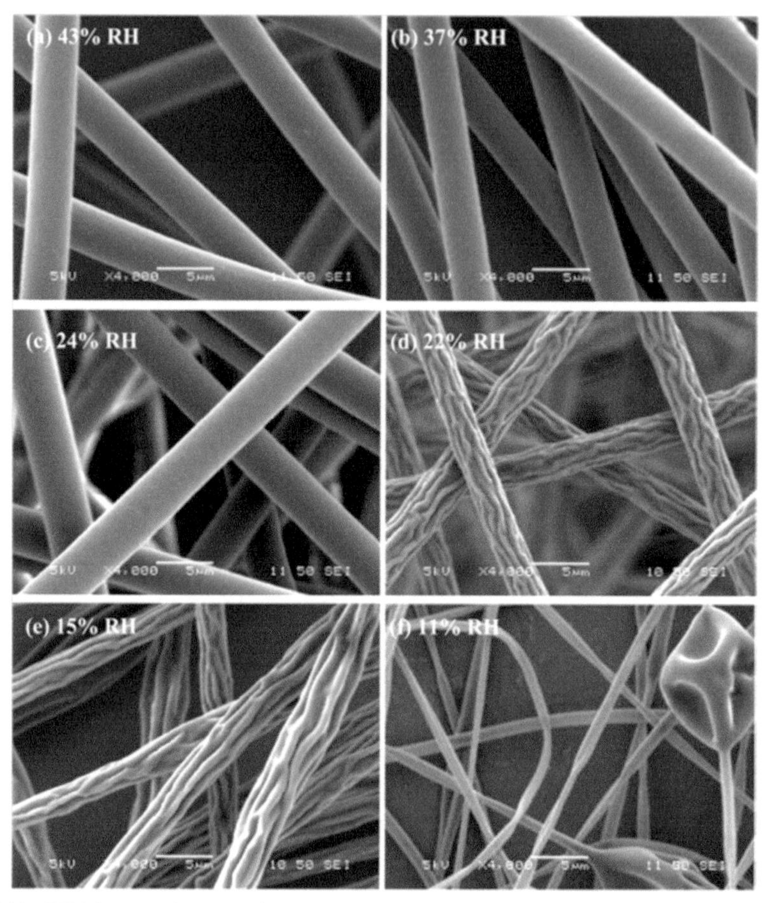

Fig. 4.10: SEM images of as-spun fibers electrospun from 30 wt% PS/DMF solutions under relative humidity ranging from 11 to 43%.

the transition from uniform wrinkled fibers to beads-on-string structures. Casper et al. reported that electrospun PS/tetrahydrofuran had a smooth surface until the relative humidity was above 31% RH. At either 50% or 70% RH, electrospun PS fibers gave regular pores, and those pores were observed only on the surface (Casper et al. 2004). In Kongkhlang's case (Kongkhlang et al. 2008), when maintaining the voltage at 4–10 kV and changing the relative humidity from 75% to 55%, the pore-riddled structure gradually changed to the regular-pore structure, demonstrating that the synergistic effects of electrical voltage and relative humidity drastically change the fiber morphology from pore-riddled to regular pores.

Air Flow Rate

Recently, electrospun nanofiber yarn has become a topic of considerable research interest because of potential applications in micro-electronics, photonics, and biomaterials. Li et al. (Li et al. 2012, Mcclure et al. 2012) made a homemade electrospinning set-up employing the funnel-shape collector with high-speed air inside flow to prepare PAN nanofiber yarn. The surrounding air around the spinneret has an important influence on the morphology of electrospun fibers. During electrospinning, the air flowing velocity in the funnel became bigger because of the gradual contractive cross-section of the funnel. Electrospun jet entered the funnel instantly after it ejected from the spinneret and formed aligned fiber along the air flowing direction. The effects of the air could be summarized as follows: initially taking the nanofibers into the collector and arranging nanofibers, then converging nanofibers, finally twisting nanofibers. During the whole process it took nanofibers moving forward and stretched nanofibers. The fibers without any airflow assist were curled, and the fibers collected by air assisted showed almost no curl, indicating that airflow actually plays a drafting role in the fiber falling, because in the initial stage of the jet, the jet velocity is slower. At this time, the auxiliary air flow is added into the inner cone nozzle, and the velocity of the jet can be accelerated by the friction between the air flow and the jet, so that the fiber becomes thinner. Furthermore, when the airflow velocity is small, the jet of the inner cone nozzle is stable, and the jet along the axial distributes uniformly. Besides, the number of the Taylor Cone is not significantly reduced compared to the phenomenon of no airflow auxiliary. However, when the air flow becomes larger, the number of jet nozzle on the inner cone surface is reduced, and the distribution uniformity of the jet circumferential becomes worse with a small circular arc no jet. It is found that due to the excessive velocity of air, the melt on the inner cone is blown away, and there are partially fan-shaped areas in the inner cone surface without melt, which leads to no jet on the upper part of the inner cone nozzle. Therefore, the velocity of air flow should not be too large during the process of electrospinning. In addition, the diameter of the fibers decreases as the velocity of air flow increases because the airflow velocity becomes bigger. The bigger speed of the jet in the initial stage can be achieved by the frictional drag effect, which can make the fiber finer. Moreover, the pore diameter also displays a dependence on air-flow velocity, because while the electrostatic field caused a decrease in fiber diameter at the sites of perforation, air-flow velocity affected the way in which those fibers were deposited at the sites of perforation and created larger pores at lower airflow rates. It was also found that when air-flow was introduced into the electrospinning process, there was a significant increase in permeability at lower air-flow velocity. Besides, it is evidence that the mechanical properties of the air-flow graft, under similar electrospinning conditions, decreased when air-flow was introduced into the system compared to the no air-flow condition, but then began to gradually increase as air-flow was increased.

Vacuum Degree

Electrospinning generally uses solvent-based solutions under atmosphere condition, which is easily operated. Rangkupan and Reneker (Rangkupan and Reneker 2003) planned to prove the concept of producing a very fine fiber in space using electrospinning process and tried to produced sub-micron fibers from molten polypropylene. Melt-electrospinning complements the solution-electrospinning for polymers which are difficult to dissolve. Additionally, melt-electrospinning can eliminate the cost associated with the removal and recovery of the solvents, and any health risks associated with solvents. Further study about the melt-electrospinning process, from both scientific and engineering standpoints can produce valuable information. A polymer melt has higher viscosity and lower electrical conductivity than a polymer solution. Therefore, in the vacuum system, they took advantage of much higher electric field compared to in air condition, enable to exert larger electric forces to fine-draw the fluid jet. The electric field strength between the spinneret and the collector was varied from 100 to 3000 kV/m. The behavior of a molten charged jet in vacuum was expected to be similar to the behavior of a molten charged jet electrospun in air, since the electrospinning processes in vacuum and in air were essentially identical, with the main difference in the solidification of the charged jets. Polypropylene was successfully electrospun into fine fibers with diameter in the range of 300 nm to 30 μm with a large fiber diameter variation. Morphological evidence also indicated that a molten charged jet underwent the electrically-driven bending instability after it was ejected from the spinneret.

Acknowledgements

The authors wish to acknowledge great help and support from Qiqi Qin, Lanlan Fan, Gang Wang, Zongjie Li, Nanping Deng, Liyuan Wang, Jingge Ju, Xinghai Zhou, Lu Gao, Yan Hao, Huiru Ren, Weicui Liu, and Huijuan Zhao.

References

Al-Qadhi, M., Merah, N., Matin, A., Abu-Dheir, N., Khaled, M. and Youcef-Toumi, K. 2015. Preparation of superhydrophobic and self-cleaning polysulfone non-wovens by electrospinning: influence of process parameters on morphology and hydrophobicity. Journal of Polymer Research 22(11): 1–9.

Arayanarakul, K., Choktaweesap, N., Aht-ong, D., Meechaisue, C. and Supaphol, P. 2006. Effects of poly(ethylene glycol), inorganic salt, sodium dodecyl sulfate, and solvent system on electrospinning of poly(ethylene oxide). Macromolecular Materials and Engineering 291(6): 581–591.

Baker, S. C., Atkin, N., Gunning, P. A., Granville, N., Wilson, K., Wilson, D. et al. 2006. Characterisation of electrospun polystyrene scaffolds for three-dimensional *in vitro* biological studies. Biomaterials 27(16): 3136–3146.

Bao, Q., Zhang, H., Yang, J. X., Wang, S., Tang, D. Y., Jose, R. et al. 2010. Ultrafast Photonics: Graphene-polymer nanofiber membrane for ultrafast photonics (Adv. Funct. Mater. 5/2010). Advanced Functional Materials 20(5): n–n/a.

Barakat, N. A. M., Kanjwal, M. A., Sheikh, F. A. and Kim, H. Y. 2009. Spider-net within the N6, PVA and PU electrospun nanofiber mats using salt addition: Novel strategy in the electrospinning process. Polymer 50(18): 4389–4396.

Baumgarten, P. K. 1971. Electrostatic spinning of acrylic microfibers. Journal of Colloid & Interface Science 36(1): 71–79.

Bhardwaj, N. and Kundu, S. C. 2010. Electrospinning: A fascinating fiber fabrication technique. Biotechnology Advances 28(3): 325–347.

Biber, E., Gündüz, G., Mavis, B. and Colak, U. 2010. Effects of electrospinning process parameters on nanofibers obtained from Nylon 6 and poly (ethylene-n-butyl acrylate-maleic anhydride) elastomer blends using Johnson S B statistical distribution function. Applied Physics A 99(2): 477–487.

Bosworth, L. A. and Downes, S. 2012. Acetone, a sustainable solvent for electrospinning poly(ε-Caprolactone) fibres: Effect of varying parameters and solution concentrations on fibre diameter. Journal of Polymers & the Environment 20(3): 879–886.

Casper, C. L., Stephens, J. S., Tassi, N. G., Chase, D. B. and Rabolt, J. F. 2004. Controlling surface morphology of electrospun polystyrene fibers: Effect of humidity and molecular weight in the electrospinning process. Macromolecules 37(2): 573–578.

Chang, G. and Shen, J. 2010. Fabrication of microropes via Bi-electrospinning with a rotating needle collector. Macromolecular Rapid Communications 31(24): 2151–2154.

Chang, J., Liu, Y., Heo, K., Lee, B. Y., Lee, S. W. and Lin, L. 2014. Direct-write complementary graphene field effect transistors and junctions via near-field electrospinning. Small 10(10): 1920–1925.

Chang, S. K., Baek, D. H., Gang, K. D., Lee, K. H., Um, I. C. and Park, Y. H. 2005. Characterization of gelatin nanofiber prepared from gelatin–formic acid solution. Polymer 46(14): 5094–5102.

Chen, H., Blitterswijk, C. V., Mota, C., Wieringa, P. A. and Moroni, L. 2017. Direct writing electrospinning of scaffolds with multi-dimensional fiber architecture for hierarchical tissue engineering. Acs Applied Materials & Interfaces 9(44): 38187–38200.

Chen, W., Huang, D. A., Chen, H. C., Shie, T. Y., Hsieh, C. H., Liao, J. D. et al. 2009. Fabrication of polycrystalline ZnO nanotubes from the electrospinning of Zn2+/Poly(acrylic acid). Crystal Growth & Design 9(9): 4070–4077.

Chew, S. Y., Wen, J., Yim, E. K. and Leong, K. W. 2005. Sustained release of proteins from electrospun biodegradable fibers. Biomacromolecules 6(4): 2017–2024.

Cork, J., Whittaker, A. K., Cooper-White, J. J. and Grøndahl, L. 2017. Electrospinning and mechanical properties of P(TMC-co-LLA) elastomers. Journal of Materials Chemistry B 5(12).

Deitzel, J. M., Kleinmeyer, J., Harris, D. and Tan, N. C. B. 2001. The effect of processing variables on the morphology of electrospun nanofibers and textiles. Polymer 42(1): 261–272.

Demir, M. M., Yilgor, I., Yilgor, E. and Erman, B. 2002. Electrospinning of polyurethane fibers. Polymer 43(11): 3303–3309.

Ding, B., Kim, H. Y., Lee, S. C., Shao, C. L., Lee, D. R., Park, S. J. et al. 2002. Preparation and characterization of a nanoscale poly(vinyl alcohol) fiber aggregate produced by an electrospinning method. Journal of Polymer Science Part B Polymer Physics 40(13): 1261–1268.

Ding, B. and Yu, J. 2011. Electrospinning and nanofibers. China Textile & Apparel Press, Beijing, China.

Dong, I., Kim, K. W., Gong, H. C., Kim, H. Y., Lee, K. H. and Bhattarai, N. 2006. Mechanical behaviors and characterization of electrospun polysulfone/polyurethane blend nonwovens. Macromolecular Research 14(3): 331–337.

Dong, Z., Kennedy, S. J. and Wu, Y. 2011. Electrospinning materials for energy-related applications and devices. Journal of Power Sources 196(11): 4886–4904.

Doshi, J. and Reneker, D. H. 2002. Electrospinning process and applications of electrospun fibers. Industry Applications Society Meeting 3: 1698–1703.

Doustgani, A. 2015. Optimization of mechanical and structural properties of PVA nanofibers. Journal of Industrial Textiles 46(3).

Flory, P. J. 1954. Book Reviews: Principles of Polymer Chemistry. Scientific Monthly 79.

Fong, H., Chun, I. and Reneker, D. H. 1999. Beaded nanofibers formed during electrospinning. Polymer 40(16): 4585–4592.

Fridrikh, S. V., Yu, J. H., Brenner, M. P. and Rutledge, G. C. 2003. Controlling the fiber diameter during electrospinning. Physical Review Letters 90(14): 144502.

Ghelich, R., Keyanpour-Rad, M., Youzbashi, A. A. and Khakpour, Z. 2014. Comparative study on structural properties of NiOâ GDC nanocomposites fabricated via electrospinning and gel combustion processes. Materials Research Innovations 19(1): 44–50.

Gogotsi, Y. 2006. Nanotubes and nanofibers. Small 1(1): 91.

Gu, S. Y., Ren, J. and Vancso, G. J. 2005. Process optimization and empirical modeling for electrospun polyacrylonitrile (PAN) nanofiber precursor of carbon nanofibers. European Polymer Journal 41(11): 2559–2568.

Gupta, P. and Wilkes, G. L. 2003. Some investigations on the fiber formation by utilizing a side-by-side bicomponent electrospinning approach. Polymer 44(20): 6353–6359.

Gupta, P., Elkins, C., Long, T. E. and Wilkes, G. L. 2005. Electrospinning of linear homopolymers of poly(methyl methacrylate): exploring relationships between fiber formation, viscosity, molecular weight and concentration in a good solvent. Polymer 46(13): 4799–4810.

Heikkil Auml, P., Söderlund, L., Uusimäki, J., Kettunen, L. and Harlin, A. 2010. Exploitation of electric field in controlling of nanofiber spinning process. Polymer Engineering & Science 47(12): 2065–2074.
Hekmati, A. H., Rashidi, A., Ghazisaeidi, R. and Drean, J. Y. 2013. Effect of needle length, electrospinning distance, and solution concentration on morphological properties of polyamide-6 electrospun nanowebs. Textile Research Journal 83(14): 1452–1466.
Henriques, C., Vidinha, R., Botequim, D., Borges, J. P. and Silva, J. A. 2009. A systematic study of solution and processing parameters on nanofiber morphology using a new electrospinning apparatus. Journal of Nanoscience & Nanotechnology 9(6): 3535.
Hong, P. D., Chou, C. M. and He, C. H. 2001. Solvent effects on aggregation behavior of polyvinyl alcohol solutions. Polymer 42(14): 6105–6112.
Huang, Z. M., Zhang, Y. Z., Kotaki, M. and Ramakrishna, S. 2003. A review on polymer nanofibers by electrospinning and their applications in nanocomposites. Composites Science & Technology 63(15): 2223–2253.
Jarusuwannapoom, T., Hongrojjanawiwat, W., Jitjaicham, S., Wannatong, L., Nithitanakul, M., Pattamaprom, C. et al. 2005. Effect of solvents on electro-spinnability of polystyrene solutions and morphological appearance of resulting electrospun polystyrene fibers. European Polymer Journal 41(3): 409–421.
Jiang, G., Zhang, S. and Qin, X. 2013. High throughput of quality nanofibers via one stepped pyramid-shaped spinneret. Materials Letters 106(106): 56–58.
Ju, J., Kang, W., Li, L., He, H., Qiao, C. and Cheng, B. 2016. Preparation of poly (tetrafluoroethylene) nanofiber film by electro-blown spinning method. Materials Letters 171: 236–239.
Ju, J., Shi, Z., Fan, L., Liang, Y., Kang, W. and Cheng, B. 2017. Preparation of elastomeric tree-like nanofiber membranes using thermoplastic polyurethane by one-step electrospinning.
Kanafchian, M., Valizadeh, M. and Haghi, A. K. 2011a. Fabrication of nanostructured and multicompartmental fabrics based on electrospun nanofibers. Korean Journal of Chemical Engineering 28(3): 763–769.
Kanafchian, M., Valizadeh, M. and Haghi, A. K. 2011b. Prediction of nanofiber diameter for improvements in incorporation of multilayer electrospun nanofibers. Korean Journal of Chemical Engineering 28(3): 751–755.
Kancheva, M., Toncheva, A., Manolova, N. and Rashkov, I. 2014. Advanced centrifugal electrospinning setup. Materials Letters 136(136): 150–152.
Kanehata, M., Ding, B. and Shiratori, S. 2007. Nanoporous ultra-high specific surface inorganic fibres. Nanotechnology 18(31): 315602.
Kim, B., Park, H., Lee, S. H. and Sigmund, W. M. 2005. Poly(acrylic acid) nanofibers by electrospinning. Materials Letters 59(7): 829–832.
Kongkhlang, T., Kotaki, M., Kousaka, Y., Umemura, T., Nakaya, D. and Chirachanchai, S. 2008. Electrospun polyoxymethylene: spinning conditions and its consequent nanoporous nanofiber. Macromolecules 41(13): 4746–4752.
Koski, A., Yim, K. and Shivkumar, S. 2004. Effect of molecular weight on fibrous PVA produced by electrospinning. Materials Letters 58(3): 493–497.
Lee, K. H., Kim, H. Y., Khil, M. S., Ra, Y. M. and Lee, D. R. 2003. Characterization of nano-structured poly(ε-caprolactone) nonwoven mats via electrospinning. Polymer 44(4): 1287–1294.
Li, D. and Xia, Y. 2010. Electrospinning of nanofibers: reinventing the wheel? Advanced Materials 16(14): 1151–1170.
Li, D., Wang, Y. and Xia, Y. 2010. Electrospinning nanofibers as uniaxially aligned arrays and layer-by-layer stacked films. Advanced Materials 16(4): 361–366.
Li, D., Chen, W., Sun, B., Li, H., Wu, T., Ke, Q. et al. 2016. A comparison of nanoscale and multiscale PCL/gelatin scaffolds prepared by disc-electrospinning. Colloids Surf B Biointerfaces 146: 632–641.
Li, L., Jiang, Z., Xu, J. and Fang, T. 2014. Predicting poly(vinyl pyrrolidone)'s solubility parameter and systematic investigation of the parameters of electrospinning with response surface methodology. Journal of Applied Polymer Science 131(11).
Li, N., Hui, Q., Xue, H. and Xiong, J. 2012. Electrospun polyacrylonitrile nanofiber yarn prepared by funnel-shape collector. Materials Letters 79(23): 245–247.

Li, Z., Xu, Y. Z., Fan, L. L., Kang, W. M. and Cheng, B. W. 2016. Fabrication of polyvinylidene fluoride tree-like nanofiber via one-step electrospinning. Materials & Design 92: 95–101.
Liu, H. Y., Kong, H. Y., Wang, M. Z. and He, J. H. 2014. Lightning-like charged jet cascade in bubble electrospinning with ultrasonic vibration. Journal of Nano Research 27(9): 111–119.
Liu, Y. and He, J. H. 2007. Bubble electrospinning for mass production of nanofibers. International Journal of Nonlinear Sciences & Numerical Simulation 8(3): 393–396.
Liu, Y., Dong, L., Fan, J., Wang, R. and Yu, J. Y. 2011. Effect of applied voltage on diameter and morphology of ultrafine fibers in bubble electrospinning. Journal of Applied Polymer Science 120(1): 592–598.
Liu, Y., Liang, W., Shou, W., Su, Y. and Wang, R. 2013. Effect of temperature on the crater-like electrospinning process. Heat Transfer Research 44(5): 447–454.
Lu, B., Wang, Y., Liu, Y., Duan, H., Zhou, J., Zhang, Z et al. 2010. Superhigh-throughput needleless electrospinning using a rotary cone as spinneret. Small 6(15): 1612.
Lu, X., Zhao, Y. and Wang, C. 2010. Fabrication of PbS nanoparticles in polymer-fiber matrices by electrospinning. Advanced Materials 17(20): 2485–2488.
Matabola, K. P. and Moutloali, R. M. 2013. The influence of electrospinning parameters on the morphology and diameter of poly(vinyledene fluoride) nanofibers-effect of sodium chloride. Journal of Materials Science 48(16): 5475–5482.
Mcclure, M. J., Wolfe, P. S., Simpson, D. G., Sell, S. A. and Bowlin, G. L. 2012. The use of air-flow impedance to control fiber deposition patterns during electrospinning. Biomaterials 33(3): 771–779.
Mckee, M. G. and Long, T. E. 2006. Phospholipid nonwoven electrospun membranes. Science 311(5759): 353–355.
Mit-uppatham, C., Nithitanakul, M. and Supaphol, P. 2004. Ultrafine electrospun polyamide-6 fibers: effect of solution conditions on morphology and average fiber diameter. Macromolecular Chemistry & Physics 205(17): 2327–2338.
Mo, X. X. C., Kotaki, M. and Ramakrishna, S. 2004. Electrospun P(LLA-CL) nanofiber: a biomimetic extracellular matrix for smooth muscle cell and endothelial cell proliferation. Biomaterials 25(10): 1883–1890.
Moon, S., Gil, M. and Lee, K. J. 2017. Syringeless electrospinning toward versatile fabrication of nanofiber web. Scientific Reports 7: 41424.
Mottaghitalab, V. and Haghi, A. K. 2011. A study on electrospinning of polyacrylonitrile nanofibers. Korean Journal of Chemical Engineering 28(1): 114–118.
Narasimhan, V. C. 2016. A novel dual layered biodegradable scaffold system through electrospinning for the simultaneous delivery of hydrophilic and hydrophobic drugs for application in bone tissue engineering. Tissue Engineering Part A 22: 138–138.
Ojha, S. S., Afshari, M., Kotek, R. and Gorga, R. E. 2010. Morphology of electrospun nylon-6 nanofibers as a function of molecular weight and processing parameters. Journal of Applied Polymer Science 108(1): 308–319.
Ostermann, R., Li, D., Yin, Y., Mccann, J. T. and Xia, Y. 2006. V_2O_5 nanorods on TiO_2 nanofibers: a new class of hierarchical nanostructures enabled by electrospinning and calcination. Nano Letters 6(6): 1297–1302.
Ou, K. L., Chen, C. S., Lin, L. H., Lu, J. C., Shu, Y. C., Tseng, W. C. et al. 2011. Membranes of epitaxial-like packed, super aligned electrospun micron hollow poly(l-lactic acid) (PLLA) fibers. European Polymer Journal 47(5): 882–892.
Pai, C. L., Boyce, M. C. and Rutledge, G. C. 2010. Morphology of porous and wrinkled fibers of polystyrene electrospun from dimethylformamide. Macromolecules 42(6): 2102–2114.
Patel, A. C., Li, S. X., Wang, C., Zhang, W. J. and Wei, Y. 2007. Electrospinning of porous silica nanofibers containing silver nanoparticles for catalytic applications. Chemistry of Materials 19(6): 1231–1238.
Qin, X. H., Yang, E. L., Li, N. and Wang, S. Y. 2007. Effect of different salts on electrospinning of polyacrylonitrile (PAN) polymer solution. Journal of Applied Polymer Science 103(6): 3865–3870.
Ramakrishna, S., Fujihara, K., Teo, W. E., Lim, T. C. and Ma, Z. 2005. An Introduction to Electrospinning and Nanofibers: World Scientific.
Rangkupan, R. and Reneker, D. H. 2003. Electrospinning process of molten polypropylene in vacuum. Journal of Metals.

Shao, H., Fang, J., Wang, H., Lang, C. and Lin, T. 2015. Robust mechanical-to-electrical energy conversion from short-distance electrospun poly(vinylidene fluoride) fiber webs. Acs Applied Materials & Interfaces 7(40): 22551.

Shenoy, S. L., Bates, W. D., Frisch, H. L. and Wnek, G. 2005. Role of chain entanglements on fiber formation during electrospinning of polymer solutions: good solvent, non-specific polymer–polymer interaction limit. Polymer 46(10): 3372–3384.

Shuakat, M. N. and Lin, T. 2015. Highly-twisted, continuous nanofibre yarns prepared by a hybrid needle-needleless electrospinning. Rsc Advances 5(43): 33930–33937.

Smit, E., Bűttner, U. and Sanderson, R. D. 2005. Continuous yarns from electrospun fibers. Polymer 46(8): 2419–2423.

Su, Y., Lu, B., Xie, Y., Ma, Z., Liu, L., Zhao, H. et al. 2011. Temperature effect on electrospinning of nanobelts: the case of hafnium oxide. Nanotechnology 22(28): 285609.

Subbiah, T., Bhat, G. S., Tock, R. W., Parameswaran, S. and Ramkumar, S. S. 2010. Electrospinning of nanofibers. Journal of Applied Polymer Science 96(2): 557–569.

Sui, X. M., Wiesel, E. and Wagner, H. D. 2011. Enhanced mechanical properties of electrospun nano-fibers through NaCl mediation. Journal of Nanoscience and Nanotechnology 11(9): 7931–7936.

Tacx, J. C. J. F., Schoffeleers, H. M., Brands, A. G. M. and Teuwen, L. 2000. Dissolution behavior and solution properties of polyvinylalcohol as determined by viscometry and light scattering in DMSO, ethyleneglycol and water. Polymer 41(3): 947–957.

Tao, J. 2003. Effects of molecular weight and solution concentration on electrospinning of PVA.

Teo, W. E., Gopal, R., Ramaseshan, R., Fujihara, K. and Ramakrishna, S. 2007. A dynamic liquid support system for continuous electrospun yarn fabrication. Polymer 48(12): 3400–3405.

Thoppey, N. M., Bochinski, J. R., Clarke, L. I. and Gorga, R. E. 2010. Unconfined fluid electrospun into high quality nanofibers from a plate edge. Polymer 51(21): 4928–4936.

Thoppey, N. M., Gorga, R. E., Bochinski, J. R. and Clarke, L. I. 2012. Effect of solution parameters on spontaneous jet formation and throughput in edge electrospinning from a fluid-filled bowl. Macromolecules 45(16): 6527–6537.

Tripatanasuwan, S., Zhong, Z. and Reneker, D. H. 2007. Effect of evaporation and solidification of the charged jet in electrospinning of poly(ethylene oxide) aqueous solution. Polymer 48(19): 5742–5746.

Vrieze, S. D., Camp, T. V., Nelvig, A., Hagström, B., Westbroek, P. and Clerck, K. D. 2009. The effect of temperature and humidity on electrospinning. Journal of Materials Science 44(5): 1357–1362.

Wang, H., Zeng, S., Yu, C., Wang, X. and Xia, D. 2010. Preparation and characterization of polypyrrole/TiO_2 coaxial nanocables. Macromolecular Rapid Communications 27(6): 430–434.

Wang, N., Wang, X. F., Ding, B., Yu, J. Y. and Sun, G. 2012. Tunable fabrication of three-dimensional polyamide-66 nano-fiber/nets for high efficiency fine particulate filtration. Journal of Materials Chemistry 22(4): 1445–1452.

Wang, T. and Kumar, S. 2010. Electrospinning of polyacrylonitrile nanofibers. Journal of Applied Polymer Science 102(2): 1023–1029.

Wang, X., Ding, B., Yu, J. and Yang, J. 2011. Large-scale fabrication of two-dimensional spider-web-like gelatin nano-nets via electro-netting. Colloids and Surfaces B-Biointerfaces 86(2): 345–352.

Wang, X., Niu, H., Wang, X. and Lin, T. 2012. Needleless electrospinning of uniform nanofibers using spiral coil spinnerets. Journal of Nanomaterials 2012(10): 3.

Wang, X. F., Ding, B., Yu, J. Y., Si, Y., Yang, S. B. and Sun, G. 2011. Electro-netting: Fabrication of two-dimensional nano-nets for highly sensitive trimethylamine sensing. Nanoscale 3(3): 911–915.

Yalcinkaya, F. 2015. Effect of current on polymer jet in electrospinning process. Tekstil Ve Konfeksiyon 25(3): 201–206.

Yan, J., Choi, J. H. and Jeong, Y. G. 2017. Freestanding supercapacitor electrode applications of carbon nanofibers based on polyacrylonitrile and polyhedral oligomeric silsesquioxane. Materials & Design, 139.

Yang, Q. B., Li, D. M., Hong, Y. L., Li, Z. Y., Wang, C., Qiu, S. L. et al. 2003. Preparation and characterization of a pan nanofibre containing ag nanoparticles via electrospinning. Synthetic Metals 137(1): 973–974.

Youn, D. H., Yu, Y. J., Jin, S. C., Park, N. M., Sun, J. Y., Lee, I. et al. 2016. Transparent conducting films of silver hybrid films formed by near-field electrospinning. Materials Letters 185: 139–142.

Yu, Z., Lin, Y., Huang, W., Zhuang, M., Hong, Y., Zheng, G. et al. 2014. Multi spinnerets electrospinning with assistant sheath gas. IEEE International Conference on Nano/micro Engineered and Molecular Systems, 64–67.

Yuan, D. X., Zhang, Y., Dong, C. and Sheng, J. 2004. Morphology of ultrafine polysulfone fibers prepared by electrospinning. Polymer International 53(11): 1704–1710.

Zeng, J., Hou, H., Schaper, A., Wendorff, J. H. and Greiner, A. 2003. Poly-L-lactide nanofibers by electrospinning—Influence of solution viscosity and electrical conductivity on fiber diameter and fiber morphology. E-Polymers 3(1): 102–110.

Zeng, J., Hou, H., Wendorff, J. H. and Greiner, A. 2005. Poly(vinyl alcohol) nanofibres by electrospinning: influence of molecular weight on fibre shape. E-Polymers 5(1): 387–393.

Zhang, C., Yuana, X., Han, Y. and Sheng, J. 2005. Study on morphology of electrospun poly(vinyl alcohol) mats. European Polymer Journal 41(3): 423–432.

Ziabari, M., Mottaghitalab, V. and Haghi, A. K. 2008. Simulated image of electrospun nonwoven web of PVA and corresponding nanofiber diameter distribution. Korean Journal of Chemical Engineering 25(4): 919–922.

Zussman, E., Theron, A. and Yarin, A. L. 2003. Formation of nanofiber crossbars in electrospinning. Applied Physics Letters 82(6): 973–975.

Chapter 5

Theoretical Simulation of Electrospinning Process:

Magnitude, Distribution and Improvement of Electric Field Intensity during Electrospinning

Yanbo Liu,[1,2,*] *Yong Liu*,[2,*] *Wenxiu Yang*,[3] *Jian Liu*[4] and *Daxiang Yang*[5]

Introduction

As addressed in the previous chapters of this book, a variety of electrospinning technologies based on different principles have emerged with the rapid development of materials science and nanotechnology. The currently existing electrospinning technologies are usually classified into two subcategories, i.e., needle electrospinning and needleless electrospinning.

Features of Needle Electrospinning

During needle electrospinning process, needles or nozzles are employed to generate Taylor cones from the liquid drops on the top of the needles or nozzles, resulting in nanofiber web with relatively condensed and uniform web structure. However, its nanofiber productivity is rather low compared to that of needleless electrospinning. Besides, needle-blocking (Fig. 5.1) and End-effect phenomena (Theron et al. 2005) (Fig. 5.2) are the typical drawbacks of needle electrospinning process.

[1] School of Textile Science and Engineering, Wuhan Textile University, 1 Yangguang Avenue, Jiangxia District, Wuhan City, Hubei Province 430200, China.
[2] School of Textiles, Tianjin Polytechnic University, 399 Western Binshui Avenue, Tianjin 300387, China.
[3] College of Textile and Garment, Hebei University of Science and Technology, 26 Yuxiang Street, Yuhua District, Shijiazhuang City, 050080, China.
[4] Tianjin Key Laboratory of Modern Technology & Equipment, Tianjin, 300387, China.
[5] Chongqing ChinaNano Sci & Tech Co. Ltd., Immigrant Eco-Industrial Park, Fengjie County, Chongqing City 404677, China.
* Corresponding authors: yanboliu@gmail.com; liuyong@tjpu.edu.cn

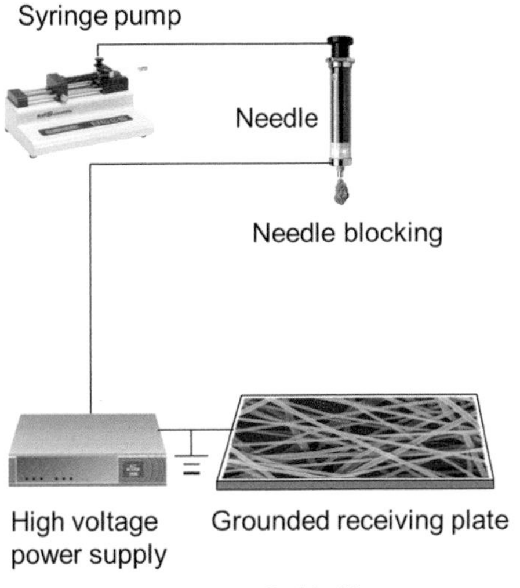

Fig. 5.1: Needle-blocking.

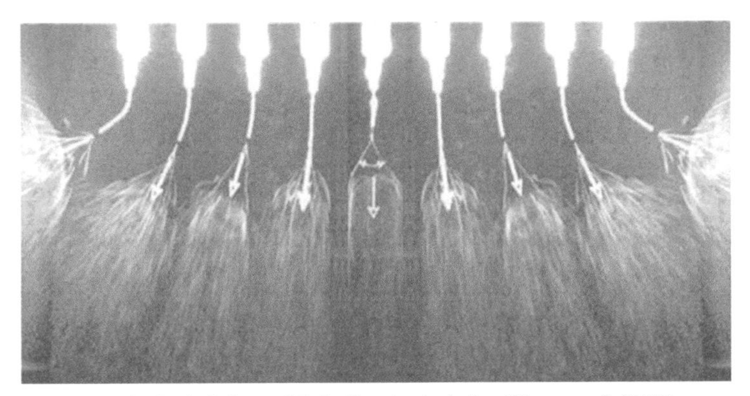

Fig. 5.2: Optical photo of End-effect (revised after (Theron et al. 2005)).

It can be obviously found from Taylor's Equation, as expressed in Eq. 5.1 (Taylor 1964, 1966, 1969), that the main factors influencing on electrospinning process include applied voltage U, surface tension γ of the spinning solution, needle length L, needle radius R, and the receiving distance H. Surface tension γ is closely related to another solution characteristic parameter, i.e., viscosity or concentration.

Previously, Taylor (Taylor 1966, 1969) determined the critical value of U, U_C, at which a fairly conducting and viscous fluid is drawn from the tip of the tube with length of L and outer radius R:

$$U_c^2 = \frac{4H^2}{L^2}\left(ln\frac{2L}{R} - \frac{3}{2}\right)(1.30\pi\gamma R)(0.09) \tag{5.1}$$

where H is the distance from the needle tip to the collector, and γ denotes the surface tension. The factor 0.09 was inserted to predict the voltage in kV, and 1.30 is the result of 2cos49.3°.

Later, Taylor (Taylor 1964) determined the critical value of voltage U_c at which the jets or drops with surface tension, γ, can appear from the tube of Zeleny apparatus (Zeleny 1914) based on the theory of electrically driven jets, as Eq. 5.2:

$$U_c^2 = 4\left(ln\frac{2h}{R}\right)(1.30\pi\gamma R)(0.09) \tag{5.2}$$

However, practical experiments indicated that the key process parameters for needle electrospinning should also include the solution concentration, feed rate, and conductivity, etc. In addition, the critical spinning voltage for a multiple needle electrospinning process cannot be expressed with a simple formula like Eq. 5.2, due to the complex process and End-effect phenomenon caused by the interaction between electric field force and Coulombic force among the needles and jets, but it can be predicated with finite element analysis method, i.e., by means of the established or build-in constitutive equations of electric field intensity, and the simulation results can be used to guide the practical electrospinning experiments or trials, especially the design and manufacture of electrospinning machines, including both needle electrospinning and needleless electrospinning.

Features of Needleless Electrospinning

Most needleless electrospinning methods are based on the principle of self-organization of free liquid on the surface of spinning electrode, where multiple Taylor cones are generated at multiple locations having equal spacing (i.e., wave length λ), and the spacing can be predicted with Eq. 5.3 (Lukas et al. 2008):

$$\lambda = \frac{12\pi\gamma}{2\varepsilon E_0^2 + \sqrt{(2\varepsilon E_0^2)^2 - 12\gamma\rho g}} \tag{5.3}$$

where γ, ρ, g, and ε stands for surface tension, liquid density, gravity, and absolute dielectric constant (permittivity) of vacuum medium, respectively; E_0 represents the applied voltage, and the critical spinning voltage E_c, at which the stable waves can be formed, is expressed as Eq. 5.4 (Lukas et al. 2008):

$$E_c = \sqrt[4]{4\gamma\rho g / \varepsilon^2} \tag{5.4}$$

These two equations can be deduced based on Landau's dispersion law. Equation 5.3 indicates that the distance between the neighboring jets, i.e., the jet spacing, is right along the wave length λ of liquid turbulence waves at corresponding field intensity when the jets are stable, and the wave length λ decreases with the increasing field intensity, which is depicted in Fig. 5.3.

This process has higher productivity, no needle-blocking issue, and less serious End-effect phenomenon than needle electrospinning, however, this process is unstable and the viscosity of the liquid in the open-air solution reservoir is inconsistent because the solution tends to become more concentrated with the solvent gradually volatilizing.

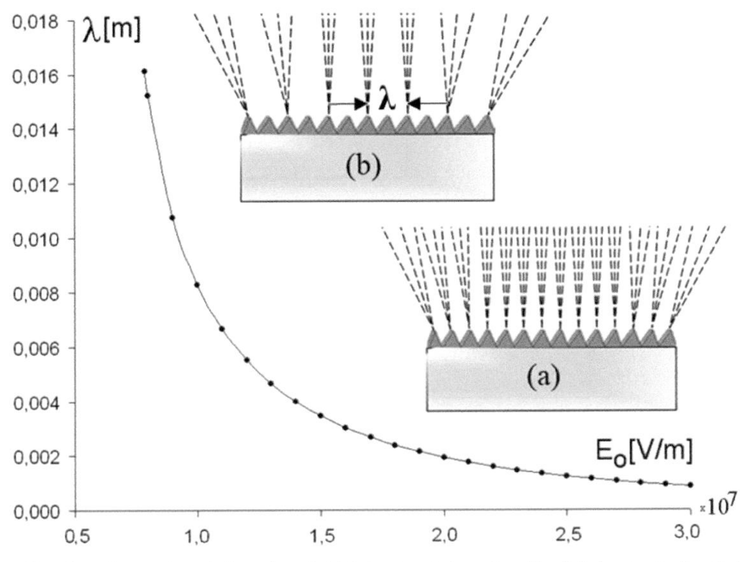

Fig. 5.3: The plot of wave length λ against field intensity E_0 based on Eq. 5.3 (revised after Ref. (Lukas et al. 2008)).

Another problem with needleless electrospinning process is that the solution on the surface of the spinning electrode also faces the same issue as the solution in the reservoir, i.e., the viscosity of the solution on the spinneret is getting thicker and thicker with the solvent evaporating, and hence the polymer will finally consolidate on the surface of the electrode, resulting in reduced electric filed intensity due to the increased diameter of the spinneret, which also leads to an unstable process in the needleless electrospinning. Besides, the needleless electrospinning usually generates fibrous webs with less density than a needle electrospinning process.

Till date, the most frequently used nanofiber manufacture method in laboratory or industry is still the needle electrospinning technology due to its setup simplicity, product uniformity, process controllability, and ease for shift of product category. Therefore, multineedle electrospinning tends to be used to prepare nanofiber products at large scale towards applications in fields such as lithium ion battery separators, biosensors, wound dressings, carriers for controlled release of medicine, tissue engineering scaffolds, and so on. Needleless electrospinning method, however, is only employed in enterprises where nanofiber materials are massively produced for commercial applications in fields, such as gas or liquid filtration, water-proof & breathable membrane, etc., where the structural uniformity is not strictly required for electrospun membrane.

Although various nanofibrous materials have been prepared based on different electrospinning methods and corresponding electrospinning setups towards their applications in many important areas, most products are still at the stage of laboratory trials or development, and only a few of them have been commercialized in some special areas, such as industrial air filtration medium (Donaldson, 2018), personal protective mask (Zhongke Best (Xiamen) Environmental Protection Technology Co.), lithium ion battery separator (Jiangxi Advanced Materials Nanofiber Sci & Tech Co. Ltd, 2018), etc. The barrier against further commercialization of electrospinning technology is either the low production rate or productivity of nanofiber web, which is as low as 0.1–1.0 g/hr for single needle electrospinning (Tang et al. 2010), or the inconsistent process of the needleless electrospinning. Then multiple needles were considered to electrospin more nanofibers simultaneously to increase the productivity of the nanofibrous web; however, it was soon found that multineedle electrospinning was not capable of producing nanofibers as many as theoretically expected due to the existence of the End-effect phenomenon (Fig. 5.1).

End-effect vs Uniformity and Productivity

It was reported that electrical shielding in multineedle electrospinning prevents the formation of Taylor cones in the central needle and deflects the jets located at the periphery, and the deflection of spinning jets has been observed from the edge needles in the array of multi-needle electrospinning equipment. This phenomenon is referred to as End-effect (Theron et al. 2005). As shown in Fig. 5.1, it could be observed that the center needle sprays polymer jets almost vertically towards the fiber receiver, while jets from the end needles travelled towards the lateral sides rather than directly to the receiver, with the jets from the outmost needles showing the most severe skew, and this inclined spraying tendency of the spinning jets increased with the increasing number of needles. This End-effect phenomenon is attributed to the unbalanced sum of Coulombic force on the needles far away from the central needle, and the balanced Coulombic force on the central needle.

As the End-effect phenomenon shown in Fig. 5.1 goes severe, some fibers are observed to fly slantwise or even horizontally towards other objects, not depositing directly on the receiver (counter electrode). Therefore, the degree of End-effect has important influence on the uniformity of fiber structure and the corresponding web structure due to the resultant electric field intensity and distribution during massive electrospinning process. The nanofiber productivity is directly related to the severity degree of the End-effect occurring during the massive electrospinning process.

The End-effect in non-capillary electrospinning process is not as serious as that in capillary electrospinning and Lucas group (Pokorny et al. 2013) has provided the Eq. 5.3 to judge where the Taylor cones could be generated on the free surface of the spinning electrode, although, it is more difficult to control due to the fact that less parameters can be altered during this electrospinning process, the spinning electrode is actually an integrate bulk metal, and the polymer solution could only be supplied to the rotating electrode at the same feed rate with the rotation of the spinning electrode.

Therefore, it is quite important to understand the electric field and its distribution during massive electrospinning process, including both multineedle electrospinning and needleless electrospinning, and hence a few researchers (Niu et al. 2012a, b, Li 2012, Ma 2016, Liu et al. 2013, 2017a, b, Zhang 2017, Zhang 2013, Guo et al. 2011a, b, 2012, Li et al. 2011, Zhang and Liu 2012, Yang et al. 2016, Liu et al. 2015, 2017, Liu and Guo 2013, Guo 2012, Yang 2016, Chen 2012, Chen et al. 2014a, b,

Sun 2015, Liu and Chen 2012) started to seek the appropriate methods for simulation and analyzation of the electric field and its distribution during electrospinning, as the applications of finite element analysis software such as Ansys® and Comsol®, were extended to the area of electromagnetics, to finally find the way to modulate and modify the electric field distribution during massive electrospinning process, towards high productivity and high quality of nanofiber products.

Fundamentals of Electrostatics and Electrostatic Field

Electrospinning technology deals with many different sciences or principles such as electrostatics, rheology, hydromechanics, solvent evaporation, surface tension, solution conductivity, and dielectrics, etc. All these principles interact with and influence each other during the electrospinning process, making it a complicated process of electrohydrodynamics (EDH).

Electrostatics is a branch of classic physics dealing with the phenomena and properties of stationary or slow-moving electric charges. A static shock occurs when the surface of a material which is negatively charged with electrons, touches a positively charged material, or vice versa.

Electric charge is the most basic unit for electrostatic field and electrostatics. Atoms are made up of protons inside the nuclei and electrons outside the nuclei, which are charged positively and negatively, respectively. Electric charge is the physical property of matter that causes it to experience a force when placed in an electric field. There are two types of electric charges—positive and negative (commonly carried by protons and electrons, respectively).

The smallest unit that describes electricity is coulomb (C), which is the electric quantity that the 1A current transports in 1s time, and $1C = 1\text{A} \times 1\text{s} = 1\text{A}\bullet\text{s}$. The charge e, carried by one electron, is called the elementary charge, and is equal to $1.602 \times 10^{-19}\ C$. A proton has the same amount of positive charge as its corresponding nuclear electron, so if an object lacks electrons, it has a positive charge. If it has excessive electrons, it will be negatively charged.

The physical field that transmits the interaction between charge and charge is called an ***electric field***. The electric field force can be either the repulsion between the same polar charges or the attraction between different polar charges (like charges repel and unlike attract). ***Coulomb's law*** is used to describe the force (F) between the two charges, which is given in Eq. 5.5, where q is the charge quantity; ε_p is the absolute capacitance of the space between the two charges; d is the distance between the two charges.

$$F = \frac{q_1 q_2}{4\pi\varepsilon_p d^2} \tag{5.5}$$

Coulomb's law applies only to point charges, but in most cases, an electric field is defined as an area where an electric charge is generated by other charges around it. The magnitude of the field (F) is given by the electric field intensity as shown in

Eq. 5.6, where F, q, and E represent the ***electric field force***, electric charge, and ***electric field intensity***, respectively.

$$F = qE \tag{5.6}$$

For a positive charge, the electric field force is the same as the electric field; for a negative charge, the direction of the electric field force is opposite to that of the electric field.

The ***electric field intensity*** at a charge distance of d is given by Eq. 5.7, where ε is the dielectric constant of the material.

$$E = \frac{q}{4\pi\varepsilon d^2} \tag{5.7}$$

As known in the art, the work done by moving one charge from the reference point to the other is exactly the potential of that point. Therefore, the potential can be expressed by Eq. 5.8, where ϕ is the potential, W is the work done, and Q is the electric/charge quantity.

$$\phi = dW/dQ \tag{5.8}$$

The voltage between two points in space is equal to the potential difference between these two points, which is expressed by Eq. 5.9, where U is the potential difference, and ϕ stands for the potential.

$$U_{1,2} = \phi_2 - \phi_1 \tag{5.9}$$

The scalar potential of the electrostatic field is called the ***(electric) potential***, or electrostatic potential. In electric field, the ratio of the electric potential energy of the point charge to its carried electric quantity is referred to as the potential of this point, which is usually called ϕ. The electric potential is the physical quantity of an electric field from an energy point of view, whereas electric field intensity is a description of an electric field from a force's point of view. An electric potential difference can produce an electric current in a closed circuit (an insulator, such as air, can also become a conductor when the potential difference is considerable).

Coulomb's law can easily describe the interaction between two point-charges, but in a real electric field, there is often a large amount of charge. In this case, it is necessary to use the electric field and potential to further describe the interaction between the charges.

The magnitude and direction of the ***electric field force*** on the charge in the ***electric field*** $\vec{\mathbf{E}}$, in units of newtons per coulomb or volts per meter, can be expressed by the ***electric field vector***, but the ***electric field line*** is more commonly used than the electric field vector due to its visualization feature. As shown in Fig. 5.4, the electric field line arrows indicate the direction of the electric field, while the tangent direction at a certain point in the electric field line shows the direction of the electric field at that point, which starts from the positive charge and ends in the negative charge. The ***equipotential surface*** can also represent the electric field, which is the surface formed by all points having the same potential in the field, with the electric field lines at all points perpendicular to the equipotential surface.

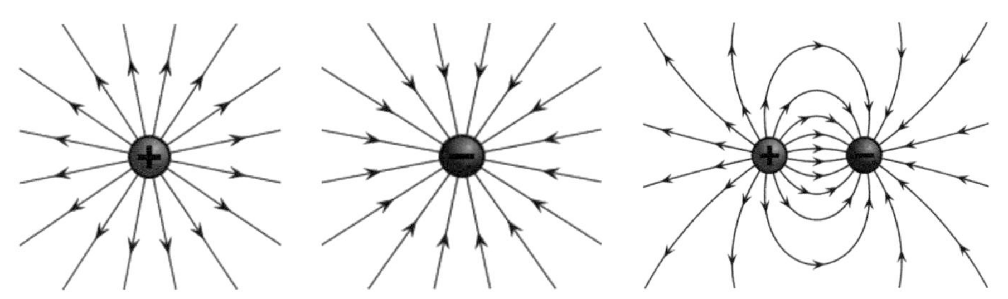

Fig. 5.4: Electric field induced by a positive electric charge (left) and a field induced by a negative electric charge (middle), as well as the electric field lines between a positive charge and a negative charge (right) (https://en.wikipedia.org/wiki/Electric_charge, as retrieved on April 11, 2018).

In the case of electrospinning, when the electrospun nanofibers are deposited on the receiving device, there will be a lot of residual charge on the nanofiber formed web, which may have a negative effect on the electrospinning process, especially when a large amount of electric charge is collected on the receiving device. The polymers used in electrospinning are mostly good insulating materials, and hence the charge on the fibers is difficult to disperse.

The ***charge density*** on the surface of the nanofiber web (that is, the amount of charge per unit area) can be expressed as Eq. 5.10, where σ_e, Q_c, and A represent charge density, total charge, and total surface area, respectively. Therefore, the total charge on an arbitrary surface can be given in Eq. 5.11.

$$\sigma_e = \frac{dQ_c}{dA} \tag{5.10}$$

$$Q_c = \int_A \sigma_e dA \tag{5.11}$$

For a certain kind of insulating material, the surface potential usually changes with different points on the material surface, and therefore it is impossible to decide the voltage on the bulk material of the insulating polymer due to the unknown point potential inside the insulating polymer material. The charges will establish an electric field on the surface of the insulating material, when it is introduced into the insulating material, just like what it will do inside the insulating material. The charges repel each other because they carry like charges, resulting in the movement of the charges, and achieving a balance where the total force on any charge is equal to nought. Consequently, the electric field intensity inside the insulator is zero due to the charge moving from inside the insulator towards the surface, and hence the charges on the surface of the polymer material will adjust its direction till its electric field is perpendicular to the surface of the material. Therefore, it is easy for charge to be collected on the point that protrudes from the material surface, resulting in stronger electric field intensity on this point.

Dielectrics vs Polarization in Electrostatic Field

Dielectrics. The so-called dielectric is an insulator, and the number of its internal free electrons is so small that it can be ignored. The charged particles in the dielectric are bound by the internal forces of the atoms, the internal forces of the molecules, or the intermolecular forces, which are called bound charges. If there is an electric field E, this charged particle can make small movement within the range of the molecule.

Every material has a dielectric constant ε, which is the ratio of the field without the dielectric (E_0) to the net field (E) with the dielectric: $\varepsilon = E_0/E$, where E is always less than or equal to E_0, so the dielectric constant is greater than or equal to 1. The larger the dielectric constant, the more charge can be stored in the parallel-plate capacitor. If a metal was used for the dielectric instead of an insulator, then the field inside the metal would be zero, corresponding to an infinite dielectric constant. The entire space between the capacitor plates are usually filled with the dielectric. However, if a metal did that it would short out the capacitor—that's why insulators are used instead of metals.

The electric charge on a capacitor plate or the charge transferred in static electricity are usually classified as "free charge". In contrast, "bound charge" means the charge of the polarized dielectric. Since all materials are polarizable to some extent, when such dielectrics are placed in an external electric field, the electrons remain bound to their respective atoms, but shift a microscopic distance in response to the field, resulting in the consequence that they're more on one side of the atom than the other. All these microscopic displacements add up to give a macroscopic net charge distribution, which constitutes the "bound charge".

Dielectric Polarization. The dielectric in an applied electric field is different from that of a conductor. The charge is bound because the internal force between the electron and the nucleus in the dielectric is quite large. The applied electric field will cause the molecules or atoms of the dielectric to form an electric dipole, which will show positive and negative bound charges on the surface of the dielectric, and this phenomenon is known as polarization, as shown in Fig. 5.5. After polarization, the electric field established by the bound charge in the dielectric is generally reduced by the applied electric field, and the surface of the dielectric is generally not equipotential (Yang et al. 2009).

The dielectrics are generally in the forms of gas, liquid, and solid, with its molecular structure being either polar or non-polar. Relative displacement occurs to the center where the positive and negative charges play a role in the effect of the

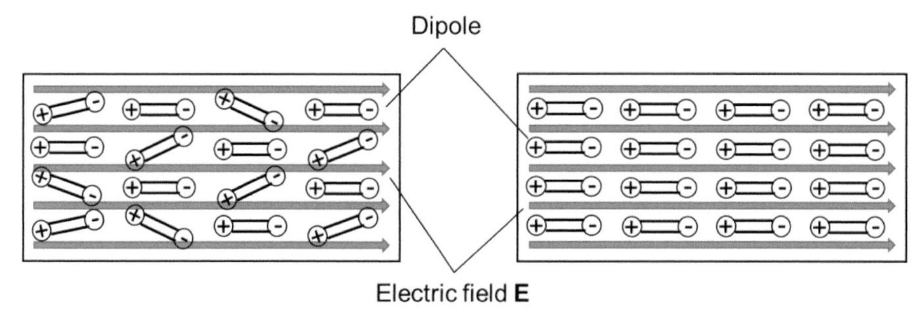

Fig. 5.5: Polarization of electric dipoles in electric field: the black "dog-bones" with plus and minus signs at the two edges indicate the electric dipoles; the orange arrows show the directions of the electric field E.

externally applied electric field, causing the directional change in the electric dipole moment of polar molecules, and then the sum of their equivalent dipole moment vectors is no longer zero, i.e., polarization occurs.

The degree of polarization of the dielectric can be expressed by the vector of polarization intensity P, which is defined as the electric dipole moment per unit volume formed after polarization, and expressed by Eq. 5.12:

$$P = \lim_{\Delta V \to 0} \frac{\Sigma P}{\Delta V} \tag{5.12}$$

In addition, the polarization intensity P is related to the electric field intensity E, and E is the actual electric field intensity in the dielectric, which is the sum of the electric field intensity caused by the applied electric field and the electric field established by bound charges.

The relationship between P and E for most dielectrics does not change with the direction of E, and the direction of P and E is the same. Such a dielectric is called isotropic dielectric, and linear dielectric if P is proportional to E. The relationship between P and E may be expressed by Eq. 5.13, where χ is the relative polarizability of the dielectric.

$$P = \chi \varepsilon_0 E \tag{5.13}$$

The effect of the dielectric on the electric field can be considered by the additional effect generated by the bound charge after the dielectric is polarized. As a result, the electric fields with electric dielectrics can be regarded to be jointly generated in a vacuum space by the free charge and the bound charge due to polarization.

The Gauss Flux Theorem. In physics, Gauss's law, also known as Gauss's flux theorem, is a law associating the distribution of electric charge with the resulting electric field. Gauss's law is one of Maxwell's four Eqs., forming the basis of classical electrodynamics, which can be used to derive Coulomb's law, and vice versa (Halliday and Resnick 1970). Gauss's law is essentially equivalent to the inverse-square Coulomb's law.

Gauss' law states that "the total electric flux through any closed surface in free space of any shape drawn in an electric field is proportional to the total electric charge enclosed by the surface." Gauss's law also says that "The net electric flux through any hypothetical closed surface is equal to $1/\varepsilon$ times the net electric charge within that closed surface (Giancoli 2008)."

The law can be mathematically expressed with vector calculus in integral form and differential form. Both are equivalent since they are associated by the divergence theorem, also referred to as Gauss's theorem. Each of these forms in turn can also be expressed by two ways: in terms of a relation between the electric field E and the net electric charge, or in terms of the electric displacement field D and the free electric charge Q (Grant and Phillips 2008).

(1) Integral form. Gauss's law can be stated using either the electric field E or the electric displacement field D, it may be expressed as Eq. 5.14:

$$\Phi_E = \frac{Q}{\varepsilon_0} \tag{5.14}$$

where Φ_E is the electric flux through a closed surface S enclosing any volume V, Q is the total charge enclosed within V, and ε_0 is the electric constant. The electric flux Φ_E is defined as a surface integral of the electric field, as expressed by Eq. 5.15:

$$\Phi_E = \oiint_S \boldsymbol{E} \cdot d\boldsymbol{A} \tag{5.15}$$

where E is the electric field, dA is a vector representing an infinitesimal element of area of the surface, and · represents the dot product of two vectors.

Although microscopically all charge is fundamentally the same, there are often practical reasons for needing to treat bound charge differently from free charge. The result is that the more fundamental Gauss's law, in terms of E (the electric field), is sometimes put into the equivalent form below, which is in terms of D and the free charge only, and expressed as Eq. 5.16:

$$\Phi_D = Q_{free} \tag{5.16}$$

where Φ_D is the D-field flux through a surface S which encloses a volume V, and Q_{free} is the free charge contained in V. The flux Φ_D is defined analogously to the flux Φ_E of the electric field E through S, as expressed by Eq. 5.17:

$$\Phi_D = \oiint_S \boldsymbol{D} \cdot d\boldsymbol{A} \tag{5.17}$$

It indicates that, in the case of electrostatic field, the flux of the electric displacement D from an arbitrary closed surface S is only related to the free charge surrounding the closed surface, having nothing to do with the polarization charge, i.e., the bound charge, no matter whether the dielectric distribution is uniform or not. However, it cannot be inferred that the distribution of D is independent of the distribution of the dielectric. In general, the distribution of D will be changed if the distribution of the dielectric is changed in the electrostatic field, even if Q, the total amount of free charge, is unchanged.

If the electric field is known everywhere, it is possible to use Gauss's law to obtain the distribution of electric charge: the charge in any given region can be calculated by integrating the electric field to obtain the flux.

The reverse problem, i.e., when the electric charge distribution is known and the electric field needs to be known, however, is much more difficult, as the net flux through a given surface gives little information regarding the electric field and can go in and out of the surface in arbitrarily complicated patterns.

An exception is that once there is some symmetry in the problem, which allows the electric field to pass through the surface in a uniform way, and then the field itself can be deduced at every point if the net flux is known. As known in the art, the geometrical structure of all the solid materials employed in electrospinning process including the needles, spinnerets, electrodes, receivers, etc., are symmetric cylindrically, planarly, or spherically. Therefore, electric field intensity can be calculated using finite element analysis software based on the Gauss's law.

(2) Differential form. The Divergence Theorem allows Gauss's Law to be written in differential form, as expressed by Eq. 5.18:

$$\nabla \cdot \boldsymbol{E} = \frac{\rho}{\varepsilon_0} \tag{5.18}$$

where $\nabla \cdot \boldsymbol{E}$ is the divergence of the electric field (divergence is a vector operator that produces a scalar field), ε_0 is the dielectric constant, and ρ is the total electric charge density (charge quantity per unit volume).

Another differential form of Gauss's law in terms of $\boldsymbol{D}$ and free electric charge density is expressed as Eq. 5.19:

$$\nabla \cdot \boldsymbol{D} = \rho_{free} \tag{5.19}$$

where $\nabla \cdot \boldsymbol{D}$ is the divergence of the electric displacement field $\boldsymbol{D}$, and ρ_{free} is the free electric charge density.

Gauss's Law vs Coulomb's Law. Gauss's law can be proven from Coulomb's law based on the superposition principle, which states that the resulting field is the vector sum of fields generated by each particle (or the integral, if the charges are distributed smoothly in space). Coulomb's law only applies to stationary charges, whereas Gauss's law does hold for moving charges, and in this respect Gauss's law is more general than Coulomb's law. Coulomb's law can be proven from Gauss's law, assuming that the electric field from a point charge is spherically symmetric (this assumption, like Coulomb's law itself, is exactly true if the charge is stationary, and approximately true if the charge is in motion).

Poisson and Laplace Equations. The definition of electrostatic potential, combined with the differential form of Gauss's law (above), provides a relationship between the potential Φ and the charge density ρ, which is expressed by Eq. 5.20:

$$\nabla^2 \Phi = -\frac{\rho}{\varepsilon_0} \tag{5.20}$$

This equation is a form of Poisson's equation. In the absence of unpaired electric charge, the equation becomes Laplace's equation, which is expressed by Eq. 5.21:

$$\nabla^2 \Phi = 0 \tag{5.21}$$

Due to the irrotational feature of electric field, it is possible to express the electric field as the gradient of a scalar function, Φ, called the electrostatic potential (also known as the voltage). An electric field, E, points from regions of high electric potential to regions of low electric potential, expressed mathematically as Eq. 5.22:

$$\vec{E} = -\vec{\nabla} \Phi \tag{5.22}$$

The gradient theorem can be used to establish that the electrostatic potential is the amount of work per unit charge required to move a charge from point a to point b with the following line integral, Eq. 5.23:

$$-\int_a^b \vec{E} \cdot d\vec{l} = \text{Ø}(\vec{b}) - \text{Ø}(\vec{a}) \tag{5.23}$$

It is observed from the equations listed above that the electric potential is constant in any region for which the electric field vanishes (such as occurs inside a conducting object).

Constitutive Equations of Electric Field

Maxwell's equations construct the basis of the macroscopic electromagnetic field and the basis of the finite element analysis of high voltage electric field in electrostatic spinning system (Grant et al. 2008, Zhang 2008, Bennet 1974). The involved differential equations are given in Eqs. 5.24~5.27, where the $\boldsymbol{H}$, $\boldsymbol{B}$, $\boldsymbol{E}$, and $\boldsymbol{D}$ are magnetic field intensity vector, magnetic flux density vector, electric field intensity vector, and electric displacement vector, respectively, and $\boldsymbol{J}$, ρ, and ∇ are current density vector, free charge density, and Hamilton operator, respectively.

$$\nabla \times \boldsymbol{H} = \boldsymbol{J} + \frac{\partial \boldsymbol{D}}{\partial t} \tag{5.24}$$

$$\nabla \times \boldsymbol{E} = -\frac{\partial \boldsymbol{B}}{\partial t} \tag{5.25}$$

$$\nabla \cdot \boldsymbol{D} = \rho \tag{5.26}$$

$$\nabla \cdot \boldsymbol{B} = \boldsymbol{D} \tag{5.27}$$

The constitutive equations to be used in investigation of electromagnetic field in the isotropic medium are listed in Eqs. 5.28–5.30, respectively, where ε, μ, and σ are dielectric constant, permeability, and conductivity, respectively, which are constants in the linear, uniform, and isotropic medium.

$$\boldsymbol{D} = \varepsilon \boldsymbol{E} \tag{5.28}$$

$$\boldsymbol{B} = \mu \boldsymbol{H} \tag{5.29}$$

$$\boldsymbol{J} = \sigma \boldsymbol{E} \tag{5.30}$$

In actual electrospinning process, only a very small amount of dynamic current exists in the entire circuit, therefore, the effect of the magnetic field on the spinning process is negligible, and only the influence from electric field will be considered during the electrospinning process, therefore, the differential equations discussed previously become a set of equations (Eqs. 5.31 and 5.32), and then, the ***constitutive equation for electrostatic field*** becomes Eq. 5.33:

$$\nabla \times \boldsymbol{E} = 0 \tag{5.31}$$

$$\nabla \cdot \boldsymbol{D} = \rho \tag{5.32}$$

$$\boldsymbol{D} = \varepsilon \boldsymbol{E} \tag{5.33}$$

Field Superposition Theory

In physics and systems theory, the superposition principle, also known as superposition property, states that, for all linear systems, the net response at a given place and time caused by multiple stimuli is the sum of the responses which would have been caused by each stimulus individually. If the system is additive and homogeneous, the superposition principle can be applied. If a homogeneous system $F(ax) = aF(x)$, and an additive system satisfies $F(x_1 + x_2) = F(x_1) + F(x_2)$, where a is a scalar, then a system that simultaneously has the property of homogeneity and additivity satisfies $F(a_1x_1 + a_2x_2) = a_1F(x_1) + a_2F(x_2)$, where a_1 and a_2 are the scalars. Therefore, the electrostatic field to be discussed in the current chapter, which is of the property of homogeneity and additivity at the same time, abides by the operation principle addressed above. This principle has many applications in physics and engineering because many physical systems can be modeled as linear systems (Wikipedia 2018b). Note that when vectors or vector fields are involved, a superposition is interpreted as a vector sum.

In physics, Maxwell's equations imply that the distributions of charges and currents are related to the electric and magnetic fields by a linear transformation. Thus, the superposition principle can be used to simplify the computation of fields which arise from given charge and current distribution.

In the case of electric fields, they satisfy the superposition principle, because Maxwell's equations are linear. As a result, if $\mathbf{E}_1$ and $\mathbf{E}_2$ are the electric fields generated from the distribution of charges ρ_1 and ρ_2, a distribution of charges $(\rho_1 + \rho_2)$ will create an electric field $(\mathbf{E}_1 + \mathbf{E}_2)$. If charges $q_1, q_2, \ldots, q_n$ are stationary in space at $r_1, r_2, \ldots, r_n$, without the existence of currents, the superposition principle states that the resulting field is the sum of fields generated by each point charge, as described by Coulomb's law, Eq. 5.34:

$$\mathbf{E}(r) = \Sigma_{i=1}^{N} \mathbf{E}_i(r) = \frac{1}{4\pi\varepsilon_0} \Sigma_{i=1}^{N} q_i \frac{r - r_i}{\left|r - r_i\right|^3} \tag{5.34}$$

Therefore, the electric Field Superposition Theory states that the total electric field intensity vector at an arbitrary point in an electrostatic field equals to the vector sum of the electric field intensities generated by all the individual point charges existing independently (Grant et al. 2008, Bennet 1974). This theory could also be applied to the case of electrospinning, because any charged system could be viewed as a set of numerous point charges.

In the current chapter, finite element analysis method is used to simulate and understand the distribution of the vector sum of electric field intensity (referred to as **field intensity** for short in the following sections of this chapter), and hence the distribution of electric field force across the polymer jets during massive electrospinning.

Principles of Electric Field Simulation in Electrospinning using FEA

Finite Element Method (FEM)

The finite element method (FEM), which was proposed by R. W. Clough in 1960, is a numerical method for solving problems of engineering and mathematical physics. The FEM formulation of the problem results in a system of algebraic equations. In order to solve the problem, the FEM yields approximate values of the unknowns at discrete number of points over the domain (Logan and Chaudhry 2011), by subdividing a large problem into smaller, simpler parts that are called *finite elements*. The simple equations that model these finite elements are then assembled into a larger system of equations that models the entire problem. FEM then uses variational methods from the calculus of variations to approximate a solution through minimizing an associated error function (i.e., least squares approach), which typically involves the following steps:

(1) Dividing the domain of the problem into a collection of subdomains, with each subdomain represented by a set of element equations to the original problem: the element equations are very simple so that they can locally approximate the original complex equations to be studied, where the original equations are often partial differential equations (PDE). These sets of element equations (algebraic equations for steady state problems or ordinary differential equations for transient problems) are linear if the underlying PDE is linear, and vice versa.
(2) Systematically recombining all sets of element equations into a global system of equations for the final calculation. The global system of equations has known solution techniques and can be inferred from the initial values of the original problem to attain a numerical answer (Wikipedia): a global system of equations is generated from the element equations by a transformation of coordinates from the subdomains' local nodes to the domain's global nodes. This spatial transformation includes appropriate orientation adjustments as applied associated to the reference coordinate system. The process is often performed by FEM software with coordinate data generated from the subdomains. FEM is well known by its practical application, i.e., finite element analysis (FEA).

Finite Element Analysis (FEA)

FEA is a computational tool for performing engineering analysis and a good choice for analyzing problems over complicated domains. It includes the use of mesh generation techniques for dividing a complex problem into small elements, as well as the use of software program coded with FEM algorithm. In applying FEA, the complex problem is usually a physical system with the underlying physical theorem or equations expressed in either PDE or integral equations, with the divided small elements of the complex problem representing different areas in the physical system.

FEA has been widely used to simulate the electric field distribution in SNE with various configurations (Kong et al. 2007, Pan et al. 2008, Carnell et al. 2009, Yang et al. 2008, Yang et al. 2005, Deitzel et al. 2001). Historical studies indicate that electric field distribution can influence the cone formation, jet path, as well as the morphology and the size of resultant fibers (Kong et al. 2007, Pan et al. 2008, Yang et al. 2008, Yang et al. 2005, Deitzel et al. 2001). However, it is rare to see that FEA is used for analysis of multineedle electrospinning or needleless electrospinning (Wang et al. 2012). The authors of this chapter started their research on filed analysis based on FEA software such as ANSYS®, COMSOL® in 2009, and various needle and needleless electrospinning methods have been investigated or developed via systematic analyzation on field distribution during various electrospinning process. Methods for modification of field magnitude and distribution have been established to guide the design and manufacture of electrospinning equipment, which are described and discussed in detail in the current chapter.

Principles of Electric Field Simulation using FEA Softwares

The simulation and analysis of the high-voltage electric field formed in the process of electrospinning with the finite element simulation software deal with the approximation of potential function, the energy function of the element, and the calculation of electric field intensity (Ma and Wang 2009).

(1) The approximation of the potential function

Assuming that the potential function φ in each small element e is a linear function regarding r and z, i.e., the electric field is approximately uniform in each of the small element domains. This way, the potential at each point in the arbitrary element should satisfy the interpolated potential function given in Eq. 5.35:

$$\varphi = a_1 + a_2 r + a_3 z \tag{5.35}$$

Set the three nodes as i, j, and m, respectively (arranged counterclockwise) for an arbitrary element, starting with the Point i, then Eqs. 5.36 and 5.37 are true:

$$\begin{cases} \varphi_i = a_1 + a_2 r_i + a_3 z_i \\ \varphi_j = a_1 + a_2 r_j + a_3 z_j \\ \varphi_m = a_1 + a_2 r_m + a_3 z_m \end{cases} \tag{5.36}$$

$$\begin{cases} a_1 = \dfrac{1}{2S_e}(a_i\varphi_i + a_j\varphi_j + a_m\varphi_m) \\ a_2 = \dfrac{1}{2S_e}(b_i\varphi_i + b_j\varphi_j + b_m\varphi_m) \\ a_3 = \dfrac{1}{2S_e}(c_i\varphi_i + c_j\varphi_j + c_m\varphi_m) \end{cases} \tag{5.37}$$

where,

$$\begin{cases} a_i = r_i z_m - r_m z_j \\ a_j = r_m z_i - r_i z_m \\ a_m = r_i z_j - r_j z_i \end{cases} \quad \begin{cases} b_i = z_j - z_m \\ b_j = z_m - z_j \\ b_m = z_i - z_j \end{cases} \quad \begin{cases} c_i = r_m - r_j \\ c_j = r_i - r_m \\ c_m = r_j - r_i \end{cases} \tag{5.38}$$

and S_e is the area of the Element e, and Eq. 5.39 is also true:

$$S_e = \frac{1}{2}\begin{vmatrix} 1 & r_i & z_i \\ 1 & r_j & z_j \\ 1 & r_m & r_m \end{vmatrix} = \frac{1}{2}(b_i c_i - b_j c_j) \tag{5.39}$$

Then, the interpolation function of Element e is given in Eq. 5.40,

$$\varphi(x,y) = \frac{1}{2S_e}\left[\left(a_i + b_i r + c_i z\right)\varphi + (a_j + b_j r + c_j z)\varphi_j + (a_m + b_m r + c_m z)\varphi_m\right] \tag{5.40}$$

(2) The energy function of an element. This function is given in Eq. 5.41:

$$W_e = \iint \frac{\xi_e}{2}\left[\left(\frac{\partial\varphi^3}{\partial r} + \frac{\partial\varphi^2}{\partial z}\right)\cdot 2\pi d_r d_z\right] \tag{5.41}$$

Then, Eq. 5.25 is obtained based on Eq. 5.42:

$$\begin{cases} \dfrac{\partial\varphi}{\partial r} = a_2 \\ \dfrac{\partial\varphi}{\partial z} = a_3 \end{cases} \tag{5.42}$$

Equation 5.42 indicates that $\frac{\partial\varphi}{\partial r}$ and $\frac{\partial\varphi}{\partial z}$ are constants at any point in Element e, which is not related to the coordinate (r, z), and hence the W_e may be simplified as Eqs. 5.43 and 5.44:

$$W_e = \frac{\xi}{2}\cdot 2\pi \frac{\left(\sum b_s\varphi_s\right)^2 + \left(\sum c_s\varphi_s\right)^2}{4S_e^2}\iint r d_r d_z \tag{5.43}$$

$$\iint_s r d_r d_z = \frac{r_i + r_j + r_m}{3} S_e = r_e S_e \tag{5.44}$$

where r is the distance from the center of the Element e to z axis, then W_e is given by Eq. 5.45:

$$W_e = \frac{1}{2}\cdot\frac{2\pi\xi_e r_e}{4S_e}\left[\left(\sum b_x\varphi_s\right)^2 + \left(\sum c_s\varphi_s\right)^2\right] \tag{5.45}$$

(3) Calculation of electric field intensity. The electric field intensity is calculated according to Eq. 5.46, where Eq. 5.47 is also true,

$$\vec{E} = -\nabla\varphi = -\frac{d\varphi}{dr}\cdot\vec{e^r} - \frac{d\varphi}{dz}\cdot\vec{e^r} = E_{re}\vec{e_r} + E_{ze}\vec{e_r} \tag{5.46}$$

$$\begin{cases} E_{re} = -\dfrac{\partial \varphi}{\partial r} = -\dfrac{1}{2S_e}\left(\sum\limits_{s=i,j,m} b_s \varphi_s\right) \\ E_{ze} = -\dfrac{\partial \varphi}{\partial z} = -\dfrac{1}{2S_e}\left(\sum\limits_{s=i,j,m} c_s \varphi_s\right) \end{cases} \tag{5.47}$$

and then the absolute value of the electric field intensity can be calculated according to Eq. 5.48:

$$E = \sqrt{E_{re}^2 + E_{ze}^2} = \sqrt{\frac{\partial \varphi}{\partial r} + \frac{\partial \varphi}{\partial z}} \tag{5.48}$$

Introduction to FEA Software

The FEA based software ANSYS® and COMSOL® can be used for simulation of electric field in electrospinning. The authors of the current chapter have performed systematic simulation on field intensity and its distribution during various type of electrospinning process using ANSYS® and/or COMSOL®. Furthermore, different methods for improvement of field intensity and its distribution towards nanofiber production at increased P/C and uniform nanofiber web structure have been proposed and analyzed again with the finite element software. In order to clearly understand the simulation process and results, a brief introduction is given about the two pieces of software, i.e., ANSYS® and COMSOL® before the finite element simulation on electric field is reported in this chapter.

ANSYS®. ANSYS Inc., USA was established by Dr. John Swanson in 1970 and the finite element analysis software, ANSYS®, was developed since then (Luo and Wang 2002), which allows its analysis ranging from linear area to nonlinear area, and from a single field to coupled multiple fields. It possesses ample element library, material model library, and solution solvers, ensuring highly effective solutions to various problems of structures or fields (Zhao and Gao 2004), enabling the simulation of electric field within 2D plane or 3D space via 2D or 3D modeling under various virtual conditions using its AC/DC module, in terms of Vector Sum of field intensity.

The modular structure in ANSYS® software gives an opportunity for taking only needed features, and it can work integrated with other typical engineering software by adding CAD and FEA connection modules, importing CAD data, and building a geometry with its "preprocessor" component. Similarly, in the same preprocessor, finite element model (*a.k.a.* mesh) required for computation is generated. After defining the parameters of elements and models, the geometry modes are set up, meshing is performed for the established geometry models, physics are solved using the solution of JCG, and then results can be viewed numerically and graphically via diagrams of vector, cloud, animation and chart, and so on.

COMSOL®. COMSOL® Multiphysics (Pryor 2011) is a finite element analysis software package based on PDE (partially differential Eq.). Compared to ANSYS®, COMSOL® facilitates defining customized geometry, meshing, specifying our own physics, solving, and visualizing results. Model set-up using COMSOL® is very fast, thanks to a number of predefined physics interfaces.

COMSOL has embedded CAD modeling tools, build-in physic models, as well as the importing function of third party CAD. It goes through shorter steps during finite element analysis process than ANSYS, which are, preprocessor (defining geometry model; specifying physic properties; meshing), solutions (solving physics), and postprocessor (result visualization).

There is no need for defining or selecting elements, and no command flow is required during finite element analysis, resulting in a faster and more simple simulation process than ANSYS®. Users can select predefined multiphysics in this software package, specify the partial differential equations (PDEs), and link them with other equations or physics. Material properties, source terms, and boundary conditions can all be arbitrary functions of the dependent variables, which can be defined or adjusted conveniently during the simulation process.

The simulation of electrostatic field formed during electrospinning process follows the Poisson's Law (Eq. 5.49), a partial differential Eq. (PDE):

$$-\nabla d\varepsilon_0\varepsilon_r \nabla V = d\rho \tag{5.49}$$

where ε_0 is the relative permittivity of vacuum, ε_r is the relative permittivity of the medium, ρ is the space charge density, and V is the electric potential. The PDE (Eq. 5.32) is the basis for setting each subdomain.

For simulation of electric field in electrospinning, the space charge density ρ is set to 0 C/m^3, and the relative permittivity of the metal (steel) based receiver and spinning electrode is set to 1.5. Consequently, the corresponding values of field intensity and electric displacement vector can be obtained, based on Eqs. 5.50 and 5.51, where E is the electric field intensity; V is the potential energy; D is the electric flux density; ε_0 and ε_r are the dielectric constant of the vacuum and the relative dielectric constant of the medium, respectively.

$$E = -\nabla V \tag{5.50}$$

$$D = \varepsilon_0\varepsilon_r E \tag{5.51}$$

Apart from the numerical values set in subdomain, determining the boundary conditions is another important step towards the simulation of electric field and every boundary should be constrained based on Eq. 5.52, where n is the normal vector of interface, D is the dielectric flux density, and ρ_s is the density of surface charge.

$$-nD = \rho_s \tag{5.52}$$

In the current study, ambient space is confined to a given area in 2D or a given space in 3D during simulation process, and the electrospinning model is placed in an open space. Therefore, zero charge symmetry is set as the boundary condition for the six planes of the air model in electrospinning, aiming to reach the aim of the infinity of surroundings corresponding to Eq. 5.53:

$$-nD = 0 \tag{5.53}$$

When no free charges exist on the interface between two dielectrics, for instance, $\rho_s = 0$, i.e., $n (D_1 - D_2) = 0$, the conditions of boundary between the electrospinning spinneret and atmosphere are continuity, which is expressed as Eq. 5.54:

$$-nD_1 = -nD_2 = 0 \tag{5.54}$$

Assumptions for Simulating Electric Field using FEA

In a real electrospinning process, the charged dynamic electrospinning jets interact with the electric field formed by the high voltage power source, and hence the electric field is changing all the time. Plus, factors such as charge density, gravity, etc., also change from time to time due to the constantly splitting and whipping of the spinning jets, resulting in a very complicated process of electrospinning.

It is impossible to make the model completely consistent with the actual situation when the finite element model is established, therefore, in order to make the model as close to the actual situation as possible, and to make the simulation process proceed smoothly, the following assumptions should be acknowledged when a finite element simulation method is employed, which are:

(1) The dielectric constant used in all models is a constant, which does not change with the change of electric field.
(2) The polymer spinning solution is induced by high pressure electric field to generate spinning jets, ignoring the change of charge in the spinning jet and its effect on the whole electric field.
(3) Ignore the effect on electric field from other components of electrospinning equipment.
(4) The effects from other factors, such as temperature and humidity, on the polymer solution are ignored and the properties of all polymer solutions are considered to be the same.

Analysis Steps for Field Simulation using FEA Software

General Steps. The steps for electric field simulation with FEA software, including ANSYS and COMOSOL® are presented in the following steps, with tiny difference between the two kinds of software (Ma et al. 2009, Dang et al. 2010):

(1) Establish the geometric model and then export it to the FEA software, i.e., ANSYS® or COMSOL®.
(2) Set Multiphysics: set the physical field as electrostatic field, via selecting ES mode in AC/DC module.
(3) Set parameters for the model: define the properties of materials, equipment and boundary conditions.
(4) Create mesh for the established model: create mesh to obtain the mesh model with nodes and elements.
(5) Define the physics: set the infinitely far boundaries and apply loads.
(6) Solve the physics.

(7) Perform post-treatment: visualize the analysis results via Plot Results and Deformed Shape; analyze, treat, and evaluate the results.

It should be remembered that for simulation with ANSYS®, selection and definition of element type are needed between the Steps (2) and (3), and Steps (1) ~ (4) are usually called Preprocessor. Some key steps for FEA simulation using ANSYS are described furthermore, for clearer understanding of the simulation steps of electrostatic field with ANSYS®. Similar operations will be followed in case of COMSOL®.

Key Steps for Electric Field Simulation using ANSYS® Software

The key steps for simulation of electrospinning process using the software ANSYS® are described below, but the key steps for COMSOL® will not be described in detail, which are much simpler than those of ANSYS, with which no definition of elements is needed and no command flow is necessary.

Modeling of Electrospinning Process. The single needle electrospinning is the mostly used process for nanofiber preparation in laboratories worldwide, which is depicted in Fig. 5.6 for the convenience of FEA, mainly including a syringe (spinning electrode), micropump (solution supplier), DC power source, and metal receiver (counter electrode).

(1) Simplify the model in 2D plane

The model for finite element analysis of field intensity during single needle electrospinning is simplified and shown in Fig. 5.7. For a clearer view of the pictures based on the 2D models in this chapter, only the spinning electrode(s) and its (their) field distribution results will be displayed in the relevant figures, while the collector/receiver section will not occur in the figures due to the large magnifications.

(2) Simplify the model in 3D space

The 3D model of electrospinning process for FEA is simplified and shown in Fig. 5.8, mainly including the square electrospinning head (spinneret), the square receiving plate, and the cubic air, as well as needles, if necessary.

Selection of Element Type. Electrostatic field analysis using ANSYS is based on the Poisson's equations, during which the main unknown variable (node degree of freedom) is the scalar electric potential (voltage), and other physical variables can be derived from the node potential.

The ANSYS elements that can be used for electric field analysis include the PLANE121 (2D solid element), SOLID122 or SOLID123 (3D solid element), as well as some special elements such as MATRIX50 (Maxwell force label element), INFIN111 (infinite plane label element), etc.

The electrospinning device involved in this chapter is featured with the steel receiver on the bottom or the top, the steel spinneret located on the top or bottom, and open space between the spinneret and receiver, resulting in an open electric field space established by all the components. Therefore, the infinite plane label element of INFIN111 should be used to create the abstract model of the electrospinning

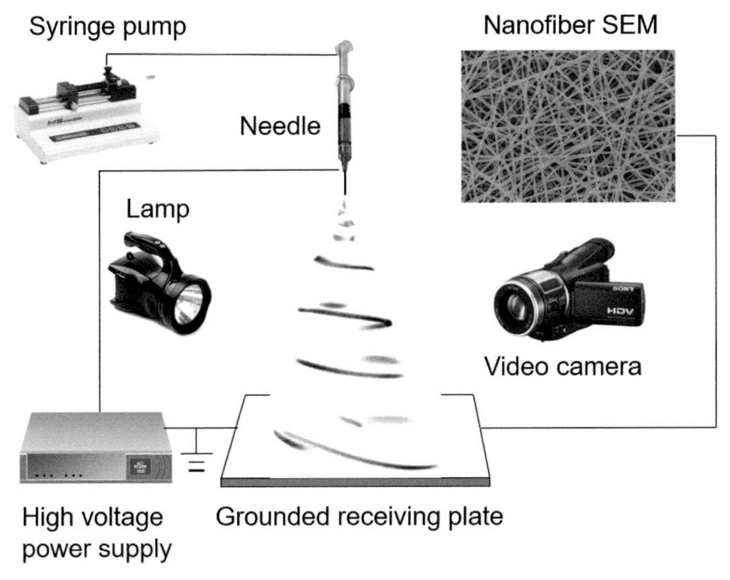

Fig. 5.6: Schematic of single needle electrospinning process.

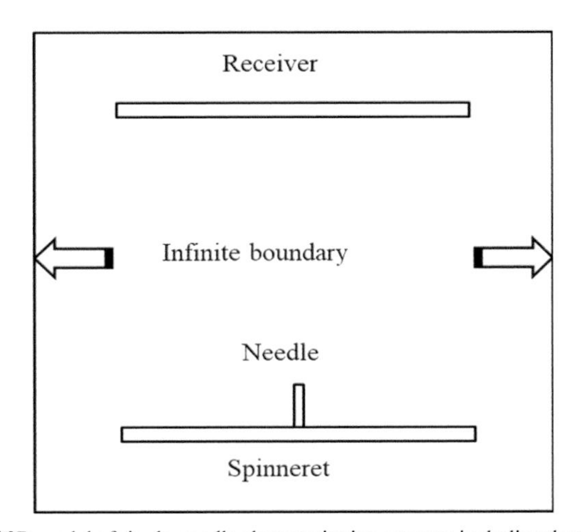

Fig. 5.7: Simplified 2D model of single-needle electrospinning process: including the spinneret, needle(s), receiver, and infinite boundary (air) only.

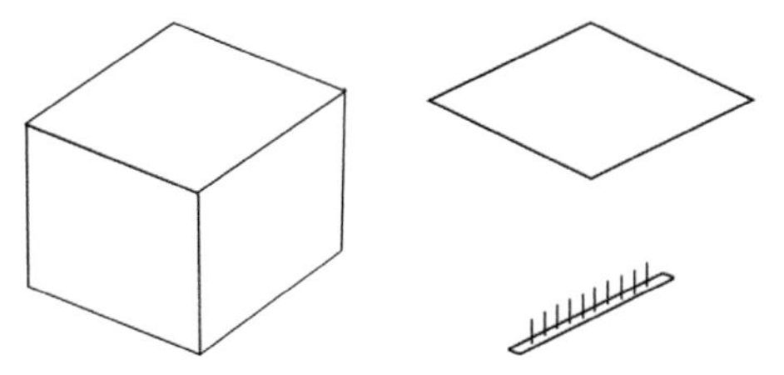

Fig. 5.8: Simplified models of multineedle electrospinning system in 3D space: (a) air; (b) receiver (the upper) and the spinneret (the lower).

device, combined with the 2D plane element of PLANE121, and the 3D element of SOLID122, towards the FEA for electric field with ANSYS®.

Select Elements and Set Parameters for the Established Model. After the electrospinning model is established and simplified, the optimal elements are selected for simulating electric field in electrospinning:

(1) PLANE121 is usually selected for 2D electrostatic field analysis, which is a 2D electrostatic solid element of an 8-node quadrilateral shape, a charge-based electric element, with one degree of freedom, i.e., voltage at each node, as shown in Fig. 5.9. This element is based on the electric scalar potential formulation and has compatible voltage shape, and hence is well suited to model curved boundaries.
(2) SOLID122 is selected for 3D electrostatic field analysis, which is a 3D electrostatic solid element having the shape of a 20-node hexahedron with voltage as its one degree of freedom at each node (Fig. 5.10); SOLID122 element has compatible voltage shapes and are well suited to model curved boundaries, and it can tolerate irregular shapes without much loss of accuracy.
(3) INFIN111 is a 3D Infinite Solid element, a special element of hexahedron shape having 8 or 20 nodes with scalar electric potential as the degree of freedom, as shown in Fig. 5.11, which models an open boundary of a 3D unbounded field problem. A single layer of INFIN111 elements is used to represent an exterior sub-domain of semi-infinite extent, modeling the effect of far-field decay in electrostatic field analysis during the electrospinning process.

The degree of freedom of all the three elements is electric potential (voltage), therefore the corresponding reaction solution is needed, which is Electric Charge F label = CHRG.

Set Properties for Materials and Equipment. The material properties are mainly concerned with the dielectric constants of the materials involved in the electrospinning process, which is a physical quantity related to the dielectric, and is divided into two kinds, i.e., the relative dielectric constant ε and the absolute/vacuum dielectric constant ε_0 (usually $\varepsilon_0 = 1$), where ε is a dimensionless ratio. Generally speaking, the substance with $\varepsilon \neq 1$ is called a dielectric, and usually ε is greater than 1 for

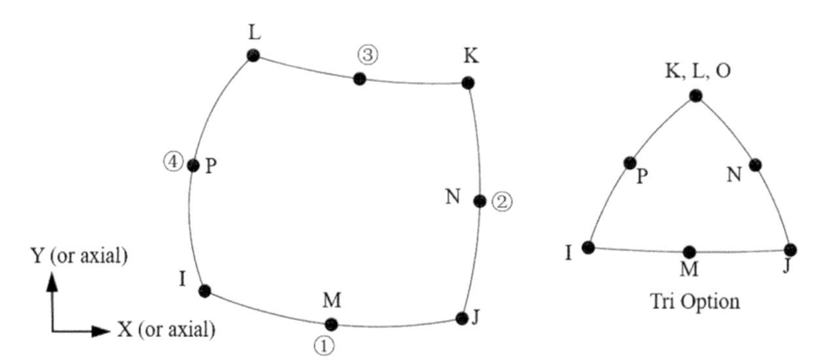

Fig. 5.9: PLANE121 Geometry: the left is the coordinate system, and the middle and right are the geometry and node locations (https://www.sharcnet.ca/Software/Ansys/16.2.3/en-us/help/ans_elem/Hlp_E_PLANE121.html, as retrieved on April 9, 2018).

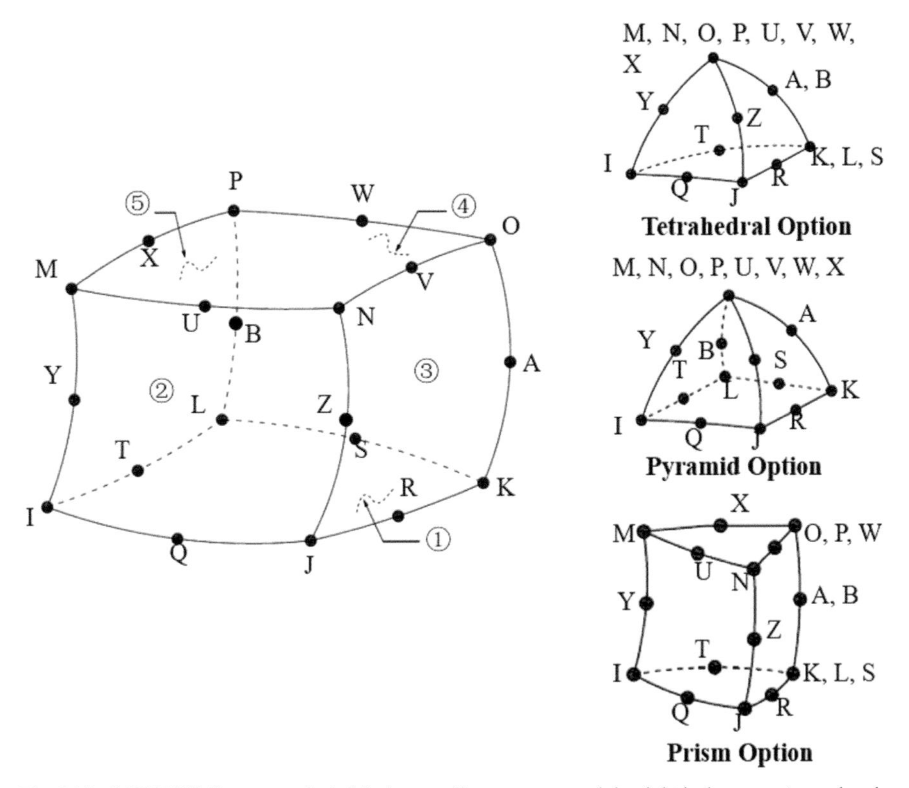

Fig. 5.10: SOLID122 Geometry: the left is the coordinate system, and the right is the geometry and node locations (https://www.sharcnet.ca/Software/Ansys/16.2.3/en-us/help/ans_elem/Hlp_E_SOLID122.html, as retrieved on April 9, 2018).

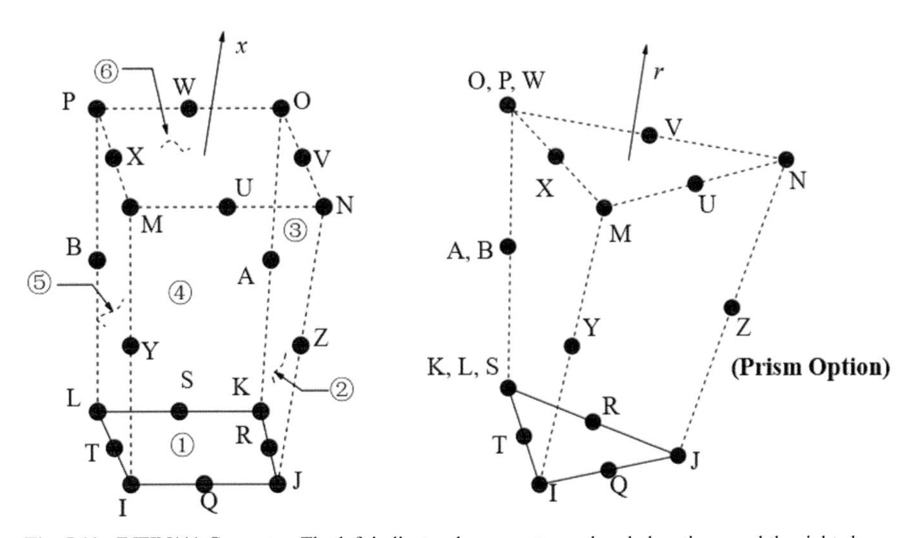

Fig. 5.11: INFIN111 Geometry: The left indicates the geometry and node locations, and the right shows the coordinate system (https://www.sharcnet.ca/Software/Ansys/16.2.3/en-us/help/ans_elem/Hlp_E_INFIN111.html#elem111table2, as retrieved on April 9, 2018).

all substances, so they are all dielectrics. The dielectric constants for spinnerets, electrodes or receivers, and air medium could be set to 2, 2, and 1, respectively (Xue 2007).

Meshing. Triangle element or quadrilateral element can be used for mesh creation in 2D plane, while hexahedron element or tetrahedron element can be used for meshing 3D object. In order to achieve accurate mesh creation, the combined form of free meshing and mapping meshing is adopted for meshing the planar object and spatial object.

Confining Infinity and External Boundaries. In order to thoroughly describe the distribution of electric fields, the boundary of the physical domain needs to be defined as infinity, namely, the infinite surface mark (INF) needs to be loaded on the boundaries of physical domain, which is not a real load, but only a label.

Applying Loads and Setting Bound Conditions. Loads such as voltage (VOLT), charge density (CHRG), and charge density of surface (CHRGS), are the typical loads used in electrostatic field analysis. However, as known in the art, applied voltage is one of the main parameters in the electrospinning process, and the electrospinning is usually carried out at high voltages up to hundreds of thousands of volts, therefore, the typical load during the analysis on electrostatic field should be in terms of voltage (VOLT) combined with the corresponding Infinite Label (INF).

Solutions and Post-treatments. Solve the physics using the solver JCG. Present the results of electric field intensity with vector diagram and/or nephogram, since the electric field intensity during electrospinning process is actually the vector sum of the field intensity from all the component points/dots on the electrospinning equipment.

In the following sections, ANSYS® and/or COMSOL® will be used to simulate the electric field intensity, to understand the magnitude of electric field and its distribution in needle or needleless electrospinning, and hence to propose effective methods for improvement of electric field distribution, distribution of spinning jets, and nanofiber web uniformity during electrospinning processes of different types.

As known in the art, the main factors influencing field intensity and its distribution, and hence the electrospinning process, are classified into three categories, which are: (1) characteristics of polymer, solvent, and their solution, such as concentration, viscosity, surface tension, electric conductivity, relative dielectric constant (relative permittivity), polarity, pH value, etc.; (2) process variables, e.g., solution feed rate, applied voltage, receiving distance, and equipment structure parameters, including both spinneret (spinning electrode) and receiver (receiving electrode); (3) environmental variables including wind rate, temperature, humidity, vacuum and inert gas, and so on. In order to simplify the electrospinning process model towards quick and easy process modeling and simulation for field intensity, only the following parameters will be considered as the key factors for electrospinning process, and they are: applied voltage, receiving distance, geometric structure of spinneret and receiver.

Firstly, the simulation is carried out in 2D plane with the applied voltage (VOLT) as the load. Steel is used as the raw materials for the two electrodes, 12 kV is applied to the spinning electrode (spinning head or spinneret), and the counter electrode (receiver) is set grounded.

The ranges of the parameters involved in FEA simulation of electric field of needle electrospinning process using ANSYS software are displayed in Table 5.1, and other parameters will remain constants when one of the variables is changing within its range, as listed in Table 5.2.

The dielectric constant of metal, steel for instance, is hard to determine because it seems to be infinity according to its definition $\varepsilon = E_0/E$. However, the ε value of a metal, like steel, cannot be calculated from this formula as it is not true anymore when $E = 0$. Actually, the value of ε can be deduced based on other dielectric constant values listed in any physics related references, as long as it is large enough. For example, the dielectric constants of vacuum, dry air, PTFE (Polytetrafluoroethylene), paper, and water are 1.00000, 1.00059, 2.55, 3.6, and 80, respectively, and it is easy to be observed that the dielectric constants increase with the increase in electric conductivity. Therefore, we finally set the dielectric constant of metal (steel) used in the current chapter as 1 or 1.5 for easy and simple calculation of electric field intensity, as we found that the value of dielectric constant of a metal spinneret or receiver did not change the simulation results significantly, and the relative values of simulated field intensity would remain the same even though different values of ε are used in field simulation. This is possibly because as the steel electrodes are not filled between two metal plates in the case of electrospinning, they are actually used as the two parallel plates themselves, resulting in a larger but limited value of ε.

Table 5.1: Range of the variables for simulation on needle electrospinning process.

Parameter	Applied voltage	Needle length	Needle diameter	Receiving distance	Needle spacing	Needle number
Range	4–40 kV	1–3 cm	0.8–2 mm	15–30 cm	0.6–6 cm	1–5

Table 5.2: Model parameters for 2D simulation of electric field in needle electrospinning.

Element type	Needle diameter	Needle length	Relative permittivity		
2D Plane 121	0.1 cm	3 cm	Air (ε_0)	Steel	PTFE
			1.00059	1.0/1.5	2.55
Side length of infinite plane	Spinneret dimension	Applied voltage	Receiver dimension	Receiving distance	Needle spacing
80 cm	6 cm × 1 cm	12 kV	20 cm × 0.6 cm	20 cm	1 cm

Electric Field Distribution in Needle Electrospinning Process

As for needle electrospinning, the influence on field intensity E and distribution CV% from key parameters, including applied voltage, receiving distance, needle length and diameter, needle number (the number of needles), and needle spacing (distance between the neighboring needles) are to be simulated with the FEA software, ANSYS, and/or COMSOL, and the results will be discussed to propose methods for improvement of field intensity and its distribution, and hence to minimize the End-effect in multineedle electrospinning process.

Field Intensity Distribution in Single Needle Electrospinning

The values of 2D field intensity simulations were obtained by FEA software, ANSYS®, using the parameters listed in Tables 5.1–5.2, following the simulation steps addressed previously, as shown in Fig. 5.12.

The left image in Fig. 5.12 is the complete diagram of the electrospinning model, including the needle and receiver; the right diagram only shows the zoom-in of the needle and this zoom-in method will be used for the remaining part of the current chapter. For a close-up picture of the field intensity distribution along and across the whole needle, to facilitate the following analysis on the simulation results.

It is found from Fig. 5.12 that the strongest field intensity (more red color area) occurs on the needle top, especially bilateral pointed edges, while the needle bottom shows weaker field intensity (less red color area), and the needle body displays the weakest field intensity (dark blue color), under the condition of being without the spinneret (spinning plate). This can be explained by the definition and superposition principle of electric field, as addressed before. The needle top shows stranger field intensity than that of the needle bottom, and this can be attributed to the nearer distance between needle top and receiving plate, which results in higher electric field intensity based on the definition of electric field definition and field superposition principle. Both the needle top and needle bottom display stronger field intensity than the needle body, which is because there are bilateral pointed edges (tips) on needle top and needle bottom, respectively, which lead to higher field intensity than the needle body, where no points exist. Pointed edges have a higher charge density than the planar zone, which can be inferred based on the Eq. 5.55 below, which is another expression of electric field intensity E:

$$E = \frac{1}{4\pi\varepsilon_0} \int \frac{\rho(r)\hat{r}}{r^2} dV \tag{5.55}$$

where $\rho(r)$ is electric charge density, the function regarding location, $\hat{r}$ is the unit vector pointed from the source dV to the field point, and r is the distance between the source dV and field point.

It is easy to understand based on Eq. 5.55, Eq. 5.10 and Eq. 5.34, that the increase in electric charge density will cause the proportional increase in electric

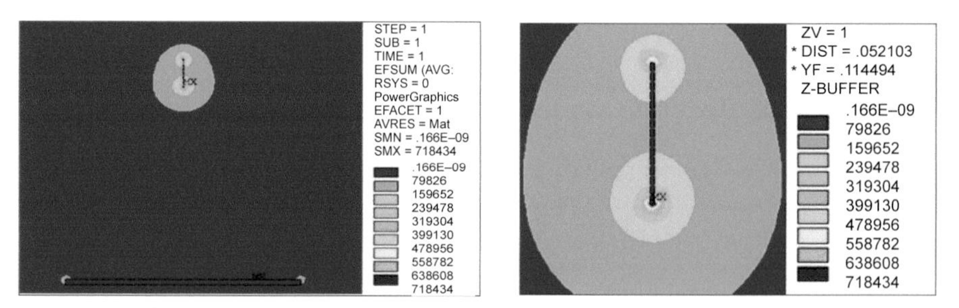

Fig. 5.12: Nephogram of field intensity distribution in single needle electrospinning process: the left shows the entire simulation result, and the right only displays the enlarged part of the needle, with the red region standing for stronger field, and blue regime representing weaker field (Guo 2012).

field intensity, and the needle body has weaker field intensity than needle top and needle bottom, and needle top has higher field intensity than needle bottom.

Field Distribution and End-effect in Multineedle Electrospinning

The multineedle electrospinning process (two to five needles) are to be discussed for field intensity simulation using ANSYS software in a 2D plane. The room-in diagrams of the simulation results are shown in Fig. 5.13, where the peak value of field intensity (peak intensity for short) is defined as the maximal field intensity at the bilateral outside pointed edge(s) of the needle top in the linear needle array, for easy later comparison between the multineedle electrospinning processes with different numbers of needles. These peak values can be obtained by ANSYS software from the simulation results.

The peak values of field intensity are 4.83e5, 4.46e5, 3.89e5, and 3.77e5 for two, three, four, and five needle electrospinning processes, respectively, when other parameters remain unchanged, which decrease with the increasing number of needles. This result also can be observed from the fact that the areas of the red regions on the top of the needles located at two outer ends reduce with the increasing number of needles. This is because the net amount of charges distributed on each needle decreases as the number of needles increases, and then the charge density, and hence the field intensity, decreases thereby.

It is also found through further observation on Figs. 5.12–5.13 that the peak intensity has different values on the top of different needles, although all the needles are linearly aligned in the same row. The nearer to the two ends of the row, the stronger the peak intensity of the needle. The peak intensity occurs at the outer pointed edges of the top of the needles located at two ends of the row. The closer to the middle of the row, the weaker the peak intensity of the needle, and the peak intensity occurs on the single needle or two needles located at the middle of the needle row, depending on whether the number of the needles is odd or even.

This is because the needle(s) in the middle of the needle row have equal Coulombic force from neighboring needles on the left side and right side, due to the symmetric location of the middle needle, and they have the same magnitude but opposite directions, so the effects from the Coulombic force of bilateral needles

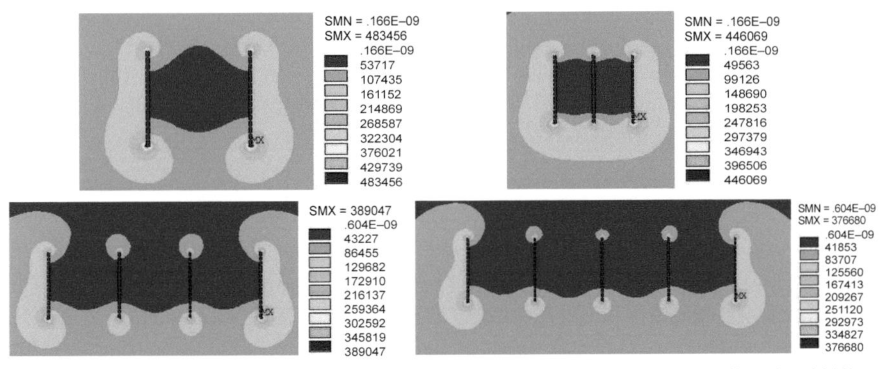

Fig. 5.13: Field intensity in multineedle electrospinning process with two–five needles (Guo 2012).

counteract each other, and no superposition effect of field intensity occurs on the middle needle(s), resulting in an unchanged field intensity on the middle needle(s), and therefore the spinning jet(s) from the middle needle(s) can remain in the direction perpendicular to the receiver with unchanged field intensity.

However, for a needle located near the two ends of the needle row, the asymmetry caused by unequal number of needles on the left side and the right side of the needle, the Coulombic force on the two sides of the needle after superposition is unbalanced, and therefore, the Coulombic force formed by more needles on one side of the needle will push the jets from the needle to traverse towards the other side, resulting in the "End-effect", as shown in Fig. 5.2 (optical image) and Fig. 5.13 (FEA results).

Actually, Fig. 5.13 can be used to explain the principle of End-effect phenomenon displayed in Fig. 5.2. The results from field intensity simulation of multineedle electrospinning indicated that the peak values of field intensity tend to decrease with the increasing number of needles, and needles at different locations show different field intensity values, with the needles located at the two ends of a needle row having the strongest field intensity, which may cause the jets from the needles located at two ends of a needle row to fly slantwise, not directly towards the receiver, resulting in poor nanofiber productivity.

This End-effect phenomenon may get worse if more needles are used in multineedle electrospinning, which may cause the thin region in the middle part of the web where the field intensity of the needles get restrained, and nanofibers to fly outwards at the two ends of the needle row, leading to the uneven web structure, waste of raw materials, and increased production cost.

High P/C of multineedle electrospinning, uniform structure of nanofiber web cannot be achieved if the End-effect issue is not solved. Hereby, parameters governing multineedle electrospinning process have been systematically investigated by the authors, and finally the methods for improving field intensity and its distribution were proposed and demonstrated.

Electric Field Intensity vs Applied Voltage

The correlation between peak value of field intensity and applied voltage is investigated by changing the voltage at 4 kV of variation gradient, and the simulation results are plotted in Fig. 5.14, from which it is observed that, for electrospinning processes with the same number of needles, peak value of field intensity increases with increasing applied voltage and decreases with increasing number of needles when the applied voltage remains the same. Since the required spinning voltage, namely, Vc, is a certain value, therefore, one needs to elevate the applied voltage to reach critical field intensity and to initiate electrospinning.

Electric Field Intensity vs Receiving Distance

Under certain voltage, increasing electrospinning distance (receiving distance) is equivalent to reducing the voltage value. Therefore, the change of receiving distance will directly cause the field intensity to change. Figure 5.15 shows that, for single needle electrospinning, the peak value of field intensity decreases dramatically with

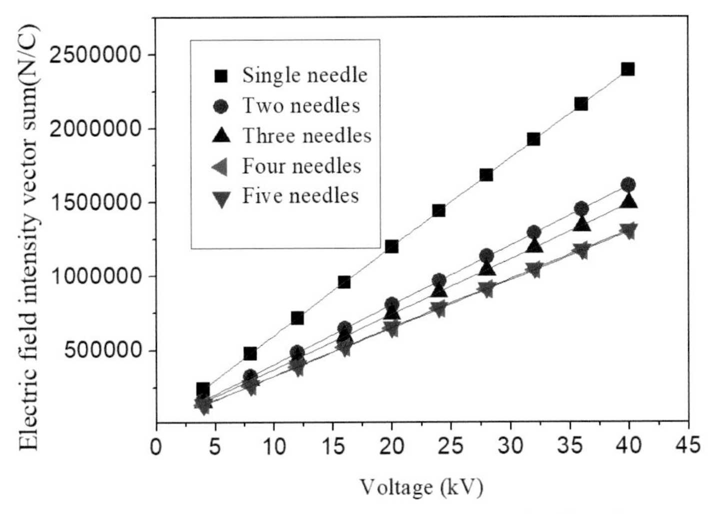

Fig. 5.14: Effect of applied voltage on peak value of field intensity.

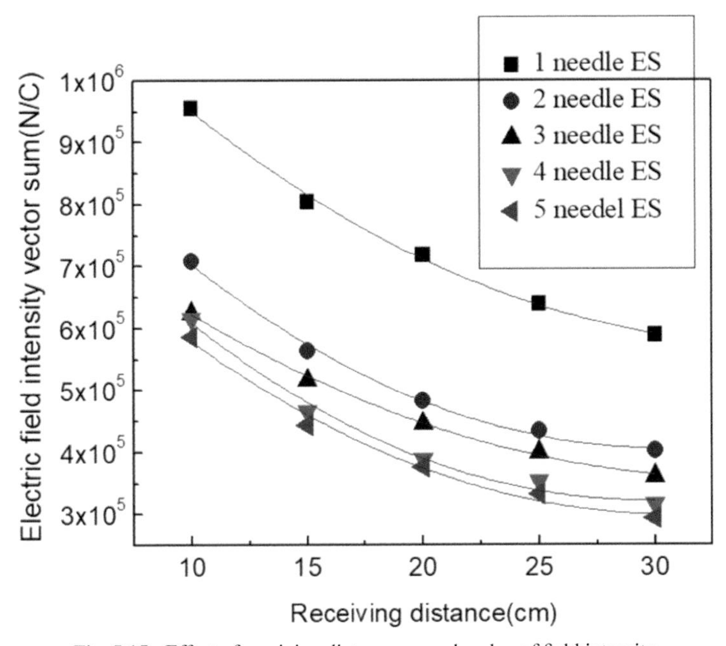

Fig. 5.15: Effect of receiving distance on peak value of field intensity.

the increase of receiving, while during multineedle electrospinning process, the peak value of field intensity decreases slowly with the increasing receiving distance, which is because there existed interference among the electrostatic fields of the multiple needles and the increase in field intensity gets suppressed (single needle has greater field intensity and hence it changes rapidly, while multiple needles have weaker field intensity and hence it changes slowly).

Multineedle electrospinning can be used to produce nanofiber materials at a large scale after appropriate measures are taken to eliminate or reduce the mutual interference between multiple jets. The increase in receiving distance may reduce field intensity, which does not favor the high productivity of nanofibers, but chances for the solvents to vaporize and for whipping instability of the spinning jets are increased, and hence the probability for production of smaller size of nanofibers is increased. Considering that increase in receiving distance usually does not dramatically decrease the field intensity in multineedle electrospinning, greater receiving distance can be accepted to create more chances for solvents to volatilize and for smaller nanofibers to be produced, meanwhile reducing the opportunity for adhesion to occur among fibers.

Electric Field Intensity vs Needle Spacing

It is observed from Fig. 5.16 that the appropriate distance between neighboring needles facilitates the effective improvement on productivity and efficiency of multineedle electrospinning. The increase in needle spacing causes the peak value of field intensity to decrease first and then increase, which is because the Coulombic force decreases initially with the increasing needle spacing, and hence the vector sum of field intensity after superposition decreases. However, when needle spacing is increased to a certain degree, e.g., above 5 cm, the Coulombic force between needles is reduced to a certain degree, the interaction between needles is too weak to be obvious, the needles located at two ends of the needle row are similar to single needles, respectively, their peak values of field intensity are hardly influenced by those from neighboring needles, and hence go up instead. If ANSYS® is continually used to simulate the field intensity at further longer needle spacing, then the field intensity is expected to approach the peak value at single needle state of the same electrospinning conditions.

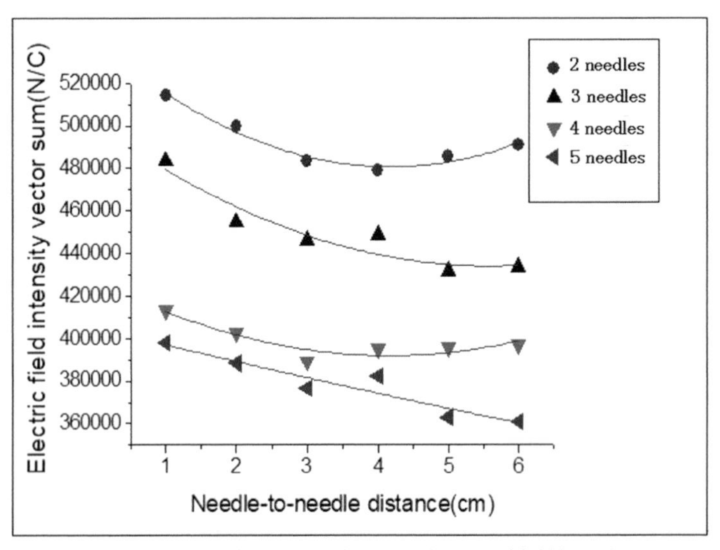

Fig. 5.16: Effect of needle spacing on peak value of field intensity.

Electric Field Intensity vs Needle Diameter

Figure 5.17 shows the results of the effect of needle diameter on the peak value of field intensity during multineedle electrospinning. It is found by observation of Fig. 5.17 that the increase in the needle diameter results in reduced charge density and decreased peak value of field intensity, based on Coulomb's law. Unfortunately, however, the increase in needle diameter and decrease in field intensity tend to form liquid drops rather than continuous spinning jets during multineedle electrospinning at a constant feed rate, due to the reduced hold force from the needle top on the meniscoid liquid drop, or lead to relatively large size of electrospun nanofibers obtained. Meanwhile, increasing the needle diameter at unchanged receiving distance and solution feed rate, favors the fast vaporization of the solvent, and hence the complete consolidation of the nanofibers.

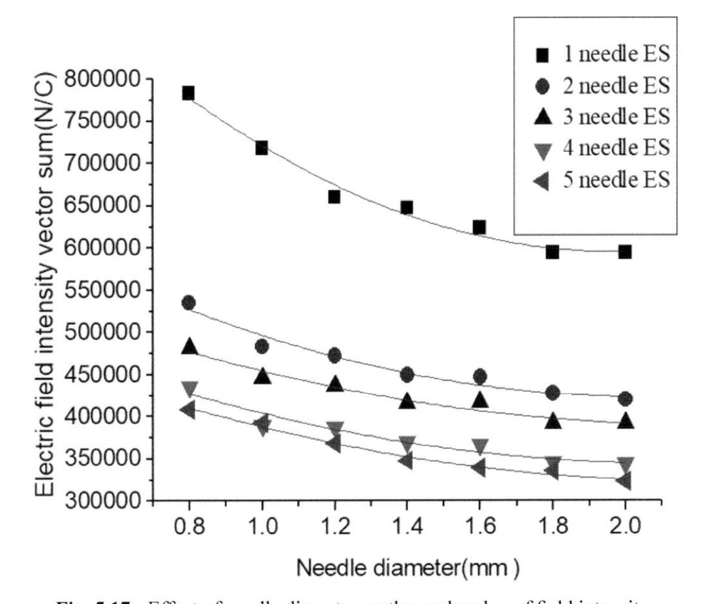

Fig. 5.17: Effect of needle diameter on the peak value of field intensity.

Electric Field Intensity vs Needle Length

The dimension of the metal needles will affect the magnitude and distribution of the field intensity, and hence influence the applied voltage and the diameter of the resultant fibers. The effect of the needle length on the peak value of the field intensity at different number of needles was simulated by ANSYS®, and the results were shown in Fig. 5.18. It is found from the results that the field intensity decreases with the increasing needle length, which can be attributed to the fact that, under the same electrospinning conditions where the applied voltage and receiving distance remain constants, longer needle length indicates larger surface area, and hence less charge density, resulting in less Coulomb's force acting on the top of the needles, and hence smaller field intensity, based on Coulombic law. The applied voltage

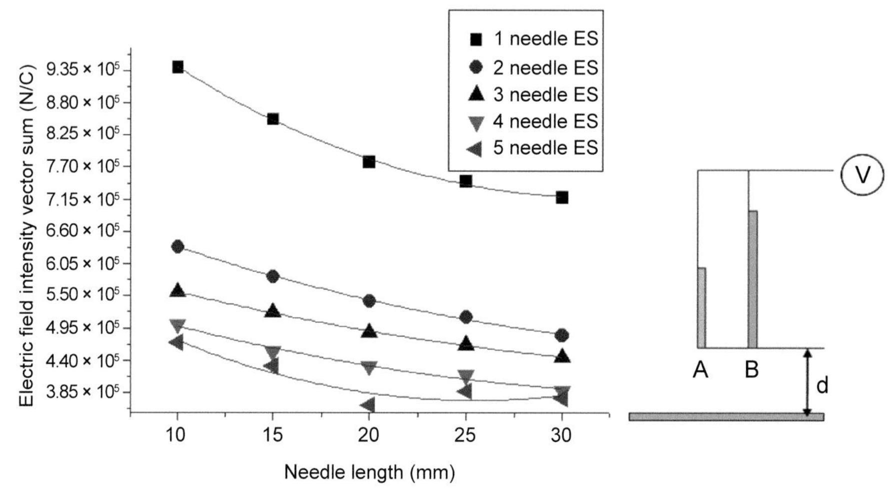

Fig. 5.18: Effect of needle length on peak value of field intensity (the right is the model of two needle electrospinning, indicating the constant receiving distance when needle length is changed: A and B stand for the first and second needles, respectively, V represents the applied voltage, and d is the receiving distance, which can be used in the case that more than two needles are used in electrospinning process).

must be increased to make up for the smaller field intensity caused by the longer needles, resulting in the energy cost thereof. Additionally, it is easy for nozzle blocking phenomenon to occur to longer needles, leading to the increase in nozzle clearing cost and decrease in productivity of nanofiber web, thereby not favoring the continuity of the electrospinning process.

Therefore, needles with as short a length as possible are recommended to be used in multineedle electrospinning process, which can increase electrospinning efficiency and mitigate the problem of needle blocking.

Field Intensity vs the Number of Needles

Figure 5.13 shows the nephogram of field intensity distribution in multineedle electrospinning processes having different number of needles. It can be clearly observed that the red area on the top of the end needles arranged in a row reduces with the increasing number of needles, indicating that more needles result in less field intensity on each needle, if other parameters remain unchanged when the number of needle increase. Figures 5.14–5.18 show the same trends that more needles result in less field intensity on each needle in terms of peak value of field intensity, which is because the charge density on each needle decreases with the increasing number of needles when the same voltage is applied to different number of needles. Smaller charge density leads to less field intensity based on Eq. 5.55.

Electric Field Distribution in Needleless Electrospinning

The processes of needleless electrospinning have many different categories; therefore, it is impossible to simulate the field intensity during needleless electrospinning without specifying the specific type of the needleless electrospinning. Here we take bubble electrospinning as a sample to investigate the field intensity distribution in needleless electrospinning.

During bubble electrospinning, multiple jets can be easily ejected from the bubble surface of the polymer solution. Before bursting, the bubble having spherical shape has been stretched to a cone shape under the action of electrostatic field force. Generally, the spherical shape of the bubble, which is going to change into a cone shape, is called the initial state of the charged bubble. The bubble cone shape is referred to as the critical state, and is going to generate jets when the applied voltage is greater than the critical voltage.

Electric Field Distribution in Single-bubble Electrospinning

In bubble electrospinning, a polymer bubble, which is surrounded by the air, plays a key role in jet generation (Kong and He 2013, Liu and He 2007). In order to understand the deformation of the bubble in electrical field, Reddy and Esmaeeli's model (Reddy and Esmaeeli 2009) can be adopted here and revised for this process.

For simplicity, some assumptions are considered—that the electrical field is uniform in this process and the polymer solution is a perfect dielectric. In the initial state, the bubble can be assumed as spherical and it is symmetrical around the central axis in the rectangular coordinate in two-dimensional condition. Consider a spherical bubble with radius a, surrounded by air of infinite boundary, and exposed to a uniform electrical field, as shown in Fig. 5.19. The electrical field intensity distribution at the interface between the bubble and air can be expressed with the following equations.

$$E_{or} = E_{\infty}\frac{2S}{S+1}\cos\theta, \tag{5.56}$$

$$E_{o\theta} = -E_{\infty}\frac{2}{S+1}\sin\theta, \tag{5.57}$$

$$S = \frac{\varepsilon_i}{\varepsilon_o}. \tag{5.58}$$

where ε_i and ε_0 are the relative permittivities of the bubble and the air, respectively, θ is the polar angle, and r is the radial distance from the origin in the cylindrical coordinates.

The Eqs. 5.56 and 5.57 based on polar coordinate system can be expressed in rectangular coordinate system by coordinate conversion. The directional components of the electrical field intensity along y-axis and x-axis are E_x and E_y, respectively, in this case, with y ranging from 0 to a, and x ranging from $-a$ to a, in Eqs. 5.59–5.60:

$$E_y = E_{\infty}\frac{(2y^2 - a^2)(S-1) + (S+1)a^2}{a^2(S+1)}, \tag{5.59}$$

$$E_x = E_{\infty}\frac{2x\sqrt{a^2 - x^2}(S-1)}{a^2(S+1)}, \tag{5.60}$$

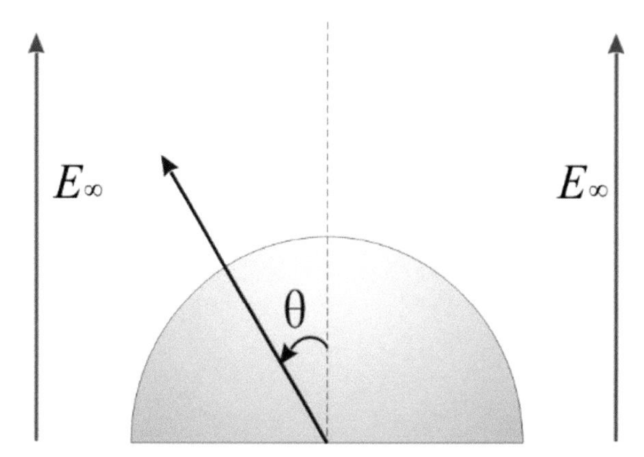

Fig. 5.19: The geometric schematic of a 2D bubble surrounded by air in an electrical field in polar coordinate system.

For an assumption of uniform electric field, the calculation of *E* can be simplified as Eq. 5.61:

$$E_\infty = \frac{U}{d}, \tag{5.61}$$

where *U* is the applied voltage in bubble electrospinning and *d* is the distance between the metal electrode (needle top) and the surface of grounded collector (counter electrode).

The bubble in an electric field is subjected to the drawing action from electric field and to be stretched to a cone shape from a spherical shape (Liu et al. 2007, Liu et al. 2008, Liu et al. 2009). The calculation of *E* can be further simplified assuming that the bubble is in elliptical shape at the critical electrospinning state during bubble electrospinning process, and hence it is possible to define the component of *E* along *y*-axis, E_y, and the component of *E* along *x*-axis, E_x (Dong et al. 2010), which can be expressed with the following Eqs. (5.62–5.70):

$$E_y = \frac{\partial \phi_z}{\partial y} = E_{\zeta\infty}[\frac{2}{c} - (1 + \frac{S-1}{S+1}\rho_o^2)(\frac{1}{c} - \frac{\sqrt{c^2 f_3 + f_1(f_1 + f_2)}}{\sqrt{2}cf_2})] \tag{5.62}$$

$$E_x = \frac{\partial \phi_z}{\partial x} = E_{\zeta\infty}(1 + \frac{S-1}{S+1}\rho_o^2)\frac{\sqrt{2}cxy}{f_2\sqrt{c^2 f_3 + f_1(f_1 + f_2)}} \tag{5.63}$$

$$f_3(x, y) = x^2 - y^2 \tag{5.64}$$

where E_{sy} and E_{sx} are the electrical field intensity along the elliptical bubble surface. and can be expressed as:

$$E_{sy} = E_{\zeta\infty}[\frac{2}{c} - (1 + \frac{S-1}{S+1}\frac{c^2}{(a+b)^2})(\frac{1}{c} - \frac{\sqrt{c^2 f_{3y} + f_{1y}(f_{1y} + f_{2y})}}{\sqrt{2}cf_{2y}})] \tag{5.65}$$

$$E_{sx} = E_{\zeta\infty}[\frac{2}{c} - (1 + \frac{S-1}{S+1}\frac{c^2}{(a+b)^2})(\frac{1}{c} - \frac{\sqrt{c^2 f_{3x} + f_{1x}(f_{1x} + f_{2x})}}{\sqrt{2}cf_{2x}})] \quad (5.66)$$

$$f_{\mu y} = f_{\mu}(\frac{a^2(b^2 - y^2)}{b^2}, y) \quad (5.67)$$

$$f_{\mu x} = f_{\mu}(x, \frac{b^2(a^2 - x^2)}{a^2}) \quad (5.68)$$

where ($\mu = 1,2,3$). The equations below will be true if a uniform electric field is assumed:

$$E_{\zeta\infty} = \frac{U}{d_{\zeta}} \quad (5.69)$$

$$d_z = \frac{c}{2}(d_{\zeta} + \frac{1}{d_{\zeta}}) \quad (5.70)$$

where U is the applied voltage in bubble electrospinning, d_{ζ} and d_z respectively present the distance between the metal electrode and the grounded collector in the ζ and z plane (Dong et al. 2010).

In order to simulate the electrical field distribution, ANSYS® software package was used to analyze the distribution of electrical field intensity in bubble electrospinning. For meshing the calculational domain, PLANE121, which is a 2D, 8-node, charge-based electric element, was used in the bubble and its vicinity. This element has one degree of freedom, voltage, at each node. The values of those parameters were designated and shown in Table 5.3.

Electrical Field Distribution on Bubble Surface at the Initial State of Bubble Electrospinning. The FEA results of field intensity component in *y*-axis on the bubble's surface and at its vicinity at the initial state of bubble electrospinning are shown in Fig. 5.20 in the form of nephogram in the whole area, where the red region on the top of the bubble stands for the maximal value of the electrical field intensity, and the dark regions represent where the field intensity are minimal. Figure 5.20 indicates that *y*-axis electrical field intensity, E_y, decreases gradually from the top to the bottom of the bubble. Figure 5.21 shows the FEA results of *x*-axis electrical field intensity on bubble surface and at its vicinity at the initial state of bubble electrospinning, in the form of nephogram in the whole area, and it can be easily observed that the electrical field intensity is symmetrically distributed on the bubble and at its vicinity due to the symmetric nature of the bubble shape.

The directional components of E, E_y, and E_x can be written as the Eqs. 5.71–5.72, according to Eqs. 5.56–5.57 and Table 5.3,

$$E_y = 0.0952381 + 18095.2y^2, \quad (5.71)$$

where the range of y is from 0 to 0.01, and,

$$E_{\mathrm{x}} = 5.42857 \times 10^{9}\ \mathrm{x}\ \sqrt{0.0001 - \mathrm{x}^{2}}, \tag{5.72}$$

where the range of x is from –0.01 to 0.01.

The graphical expressions of Eqs. 5.62 and 5.63 are shown in Figs. 5.21 and 5.22, respectively, where the scatter curves are obtained through plotting E_x and E_y values of nodes on the bubble's surface from the numerical simulation results.

The electric field force can be expressed as $\boldsymbol{F} = \boldsymbol{Eq}$, according to Coulomb's law, where q is the surface charge per unit length. It could be concluded that the maximum of electric field force occurs at the top of the bubble, based on the electric field distribution on the bubble. The electric field force is symmetrical with respect to the vertical position along x-axis direction, which can be used to explain the reason why the bubble could be stretched from a spherical shape to a cone shape in bubble electrospinning process.

Electrical Field Distribution on Bubble Surface at the Critical State of Bubble Electrospinning. The bubble deforms its shape from sphere to cone at the critical state of bubble electrospinning. The corresponding FEA results are shown in Figs. 5.23 and 5.24, which represent the zoom-out nephograms of the electric field distribution along y-axis and x-axis, respectively, and the red region at the top of the bubble stands for the area where the highest field intensity was achieved.

As can be observed from Fig. 5.23, the electric field intensity component along y-axis decreases gradually from the top to the bottom of the bubble, however, the field intensity increases on the top of the bubble suddenly and precipitously. Figure 5.24 displays the symmetric distribution of field intensity along x-axis, indicating no significant effect on the bubble electrospinning process.

The 2D FEA numerical results corresponding to the nephograms in Figs. 5.23 and 5.24 are shown in Figs. 5.25 and 5.26, respectively, where the dotted curves are

Table 5.3: Key parameters used in calculation of E.

ε_i	ε_o	a [m]	U [V]	d [m]
20	1	0.01	30000	0.1

(1) Electrical field E_y in y-axis

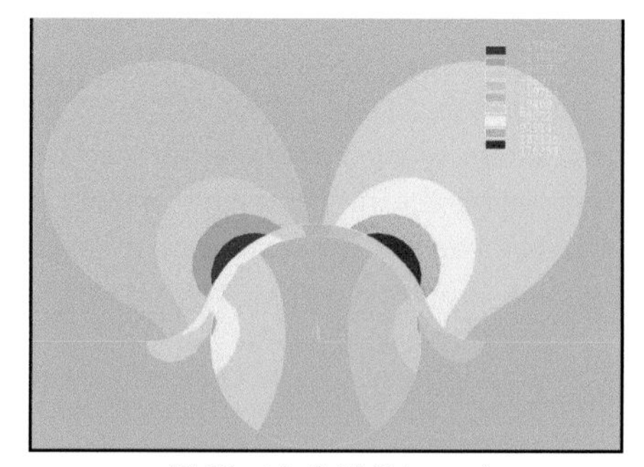

(2) Electric field E_x in x-axis

Fig. 5.20: The nephograms of electric field E.

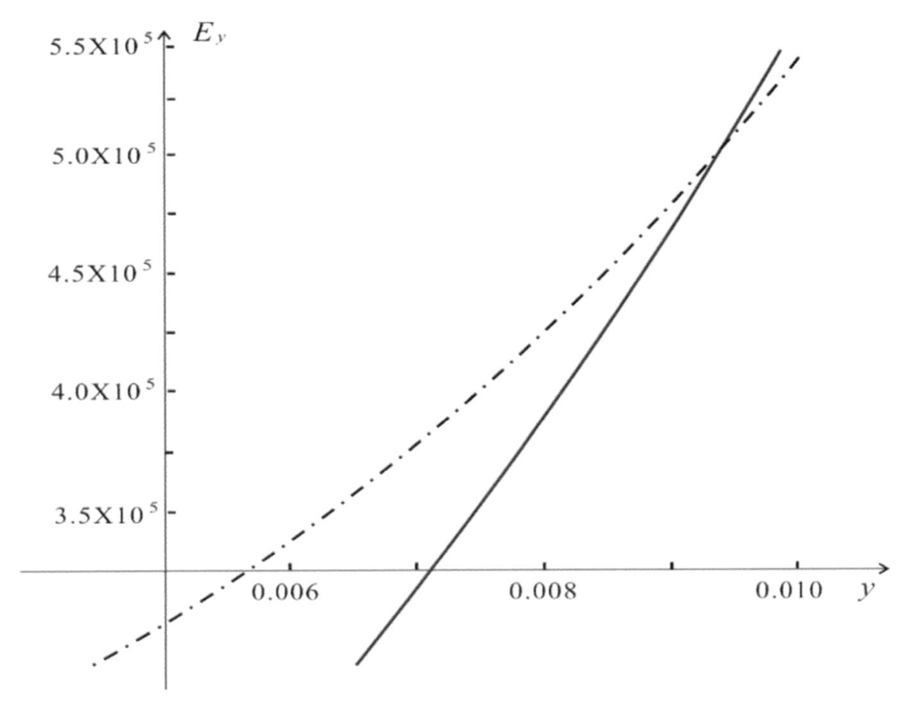

Fig. 5.21: The graphical expression of Eq. 5.62: the distribution of the electric field component E_y, where the y-axis stands for the longitudinal radius, the dotted curve represents the FEA results, and the solid curve represents the results from Eq. (5.71).

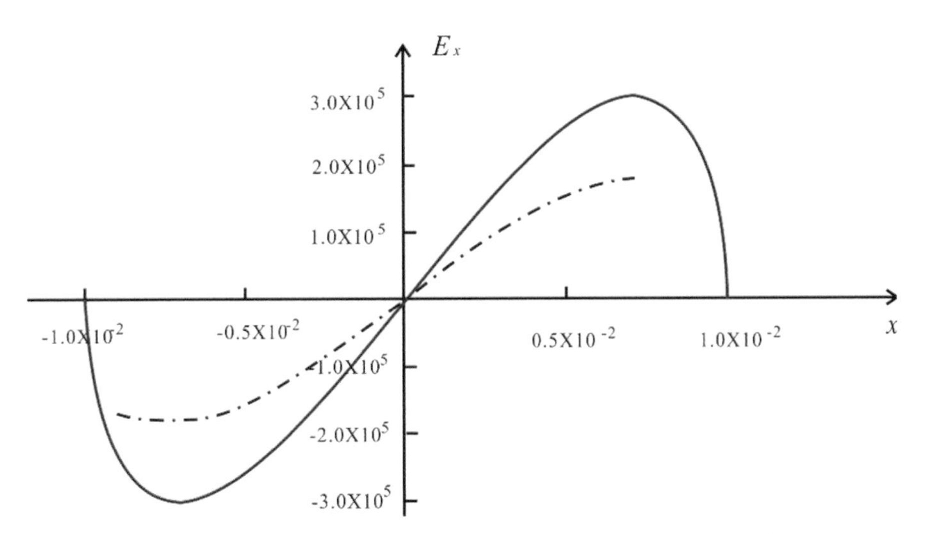

Fig. 5.22: The graphical expression of Eq. 5.63: the spatial distribution of the electric field component E_x, where the x-axis stands for the radial radius, the dotted curve represents the FEA results, and the solid curve represents the results from Eq. (5.72).

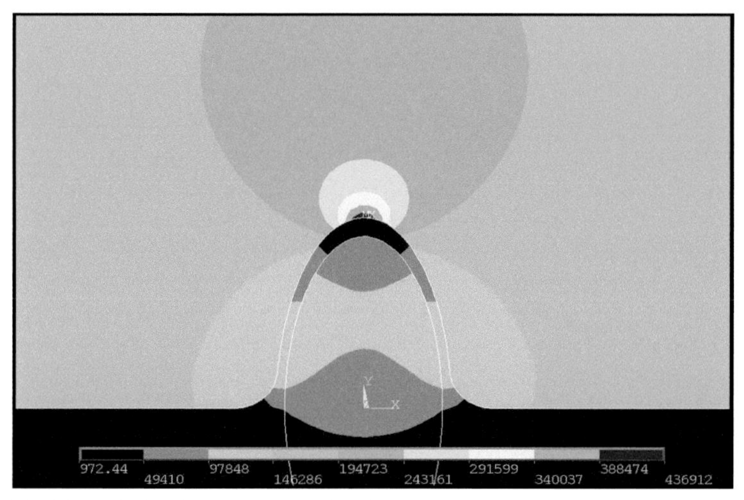

Fig. 5.23: The 2D FEA nephogram of E_y, the component of field intensity E in y-axis.

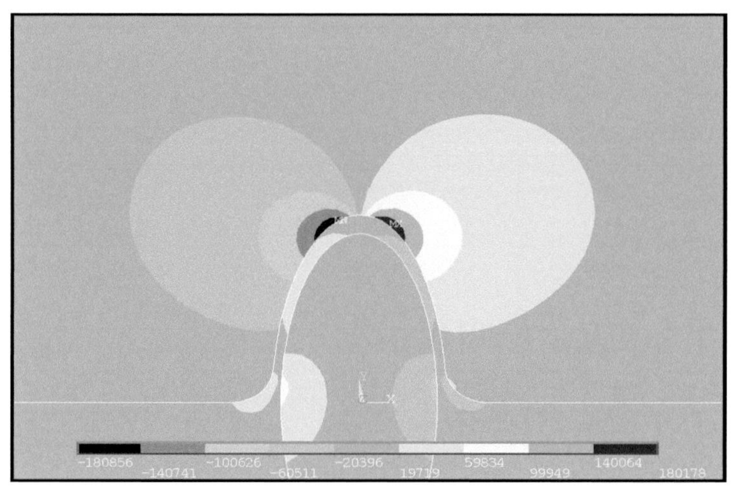

Fig. 5.24: The 2D FEA nephogram of E_x, the component of field intensity E in x-axis.

obtained by plotting E_y and E_x values of the nodes on the bubble surface based on the 2D FEA results. The vertical axes in the nephograms of Figs. 5.25 and 5.26 represent the electrical field intensity (E) along y-axis and x-axis respectively, while the y-axis and x-axis in Figs. 5.25 and 5.26 represent the directions of semi-major axis and semi-minor axis of the elliptical bubble sphere, respectively. It is easily found that the field intensity distribution is consistent with the FEA results.

It can be concluded that the maximal electric field force occurs on the top of the bubble based on the electric field intensity distribution on the bubble, and this can explain why the first jet occurs on the top of the bubble surface when the electric field force exceeds the sum of surface tension and viscous resistance.

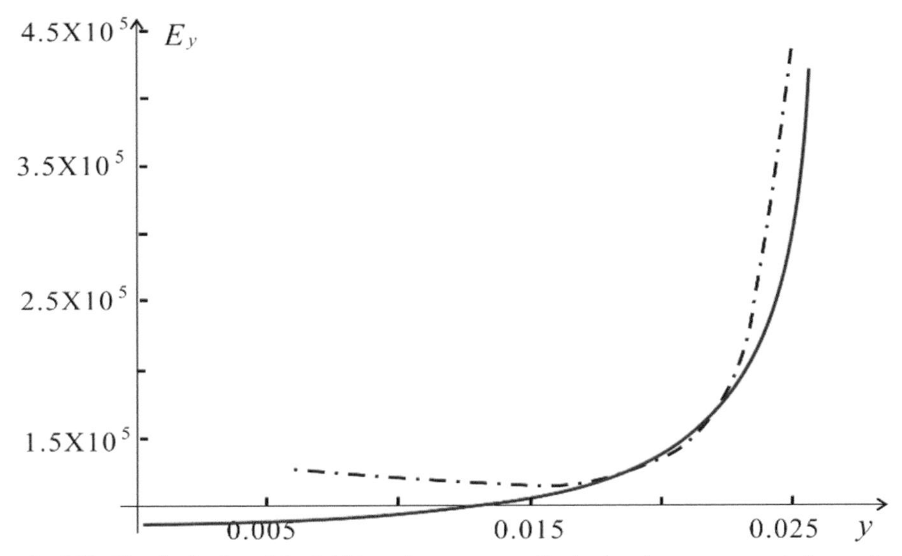

Fig. 5.25: The distribution of the field intensity component E_y: the dotted curve represents the results from FEA and the solid curve is obtained based on Eq. (5.71).

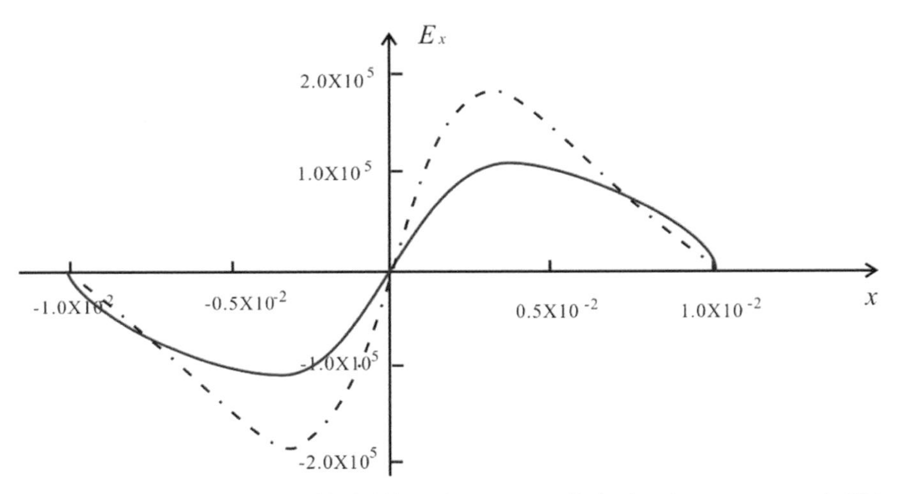

Fig. 5.26: The spatial distribution of the field intensity component E_x: the dotted curve represents the FEA results, and the solid curve is obtained from Eq. (5.72).

Electric Field Distribution in Multi-bubble Electrospinning

Figure 5.27 shows the results of the electric field distribution in multi-bubble electrospinning process. It can be found that the distribution of electric field intensity is symmetric with regard to the central axis of the linear bubble array when the number of bubble increases from 3 to 5, and the symmetry about the central axis is generated by the balanced mutual repellency between the adjacent charged bubbles on both sides of the central bubble due to Coulombic force action, which has influence on the envelope shape and path of the jets (Shou et al. 2011).

The directions of the electric field force generated by the central bubble are vertically upward, while the directions of the field intensity of the bubbles on both the left end and the right end are pointed outwards, slanting upwards, as shown in Fig. 5.28. This phenomenon has been defined as End-effect previously. The mode of field intensity distribution in multibubble electrospinning resembles the mode in multineedle electrospinning (Theron et al. 2005), and also resembles the mode in other needleless electrospinning, which are based on the principle of liquid self-reorganization on free surface.

Despite the symmetry, it is clear that the mutual interference can influence the behavior of bubbles and the paths of jets during bubble electrospinning. Hence, bubble electrospinning has a similar mode of electric field distribution to multineedle electrospinning, and End-effect will occur in large-scale electrospinning process as long as the spun jets are linearly aligned in the electric field. Effective measures are also needed in multibubble electrospinning and other needleless electrospinning processes, wherever the End-effect exists and remains an issue.

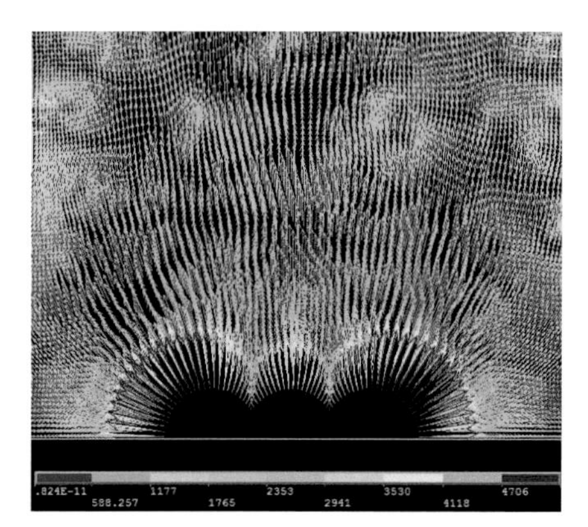

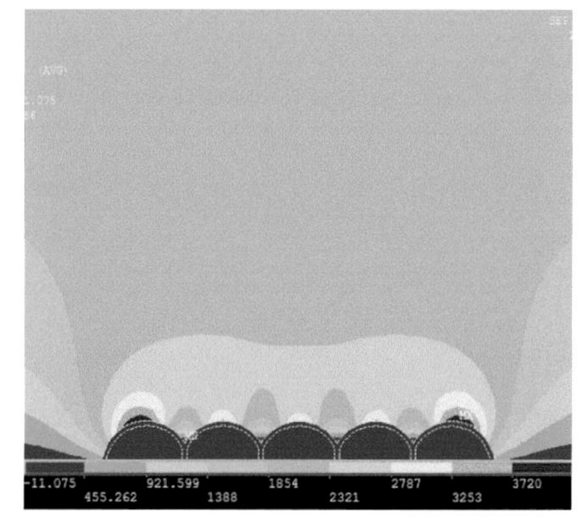

Fig. 5.27: Numerical vector plots of the field intensity in multi-bubble electrospinning with different number of bubbles: the left process has three bubbles and the right has five bubbles.

Fig. 5.28: Photo showing the mutual interference of bubbles and jets in bubble electrospinning.

Improvement on Electric Field and Distribution in Needle Electrospinning

Previous analyses on field intensity of needle electrospinning based on FEA indicated that electric field intensity distribution is not uniform on a single needle and multineedles in electric field. For single needle electrospinning, the field intensity on the top of the needle is greater than that at the bottom of the needle. For multineedle electrospinning, field intensity on the ends of the needle row is greater than that on the inner side of the needle row, especially in the middle of the needle row, which is called the End-effect. This causes the middle needles to generate no or weak spinning jets which fly directly towards the receiver, if any, while the edge needles produce strong spinning jets which fly slantways due to the End-effect.

This phenomenon can be improved by increasing the field intensity of the middle needles and decreasing the field intensity of the end needles, minimizing the difference in magnitude and direction of field intensity between the needles in the middle of the needle row and the needles in the ends of the needle row, so that relatively uniform distribution of field intensity across the needle row can be achieved. The authors have taken measures, such as using different voltages, different needle length, different needle spacing, different needle layout, and auxiliary electric field, e.g., needle casing, needle separator plate, or using plastic spinneret instead of metal spinning head, etc., which have been proved to be effective for field intensity improvement during multineedle electrospinning. All these methods can be used to guide the control of either the magnitude or the uniformity of field intensity to minimize the End-effect.

Using Plastic Needle Plate

Needle plate is usually not used in the conventional single-needle electrospinning process or a metal plate is used to plant needles in multineedle electrospinning for a simple and easy electrospinning setup. A needle plate is necessary for multineedle electrospinning to support the needles and align the needles, making the receiving distance, needle location, needle alignment adjustable. Considering that metal plate will share the charges from power supply and hence reduce the field intensity, a plastic plate is recommended as needle plate. The nephograms for field intensity of five-needle electrospinning models without and with plastic (PTFE) needle plate based on ANSYS® simulation in 2D plane are displayed in Fig. 5.29 and Fig. 5.30, respectively.

The resulting peak values of field intensity of the electrospinning models with/without plastic needle plate are 3.91e5 N/C and 3.77e5 N/C, respectively, indicating that using plastic (PTFE) needle plate can obtain relatively strong field intensity. Plastic (PTFE) needle plate will be used in FEA simulation and analysis for field intensity based on different types of needle electrospinning processes in the remaining part of this section. End-effect can be observed in both the multineedle electrospinning processes with/without the plastic needle plate. Field intensity shows symmetric distribution about the middle needle(s), which can also be found from the previous simulation results in this chapter.

Since previous FEA simulation results indicated that the plastic plate where the needles are mounted facilitated the improvement of field intensity in five-needle electrospinning, it is necessary for further FEA to be employed for investigation of the influence from the specification of the plastic plate on field intensity distribution. FEA on two-, four-, six-, and eight-needle electrospinning was performed respectively, based on constant needle length and needle spacing, as well as the same parameters of electrospinning process with different lengths of spinning plate and edge length. The model of 6-needle electrospinning process is shown in Fig. 5.31, where Edge length indicates the distance between the edge needle and the edge of the plate on the same side, and Spinneret means the plastic plate as addressed before.

Figure 5.32 showed the field intensity in terms of nephogram from FEA results of six-needle electrospinning, and Fig. 5.33 displayed the FEA results for two-, four-, six-, and eight-needle electrospinning process. It is easy to observe that the peak value of field intensity initially increased, then decreased, and finally increased again with the increase in edge length of the plastic plate, therefore the appropriate value of edge length should be either the smallest or the largest, and field intensity reached its minimum when edge length was 3 cm *c.a.* According to the previous definition for peak value of field intensity, it means that, when edge length is 3 cm,

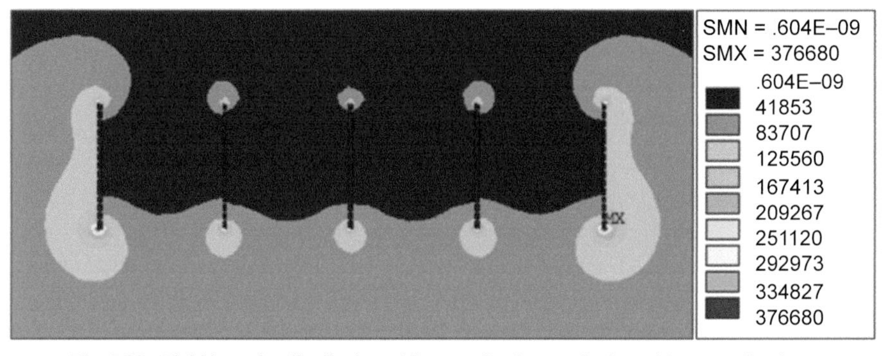

Fig. 5.29: Field intensity distribution of five-needle electrospinning without needle plate.

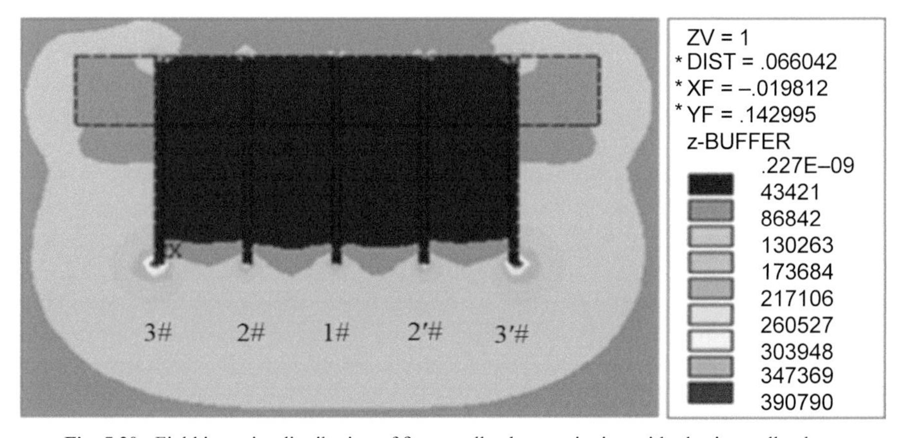

Fig. 5.30: Field intensity distribution of five-needle electrospinning with plastic needle plate.

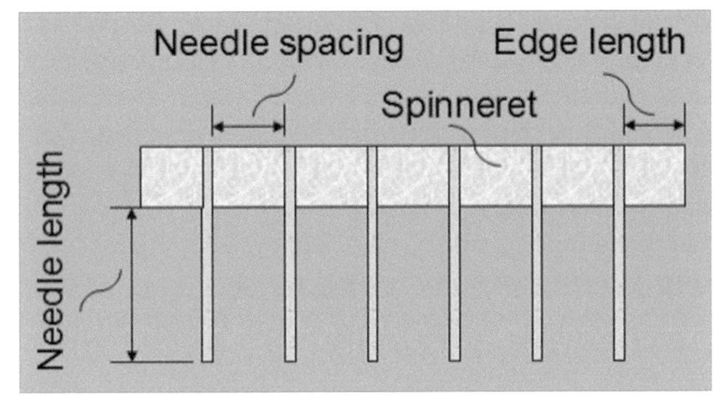

Fig. 5.31: Model of electrospinneret.

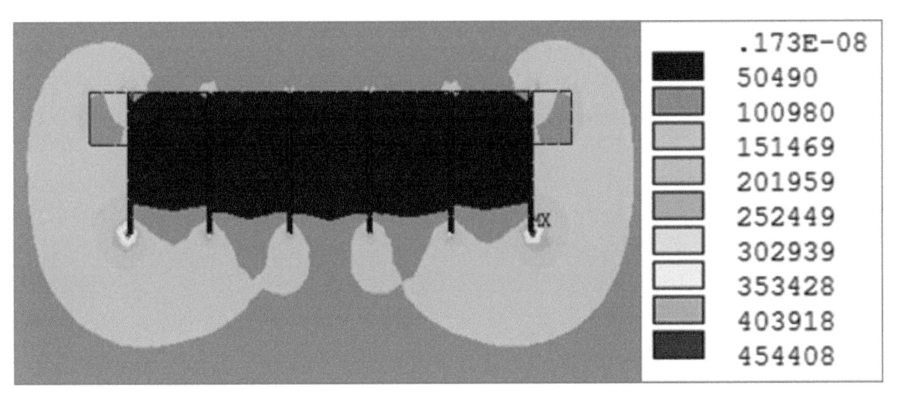

Fig. 5.32: Field intensity in six-needle electrospinning.

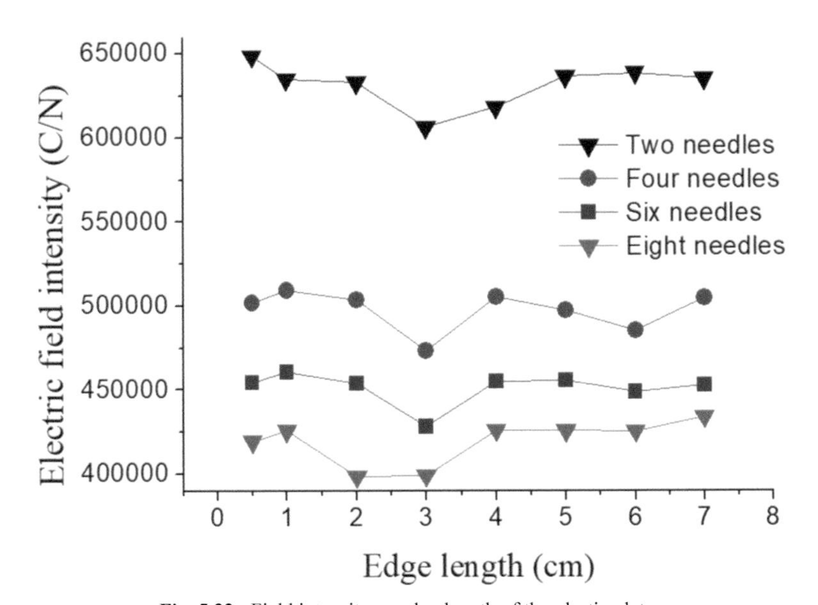

Fig. 5.33: Field intensity vs edge length of the plastic plate.

the field intensity of edge needles is the smallest, meaning the filed intensity of other needles are relatively large, and the distribution on field intensity among all the needles is relatively uniform, as the electric field is conservative, and total number of charges does not change during electrospinning under the same conditions. It was also found that more needles led to less field intensity, which was consistent with the results obtained previously.

Using Different Needle Length along the Needle Row

It has been concluded from the previous FEA on field intensity of needle electrospinning process that the field intensity decreases with the increase in needle spacing, and the middle needles show relatively weak field intensity compared to the end needles in a needle row where the needles are lined up, if the needles have the identical length and the top sides of all the needles are linearly arranged.

This situation that field intensity is not distributed uniformly along a needle row is expected to be improved by using decreasing needle length from the middle to the ends of the needle row. The results from preliminary FEA simulation on five-needle electrospinning process model indicated that extending the length of the middle needles (i.e., reducing the receiving distance) or reducing the length of the end needles (i.e., extending the receiving distance) could distribute the field intensity uniformly from the middle to the ends along a needle row.

FEA simulations based on unequal needle length but equal difference in needle length showed that the optimal field intensity distribution was obtained when the length of the middle needle was 6.0 mm longer than that of the end needles, with the length of the middle needle being 3.0 cm. It is found based on the comparison of Fig. 5.34 to Fig. 5.33, the needles No. 2# and No. 2'# showed increased field intensity by 4e4 N/C, and their nephogram color changed from green (Fig. 5.33) to yellow (Fig. 5.34), and the color of the nephogram for needle No. 1# changed from green to orange, indicating the increase in field intensity of the three needles, while the field intensity of the end needles decreased correspondingly, leading to relatively uniform field intensity distribution along the needle row, and hence the End-effect could be fairly controlled this way.

However, the results from using equal difference in needle length during five-needle electrospinning did not generate satisfactory field intensity distribution along the five needles in a row. Therefore, additional FEA work is necessary to further increase the uniformity of field intensity distribution using unequal needle length and unequal difference in needle length. The model for five-needle electrospinning process with unequal difference in needle length is shown in Fig. 5.35, where D1 stands for the difference in needle length between 2# and 3# needles, and D2 represents the difference in needle length between 1# and 2# needles.

The FEA was performed to obtain the field intensity when the value of (D2-D1) is 0.2 cm, 0.3 cm, 0.4 cm, 0.5 cm, 0.6 cm, 0.7 cm, when other conditions remained as the same as before, and the optimal result was found when the value of (D2-D1) is 0.6 cm, at which the peak values of field intensity of the five needles are 9.67e5, 9.31e5, 9.40e5, 9.31e5, 9.67e5 N/C, respectively, from left to right. Figure 5.36 shows the FEA simulation result for five-needle electrospinning when (D2-D1) is 0.6 cm.

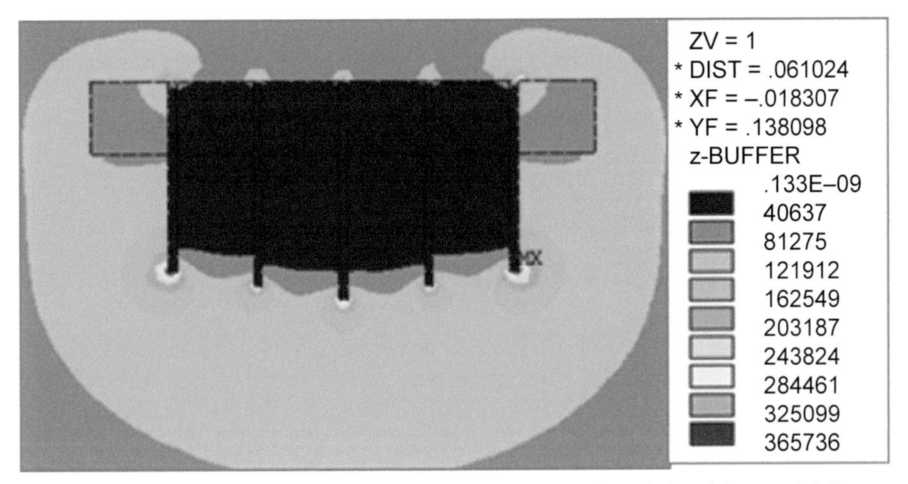

Fig. 5.34: Field intensity in five-needle electrospinning with unequal needle length but equal difference in needle length between neighboring needles.

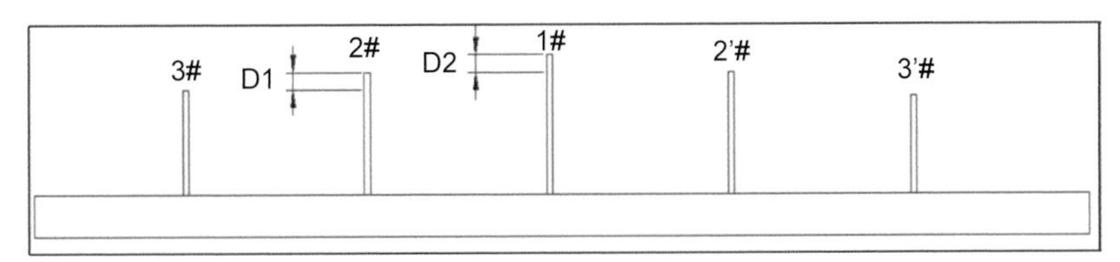

Fig. 5.35: Field intensity in five-needle electrospinning with unequal difference in needle length.

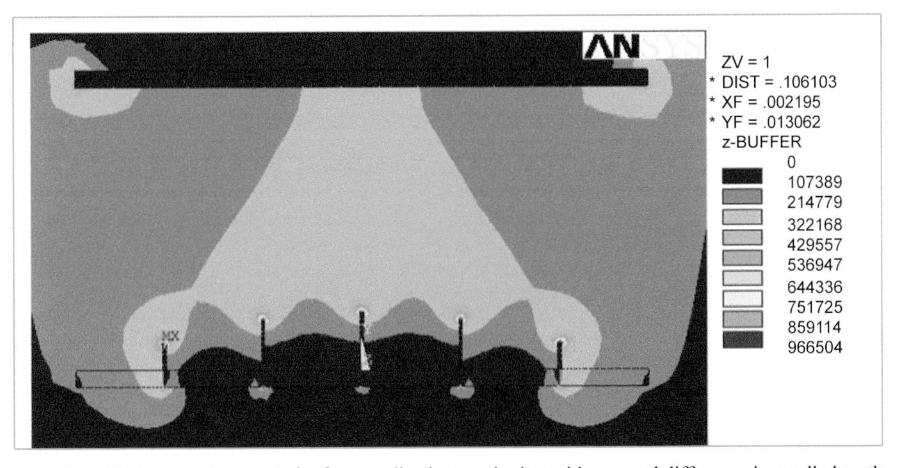

Fig. 5.36: FEA simulation result for five-needle electrospinning with unequal difference in needle length when (D2-D1) is 0.6 cm.

The FEA simulation results indicated that using unequal needle length, particularly using unequal difference in needle length can improve the uniformity of field distribution, i.e., the End-effect phenomenon. However, greater field intensity usually leads to greater non-uniformity of field intensity distribution, therefore, strongest field intensity and most uniform field intensity distribution cannot be achieved at the same time, the latter being more important than the former. The moderate field intensity but uniform field intensity distribution is more welcome for practical large-scale electrospinning process.

Applying Different Voltages along the Needle Row

Generally, the voltage is indirectly applied to different needles in the multineedle electrospinning process, taking the needles as an assembled entirety through a metal needle plate, namely, the voltage is firstly applied to the metal needle plate, and then the voltage is applied to the needles planted on plate indirectly. This method of voltage application is simple, easy for operation, however, the way the voltage is applied to the needles makes all the needles aligned in a row share the same voltage, resulting in different peak values of field intensity along the needle row. Meanwhile, the metal needle plate will share the charges from the power source and lower the magnitude of field intensity of all the needles, causing the increase in the cost of nanofiber production, not favoring electrospinning at low energy cost.

Independently applying voltage may avoid the energy waste resulted from the charge share of the metal needle plate, and different voltage values can be applied to different needles, which facilitates the field intensity distribution in a relatively uniform way. Higher voltages can be applied to the middle needles, or lower voltages can be applied to the end needles to balance the peak values of field intensity on the corresponding needles, resulting in relatively uniform distribution of the field intensity along the needle row. For example, 12 kV voltage was applied to the end needles, i.e., No. 3# and No. 3′#, and 12.6 kV was applied to the two needles adjacent to the two end needles, i.e., No. 2# and No. 2′#, and 12.8 kV was applied to the middle needle, i.e., No. 1#. The prototype electrospinning model is depicted in Fig. 5.30. The FEA simulation results are shown in Fig. 5.37, where the red color occurs

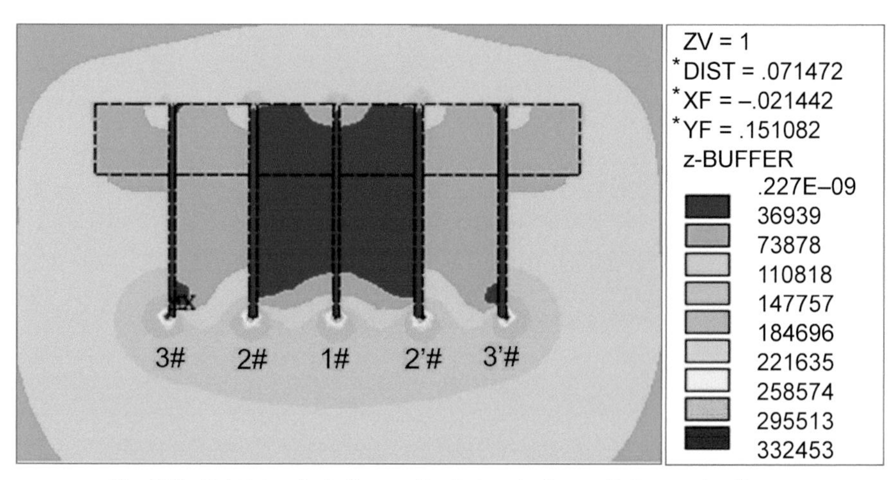

Fig. 5.37: Field intensity in five-needle electrospinning applied unequal voltages.

on the top of all the five needles, indicating uniform field intensity distribution along the five needles, at the cost that the peak value of the field intensity decreased from 3.91e5 N/C to 3.32e5 N/C.

The method of applying decreasing additional voltages to the needles from the middle to both the ends can reduce the field intensity of end needles, meanwhile increasing the field intensity of the middle needles, balancing the distribution of field intensity along the needle row, and minimizing the End-effect. This favors the generation of uniform and steady spinning jets and ensures the directions of the jets towards the fiber receiver.

Using Unequal Needle Spacing

Previous simulation results showed that the field intensity decreased with the increasing needle spacing. Therefore, End-effect was expected to be suppressed by changing the needle spacing between the end needles and their neighboring needles, and/or the needle spacing between the middle needle(s) and their neighboring needles.

Preliminary 2D FEA simulation results indicated that the field intensity of end needles increased from 3.91e5 N/C to 3.94e5 N/C when the needle spacing between the end needles and their neighboring needles were increased, as shown in Fig. 5.38, which does not favor the uniform distribution of field intensity along the needle row, however, later simulation results from increasing the distance between the middle needle and its neighboring needles showed increased uniformity of field intensity distribution, which is not shown here. Although further simulation results from adjusting needle spacing between all the needles can further optimize the uniformity of field intensity distribution along the needles, this method, however, will lead to non-uniform distribution of spinning jets across the electrospinning machine, causing non-uniform structure of electrospun web in CD (cross machine direction) due to unequal needle spacing, even if field intensity has relative uniform distribution along the needle row (CD). Therefore, using unequal needle spacing is not recommended for practical electrospinning process, unless there are no other choices or it is combined with other methods.

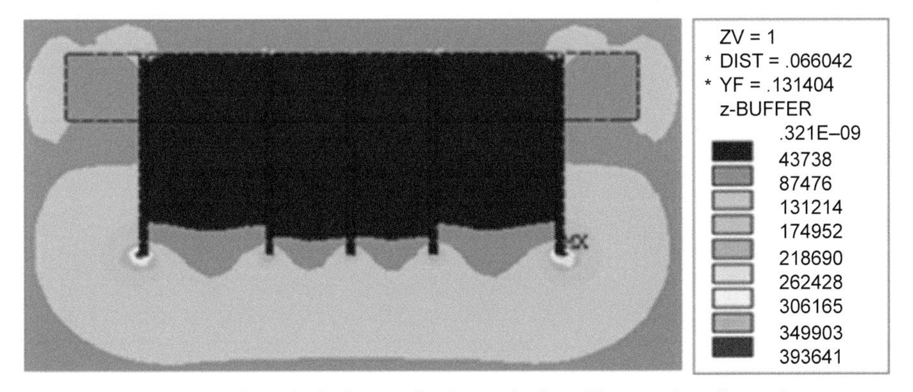

Fig. 5.38: Field intensity in five-needle electrospinning with unequal needle spacing.

Using Auxiliary Electrodes

It has been found that the uniformity of field intensity distribution in multineedle electrospinning was caused by the greater field intensity of the end needles and lesser field intensity of the middle needles, and there are strong interactions of electric field force between neighboring needles. It is hereby proposed that using additional needles, as auxiliary electrodes, at the two ends of the needle row is expected to balance the Coulombic forces on the two sides (left side and right side) of the end needles, and hence may minimize or remove the End-effect.

Voltages will be applied to the additional needles, but no solution will be fed to them as these needles are used as auxiliary electrodes only. These needles can be made of any metal materials, with or without capillary holes. The number of the additional needles can be arbitrary, depending on the required additional electric field. The same or different voltages can be applied to the additional needles, and the same or different needle spacing or dimension can be used. This method only consumes a small amount of electric energy, and hence the effect for uniformity of field intensity distribution is very cost effective.

Results from further FEA on field intensity simulation for multineedle electrospinning indicated that the more the needles are used, the more uniform distribution of field intensity can be achieved. Figure 5.39 showed the FEA simulation results for multineedle electrospinning where 13 needles were used, among which two needles at both ends, i.e., a total of four end needles, were used as auxiliary electrodes with smaller needle spacing between them than that between the inside needles, i.e., the rest nine needles between the four end needles.

It was found that, after FEA simulation many times, 3% CV of field intensity was achieved for the peak values of the field intensity of 13 needles when using shorter needle length, and smaller needle spacing for the four end needles and longer needle length and greater needle spacing for the remaining nine needles between the four end needles, resulting in an ideal distribution of field intensity distribution along the 13-needle row. The four end needles (with two at each end) functioned as

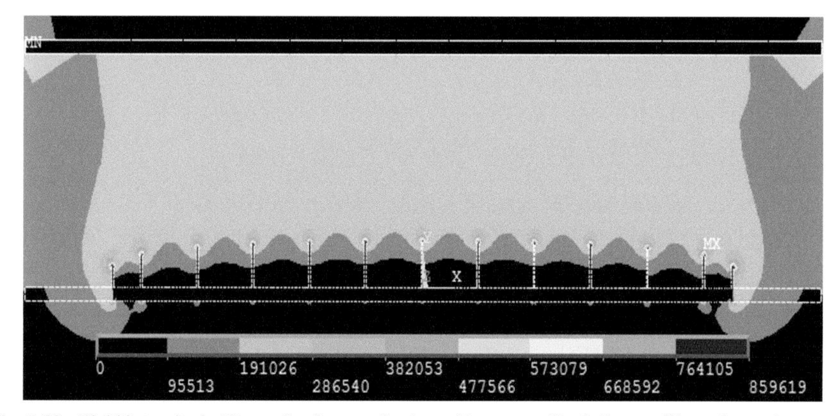

Fig. 5.39: Field intensity in 13-needle electrospinning with two needles being auxiliary electrodes at each of the two ends, where the needle spacing between the two lateral needles at each end is smaller than that of the remaining nine needles between the four end needles, and the length of the two needles at each end gradually becomes shorter and shorter.

auxiliary electrodes, providing additional, horizontal Coulombic force acting from the end needles towards the remaining nine needles between the four end needles. The web edges between the two end needles on both lateral sides should be cut off in practical large-scale electrospinning process to obtain an uniform nanofiber web structure.

Special Layouts of Needles

The effects from needle specifications and process parameters on field intensity and the ways to minimize End-effect have been investigated based on the multineedle electrospinning method, where the needles are linearly aligned in a row by means of ANSYS® software in 2D plane. Actually, needle layout with non-linear arrangement can also be considered to control End-effect during multineedle electrospinning.

The FEA on field intensity distribution in three-needle electrospinning was investigated in 3D space using ANSYS®, and the resultant vector diagram of field intensity is shown in Fig. 5.40, the front view of the vector diagram for the linearly aligned tree-needle electrospinning model, and in Fig. 5.41, the upward view of the vector diagram for the triangularly aligned three-needle electrospinning model is shown, respectively, for convenient comparison.

It is found from Fig. 5.40 that the field intensity vectors of the lateral edge of the two end needles are larger and denser than that of the middle needle, with the arrowheads being obviously deviated laterally. This is because the charges on the end needles of the linearly aligned three needles received the repellent force from the charges on the middle needle, and an unbalanced Coulombic force occurred on the charges of the end needles, resulting in the arrow directions of the field intensity vectors shown in Fig. 5.40 to be pointing tipsily rather than vertically towards the receiver. Therefore, the two end needles showed stronger field intensity than the middle needle in linearly arranged three-needle electrospinning, indicating strong End-effect in multineedle electrospinning.

However, different layouts of multineedles lead to different modes of field intensity distributions. Figure 5.41 showed the FEA result of the three-needle electrospinning method, where each of the three needles arranged triangularly received the same interactions from the other two neighboring needles, the interneedle Coulombic forces are identical to each other, and hence the field intensity distribution is uniform among the three needles. It can be foreseen that the spinning jets from the three needles will be generated uniformly among the three needles due to the uniform distribution of the field intensity among the three needles, and the peak value of field intensity in linearly aligned three-needle electrospinning (1.78e7 N/C) is greater than that in triangularly arranged three-needle electrospinning (1.75e7 N/C), due to no End-effect.

The triangular layout of three-needle electrospinning model can be extended to multirow multineedle electrospinning process, as shown in Fig. 5.43, and the End-effect in the large-scale electrospinning with needles arranged this way is expected to be effectively reduced and uniformly distributed spinning jets are expected to generate in an triangularly or rhombically aligned array, and the nanofiber web will be more uniform than that from single row multineedle electrospinning. Figure 5.44

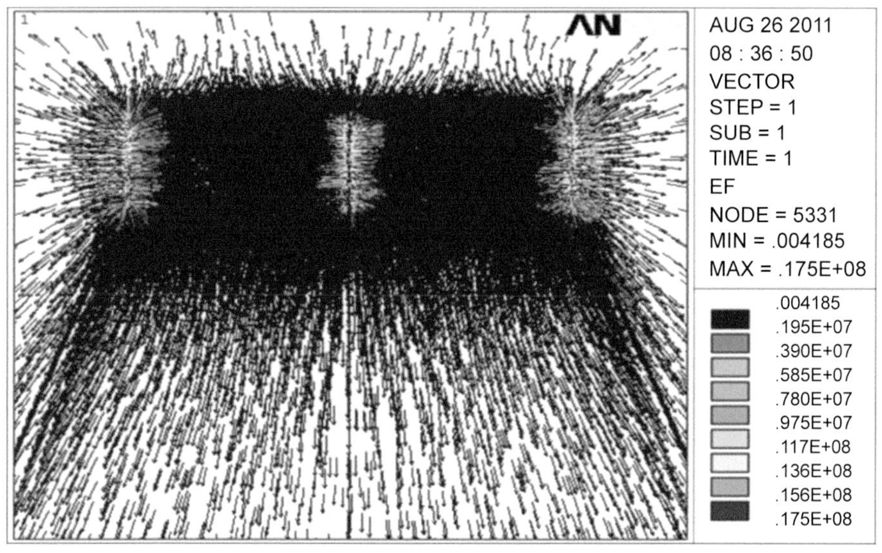

Fig. 5.40: Front view of vector diagram of field intensity distribution in three-needle electrospinning where the three needles are linearly aligned.

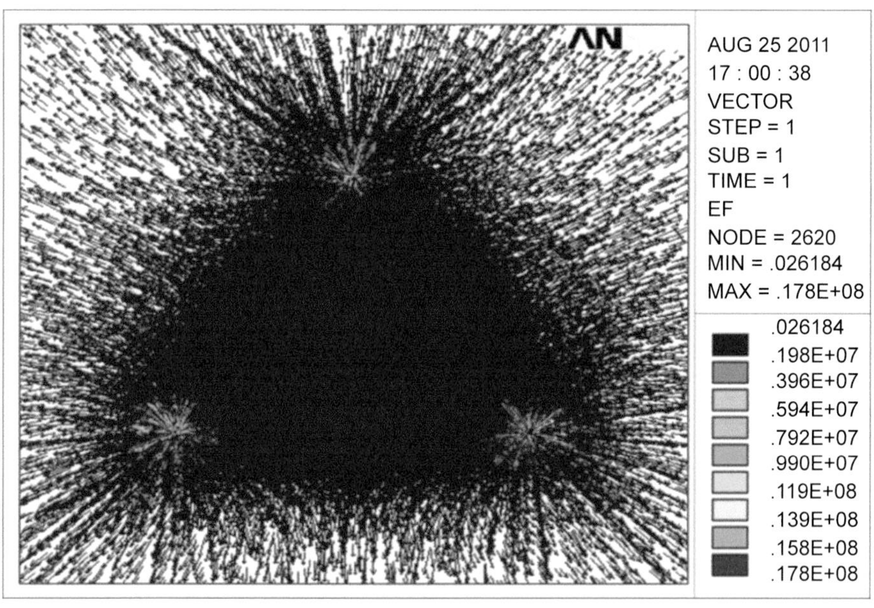

Fig. 5.41: Upward view of vector diagram of field intensity distribution in three-needle electrospinning where the three needles are triangularly aligned.

only showed three-row multineedle electrospinning model, which can be further extended to multineedle electrospinning process model with more rows and needles. The previously discussed methods for minimizing End-effect will also be suitable for this situation, and industrial scale of electrospinning can be commercialized

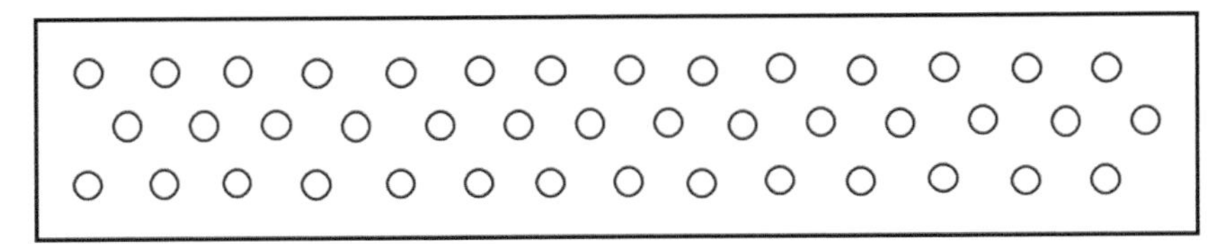

Fig. 5.42: Three row multineedle electrospinning model in which the needles are triangularly or rhombically aligned.

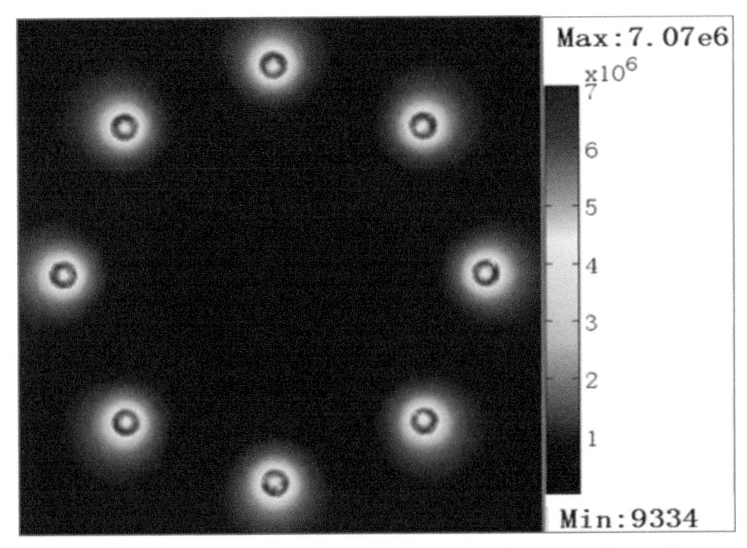

Fig. 5.43: Field intensity distribution in eight-needle electrospinning where the needles are arranged along a circle with diameter 0.01 m (Chen 2012).

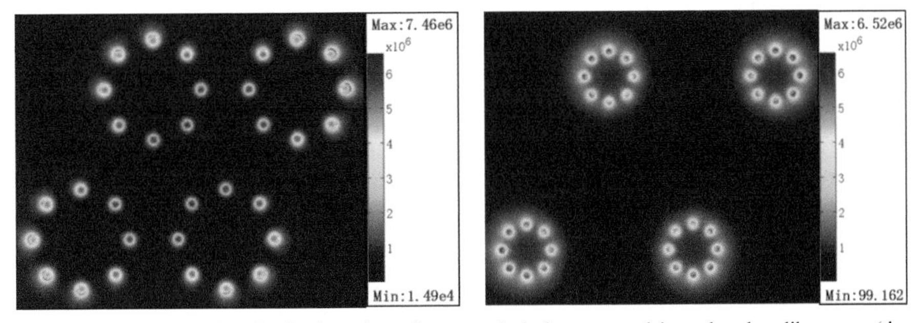

Fig. 5.44: Field intensity distribution along four round circles arranged in a rhombus-like array (the diameter of the circle was 0.01 m and the distance between the centers of horizontally aligned two circles were 0.02 m and 0.035 m, respectively for the left and right).

via combinedly using these methods for reduction of End-effect, thereby achieving uniform nanofiber web finally. Of course the multineedles can be arranged in a way so that needles in different rows have different lengths, or the neighboring needles have different needle lengths.

Additionally, the needles can also be circularly arranged along a circle in multineedle electrospinning, as proposed by historical literature (Zhou et al. 2010), to minimize the End-effect. Actually, the results of FEA on eight-needle electrospinning in 2D plane with COMSOL® where the needles are aligned along a circle (Fig. 5.43) indicated that although each needle along the circle has equal field intensity, the outer half edges of all the needles showed greater peak values of the field intensity than the inner half edges of all the needles, namely, the spinning jets are predicted to fly aslant towards outside of the circle, not vertical to the receiver. Although the End-effect in the circularly aligned multineedle electrospinning can be accepted and the field intensity is distributed uniformly along the circle, as shown in Fig. 5.43, the resulting nanofiber web cannot be accepted due to the non-uniform web structure along cross machine direction generated by unevenly distributed needles along CD (although the distance between the neighboring needles along the circle is equal, the horizontal distance between the neighboring needles varies along CD). Therefore, the spinning head on which the needles are arranged along a circle is not suitable for large-scale electrospinning, although it may be used to produce nanofiber yarns, after certain modifications.

The electrospinning process involving only one circle of multineedles is not enough for large-scale nanofiber production, where several circles of needles may be arranged in an array as shown in Fig. 5.44, taking four circles of needles for instance.

It was obviously found that the End-effect still existed when the distance between the centers of the two circles horizontally aligned were 0.02 m, and End-effect almost disappeared when that was 0.035 m. Although far distance between the circles arranged in an array facilitates the removal of End-effect and uniform distribution of field intensity, the spinning jests, and hence the resulting nanofibers, are not expected to distribute across the CD direction of the nanofiber web due to the non-uniform alignment of the needles along the CD direction.

Using Plastic Casing to Needle

In order to further investigate the methods for controlling field intensity of the needles in multineedle electrospinning process, plastic (PTFE) casing can be applied to encircle the multineedles as a “shielding” element to reduce the interaction of the field intensity between the needles. The “shielding” effect of the plastic casing is to isolate two space regions, and hence to control induction and radiation of electric field from one area to another. The electrostatic shielding principle is the basis for applying shielding measure, such as plastic casing to spinning needle, reducing the total Coulombic force of external electrostatic fields acting on the needle with casing from other needles, and increasing energy efficiency by generating the same field intensity at relatively low voltage.

It was previously found that using plastic (PTFE) plate to mount the needles into an electrospinneret not only facilitated the operation of electrospinning experiment, but also led to increased field intensity of the needles, inspired by which, the authors applied a PTFE based casing on each of the five needles, and performed FEA on the field intensity of the five-needle electrospinning in 2D plane, where the protruded length of casing from the needle top was 0 mm, 1 mm, 2 mm, 3 mm, 4 mm, and

5 mm, respectively, and the casing thickness was 0.2 mm, 0.4 mm, and 0.6 mm, respectively, with no space between needle and casing. The simulation parameters are listed in Table 5.4 and the material permittivities remained the same as listed in Table 5.2. Figures 5.45 and 5.46 showed the model and results of 2D FEA of the field intensity in five-needle electrospinning based on simulation with COMSOL software (Chen 2012).

It was found that the peak value of field intensity when each needle was sheathed with plastic casing made of PTFE (E = 1.52e6 N/C) was increased by 42.3% compared to that using the prototype without casing on the needle (E = 1.068e6 N/C), indicating that the "shielding" effect of the plastic casing caused the field intensity to be enclosed or confined in the vicinity of the needle top.

Simulation results showed that field intensity increased initially and then decreased with the increase of protruded length, which is the difference between needle length and casing length, as shown in Fig. 5.45. Thicker needle casing resulted in weaker field intensity. The maximal field intensity occurred when the thickness of needle casing was 0.2 mm and protruded length was 2 mm. Needles were wrapped completely when the protruded length was 0 mm, which lead to very strong shielding action. On the contrary, if the protruded length was longer than 2 mm, then electric field intensities would be lower due to relative weak shielding action. Thus, protruded length of 2 mm will be the optimum for real electrospinning process.

The increased field intensity due to using plastic casing for the needles was possibly because (1) the nonpolar PTFE casing prevents the needle from discharging towards the surrounding atmosphere as no space existing between the needle and

Table 5.4: Basic parameters for simulation of field intensity in five-needle electrospinning with needle casing.

Applied voltage	Needle length	Receiving distance	Needle spacing	Needle diameter	Collector dimension
20 kV	1 cm	15 cm	1 cm	1 mm	10 cm × 0.02 cm

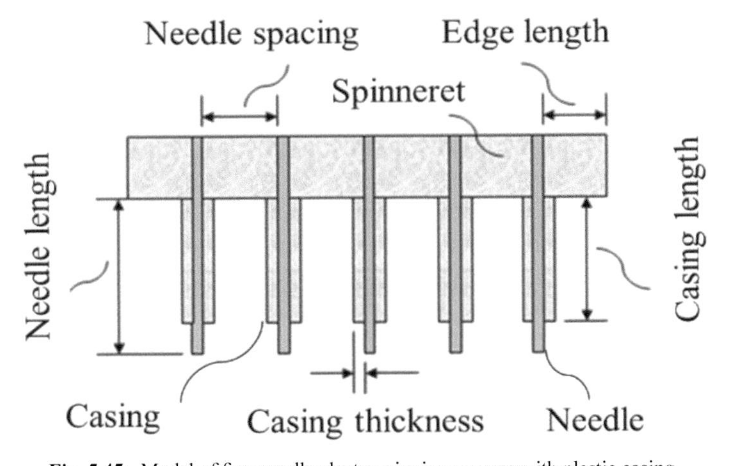

Fig. 5.45: Model of five-needle electrospinning process with plastic casing.

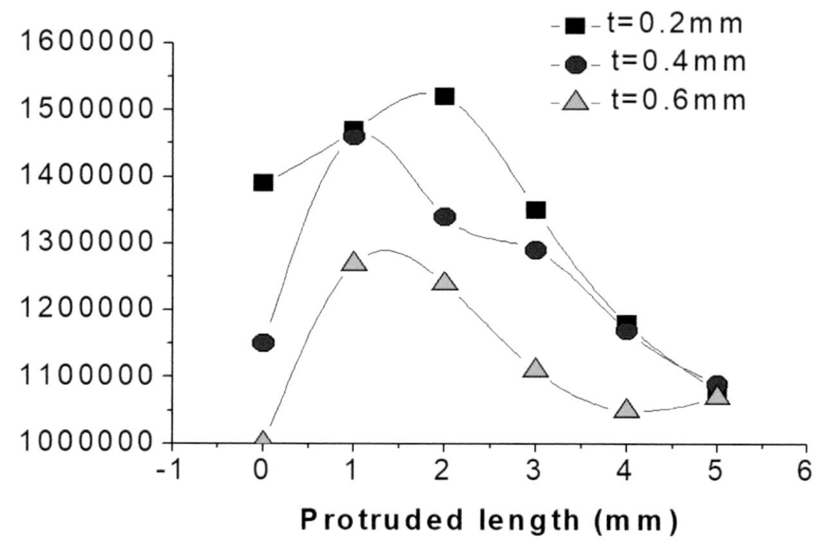

Fig. 5.46: Field intensity in five-needle electrospinning with plastic casing.

casing, and (2) the macroscopic displacement polarization occurred to the nonpolar dielectric material, the PTFE plastic in the existence of externally applied electric field, leading to redistribution of the charges on the needles, and a macroscopic electric field was generated due to the excitation of the macroscopically polarized electric charges (surface bound charges), with its direction opposite to that of the applied electric field, and therefore, the original field intensity inside the dielectric medium was reduced as a result, and hence the needle top protruded from the casing top gained more charge density than needle body after charge redistribution along the needle.

It can be concluded based on the results and analyses above that needle casing is a cost-effective method for improvement of field intensity during practical multi-needle electrospinning, with which relatively fine fibers can be produced, owing to increased field intensity and electric draw when other parameters remain the same. However, application of plastic casings to hundreds of thousands of needles is not a wise choice, instead, we may use a plastic plate containing multiple holes from which the multineedles are protruded, with the holes and the needles having the same central axis.

Using Metallic Casing to Needles

There is significant difference between the shielding effects on field intensity from metal casing and PTFE casing when they are used as shielding measures due to their different Relatively Dielectric Constants. Now that the previous FEA simulation on field intensity in multineedle electrospinning returned reasonably good results and provides potential measures for improvement of field intensity distribution, metal casing was also used in multineedle electrospinning to evaluate its effect on field intensity distribution, and controlling with parameters the same as previous process using PTFE as shield casing. The FEA results of electrospinning modeling with different metallic casing thickness values are shown in Fig. 5.47.

As can be found from Fig. 5.47, the relationship between the thickness of metal casing and the field intensity follows a similar tendency to that of the field intensity when using PTFE as casing material, and the casing thickness was 0.4 mm and 0.6 mm, i.e., the field intensity increased initially and then decreased with the increase in protruded length of needle. However, the field intensity when using steel casing was greater than using PTFE casing due to better shielding effect of metal materials than plastic materials.

The field intensity decreased with the increase in protruded length of needle when the thickness of the metal casing was reduced to 0.2 mm, with the optimal field intensity being the one when the protruded length of needle was 0 mm. This indicated that the needle top should be in line with the metal casing top to achieve the strongest field intensity when 0.2 mm thick metal casing is used as shielding measure for field intensity shielding during the multineedle electrospinning process. The protruded length of needle should be 1 mm to achieve the strongest field intensity when 0.4 mm and 0.6 mm thick metal casings are used as shielding measures for field intensity during the multineedle electrospinning process.

As shown in Fig. 5.47, the highest peak value of field intensity occurred when 0.2 mm thick steel casing was used as shielding measure with 0 mm protruded length of needle, which was 1.684e6 N/C. It was easy to know that field intensity increased by 57.68% compared to the field intensity of 1.068e6 N/C with no metal casing used. Therefore, the application of metal casing may help field intensity to reach the critical field intensity at relatively low applied voltage, indicating this is a cost-effective method and energy can be saved for industrial electrospinning. Besides, relatively fine fibers can be produced at the same voltage if metal casings are employed in the multineedle electrospinning process. Further investigations on

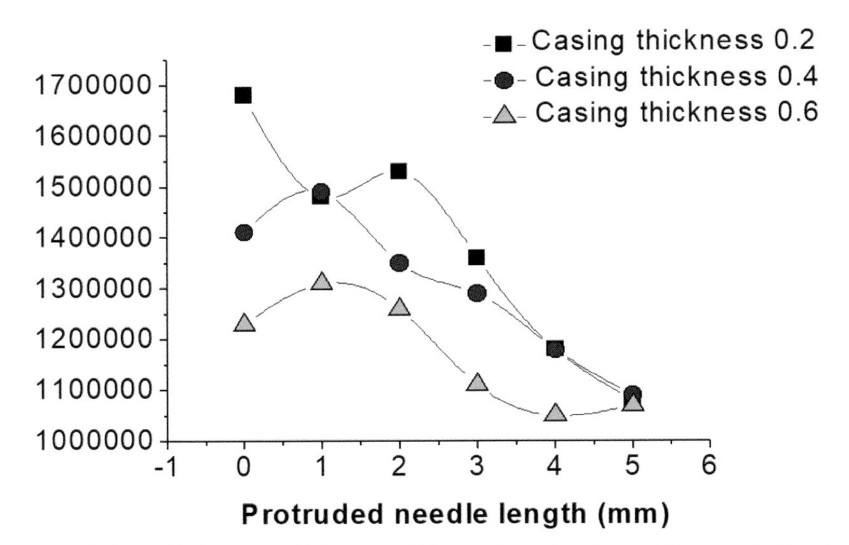

Fig. 5.47: Relationship between field intensity, thickness of metal casing, and the protruded length of the needle.

needle electrospinning process indicated that spinning jets became more stable and concentrated when metal nuts were used as auxiliary field shielding measure (Guo 2012). Therefore, application of metal casing during the needle electrospinning process has been proved to be a simple and effective method for improvement of field intensity and its distribution.

Using Plastic or Metal Sleeve to Needles

Firstly, a round PTFE sleeve was symmetrically placed around the three triangularly arranged needles to further investigate the effect of the plastic sleeve on the field intensity distribution in three-needle electrospinning, the model of which is shown in Fig. 5.48, and the FEA result was displayed in Fig. 5.49. The comparison between the field intensity with/without plastic sleeves shown in Fig. 5.50 and Fig. 5.52 indicated that the needle sleeves could (1) save energy during electrospinning, namely, the increase in field intensity from 1.78e7 to 2.02e7 N/C leads to increased electric field force at the same applied voltage, and then energy consumption will be lowered as a result; (2) effectively reduce End-effect of the three needles, leading to less density of electric field vectors outwards, as shown in Fig. 5.49 compared to Fig. 5.51; (3) result in uniform distribution of field intensity among the three needles.

Furthermore, the application of metal sleeves around symmetrically aligned needles may result in a similar effect on field intensity during the multineedle electrospinning process, including increased electric field force, prolonged stable segment of the spinning jets, and precise control over the deposition position of the formed nanofibers.

Similarly, the material types from which the sleeves are made may also influence the degree of field intensity to be improved. As a consequence, the effect of metal sleeves on field intensity distribution during three-needle electrospinning was also performed, and the results are shown in Fig. 5.50, which displayed the similar changing trend to Fig. 5.49, with higher peak value of field intensity vector (2.13e7 N/C vs 2.02e7 N/C), favoring energy saving, theoretically speaking.

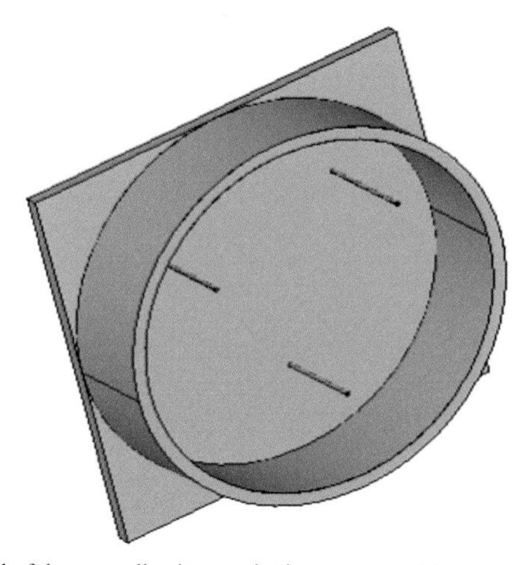

Fig. 5.48: Model of three-needle electrospinning process with round sleeves (3D view).

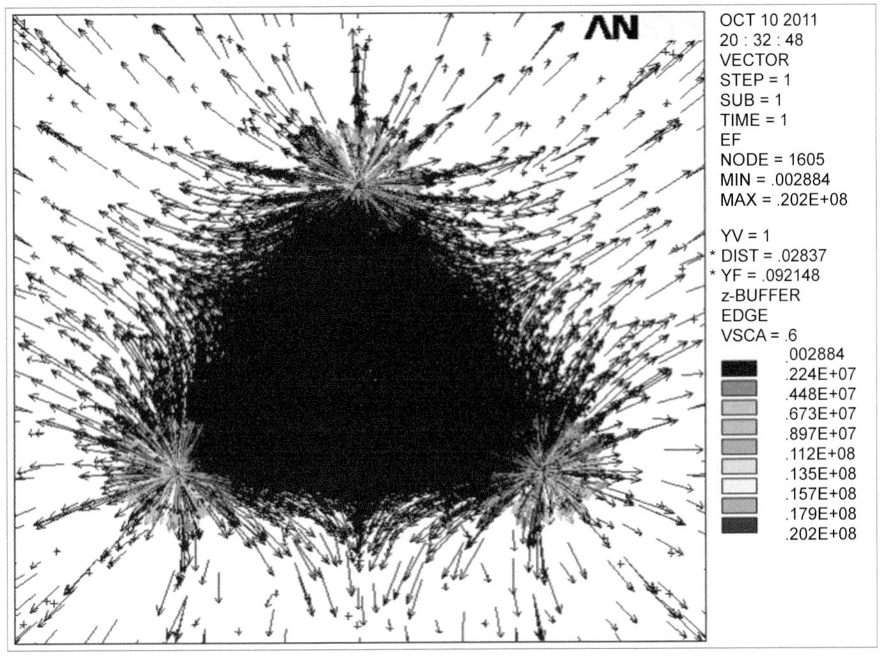

Fig. 5.49: Field intensity vectors in three-needle electrospinning with plastic sleeves (top view).

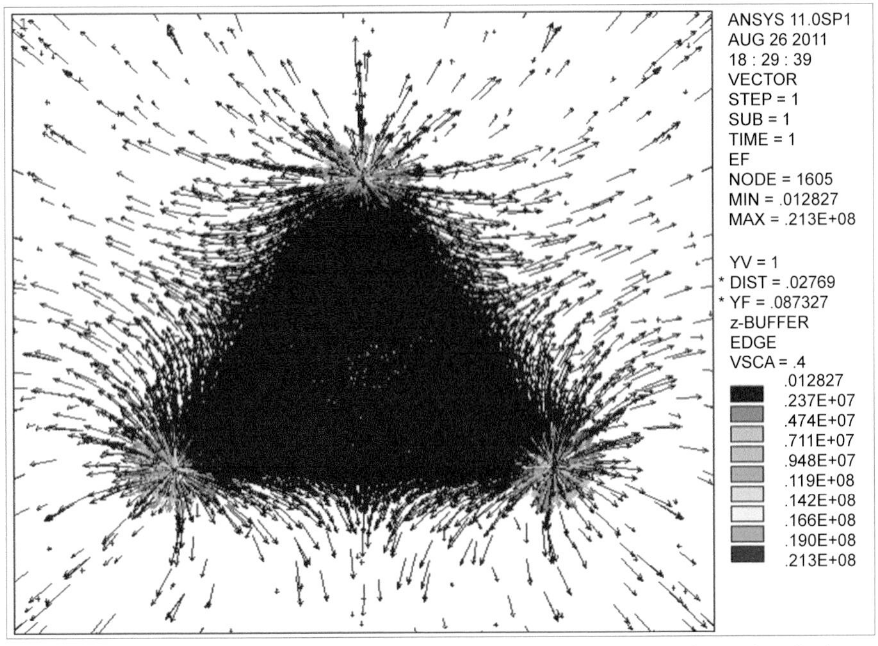

Fig. 5.50: Field intensity vectors in three-needle electrospinning with metal sleeves (top view).

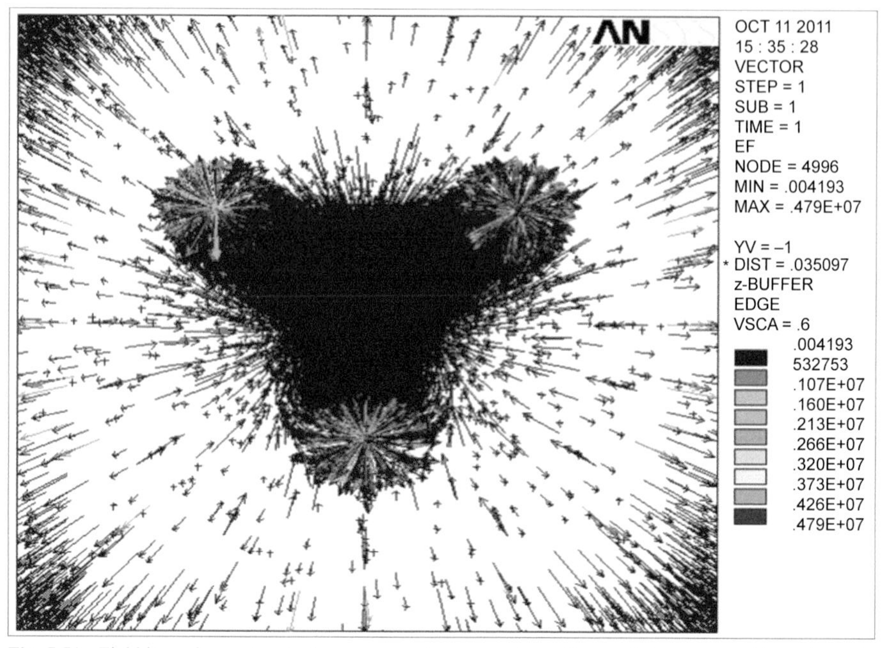

Fig. 5.51: Field intensity vectors in three-needle electrospinning with metal sleeve as auxiliary electrode (upward view).

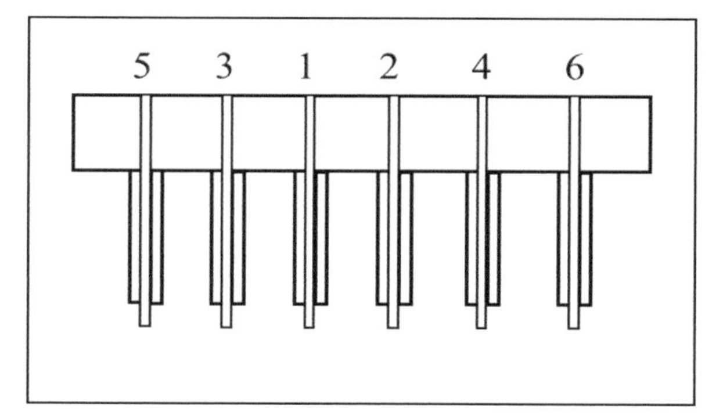

Fig. 5.52: Model of six-needle electrospinning process with PTFE casings.

Using Metal Sleeves as Auxiliary Electrodes

Now that it was found that metal sleeves and metal casings could increase field intensity and restrain End-effect phenomenon during multi-needle electrospinning process, the auxiliary metal electrode made from steel, which is also connected to an applied voltage, was further investigated to understand its effect on field intensity distribution. The FEA result based on a three-needle electrospinning model was shown in Fig. 5.51, which indicated that the application of metal sleeve as an auxiliary electrode dramatically weakened the field intensity (4.79e6 N/C < 2.13e7 N/C), and even less number of field intensity vectors were observed in Fig. 5.51.

Optimization of Parameters Controlling Field Intensity During Needle Electrospinning

Although the previously proposed methods have been theoretically demonstrated to be the effective measures for improvement of field intensity and its distribution during multineedle electrospinning by means of FEA simulation technology, some structural parameters still need to optimize towards the most reasonably structural design of the spinneret in multineedle electrospinning. The model of six-needle electrospinning process, for instance, is to be investigated to obtain the optimal structure parameters of the electrospinneret.

Optimization on Specifications of Plastic Casing. The application of casings is very beneficial to the enhancement in multi-needle electrospinning, but the specifications of the plastic casings need to be optimized, including the wall thickness, inner diameter, and length, in order to obtain the optimal quality of the electrospun nanofibers.

The simplified 2D model of six-needle electrospinning process is shown in Fig. 5.52, where each needle was applied a PTFE based plastic casing, with the needle top being 2 mm longer than the casing top, the wall thickness of the casing being 1 mm, and no space existing between the needle and the casing. The FEA results are shown in Fig. 5.53, where the peak value of field intensity was 5.91e5 N/C.

For further optimal results of the field intensity, the FEA results based on the electrospinning model shown in Fig. 5.52 with 2 cm of needle spacing, 2 mm, and 4 mm of protruded length of needle from casing and wall thickness of casing ranging from 1 mm to 7 mm were depicted in Fig. 5.54. The curves indicated that field intensity in multineedle electrospinning drops with the increase in wall thickness of the plastic casing, meaning higher shielding actions against electric field from plastic casings with thicker walls, due to weaker superposition effect of multineedles with thicker plastic casings.

It could be found in Fig. 5.54 that the two curves showed a similar tendency of field intensity with increasing wall thickness of needle casing, with the needles having shorter protruded length displaying higher field intensity, but the field intensity of the two curves tends to be almost the same when casing thickness is greater than 6 mm, indicating that the protruded needle length is no longer a key factor influencing field intensity distribution in six-needle electrospinning process. Meanwhile, it also could be observed that the field intensity values of the two curves at casing thickness of 6 mm and 7 mm have less difference, indicating that further increase in casing thickness would not have significant influence on field intensity.

Now that it was found from previous FEA work that the applied plastic casings showed great effect on field intensity in multineedle electrospinning, further investigation was performed to understand the change in field intensity before and after applying plastic casings. The FEA results were listed in Table 5.5, from which it was observed that the field intensity of each needle was increased by 30% after applying PTFE casing, implying 30% voltage or electric power could be saved when plastic casings are used in six-needle electrospinning, or finer nanofibers could be produced if plastic casings are used under the same voltage applied.

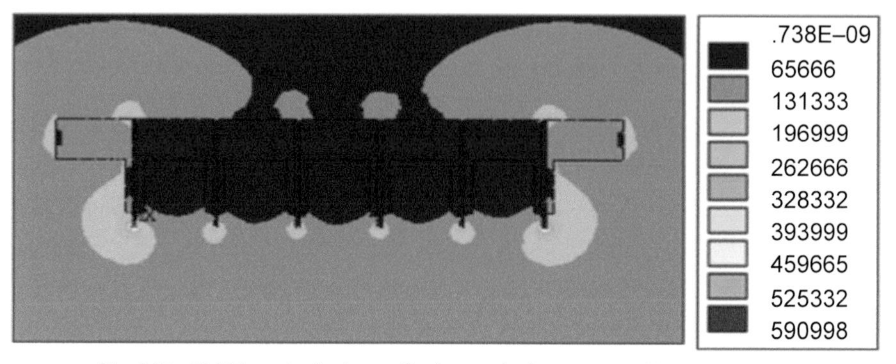

Fig. 5.53: Field intensity in six-needle electrospinning process with PTFE casings.

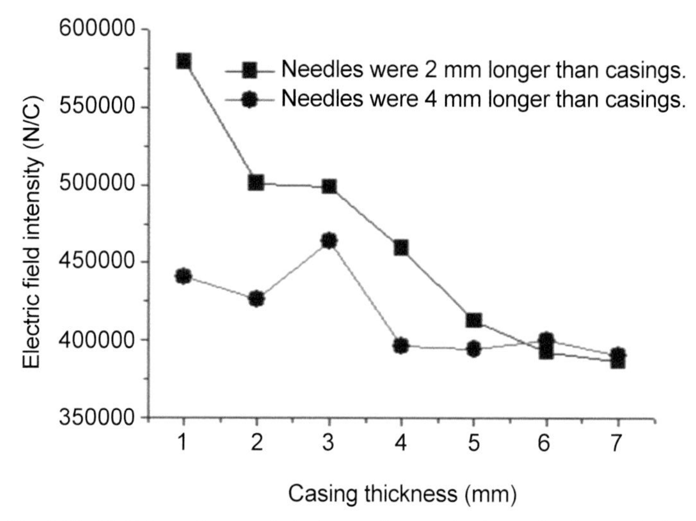

Fig. 5.54: Field intensity vs wall thickness of casing in six-needle electrospinning.

Table 5.5: Field intensity before and after applying PTFE casings in six-needle electrospinning.

Needle No.		**5,6**	**3,4**	**1,2**
Field intensity (N/C)	Without casings	454408	403918	353428
	With casings	590998	525332	459665
Change ratio of E		+30%	+30%	+30%

Optimization of Specifications of Separation Plate. Although the application of plastic casings favors the improvement of field intensity during multineedle electrospinning, a plastic casing needs to be applied to each of hundreds of thousands of needles, which is time-consuming and hard to do. The simplest way to apply a shielding electric field to each needle in large-scale needle electrospinning process is using a separation plate to wrap each needle in the corresponding hole in a plastic plate. The simplified 2D model of six-needle electrospinning process with PTFE separation plate is illustrated in Fig. 5.55, where the needles are relatively independent from each other due to the existence of the PTFE separation plate. The protruded needle length is 2 mm, and the space between each needle and its hole wall is 1 mm. One sample of the corresponding FEA results is shown in Fig. 5.56.

It could be found that the peak values of the field intensity still occurred at the end needles, however, the field intensity when using plastic separation plate was smaller than when using plastic casing (3.54e5 N/C < 5.91e5 N/C). The needle spacing remains constant in the current electrospinning model, therefore the space between the needle and the hole wall of the plastic separation plate will definitely lead to the decrease in the thickness of the separation plate. The parameters of thickness of separation plate and space between needle and hole wall of the separation plate (referred to as "Needle-to-separator gap") should be taken consideration to obtain optimal structural specifications of the spinneret design in multineedle electrospinning.

Further FEA was carried out with the values of Needle-to-separator gap ranging from 0 to 4 mm, and the results are shown in Fig. 5.57.

It was found from Fig. 5.57 that field intensity reached its maximum (5.07e5 N/C) when there was no space between needle and the corresponding hole wall (i.e., the value of Needle-to-separator gap was 0 mm), as shown in Fig. 5.58, and the field intensity decreased initially then increased with the increase in Needle-to-separator gap, and finally it did not change dramatically. Additionally, field intensity showed

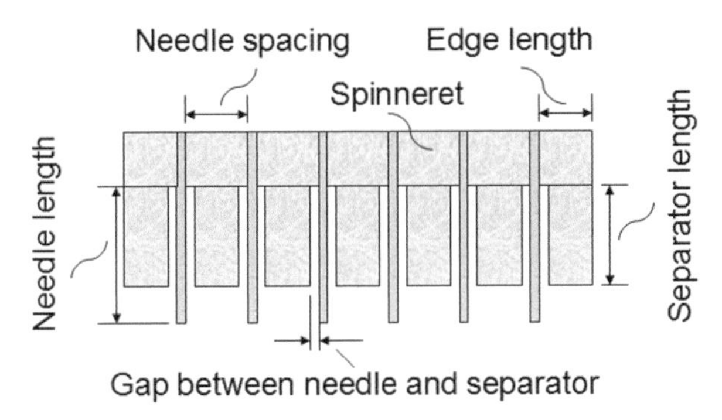

Fig. 5.55: Model of six-needle electrospinning process with PTFE separation plate.

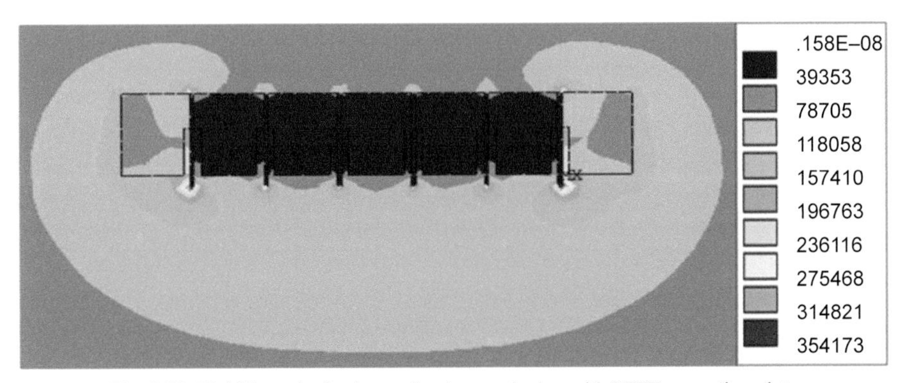

Fig. 5.56: Field intensity in six-needle electrospinning with PTFE separation plate.

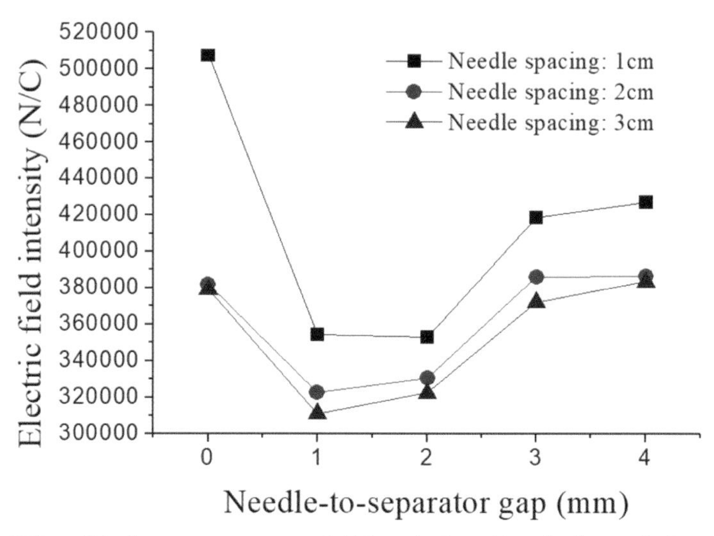

Fig. 5.57: Effect of Needle-to-separator gap on field intensity in multineedle electrospinning with PTFE separation plate.

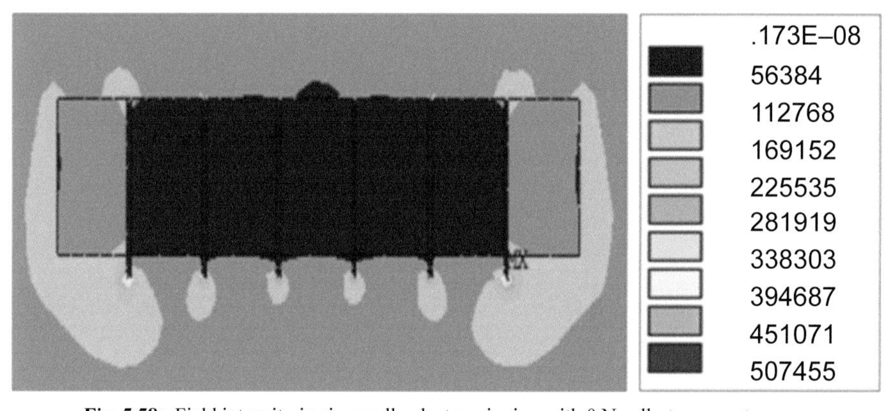

Fig. 5.58: Field intensity in six-needle electrospinning with 0 Needle-to-separator gap.

increasing tendency with the decrease in needle spacing, indicating that smaller needle spacing favored higher field intensity when plastic separation plate was used with needle spacing in the range of 1–3 cm. This result is consistent with that form Fig. 5.16, i.e., field intensity increases as the needle spacing decreases.

Further observation on Fig. 5.58 indicated that the minimal field intensity achieved when the value of Needle-to-separator gap was 1 cm, was when the gap changed from 0 to 1 cm. Further increase of the gap from 1 cm to 4 cm resulted in increase in field intensity, however, the field intensity values were still less than those with no gap between needle and the hole wall of the separation plate.

Therefore, the specifications and dimensions of the needle separation plate, such as the protruded needle length, needle-to-separator gap, the materials used, etc., can be optimized based on 2D FEA on the multineedle electrospinning model, and the results can be used to design and manufacture multineedle electrospinning devices to aim for massive production of nanofibers and their products.

Summary Remarks

So far, large-scale electrospinning is the best way to produce nanofibers for commercial applications in many areas, including gas/liquid filtration, gas/liquid separation, battery separator, supercapacitor separator, water-proof & breathable membrane, biosensor, wound dressing, tissue engineering scaffold, energy storage, and environmental protection, etc., compared to other nanofiber preparation methods. Although different electrospinning principles and technologies have been employed worldwide to produce nanofibrous non-woven webs or yarns for different commercial applications, the low productivity and low quality of the electrospun products are still the key issues needing to be solved as soon as possible to aim for real industrialization of the best nanofiber producing technology.

This chapter deals with the feasible methods for investigating and/or improving the magnitude and distribution of electric field intensity in massive electrospinning, including multineedle electrospinning and multi-bubble electrospinning by means of finite element analysis (FEA), and the results indicated that different values of needle length, needle spacing, applied voltage, and effective measures such as auxiliary metal electrode, plastic/metal sleeve, plastic/metal casing, separation plate, as well as special layout of needles, etc., can be employed to help minimize End-effect, adjust the magnitude and distribution of electric field intensity in multineedle electrospinning, in order to aim for the production of uniform electrospun web with massive needle electrospinning technology. This can be easily realized by combination of the measures addressed in the current chapter.

Acknowledgements

The authors wish to acknowledge grant support from National Natural Foundation of China (NSFC) with the approval Nos. 51373121 and 51573133. We also would like to thank the great supports from our graduate students, family members, colleagues, friends, and all other people who offered kind help and assistance during the book writing process. We are also thankful for the financial support from school projects of Wuhan Textile University, China. Last but not the least, we highly appreciate the editorial team, Taylor & Francis Group, CRC Press, for their kind support and great patience with this chapter.

References

Bennet, A. G. G. 1974. Electricity and modern physics (2nd ed.). UK: Edward Arnold.

Carnell, L. S., Siochi, E. J. and Wincheski, R. A. 2009. Electric field effects on fiber alignment using an auxiliary electrode during electrospinning. Scripta Materialia 60(6): 359–61.

Chen, W. -Y. 2012. Finite element analysis on field intensity distribution during multineedle electrospinning. B.S. Tianjin Polytechnic University, Tianjin, China.

Chen, W. -Y., Liu, Y. -B., Zhang, Z. -R. and Ma, Y. 2014a. Finite element analysis of improvement of field intensity in multi-needle electrospinning. Journal of Textile Research 35(4): 21–5+31.

Chen, W. -Y., Liu, Y. -B., Wang, Y. -Z., Shen, Y. -W. and Guo, L. -L. 2014b. Finite element analysis on electric field intensity and distribution during multi-needle electrospinning process. Journal of Textile Research 35(6): 1–6.

Dang, S. -S., Xu, Y. and Zhang, H. -S. 2010. ANSYS 12.0 multi-physical coupling field finite element analysis from entry to proficiency. Beijing: Machinery Industry Press.

Deitzel, J. M., Kleinmeyer, J. D. and Hirvonen, J. K. 2001. Controlled deposition of electrospun poly(ethylene oxide) fibers. Polymer 42(19): 8163–70.

Donaldson, Ultra-Web [cited April 19, 2018]. Available from: https://www.donaldson.com/en-us.

Dong, L., Liu, Y., Wang, R., Kang, W. -M. and Cheng, B. -W. 2010. Mathematical model of electric field distribution at the critical state in bubble electrospinning. Textile Bioengineering & Informatics Symposium 3(2): 117–20.

Giancoli, D. C. 2008. Physics for scientists and engineers with modern physics (4th ed.): Pearson/Prentice Hall.

Grant, I. S. and Phillips, W. R. 2008. Electromagnetism. manchester physics (2nd ed.): John Wiley & Sons.

Guo, L. -L., Liu, Y. -B. and Chen, W. -Y. 2011a. Simulation on process control of multi-needle electrospinning. Journal of Modern Textile Science and Engineering 2(1): 12–22.

Guo, L. -L., Liu, Y. -B. and Zheng, Y. 2011b. Simulation about multi-needle electrospinning based on finite element method. Advanced Materials Research 332-334: 2157–60.

Guo, L. -L. 2012. Fundamentally theoretical study on large-scale electrospun nanofiber technology. M.S. Tianjin Polytechnic University, Tianjin, China.

Guo, L. -L., Liu, Y. -B., Zhang, Z. -R. and Ma, Y. 2012. ANSYS simulation of homogeneous field intensity during multi-needle electrospinning. Journal of Tianjin Polytechnic University 31(2): 23–6.

Halliday, D. and Resnick, R. 1970. Fundamentals of physics. New Jersey: John Wiley & Sons.

Jiangxi Advanced Nanofiber S&T Co., Ltd. PI nanofiber based lithium ion battery separators. [cited April 19, 2018]. Available from: www.hinanofiber.com.

Kong, C. S., Lee, T. H., Lee, S. H. Kim, H. S. 2007. Nano-web formation by the electrospinning at various electric fields. Journal of Materials Science 42(19): 8106–12.

Kong, H. -Y. and He, J. -H. 2013. A modified bubble electrospinning for fabrication of nanofibers. Journal of Nano Research 23(35): 125–8.

Li, P. -C., Liu, Y. -B. and Chen, W. -Y. 2011. Research on sawtooth type of needleless electrostatic spinning. Journal of Modern Textile Science and Engineering 2(1): 23–32.

Li, P. -C. 2012. Study on sawtooth-based electrospun nanofiber technology. M.S. Tianjin Polytechnic University, Tianjin, China.

Liu, J., Liu, Y. -B., Jiang, X. -M. and Ma, Y. 2015. Improvement of electrospun membrane uniformity by noncircular gear traverse mechanism. Journal of Textile Research 36(7): 116–20.

Liu, J., Liu, Y. -B., Bukhari, S. H., Ren, Q. and Wei, C. -H. 2017. Electric field simulation and optimization on solid-core needles of electrospinning device. Chemical Journal of Chinese University 38(6): 1011–7.

Liu, Y. and He, J. -H. 2007. Bubble electrospinning for mass production of nanofibers. International Journal of Nonlinear Sciences and Numerical Simulation 8(3): 393–6.

Liu, Y., He, J. -H., Xu, L. and Yu, J. -Y. 2008. The principle of bubble electrospinning and its experimental verification. Journal of Polymer Engineering 28(1-2): 55–66.

Liu, Y., He, J. -H., Wang, R., Xu, L. and Yu, J. -Y. 2009. Multiple jets from a polymer bubble in electrospinning. Proceedings of the Fiber Society 2009 Spring Conference. Shanghai, China.

Liu, Y. -B. and Chen, W. -Y. 2012. Finite element analysis for the multi-needle electrospinning process. Journal of Modern Textile Science and Engineering 3(1): 15–27.

Liu, Y. -B. and Guo, L. -L. 2013. Homogeneous field intensity control during multi-needle electrospinning via finite element analysis and simulation. Journal of Nanoscience and Nanotechnology 13(2): 843–7.

Liu, Y. -B., Zhang, Z. -R. and Li, P. -C. 2013. Velmurugan thavasi. fundamental study on needleless electrospinning based on metal (card) clothing. Adv. Mater. Res. 662: 103–7.

Liu, Y. -B., Zhang, L. -G., Yang, Y. -Y. and Ren, Q. 2017a. Study on improvement of field intensity in massive electrospinning. Journal of Chengdu Textile College 34(2): 17–21.

Liu, Y. -B., Cao, H., Zhang, L. -G. and Yang, Y. -Y. 2017b. The improvement of the field strength during the electrospinning of multiple needles. Journal of Chengdu Textile College 34(2): 30–6.
Logan, D. L. and Chaudhry, K. K. 2011. A first course in the finite element method: Cengage Learning.
Lukas, D., Sarkar, A. and Pokorny, P. 2008. Self-organization of jets in electrospinning from free liquid surface: A generalized approach. J. Appl. Phys. 103(8): 084309.
Luo, G. -Z. and Wang, P. 2002. ANSYS and applications in structural analysis. Journal of Hubei Polytechnic University 18(3): 26–8.
Ma, H. and Wang, G. 2009. Basic operation guidelines and answers to frequently asked questions. Beijing: People's Communications Press.
Ma, S. -M. 2016. Study on methods for improving field intensity and its distribution during large-scale electrospinning. M.S. Tianjin Polytechnic University, Tianjin, China.
Niu, H. -T., Wang, X. -G. and Lin, T. 2012a. Needleless electrospinning: influences of fibre generator geometry. Journal of the Textile Institute 103(7): 787–94.
Niu, H. -T., Wang, X. -G. and Lin, T. 2012b. Upward needleless electrospinning of nanofibers. J. Eng. Fibers and Fabrics (07): 17–22.
Pan, C., Han, Y. -H., Dong, L., Wang, J. and Gu, Z. -Z. 2008. Electrospinning of continuous, large area, latticework fiber onto wwo-dimensional pin-array collectors. Journal of Macromolecular Science Part B 47(4): 735–42.
Pokorny, P., Kocis, L., Chvojka, J., Lukas, D. and Beran, J. 2013. Alternating current electrospinning method for preparation of nanofibrous materials. Proceedings of NANOCON. Brno, Czech Republic, EU.
Pryor, R. W. 2011. Multiphysics modeling using comsol: a first principles approach, infinity science series. Burlington,US: Jones & Bartlett Learning.
Reddy, M. N. and Esmaeeli, A. 2009. The EHD-driven fluid flow and deformation of a liquid jet by a transverse electric field. International Journal of Multiphase Flow 35(11): 1051–65.
Shou, W., Dong, L., Liu, Y. and Wang, R. 2011. Mutual interaction of the electric field in multi-bubble electrospinning. Nonlinear Sci. Lett. D 2: 119–21.
Sun, J. 2015. Finite element analysis of field intensity distribution during large-scale multineedle electrospinning. M.S. Tianjin Polytechnic University, Tianjin, China.
Tang, S., Zeng, Y. and Wang, X. 2010. Splashing needleless electrospinning of nanofibers. Polym. Eng. Sci. 50(11): 2252–7.
Taylor, G. 1964. Disintegration of water drops in an electric field. Proc. R Soc. London 280(1382): 383–97.
Taylor, G. 1966. The force exerted by an electric field on long cylindrical conductor. Proc. R Soc. London, Ser A 291(1425): 145–58.
Taylor, G. 1969. Electrically driven jets. Proc. R Soc. London 313(1515): 453–75.
Theron, S. A., Yarin, A. L., Zussman, E. and Kroll, E. 2005. Multiple jets in electrospinning: experiment and modeling. Polymer 46(9): 2889–99.
Wang, X., Niu, H. -T., Wang, X. -G. and Lin, T. 2012. Needleless electrospinning of uniform nanofibers using spiral coil spinnerets. Journal of Nanomaterials 2012(10): 1–9.
Wikipedia. Finite element method. [cited April 13, 2018a]. Available from: https://en.wikipedia.org/wiki/Finite_element_method.
Wikipedia. Superposition principle. [cited April 19, 2018b]. Available from: https://en.wikipedia.org/wiki/Superposition_principle.
Xue, R. 2007. Study on methods for calculating electric field of conductive wires of ±800 kV EHVDC transmission line. Ph.D. Chongqing University, Chongqing, China.
Yang, E. -B., Yang, H. -H. and Liu, Y. -H. 2009. Foundation and applications of engineering electromagnetic field. Beijing: Press of China Electric Power.
Yang, W. -X. 2016. Fractal electrospinning technology and its application in preparation of lithium ion battery separator. Ph.D. Tianjin Polytechnic University, Tianjin, China.
Yang, W. -X., Liu, Y. -B., Zhang, L. -G. and Cao, H. 2016. Optimal spinneret layout in Von Koch curves of fractal theory based needleless electrospinning process. AIP Adv. 6(6): 065223: 1–15.
Yang, Y., Jia, Z. -D. and Guan, Z. -C. 2005. Controlled deposition of electrospun poly(ethylene oxide) fibers via insulators. IEEE International Conference on Dielectric Liquids.

Yang, Y., Jia, Z. -D., Liu, J. -N. and Hou, L. 2008. Effect of electric field distribution uniformity on electrospinning. Journal of Applied Physics 103(10): 89.

Zeleny, J. 1914. The electrical discharge from liquid points, and a hydrostatic method of measuring the electric intensity at their surface. Phys. Rev. 3(2): 69–91.

Zhang, L. 2008. Finite element methods and application foundation of ANSYS programs. Beijing: Science Press.

Zhang, L. -G. 2017. Study on distribution and improvement of field intensity during multineedle electrospinning. M.S. Tianjin Polytechnic University, Tianjin, China.

Zhang, Z. -R. and Liu, Y. -B. 2012. Research on multi-needle electrospinning and needleless electrospinning. Journal of Modern Textile Science and Engineering 3(1): 1–8.

Zhang, Z. -R. 2013. Study on application of electrospun PBS nanofiber materials in blood filtration. M.S. Tianjin Polytechnic University, Tianjin, China.

Zhao, H. -H. and Gao, X. -J. 2004. ANSYS and its applications. Automation in Manufacture 26(5): 20–3.

Zhongke Best (Xiamen) Environmental Protection Technology Co., Ltd. [cited April 19, 2018]. Available from: www.cas-best.cn.

Zhou, F., Gong, R. and Porat, I. 2010. Polymeric nanofibers via flat spinneret electrospinning. Polymer Engineering & Science 49(12): 2475–81.

Chapter 6

Morphology and Structure of Electrospun Nanofibrous Materials

Zhaoling Li,[1,2] *Miaomiao Zhu,*[3] *Ibrahim Abdalla,*[3] *Jianyong Yu*[2] and *Bin Ding*[1,2,*]

Introduction

Electrospinning is a facile and convenient technique for producing nano-scale fibers or micro-scale fibers by using polymer-based solutions under appropriate operation parameters (Khan et al. 2013). It is capable of fabricating some commonly existing synthetic and natural polymeric fibers (Khan et al. 2013, Greiner and Wendorff 2007). Till date, there are more than one hundred kinds of polymers (Lu et al. 2009, Huang et al. 2003, Xie et al. 2008) that have been successfully constructed to continuous nanofibers or microfibers (Lu et al. 2009). The structure and morphology of electrospun nanofibers largely determine their final properties and performance. In order to develop functional nanofibers, it is very necessary to tailor the structure and control the morphology. Via regulating and optimizing both individual electrospun nanofibers and their assembles, it can greatly enrich the application performance of nanofibers in a wide fields of electronic, environment, energy, bioengineering, etc. The polymeric fibers with various morphologies (e.g., electrospun fibers with beads) (Liu et al. 2008), special patterns (Li et al. 2005b), and non-regular cross-sections (Koombhongse et al. 2001) can be created by properly controlling the parameters of electrospinning process (Doshi and Reneker 1995, Bhardwaj and Kundu 2010, Greiner and Wendorff 2007, Subbiah et al. 2005, Casper et al. 2004, Liu et al. 2008, Zhu et al. 2016).

[1] Key Laboratory of Textile Science and Technology, Ministry of Education, College of Textiles, Donghua University, Shanghai 201620, China.
[2] Innovation Center for Textile Science and Technology, Donghua University, Shanghai 200051, China.
[3] College of Materials Science and Engineering, Donghua University, Shanghai, 201620, China.
* Corresponding author: binding@dhu.edu.cn

In addition to polymeric electrospun nanofibers, polymers loaded with active agents, ceramics, nanoparticles, metals, as well as chromophores (Greiner and Wendorff 2007), can be fabricated to polymer/inorganic hybrid or composite electrospun nanofibers. These hybrid or composite nanomaterials can be generally divided into three different kinds, namely polymer/polymer, polymer/inorganic, and inorganic/inorganic composite systems (Khan et al. 2013, Lu et al. 2009). Besides, the spinning equipment and apparatus can be further modified or improved (Ding and Yu 2014) so as to prepare the differential nanofibers with special and complex architectures (Greiner and Wendorff 2007, Bhardwaj and Kundu 2010), such as solid structures (Zhang et al. 2011), Janus structures (Chen et al. 2015, Ma et al. 2015), core-shell structures (Bazilevsky et al. 2007, Sun et al. 2003), hollow structures (Zhao and Jiang 2009), 2D nano-nets (Wang et al. 2011b), 3D spiral coil nanofibers (Lin et al. 2005), and hierarchical secondary structures (Zander 2013, Sun and Xia 2004). In this chapter, different morphologies of polymeric electrospun nanofibers are first introduced, and then the effects of operation parameters on nanofibers' morphologies are systematically described. Additionally, different structures of obtained electrospun nanomaterials and their unique applications are also introduced in detail.

Polymeric Nanofibers

Electrospinning process serves as an effective and useful technique to manufacture the successive polymeric nanofibers with a few nanometers (within 50 to 500 nm (Chronakis 2005)) and several micrometers (Huang et al. 2003) for various applications. Many different categories of polymers have been successfully made into nanofibers on a large scale via electrospinning (Fang et al. 2011) that are mostly in solution and some in melt form (Huang et al. 2003). When the diameters of polymeric fibers decrease from micrometers to nanometers (Lu and Ding 2008), they often exhibit some amazing characteristics, include large surface area-to-volume ratio, small size effect (Sridhar et al. 2015), high porosity (Fang et al. 2008), interconnected fibrous network (Casper et al. 2004) and tunable network geometry (Lu and Ding 2008). These outstanding features enable the polymer nanofibers to be excellent candidates for wide and significant applications. It is worth noting that the operation variables play a vital role in the production process, which can finally determine both the morphology and structure of the resultant electrospun nanofibers (Casper et al. 2004). Great efforts have been devoted to studying the fabrication process as well as the characterization of electrospun nanofibers (Lu and Ding 2008). The morphology, diameter, and patterning of nanofibers can be properly regulated or controlled over a wide range (Ramakrishna 2005). Different morphologies and structures of polymer nanofibers can be obtained via strictly adjusting the processing conditions, such as solution parameters, apparatus parameters, and environment parameters (Ding and Yu 2014). The obtained morphologies could be divided mainly into six categories: regular round cross-sectional nanofibers, beads in electrospun polymer nanofibers, liquid drops on nanofiber mats, micropores within nanofiber surface, nanofibers with special patterns, and non-regular cross-sectional nanofibers.

All mentioned nanofibers will be described in more detail in the following sections. The current advancement and latest progress will be systematically summarized in this growing field. These newest developments can provide some useful information in designing functional nanomaterials, and can benefit the relevant researchers to facilitate the further potential applications of electrospun nanofibers.

Regular Round Cross-Sectional Nanofibers

Numerous work has been done to illustrate the morphologies and properties of electrospun nanofibers as a function of process and material parameters (Subbiah et al. 2005) by many researchers in different groups. The transformation of the polymer from solutions to electrospun nanofibers (Huang et al. 2003) can be affected by solution parameters (concentration, molecular weight, viscosity, surface tension, conductivity, surface charge density, polymer type (Li and Xia 2004b)), processing parameters (applied voltage, feed/flow rate, collector type, distance between tip to collector), and ambient parameters (humidity, temperature) (Subbiah et al. 2005, Ding and Yu 2014, Zhu et al. 2016, Bhardwaj and Kundu 2010). Each parameter significantly influences the morphology and structure of obtained electrospun nanofibers (Ding and Yu 2014). Generally, the diameter of polymer nanofibers prepared by electrospinning are uniform with normal round cross-section when we choose proper parameters. And the regular round cross-sectional nanofibers are the most common type, which possess a wide range of potential applications. However, different electrospun nanofibers of required morphologies can be appropriately fabricated by adjusting those mentioned operation parameters (Bhardwaj and Kundu 2010, Ramakrishna 2005). The following sections will describe how to obtain the polymeric nanofibers with irregular morphologies and structures.

Occurrence of Beads in Electrospinning Materials

The controllable surface morphologies of the electrospun products can be realized, such as random or aligned fibers, smooth or rough surfaces, hollow and porous morphologies, and bends or beaded structures (Megelski et al. 2002, Eda and Shivkumar 2006, Zheng et al. 2006, Li et al. 2004). The presence of beads in electrospun polymeric nanofibers is a common and widely observed phenomenon (Li and Xia 2004). However, the detection of beads on the electrospun fibers produced by the electrospinning procedure is considered one of the main weaknesses of this technique (Miyoshi et al. 2005, Yuan et al. 2004). The applied voltage, viscoelasticity, charge density, and surface tension of the solution can be the main factors to cause the occurrence of beads in the nanofibers. Liu and co-workers (Liu et al. 2008) reported that controlling the concentrations of both polymer solutions and salt additives could largely avoid the production of beads in the electrospinning process. And they also found that solvents had very significant impact on the number and morphology of beads, as well as the size of the electrospun nanofibers. Therefore, these factors largely affect the formation of beads and thus the efficiency of electrospinning technique.

Though beaded structures (Fong et al. 1999) are commonly regarded as defects in the electrospinning method, these special structures have gained great attention for wide use in superhydrophobic surfaces, catalysis (Yoon et al. 2008, Zhang et al. 2012), drug release systems, and photonic devices (Somvipart et al. 2013, Sridhar et al. 2015, Tomczak et al. 2005) due to their specific shapes, sizes, and surface morphologies (Chen et al. 2014). Small-scale nanofibers are necessarily required to prevent the formation of beads on nanofibers, therefore nanofibers with diameters less than 100 nm without beads are very important and challenging to develop more advanced technologies (Huang et al. 2006, He et al. 2007). Figure 6.1 shows the scanning electron microscopy micrographs of electrospun polybutylene succinate (PBS) polymer produced in different solvents (Liu et al. 2008). The electrospun PBS products were 'beads on a string' due to the use of 100% chloroform (CF) (Fig. 6.1a). Many microspheres were observed by using the 100% dichloromethane (DCM) solvent to produce the fibers. Conversely, the mixing solvent of CF/DCM (7/3 w/w) or CF/IPA (8/2 w/w, IPA is isopropanol) would create bigger microporous beads in the final structure, as shown in Fig. 6.1b. Besides, the fibers were achieved with few beads, and the spinning procedure was of higher quality and highest efficiency for the mixed solvent CF/CE (7/3 w/w, CE is 2-chloroethanol), as displayed in Fig. 6.1c. When the solvent was CF, CF/IPA (8/2 w/w), and CF/CE (7/3 w/w) without adding salt under same conditions, the PBS electrospun products appeared to undergo a sharp reduction in the number of beads with an increase of salt content. The effects of different solvents and polymer concentrations on electrospun products were summarized in Table 6.1.

Fig. 6.1: SEM micrographs of PBS electrospun products. The solvent was (a) CF, (b) CF/IPA (8/2 w/w), (c) CF/CE (7/3 w/w), with all other conditions being equal (Copyright 2007, reproduced with permission from Society of Chemical Industry) (Liu et al. 2008).

Table 6.1: Electrospun products for PBS in different solvents (Copyright 2007, reproduced with permission from Society of Chemical Industry) (Liu et al. 2008).

Solvent (w/w)	Polymer concentration (wt %)	Electrospun products
CF	11	Beads + few fibers
CF/DCM (7/3)	11	Microspheres + few fibers
CF/IPA (8/2)	15	Spoon-shaped beads + fibers
CF/CE (7/3)	15	Few beads + fibers
DCM/CE (5/5)	15	Beaded fibers
CF/1-CP (9/1)	15	Beaded fibers
DCM/3-CP (9/1)	15	Fibers
DCM/1-CP (9/1)	15	Fibers
CF/3-CP (9/1)	15	Fibers
DCM/CE (7/3)	15	Fibers
DCM/CE (6/4)	15	Fibers

Formation of Liquid Drops on Nanofiber Mats during Electrospinning

The presence of incomplete solvent evaporation is one of the problems in atmospheric pressure ionization interfaces. The formation of solvent assemblies and related solvent group ions could lead to such incomplete solvent evaporation. Various procedures have been advanced so as to enable the breakdown of these dusters and duster ions prior to introduction into the expansion region (Zhou and Hamburger 1996, Niessen and Tinke 1995). Besides, the droplet desolvation can be obviously improved by additional heating. Lee and co-workers (Lee et al. 2003) reported that the fiber morphology can be improved and the diameter of the filament can be decreased by adding the dimethyl formamide (DMF) to electrospinning solution. During the spinning process, the ketanserin/polyurethane (PU) structures may resemble small knots and reflect the incomplete fiber formation. In the drug separating, the solvent incomplete evaporation will cause the occurrence of inhomogeneity in the drug-laden fibers with crystalline. Additionally, Rayleigh instabilities are favored at low solution viscosity (Deitzel et al. 2001), which affect in particle formation rather than the generation of fibers. As we can see in Fig. 6.2, several small defects were renowned on the electrospun ketanserin/PU fibers, which were not observed in the itraconazole-based systems.

Formation of Micropores within Nanofiber Surface

In recent times, palladium-catalyzed cross-coupling chemistry was used to develop the micro-porous conjugated polymers with high surface areas such as poly(arylene

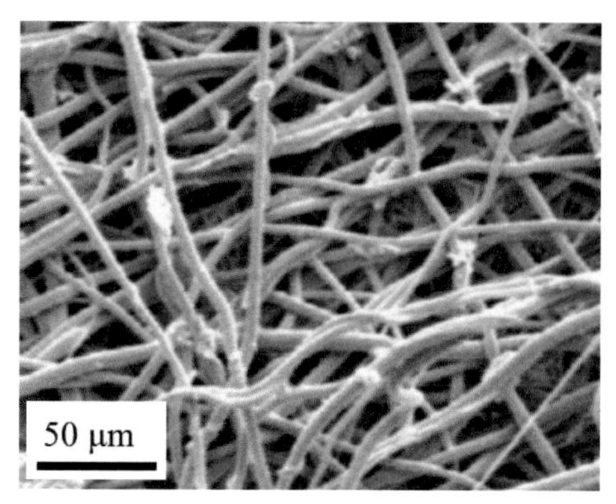

Fig. 6.2: SEM of ketanserin/PU (10% w/w) electrospun fibers at 16 kV (Copyright 2003, reproduced with permission from Elsevier B.V.) (Verreck et al. 2003).

ethynylene) and poly(phenylene) (Jiang et al. 2008b, Weber and Thomas 2008, Jiang et al. 2011, Jiang et al. 2008a). Porous carbon nanofibers and carbon nanotubes with high surface areas up to 900 m^2g^{-1} were accordingly achieved after thermal pyrolysis. Moreover, the HR-TEM was used to confirm the formation of intrinsic microporous structure in rigid carbon-rich networks (Liu et al. 2013, Chen et al. 2012). Determined by rectangular cyclic voltammograms (CV) profiles, the double layer capacitance is mainly resulting from the significant number of interconnected micropores, which can form unique three-dimension (3D) frameworks for charge storage and release (Feng et al. 2009). The enhanced capacitance behavior might originate from the large amount of sub-micropores, which were gradually accessible by the electrolyte along with the circular process (Liu et al. 2013).

Jiang and co-authors (Jiang et al. 2008b) have reported that micropore size distribution can be readily realized by simply controlling the rigid node-strut topology, particularly the average strut length. In addition, in order to signify a huge variation in surface areas and micro-pore volumes, some sorbents were properly selected. Generally, there are two different categories of sorbents, namely non-porous materials such as aerosols and graphite, the other one is micro-porous materials, such as activated carbons. Actually, activated carbons include highly micro- and mesoporous carbon materials (Nijkamp et al. 2001). The open multi-walled carbon nanotubes (MWNTs) are meso-porous, whereas open single-walled carbon nanotubes (SWNTs) are micro-porous (Eswaramoorthy et al. 1999, Mauter 2011). Most carbon nanofibers (CNFs) and carbon nanotubes (CNTs) are consisted of crystallized graphitic cylindrical layer. Conversely, one-dimensional porous carbon materials mainly contain the amorphous structures. Besides, the synthesis of most structures of carbon materials has been extensively exploited using the carbonization process of carbon precursors in the pores of suitable templates (Eswaramoorthy et al. 1999, Feng et al. 2009).

Formation of Nanofibers with Special Patterns

The electrospun nanofibers are usually collected on the nonwoven mats, in which the obtained nanofibers are randomly oriented (Li et al. 2005b, Dai et al. 2002).

Electrospinning has been widely explored for manufacturing special patterns of nanofibers (Li and Xia 2004, Srinivasan and Reneker 1995, MacDiarmid et al. 2001, Bognitzki et al. 2001, Larsen et al. 2003, Ko et al. 2003). Numerous applications require patterned constructions such as the materials with specific patterned structure or parallel orientation, and those materials may have specific biological effects on tissue regeneration (Lee et al. 2005, Zong et al. 2005). Electrospinning method has been used to prepare the uniaxially aligned nanofibers made from different raw materials. And a variety of materials such as organic polymers, ceramics, carbons, and polymer/ceramic composites were confirmed to be successfully electrospun as the uniform nanofibers (Li et al. 2003a, b). Furthermore, this technique is also versatile for producing the core/shell or hollow structure nanofibers (Li and Xia 2004, Li et al. 2004, Li et al. 2005a). Figure 6.3 displays a diagram design of fibrous tubes production by electrospinning method using three-dimension columnar collectors (w, working collector; sa, stick assistant collector; pa, plane assistant collector) (Zhang and Chang 2008). It is worth noting that controlling the shapes of collectors can effectively tailor the macroscopic structures of the tubes. Figure 6.4 presented that the fibers collected at the edge of larger insulating area electrode are straighter than that on an electrode with a smaller insulating area. Figure 6.3a–c was the dark-field optical micro-graphs representing the orientation of fibers deposited on the edges of gold electrodes with different insulating areas. The inset images,

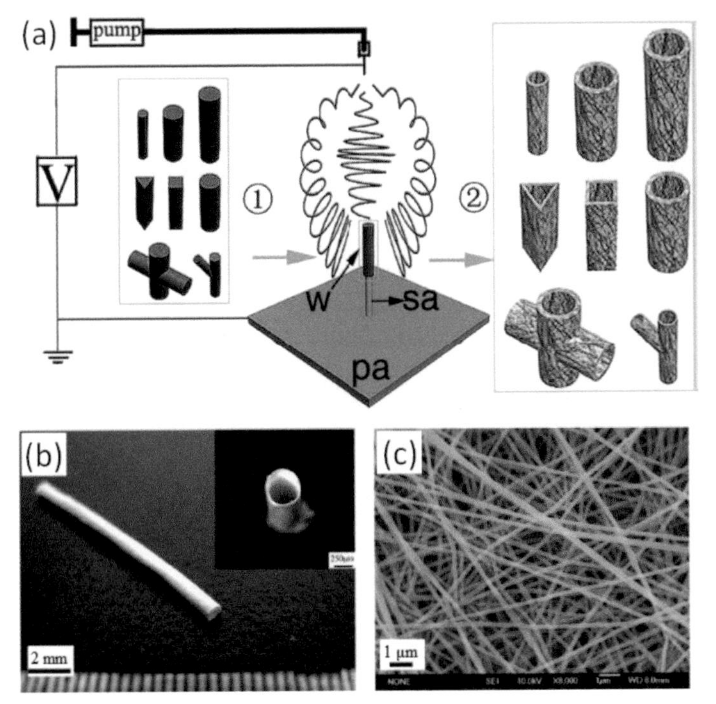

Fig. 6.3: (a) Schematic illustration of fabrication of fibrous tubes by electrospinning technique using 3D columnar collectors, (b) Fibrous tube with a diameter of 500 µm (inset is the cross-section image), and (c) SEM image of fiber assemblies of the tube (Copyright 2008, reproduced with permission from American Chemical Society) (Zhang and Chang 2008).

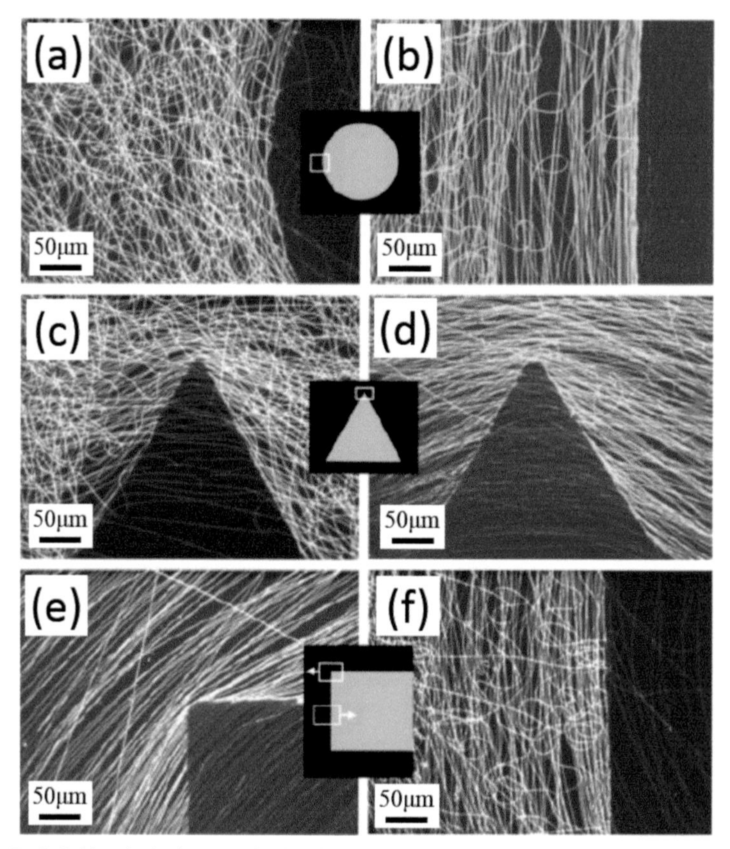

Fig. 6.4: Dark-field optical micro-graphs showing the orientation of fibers deposited on the edges of gold electrodes with different insulating areas. The insets are images taken in the transmission mode, showing the areas from which the dark-field images were captured (Copyright 2005, reproduced with permission from American Chemical Society) (Li et al. 2005b).

taken in the transmission mode, displayed the areas from which the dark-field images were captured. Besides, with an increase of the area of the insulating region, the degree of orientation became dominant (Li et al. 2005b). Li and co-workers (Li et al. 2003b) reported that the uniaxial alignment can be realized by collecting the electrospun nanofibers over a gap formed between two conductive substrates, and finally controllable structures and patterns can be successfully obtained in the as-spun mats (Zhang and Chang 2007). A patterned collector with protrusions was used to investigate the influence of electrospinning parameters on the order degree of the patterns. In addition, the voltage, feeding rate, and volume ratio of solvent largely affect the patterned structures (Fig. 6.5a–c). And a fully patterned structure could be produced under the suitable parameters (Zhang and Chang 2007, Pennella et al. 2013). Moreover, with an increase of voltage more than 15 kV, the morphology of the fibers are going to change as some fibers are adhered together, and therefore the ordered orientation of suspended crossing fibers will be changed. The ordered fiber arrangement cannot be observed when the voltage is less than 10 kV. Moreover,

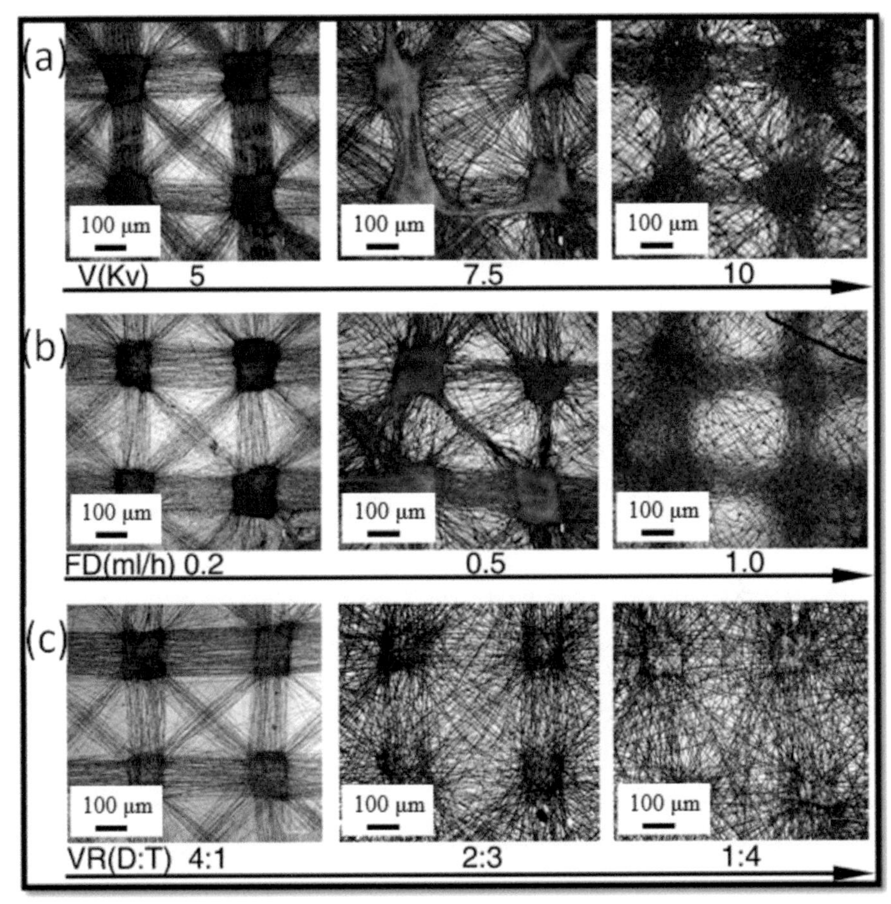

Fig. 6.5: (a) Influence of voltage on patterned structures (v = voltage). (b) Influence of feeding rate on patterned architectures (FD = feeding rate). (c) Influence of volume ratio of solvents (DMF/THF) on patterned architectures (VR = volume ratio; D = DMF; T = THF) (The scale bar is 100 μm) (Copyright 2008, reproduced with permission from American Chemical Society) (Zhang and Chang 2008).

when THF volume ratio in solvents increased, both the ordered fiber arrangement and specific fiber deposition on protrusions changed, and patterned architectures tended to disappear.

Pan and co-authors (Pan et al. 2008) used a micro-matrix collector to produce a latticework fibers pattern, and then analyzed the difference of a plane collector and 2D collector devices for the first time from an electrostatic field view. Furthermore, it is possible to make patterns of nanofibers in particular by using this simple method, because it is easy to strip the nanofibers film from the 2D collector (Katta et al. 2004), and thus it has probable applications in cell cultures. It is also imaginable to provide a common strategy for the assembly of polymer fibers measured by the location of the electrodes. The fibers pattern consisted of polyvinyl pyrrolidone (PVP)-based sub-micron fibers with diameters within a range of 910 nm to 1300 nm, which have promising applications in tissue cell cultures.

Non-Regular Cross-Sectional Nanofibers

In addition to round nanofibers, different cross-sectional shapes of thin nanofibers can be fabricated by electrospinning technique from polymer solution. Koombhongse and co-authors (Koombhongse et al. 2001) have studied different kinds of fibers such as branched fibers, flat ribbons, ribbons with other forms, and fibers that were split longitudinally from longer fibers. Lim and co-authors (Lim et al. 2006) have reported due to the hydrogen bonding of the silanol groups of silica and carbonyl groups of polyacrylamide (PAM), the silica particles were well dispersed in the PAM solution, and it can also decrease the macroscopic phase separation (Jang and Park 2002). Figure 6.6a–e exhibited SEM images of the silica/polyacrylamide (PAM) nanofibers produced at 0.9 kV/cm for different sizes of silica particles. The silica particles (300 nm) were dissolved in an aqueous PAM solution and finally obtained the electrospun PAM nanofibers, as shown in Fig. 6.6b. Alternatively, the salts soluble in organic solvent could be used to increase the electrical conductivity of the solution. Therefore, the jets could carry more current, which resulted in smooth surface and

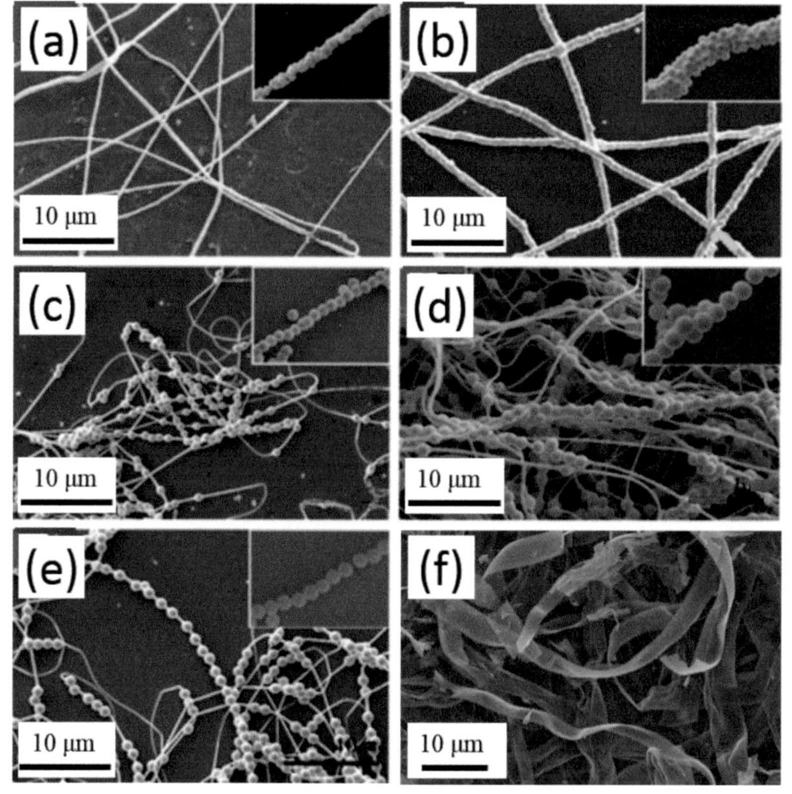

Fig. 6.6: SEM images of electrospun composite PAM nanofibers for various sizes of silica particles: (a) 100 nm, (b) 300 nm, (c) 450 nm, (d) 700 nm, and (e) 1 μm (Copyright 2006, reproduced with permission from American Chemical Society) (Lim et al. 2006). The insets show calcined 1-D silica particle assemblies and (f) SEM image of 4 wt% $AgNO_3$-doped mesoporous silica nanostructured ribbons (Khan et al. 2013).

ribbon-like morphology, as shown in Fig. 6.6f. It means that selection of solvent and inorganic salt may play an important role (Khan et al. 2013). The geometrical shapes formed by the fluid jet are often well-maintained in the shape of solid fibers using electrospinning method. Figure 6.6e showed a web made from the skin connected the two tubes (hollow or filled) and shaped after atmospheric pressure forced the skin into contact, initially at separated points. Then, interconnected forces tend to raise the area of the web by sticking together the skin from opposite sides of the tube. The formation of flat ribbons of the sort were resisted due to the stiffness of the skin and the self-repulsion of the charge on each of the small tubes, shown in Fig. 6.6d.

Figure 6.8a demonstrated the broken end of an individual tube that was fractured in tension at the temperature of liquid nitrogen. The broken end provides a cross-sectional view of folded layers of skin that possesses a thickness of a few hundred nanometers. Figure 6.8b showed different forms of the fibers, such as the flat fiber from 10% poly(ether imide) in hexafluoro-2-propanol. The section that runs parallel across the center of the Fig. 6.8b was seen from the wide side. The segment that runs from the upper left to the lower right was seen edgewise. Figure 6.8c, d presented the flat fiber of 30% polystyrene (PS) in dimethyl formamide (DMF) solvent with an enlarged image showing the irregular wrinkling. The electrospinning method is demonstrated to be a versatile and effective process for producing fibers from microscale to nanoscale. A variety of materials, such as polymers (Reneker and Chun 1996, Doshi and Reneker 1995, Bognitzki et al. 2001, Casper et al. 2004, Khil et al. 2004) inorganic, hybrid compounds nanofibers, and regular/non-regular cross-sectional nanofibers have been prepared using this technique (Ko et al. 2003, Larsen et al. 2003, Caruso et al. 2001, Dai et al. 2002, Wang et al. 2004, Lu et al. 2005a, b). The skin of the jet sometimes irregularly wrinkled in a bend, as shown in Fig. 6.9. The improvement of skins on polymer fibers spun using polymer solutions is a recognized method detected in manufacturing the textile fibers from viscose and other polymers (Cumberbirch et al. 1961, Capone 1995). Kang and co-authors (Kang et al. 2010) have synthesized Ag-doped silica nanostructured ribbons by electrospinning. Thus, the ribbon shapes with high surface and thinner thickness enabled their strong mass transition ability. As a consequence, the ribbons with high width/height ratio have a uniform and continuous morphology.

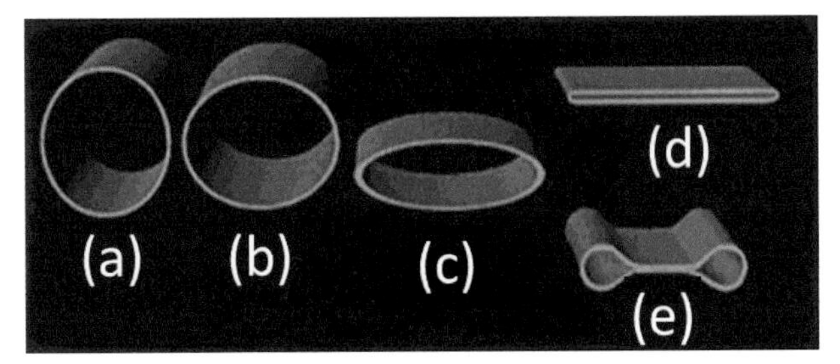

Fig. 6.7: Photographs showing a web made from the skin connected the hollow or filled tubes, the circular cross section (a, b) became elliptical and then flat. Flat ribbons are formed when Fig. (c) is followed by a flat ribbon, as shown in Fig. (d). An alternative collapse mode is a ribbon with two tubes, as shown in Fig. (e) (Copyright 2001, reproduced with permission from John Wiley & Sons, Inc.) (Koombhongse et al. 2001).

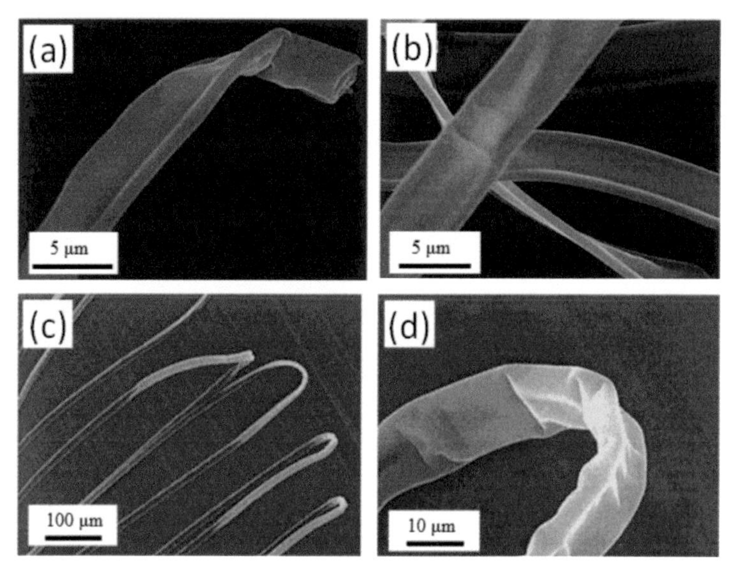

Fig. 6.8: (a) The broken end of a small tube (b) flat fiber from and (c, d) flat fiber from 30% PS in DMF solvent: bent ribbons, with an enlarged image showing the irregular wrinkling of a thin skin (Copyright 2001, reproduced with permission from John Wiley & Sons, Inc.) (Koombhongse et al. 2001).

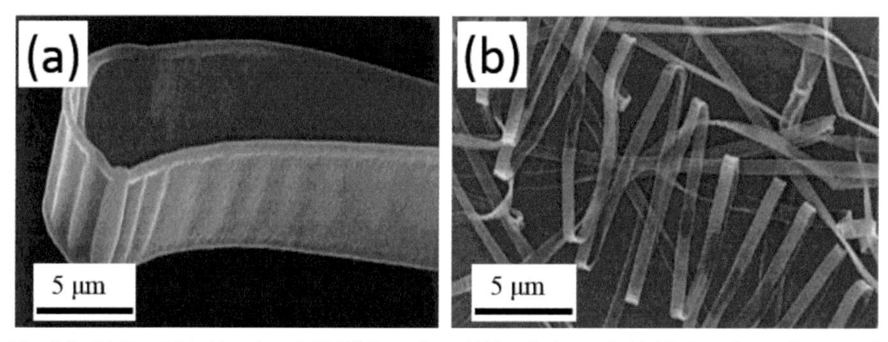

Fig. 6.9: (a) A wrinkled bend, and (b) Ribbons from 10% poly (ether imide) in hexafluoro-2-propanol (Copyright 2001, reproduced with permission from John Wiley & Sons, Inc.) (Koombhongse et al. 2001).

Differential Nanofibers

The electrospinning procedure could be modified in order to produce electrospun fibers with desired structure and features (Ding and Yu 2014). Generally, it is crucial to find out how to prepare electrospun nanofibers with different circular cross sections according to design, and the main objective of electrospinning for special applications is to fabricate various especial kinds of nanofibers (Ding and Yu 2014). What's more, other kinds of structure of polymer nanofibers may be advantageous to particular applications and functionality. Up till now, through attentively adjusting the fabricating conditions of electrospinning process, various particular structures of electrospun nanofibers have been successfully prepared and developed, such as solid structures, Janus structures, core-shell structures, hollow structures, hierarchical secondary structures, 2D nanonets, and 3D spiral coil nanofibers. Those differential nanofibers with complex architectures can be produced by distinctly special electrospinning methods (Greiner and Wendorff 2007).

Solid Structures

Electrospinning is a highly accomplished approach to prepare successive polymeric nanofibers with diameters ranging from nanoscale to microscale (Greiner and Wendorff 2007). And the solid structure is one of the most common structures in electrospun nanofibers. The solid nanofibers can be prepared easily and directly, and do not require any post treatment or modification techniques. The properties of solid electrospun nanofibers distinguish themselves greatly from those of hollow electrospun nanofibers. Among different types of electrospun nanofibers, tubular structured nanofibers (e.g., hollow nanofibers) double the surface area contrasted with the normal solid nanofibers (Zhang et al. 2011). This special characteristic is rather useful in surface related applications, such as photocatalysis (Zhang et al. 2011). However, the electrospun nanofibers with solid structures possess various forms (e.g., Janus nanofibers, hierarchical nanofibers, helical nanofibers, and nano-nets, and these structures will be described in the following sections) via different treatment, modification techniques, or prepared directly using different electrospinning apparatuses (such as the syringe needle or spinneret). These solid electrospun nanofibers have various properties, resulting in different performance that can be widely applied in many fields compared to hollow nanofibers, such as personal protective clothing (Faccini et al. 2012), tissue engineering (Li et al. 2002), filtration (Gopal et al. 2006), nano-sensors (Aussawasathien et al. 2005), and some industrial applications (Doshi and Reneker 1995).

Janus Structures

'Janus' is an ancient Roman God's name, who has two faces turning in the direction of the future and past, respectively (Wang et al. 2015). Analogously, Janus materials have two obviously remarkable chemical properties or surfaces on the two sides. Moreover, Janus electrospun nanofibers often possess versatile properties that can be used as multifunctional membranes (Wang et al. 2015). This capacitates the double-component materials having different features from each of the polymer components, for instance, one of the polymer could enhance the mechanical performance of the membranes and the other component could contribute to a desired chemical functionality and so forth (Gupta and Wilkes 2003). Till date, almost all the reported Janus nanomaterials have been created by bottom-up methods, for example, molecular self-assembly (Bhaskar and Lahann 2009, Chen et al. 2008, Starr and Andrew 2013). However, electrospinning is a convenient top-down approach for preparing nanofibers (Yu et al. 2017), and these multifunctional Janus nanofibers can be fabricated by side-by-side electrospinning (Ma et al. 2015), as shown in Fig. 6.10a (Zhou et al. 2015). It is worth noting that the side-by-side electrospinning contains some complicated interactions between rheology, electrodynamics, and fluid dynamics, and it is a big challenge to control the motion in joint of two different solutions in a side-by-side fashion under an electrical field from metal spinneret to

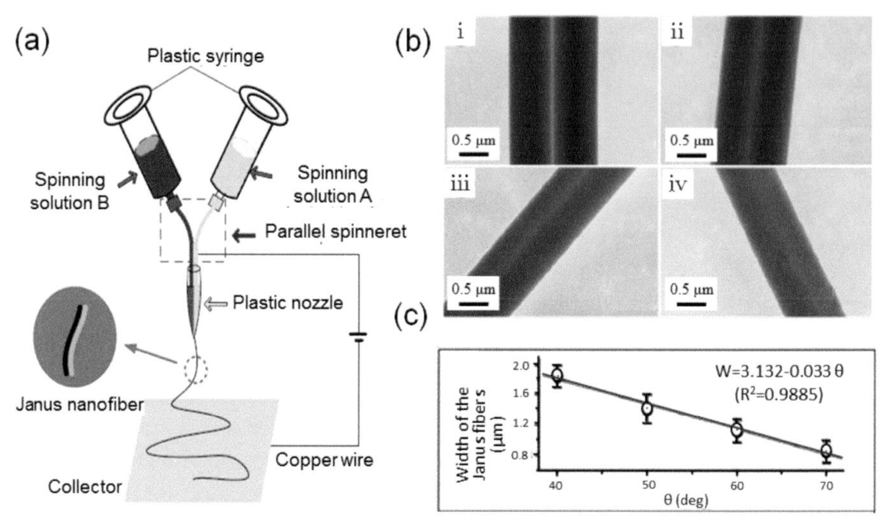

Fig. 6.10: (a) Schematic illustration of the electrospinning device for fabricating Janus nanofibers (Copyright 2015, reproduced with permission from Springer Science+Business Media New York) (Zhou et al. 2015). (b) Adjustable interfacial area and width of the Janus electrospun nanofibers by turning the port angle (Copyright 2015, reproduced with permission from Royal Society of Chemistry) (Chen et al. 2015). (i–iv) TEM images of Janus electrospun nanofibers generated by spinnerets with a port angle of 40°, 50°, 60°, and 70°, respectively. (c) The linear relationship between the port angle (θ) and the width of the Janus electrospun nanofiber (W) (Copyright 2015, reproduced with permission from Royal Society of Chemistry) (Chen et al. 2015).

collector (Chen et al. 2015). And the key point to solve this issue existing in the need to manipulate electrospinning solution with very different features. And it is highly needed that the Janus nanofibers are ejected simultaneously from the metal spinneret without separation under the electrical field (Yu et al. 2017). This implies that both conductivity and viscosity of each electrospinning polymer solution are serving crucial process parameters for double-component electrospinning process (Gupta and Wilkes 2003). What's more, a careful and reasonable spinneret design is really significant. The spinneret are supposed to provide a template for generating the desired nano-sized structure, and also it can be used to control the behavior of the polymer solutions under the high voltage (Yu et al. 2017).

Gupta and Wilkes (Gupta and Wilkes 2003) firstly reported the fabrication of side-by-side Janus electrospun nanofibers (Chen et al. 2015). They designed and prepared an electrospinning equipment where the two different polymer solutions, namely (poly(vinyl chloride)/poly(vinylidiene fluoride) and poly(vinyl chloride)/ segmented polyurethane) were electrospun synchronously in a side-by-side manner. And Bi et al. (Bi et al. 2015) have successfully fabricated magnetic-luminescent bifunctional [$CoFe_2O_4$/PVP]//[YAG: 5% Eu^{3+}/PVP] Janus nanofibers (Bi et al. 2015) and novel [YF_3:$Eu3^{3+}$/PVP]//[PANI/Fe_3O_4/PVP] (Yin et al. 2015) photoluminescence-electricity-magnetism trifunctional Janus nanofibers by a facile and convenient electrospinning technology using a side-by-side double spinneret. Compared to composite electrospun nanofibers, those Janus nanofibers possess higher fluorescent intensity, higher conductivity, and saturation magnetization. Chen et al. (Chen et al.

2015) found accidentally that the structures of the Janus electrospun nanofibers could be manipulated easily, mainly by varying the port angle in the process of optimizing and adjusting the spinneret design. The variations in Janus structure include (1) the width of fibers, (2) the surface or volume of the double sides, and (3) the interfaces of Janus nanofibers. As shown in Fig. 6.10b–c, by manipulating the port angle of the syringe needle, electrospun Janus nanofibers with adjustable structures on the basis of width, volume of each side, and the interfacial area can be properly adjusted. For all these researches, the Janus polymeric nanofibers can be successfully prepared by strictly controlling the operation parameters of the electrospinning process.

Core-Shell Structures

As discussed before (Section 6.2.1), the conventional electrospinning device often contains a common and single capillary as the syringe needle, which enables the continuous fabrication of solid polymer fibers. In many cases, there are some obstacles in conventional electrospinning with common single syringe needle to fabricate functionalized nanofibers (Greiner and Wendorff 2007). However, if we changed the single syringe needle with the coaxial syringe needles, as shown in Fig. 6.11a (Zhang et al. 2004), the core-shell electrospun nanofibers could be formed. This modified electrospinning technology can be effectively applied, which is called coaxial electrospinning (Lu et al. 2009). If the proper electrospinning process parameters are chosen, the core-shell electrospun nanofibers can be prepared with high precision from a great variety of materials including polymer and some inorganic substances (Greiner and Wendorff 2007). Therefore, the coaxial electrospinning technique provided us with a possibility to fabricate functional electrospun composite nanofibrous membranes (Lu et al. 2009). Therefore, there have been numerous efforts devoted to overcoming the drawback of single component electrospun membranes through reasonable designing of the core-shell nanofibers by coaxial electrospinning, such as poor mechanical stability (Hua et al. 2016), uncontrollable drug release rate, and so on (Sun et al. 2003). And in the past few years, there were a large number of reports on the fabrication of core-shell composite nanofibers through coaxial electrospinning, including polymer/inorganic, polymer/polymer and inorganic/inorganic composites (Lu et al. 2009).

Ma et al. (Ma et al. 2016) designed a high-strength core-shell cellulose acetate-polyamide (CA-PI, polymer/polymer) nanofibrous membranes for oil-water separation, which significantly enhanced the potential in industrial applications of electrospun membranes. Furthermore, they fabricated high strength and high flexible core-sheath CA-PI nanofibrous membranes (Fig. 6.11b) (Ma et al. 2017), which are excellently suited for an energy-efficient oil-water separation (Ma et al. 2017). In addition to polymer/polymer electrospun core-sheath nanofibers, the polymer/inorganic composite core-sheath nanofibers have also been created by coaxial electrospinning (Graeser et al. 2007, Larsen et al. 2004). For example, Martin Graeser et al. (Graeser et al. 2007) have prepared polymer core-shell fibers, which contained some salts of platinum palladium and rhodium. And the testing results showed that the salts of Pd-Rh have been incorporated into polymer nanotubes during the generation of a new catalyst system, which exhibited high activity in hydrogenation reductions.

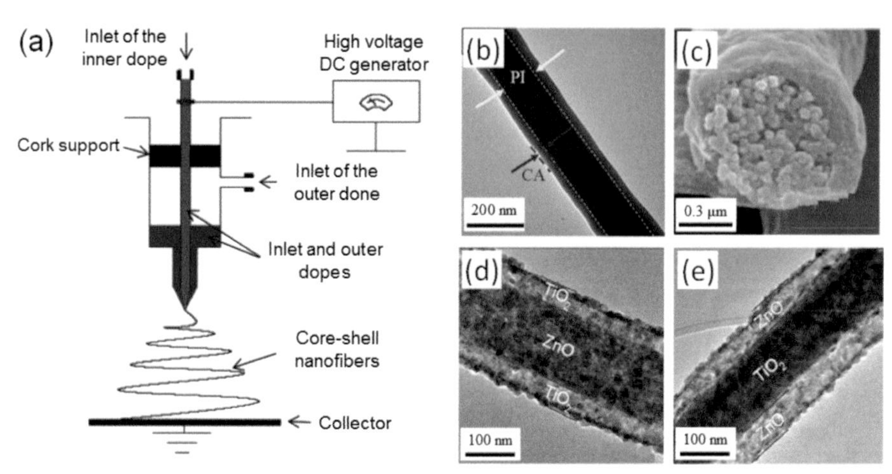

Fig. 6.11: (a) Schematic diagram of the coaxial electrospinning equipment used in preparing the core-sheath composite nanofibers (Copyright 2004, reproduced with permission from American Chemical Society) (Zhang et al. 2004). (b) TEM image of a flexible PI/CA nanofiber (Copyright 2016, reproduced with permission from Elsevier B.V.) (Ma et al. 2017). (c) A cross-sectional SEM image of a single SiNP@C composite nanofiber indicating that the full core of Si NPs is wrapped by carbon shell (Copyright 2012, reproduced with permission from American Chemical Society) (Hwang et al. 2012). (d) TEM images of the ZnO-TiO_2 and (e) TiO_2-ZnO core-shell hetero-junction nanofibers, respectively (Copyright 2014, reproduced with permission from Royal Society of Chemistry) (Kayaci et al. 2014).

What's more, coaxial electrospinning can be used in generating inorganic/inorganic composite fibers as well (Gu et al. 2007, Li et al. 2005a, Joo 2006). For example, the core-sheath fibrous structures fabricated by a coaxial electrospinning technique are scalable and feasible for SiO_2 nanoparticles based lithium ion batteries anodes with highly robust electrochemical property (Fig. 6.11c), which have been reported by Hwang et al. (Hwang et al. 2012). In addition, Kayaci and coworkers (Kayaci et al. 2014) have designed core-sheath hetero-junction ZnO/TiO_2 nanofibers in which only the 'shell' of the hetero-junction is exposed outside to take part in the photocatalysis (the structures is shown in the Fig. 6.11d–e).

Conventional coaxial electrospinning needs a core-shell nozzle attached to a double-compartment syringe or supply of two different polymer solutions by means of two separate syringe pumps and pipelines leading to a core-shell nozzle (Yarin 2011). In addition, a core-sheath electrospun polymer nanofibers is probably formed through an ordinary electrospinning device with a single needle under the condition of an emulsion which contains two polymer solutions (Bazilevsky et al. 2007). In particular, when the electrospinning solution contains two polymers, the phase separates as the solvent is evaporated (Li and Xia 2004), and this process results in a coaxial electrospun nanofibers in another way. As Fig. 6.12a shows, it is an ordinary electrospinning set-up which could be used to fabricate coaxial nanofibers through polymer emulsion. Bazilevsky et al. (Bazilevsky et al. 2007) fabricated the core-shell electrospun nanofibers which used emulsion (Fig. 6.12b) of poly(methyl methacrylate) (PMMA) and polyacrylonitrile (PAN) through a single needle. This periodical process illustrated that a PMMA/DMF droplet was dragged in the tip of a single-liquid Taylor cone which created by PAN/DMF matrix. In this case core-shell

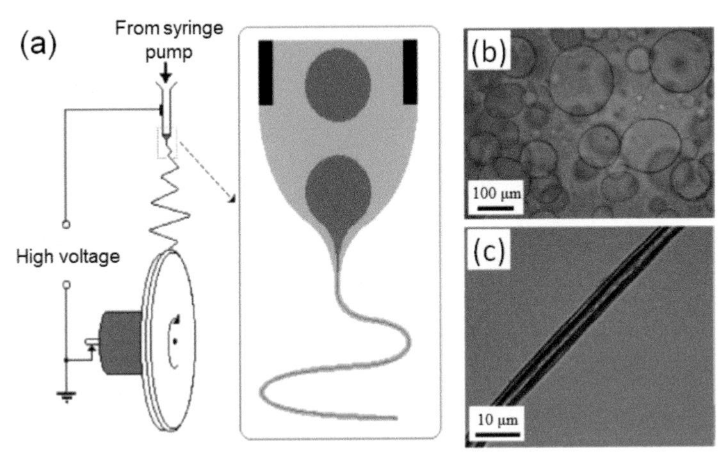

Fig. 6.12: (a) Emulsion co-electrospinning of PMMA/PAN blend in DMF using a single needle. The PMMA/DMF droplets (core) are shown in dark gray, while the PAN/DMF matrix (shell) is lighter. (b) PMMA/PAN emulsion about one day after mixing equal amounts of each polymer in DMF to generate a homogeneous emulsion. The PMMA/DMF droplets are dispersed in the surrounding PAN/DMF matrix. (c) Optical images of electrospun core-shell (PMMA-PAN) microfibers (Copyright 2007, reproduced with permission from American Chemical Society) (Bazilevsky et al. 2007).

Taylor cone at the syringe needle exit is only momentary. Therefore, in principle, the core-shell nanofibers, which are fabricated by polymer emulsion, should not possess a continuous core. However, as found by testing the as-spun PMMA/PAN nanofibers (As shown in Fig. 6.12c), there are few disruptions of the core. The reason is that there is a very strong stretching of emulsion in the coaxial electrospun jet. And the emulsion electrospinning technique is expected to be a promising approach in future biomedical applications, especially in controlled drug delivery system (Lu et al. 2009).

Hollow Structures

The electrospinning set-up can be modified to enable the fabrication of nanofibers with a differential structure, as discussed above. Many groups have demonstrated that core-sheath and hollow nanofibers with adjustable dimensions can be prepared by two or several immiscible solutions via a coaxial nozzle during electrospinning process (McCann et al. 2005, Li and Xia 2004, Xue et al. 2017). For the system shown in Fig. 6.13a (Li and Xia 2004), any substance soluble in the oil phase could be straight introduced into the interiors of nanofibers, as they were electrospun from the syringe needle (Li et al. 2005a). Li and Xia demonstrated the hollow electrospun nanofibers with controllable dimensions could be expediently prepared by electrospinning two immiscible polymer solutions via two capillary syringe or a coaxial needle (Li and Xia 2004). They have reported the approach which is using an ethanol solution (including poly(vinyl pyrrolidone) (PVP) and $Ti(OiPr)_4$) and heavy mineral oil as the materials for shell and core of the fibers. As shown in Fig. 6.13b, the long and hollow electrospun nanofibers with uniform diameters and regular circular cross-sections were successfully generated by co-electrospinning.

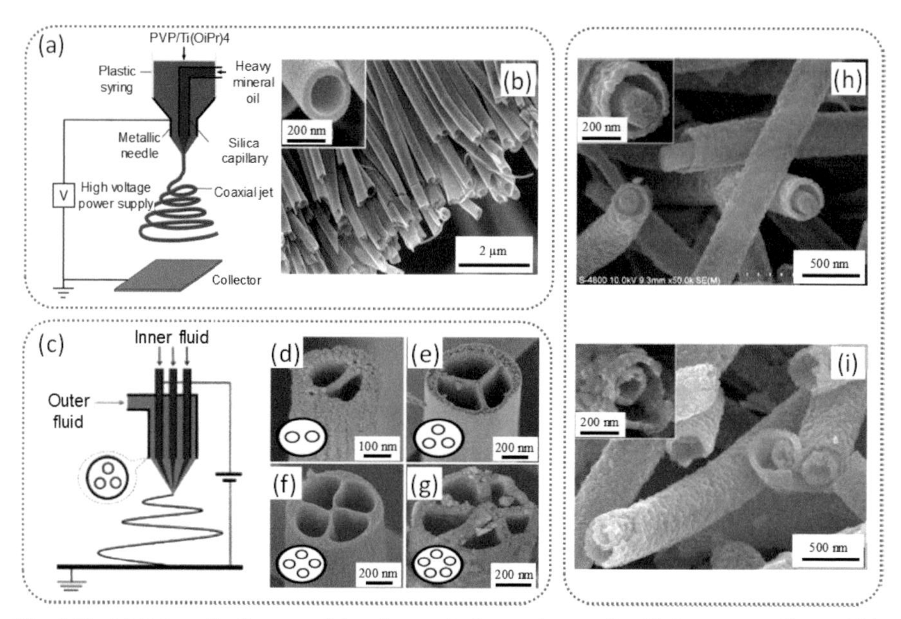

Fig. 6.13: (a) Schematic diagram of the electrospinning equipment for fabricating nanofibers which possessed a core-shell structure. And the spinneret was fabricated from two coaxial capillaries (Copyright 2004, reproduced with permission from American Chemical Society) (Li and Xia 2004). (b) SEM image of a uniaxially aligned array of electrospun hollow fibers (Copyright 2004, reproduced with permission from American Chemical Society) (Li and Xia 2004). (c) Schematic diagram for the preparation of a three-channel fiber. The outer and inner fluids (light area: titanium tetra isopropoxide solution; dark area: mineral oil) were injected to the individual syringe needles, respectively (Copyright 2007, reproduced with permission from American Chemical Society) (Zhao et al. 2007). (d–g) Corresponding to electrospun fibers with channel number from two to five (Copyright 2007, reproduced with permission from American Chemical Society) (Zhao et al. 2007). (h–i) SEM images of the as obtained γ-Fe_2O_3 fiber-in-tube and tube-in-tube, respectively (Copyright 2010, reproduced with permission from American Chemical Society) (Mou et al. 2010).

In addition to single hollow nanofibers, there have been some multilevel hollow nanofibers fabricated. The multilevel hollow fibers, such as multichannel hollow fibers, fiber-in-tube, tube-in-tube, and multi-shell usually exhibit improved electronic, catalytic, optical, mechanical, and magnetic properties compared to those of the simple hollow structures due to their inherent properties involving isolated interior structures, controllable physicochemical microenvironment, and multiple heterogeneous interfaces (Zhao and Jiang 2009). There are many methods, such as multi-fluidic compound-jet electro-hydrodynamic techniques (Chen et al. 2010), galvanic replacement reaction (Sun and Xia 2004), template method (Qin et al. 2008), and Kirkendall effect (Peng et al. 2009) to obtain the multilevel hollow fibers. Nevertheless, it is still a great challenge to set up a low cost, versatile way for large-scale production of multilevel hollow nanofibers because of their small size and structural complexity (Mou et al. 2010). Fortunately, electrospinning is a very convenient and effective non-equilibrium heat-treatment approach to prepare multilevel hollow nanofibers.

Jiang and coworkers (Zhao et al. 2007) developed a very powerful and simple multi-fluidic compound-jet electrospinning process (seen in the Fig. 6.13c) for generating biomimic multichannel fibers that have been seldom obtained by other approaches. As shown in Fig. 6.13d–g, electrospun fibers with two to five channels have been successfully fabricated. By changing the inner solutions into other functional polymers, multi-channels core-shell electrospun polymer nanofibers would be fabricated and different polymers could be integrated in the nano-sized materials without interaction. These multiple electrospun nanofibers should be of novel and improved features which do not exist in each ingredient. Consequently, the multi fluidic compound-jet electrospinning technology breaks through the limitation that could create programmable multichannel one dimension electrospun microfibers or nanofibers in a convenient and efficient way. And the multichannel hollow electrospun nanofibers is a promising candidate for various applications, such as vessels, macro/nanofluidic devices, high efficient catalysts, bio-mimic super-lightweight thermo-insulated textiles, and multicomponent drug delivery. Mou et al. (Mou et al. 2010) reported an effective and facile non-equilibrium heat-treatment method which enables the simple fabrication of γ-Fe_2O_3 (maghemite) tube-in-tube and fiber-in-tube nano-sized materials by heat-treating electrospun precursor fibers which are composed of polyvinyl pyrrolidone (PVP) and iron citrate. And the resultant γ-Fe_2O_3 fiber-in-tube (Fig. 6.13h) and tube-in-tube (Fig. 6.13i) nanomaterials may possess excellent performance in various applications, such as microreactors, catalyst supporting materials, absorbents, sensors and so forth, because of their structural features and excellent magnetic performance. Therefore, electrospinning is an important, flexible, and effective top-down method for the preparation of hollow nanofibers (McCann et al. 2005).

2D Nano-Nets

Since 2006, the novel electro-netting technique has developed very fast and overcome the bottleneck issue of continuous thinning for electrospun nanofibers, which serves as an advanced approach for creating thinner polymeric nanofibers. Two-dimensional nano-nets with extremely small diameters of ~ 20 nm and unparalleled Steiner tree pore structures, can be fabricated on mass production via electro-netting. 2D nano-nets possess weighted and ideal Steiner networks which is a clear geometric feature because of the fast phase separation process, and it obeys the minimal energy principle (Wang et al. 2011a). Owing to the 2D nano-nets' charming characteristics of small pore size, extremely small diameter, larger specific surface area, and high porosity, the nano-nets membranes exhibit remarkably enhanced performance in the application of lithium ion batteries, tissue engineering, air filtration, ultrasensitive sensors, and so forth.

Ding's group have developed a one-step approach (as shown in Fig. 6.14a–b) to controllably fabricate the 2D nano-nets with a large area density through the introduction of different additives into electro-netting solutions (Yang et al. 2011). 2D nano-nets are stacked layer-by-layer, which contain some interlinked 1D ultra-thin (~ 35 nm) nanowires. And they are widely distributed in the 3D structured membranes (seen in Fig. 6.14c). Figure 6.15 shows the typical morphologies of nano-fiber/nets membranes fabricated by different electrospun polymeric solutions. The final nanowire diameter, pore-width, and area density of the 2D nano-nets highly

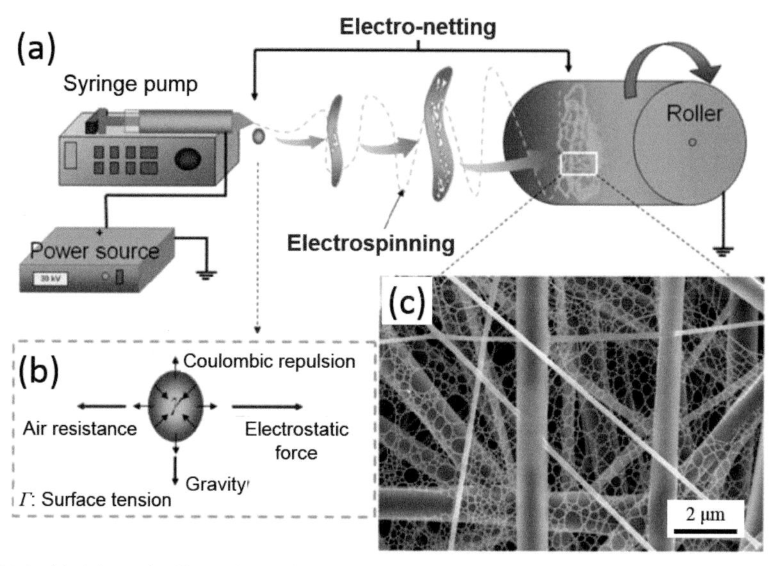

Fig. 6.14: (a) Schematic illustrations of the probable mechanism of nanonets generation during the electrospinning/electro-netting process. (b) The forces acting on the charged droplet. (c) Typical FE-SEM image of PAA/DBSA nano-nets (Copyright 2010, reproduced with permission from Royal Society of Chemistry) (Yang et al. 2011).

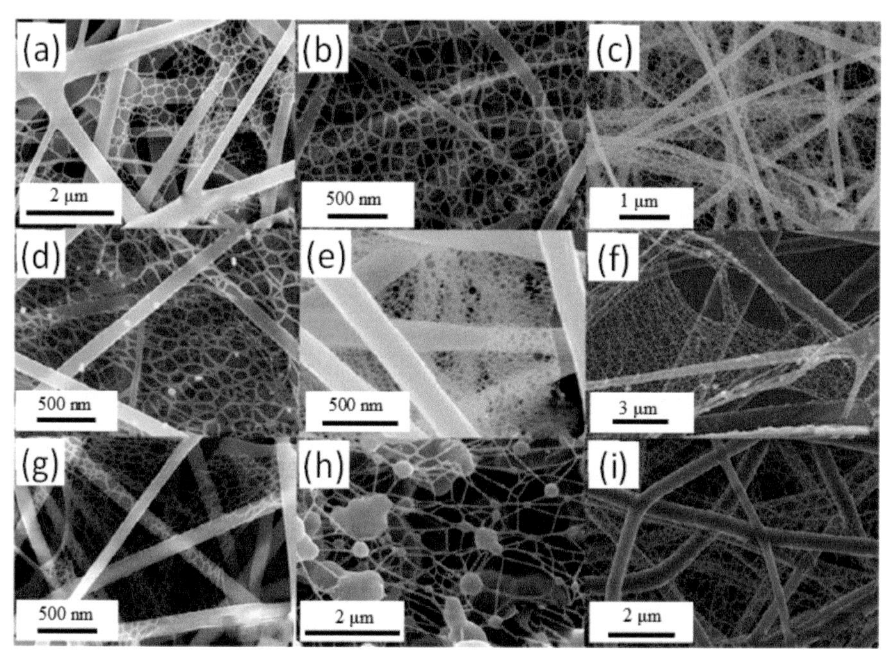

Fig. 6.15: Several typical nanonets based on different polymer fabricated in Ding's laboratory (Wang et al. 2013). (a) PAA (Wang et al. 2009); (b) PA-6 (Ding et al. 2011b); (c) PANI/PA-6 (Ding et al. 2011a); (d) PA-66; (e) PVA/ZnO (Ding et al. 2008); (f) PVA/SiO_2 (Ding et al. 2010); (g) Gelatin (Wang et al. 2011b); (h) CS (Wang et al. 2014); (i) PU (Hu et al. 2011) (Copyright 2013, reproduced with permission from Elsevier Ltd.).

rely on the kinds of additives, additives' concentration, and solvent properties. By adjusting these parameters, various polymers can be fabricated into 2D nano-nets by electro-netting technique, such as polyacrylic acid (PAA), polyamide-6 (PA-6), polyamide-66 (PA-66), poly(vinyl alcohol) (PVA), gelatin, chitosan (CS), and polyurethane (PU), which all have been successfully prepared in the laboratory.

3D Spiral Coil Nanofibers

Microscale or nanosized 3D spiral coil structures are greatly significant and have gained a lot of attention. This is mainly attributed to their promising potential applications as advanced inductive or optical components in nanoscale or microscale electromechanical and electromagnetic systems, as well as drug delivery system. However, the preparation of the helical polymeric nanofibers in a controllable and economical way is still a challenge so far. The nanofibers are generally collected in the fashion of randomly oriented media because of the unsteady bending of the highly charged jet (Reneker et al. 2000) for conventional electrospinning approach. Highly ordered and well-aligned architectures are usually needed in many applications both for helical polymer nanofibers and straight nanofibers (Theron et al. 2001, Dersch et al. 2003).

Yu and coworkers (Yu et al. 2008) developed a convenient electrospinning method, by which the aligned helical polycaprolactone (PCL) nanofibers (Fig. 6.16b) were easily prepared. As shown in Fig. 6.16a, the aligned spiral coil PCL nanofibers were fabricated on tilted glass slides by using a unique set-up. The helical PCL nanofibers were shaped by the jet buckling hitting the surface of collector. The formation of the spiral coil structures rely on the obliquity and location (distance away from the syringe needle) of tilted glass slide. Interestingly, there are some researches using a quiescent water surface as a collector presented by Reneker et al. (Han et al. 2007). The straight jet buckled as it impinged onto the surface of water and then sank into the water. As shown in Fig. 6.16c, three-dimensional helical morphologies were shaped as a coiling spring. And its diameter was around 20 mm. Shin and coworkers (Shin et al. 2006) created a convenient process for preparing single compound spiral coil nanofibers and transforming these spiral coil polymer nanofibers (Fig. 6.16e) into oriented nanofibers by introducing parallel electrodes (Fig. 6.16d), which convent a common electric field into a splitting electric field. In addition, Xin et al. (Xin et al. 2006) have fabricated a large scale of uniform 3D helical electrospun polymeric nanofibers, as shown in Fig. 6.16f. They came to the conclusion that the operating voltage, viscosity, and conductivity of electrospinning solution are the main factors that influence the formation of the spiral coil structures. In addition to single spiral coil polymer nanofibers, Kessick and Teppera (Kessick and Tepper 2004) fabricated attractive spiral coil electrospun fibers consisting of two polymers with different features. One of the polymer is conductive poly(aniline sulfonic acid), and the other is nonconductive poly(ethylene oxide).

In another approach, Lin et al. (Lin et al. 2005) have created bicomponent helical polymeric nanofibers by side-by-side electrospinning (the device diagram is shown in the Fig. 6.16g), and the microfluidic device is seen in the Fig. 6.16h as the syringe needle. Self-crimping electrospun nanofibers have been obtained, as shown in the

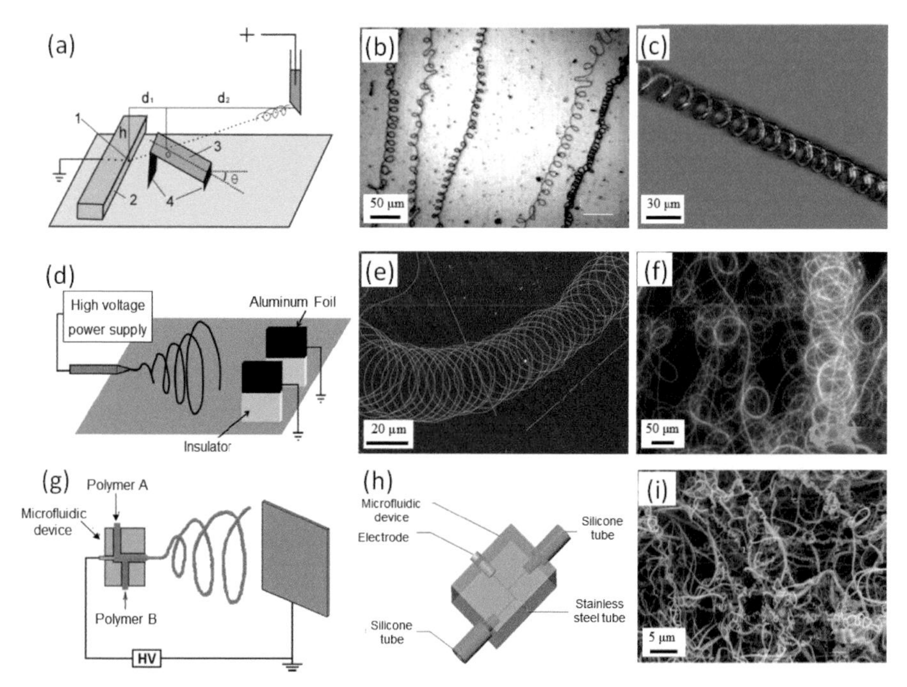

Fig. 6.16: (a) Schematic illustration of the electrospinning equipment. 1-grounded electrode wire, 2-wooden board, 3-glass slide, 4-supporter (Copyright 2008, reproduced with permission from Elsevier Ltd.) (Yu et al. 2008). (b) Optical images of the 3D spiral coil PCL electrospun fibers (Copyright 2008, reproduced with permission from Elsevier Ltd.) (Yu et al. 2008). (c) Three-dimensional buckled patterns formed after the impingement of a polystyrene jet onto a water surface (Copyright 2007, reproduced with permission from Elsevier Ltd.) (Han et al. 2007). (d) Schematic diagram of the electrospinning apparatus using parallel sub-electrodes (Copyright 2006, reproduced with permission from AIP Publishing LLC) (Shin et al. 2006). (e) SEM image showing spiral coil structured nanofibers with regular-shaped coils (Copyright 2006, reproduced with permission from AIP Publishing LLC) (Shin et al. 2006). (f) Fluorescence microscopy images of PPV/PVP electrospun fibers (Copyright 2006, reproduced with permission from AIP Publishing LLC) (Xin et al. 2006). (g) Side-by-side electrospinning set-up (Copyright 2005, reproduced with permission from WILEY-VCH Verlag GmbH & Co. KGaA, Weinheim) (Lin et al. 2005). (h) The microfluidic equipment as the electrospinning spinneret (Copyright 2005, reproduced with permission from WILEY-VCH Verlag GmbH & Co. KGaA, Weinheim) (Lin et al. 2005). (i) SEM images of PAN/PU composite helical nanofibers (Copyright 2005, reproduced with permission from WILEY-VCH Verlag GmbH & Co. KGaA, Weinheim) (Lin et al. 2005).

Fig. 6.16i. Self-crimping generates due to buckling of the compressed component in bicomponent fibers. Polyacrylonitrile (PAN) is a thermoplastic polymer and polyurethane (PU) is an elastomeric polymer, the stretching of these two polymers will result in differential shrinkage, and then lead to helical polymeric nanofibers.

Hierarchical Secondary Structures

The common smooth one-dimensional microfibers or nanofibers has usually been a limitation to various interesting applications (Zander 2013). Hierarchical integration of the ingredients and hierarchical secondary structures is required to boost the performance of electrospun fibers or lead to multifunction. Fortunately, electrospinning is an efficient and convenient technology that can incorporate and optimize the structures of nanofibers and make the nanofibers with versatile function, either through post-spinning modifications or during spinning process (Teo and Ramakrishna 2009). And various fascinating hierarchical secondary structures in microfibers or nanofibers such as nanorods, nanopillars, and nanopores can be created via electrospinning technology. Some hierarchical secondary structures may need one-step process in which the solvent parameters and humidity are carefully adjusted to generate nanometer sized pores along the fiber axis (Casper et al. 2004). And the other two-step process is, for example growing nanotubes or nanorods from nanoparticles which were introduced in electrospun polymer nanofibers or microfibers (Lai et al. 2008). It is noteworthy that these hierarchical secondary structures are distinguished greatly from smooth electrospun fibers conventionally fabricated by the electrospinning technology. Thus, the hierarchical electrospun nanofibers enable themselves to possess a great variety of special properties, including ultra-filtration capability, enhanced catalytic ability, superhydrophobicity or superhydrophilicity, improved sensing ability, and so on (Zheng et al. 2012).

6.2.7.1 Nanopores. Taking advantage of the phase separation between solvent and polymer (Bognitzki et al. 2001) may directly create highly porous electrospun fibers with increased specific surface areas (McCann et al. 2006). And the breath figures have been used to explicate the generation of pores on electrospun polymer fibers. Breath figures form due to the evaporative cooling as result of fast solvent evaporation, thus remarkably cooling the surface of the electrospinning jet. When the surface of electrospinning jet cools, the moisture in the air condenses to form droplets on the surface of electrospinning jet. As the fiber dries, the water droplets leave an imprint (i.e., pores) on the fiber (Megelski et al. 2002). Therefore, this process lead to the generation of hierarchical secondary electrospun nanofibers. Bognitzki et al. have reported the formation of nanoporous structures on electrospun poly(vinylcarbazole), polylactic acid (PLLA) and polycarbonate fibers (Bognitzki et al. 2001). These elliptical pores on the fibers axis were around 200 nm and with the longer dimension oriented (as shown in the Fig. 6.17a). The mechanism of the generation of pore structure was the fast phase separation of the solvent and polymer during the electrospinning procedure as discussed above. Simultaneously, they also discovered a phenomenon that the utilization of lower vapor pressures solvents would reduce the generation of pores on the electrospun fibers. In addition, there is another approach to fabricate porous electrospun fibers. McCann et al. (McCann et al. 2006) have reported that taking advantage of thermally induced phase separation between the polymer-poor and polymer-rich domains through the electrospinning process, electrospun nanofibers can be effectively collected by using liquid nitrogen and then removing solvent in vacuum (Fig. 6.17b). This approach is versatile because it does not need to choose dissolution of phase-separated polymers, and it can be easily used with nonvolatile solvents. Moreover, the generation of pores on electrospun fibers is not only associated with the properties of solvent itself, but also affected by humidity (Zhou 2006), according to Casper et al. (Casper et al. 2004). The features of pores become apparent when electrospinning process happens at the circumstances of more than 30% relative humidity. As shown in Fig. 6.17c–d, they found that increasing

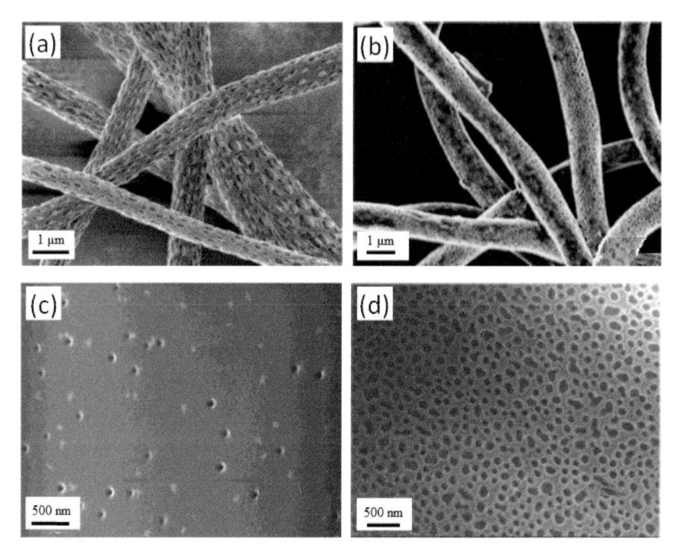

Fig. 6.17: (a) SEM micrographs of porous PLLA electrospun fibers (Copyright 2001, reproduced with permission from WILEY-VCH Verlag GmbH, Weinheim, Fed. Rep. of Germany) (Bognitzki et al. 2001). (b) Poly(ε-caprolactone) fibers obtained by electrospinning into liquid nitrogen followed by drying in vacuum (Copyright 2006, reproduced with permission from American Chemical Society) (McCann et al. 2006). (c–d) FE-SEM micrographs of PS/THF fibers electrospun under various humidity: (c) 31–38%, (d) 50–59%, respectively (Copyright 2004, reproduced with permission from American Chemical Society) (Casper et al. 2004).

humidity of circumstances during electrospinning process causes an increase in the distribution and diameter pores on the polymeric fibers (Casper et al. 2004).

6.2.7.2 Nanoprotrusions. Nanoprotrusions is another fashion of the hierarchical secondary structures of electrospun fibers, which can increase the specific surface area and roughness of the electrospun fibrous membranes and adjust its features (Zander 2013). Wan et al. (Wan et al. 2014) have reported a superhydrophobic air filtration medium with hierarchical nanostructure prepared by electrospinning (Fig. 6.18a) through addition of TiO_2 nanoparticles during electrospinning process. In their work, the hierarchical nanostructure electrospun nanofibrous membranes possess excellent filtration efficiency (99.997%). Alternatively, there is another impactful approach to create hierarchical structure nanofibers via growing nano-branches on electrospun fibers, which can efficiently increase the surface area of the fibers (Teo and Ramakrishna 2009). Cao et al. (Cao et al. 2016) have reported novel nanofibrous membranes with hierarchical nanorod-branched TiO_2 (Fig. 6.18b) and investigated their photoelectric performance. The specific method enables the nanorod branches grown on each electrospun polymer nanofibers to go through a hydrothermal reaction. The photo-electric conversion efficiency of this hierarchical composite photoanode is 46.9%, which is higher than the pristine electrospun nanofibrous membranes photoanode. What's more, Wang et al. (Wang et al. 2008) also reported an effective two-step method to fabricate novel hierarchical hetero-junctions fibers, which contained double inorganic substance (ZnO and TiO_2) by a combination of electrospinning and hydrothermal approach. In the first step, they

Fig. 6.18: (a) FE-SEM images of PSU/TiO_2 electrospun fibrous membranes (Copyright 2013, reproduced with permission from Elsevier Inc.) (Wan et al. 2014). (b) TEM image of a hierarchical electrospun fiber (Copyright 2015, reproduced with permission from Elsevier Ltd.) (Cao et al. 2016). (c) ZnO/TiO_2 hetero-junctions with ZnO nanorods grown on TiO_2 fibers (Copyright 2008, reproduced with permission from Royal Society of Chemistry) (Wang et al. 2008). (d) ZnO/TiO_2 hetero-junctions with ZnO nanoplates grown on TiO_2 fibers (Copyright 2008, reproduced with permission from Royal Society of Chemistry) (Wang et al. 2008). (e) Long, slightly curved carbon nanotubes obtained at 850°C (Copyright 2004, reproduced with permission from John Wiley & Sons, Inc.) (Hou and Reneker 2004). (f) Curved and bent carbon nanotubes obtained at 700°C (Copyright 2004, reproduced with permission from John Wiley & Sons, Inc.) (Hou and Reneker 2004).

fabricated TiO_2 fibers through electrospinning, which served as substrates to growing secondary ZnO nanostructures. Subsequently, ZnO nanostrutures were generated on the TiO_2 electrospun fibers via hydrothermal post treatment (Fig. 6.18c). Exactly as we know, the features of ZnO rely highly on their structure, including orientations, morphologies, and crystal sizes (Wang et al. 2008). Thus, ZnO/TiO_2 hierarchical hetero-junctions fibers with different ZnO structures will extend their application fields. By adjusting the synthetic process, the morphology of the ZnO/TiO_2 hierarchical hetero-junctions fibers can be further controlled, as shown in Fig. 6.18d. Hou and Reneker (Hou and Reneker 2004) reported another fascinating hierarchical structure on carbon nanofibers, as shown in Fig. 6.18e–f. Electrospun polyacrylonitrile (PAN) nanofibers were carbonized and served as templates to generate multi-walled carbon nanotubes by an iron-catalyzed growth mechanism. There are a lot of smaller and finer multi-walled fullerene tubes with a metal particle at the tip of each tube on the carbonized electrospun nanofibers. The dimensions of this interesting hierarchical secondary structure can be adjusted and controlled through the synthesis process. It is promising to produce some macroscale effects in nanoscale engineering, such as chemical interactions, mechanical, and electrical integration.

Concluding Remarks

Electrospinning technique affords us a significantly effective and convenient approach for fabricating successive nano-scale or micro-scale fibers with various and versatile morphologies and structures. In this chapter, we have described in detail the resulting morphologies and structures of some commonly existing electrospun nanofibers. By properly adjusting the polymer solution parameters (e.g., concentration, viscosity, conductivity and surface tension), processing parameters (e.g., applied voltage, tip to collector distance and feed rate) and ambient parameters (e.g., humidity and temperature), regular round cross-sectional nanofibers and non-regular round cross-sectional nanofibers can be successfully fabricated through electrospinning method. In addition to polymeric fibers, the inorganic nanofibers or polymer-inorganic hybrid nanofibers also can be obtained by electrospinning process. Besides, it is capable of creating various differential nanofibers, such as solid, Janus, core-shell, hollow, porous, hierarchical, and 1D/2D/3D composite nanomaterials via different electrospinning equipment and apparatus. These special structured nanofibers exhibit more unique properties, and can be potentially and extensively applied in many different fields.

Acknowledgements

The authors would like to acknowledge the funding support from the National Natural Science Foundation of China (Nos. 51673037, 51703022 and 51473030), the Fundamental Research Funds for the Central Universities (No. 16D310105), the "DHU Distinguished Young Professor Program".

References

Aussawasathien, D., Dong, J. -H. and Dai, L. 2005. Electrospun polymer nanofiber sensors. Synthetic Metals 154(1-3): 37–40.

Bazilevsky, Alexander V., Alexander L. Yarin and Constantine M. Megaridis. 2007. Co-electrospinning of core–shell fibers using a single-nozzle technique. Langmuir 23(5): 2311–2314.

Bhardwaj, Nandana and Subhas C. Kundu. 2010. Electrospinning: a fascinating fiber fabrication technique. Biotechnology Advances 28(3): 325–347.

Bhaskar, Srijanani and Joerg Lahann. 2009. Microstructured materials based on multicompartmental fibers. Journal of the American Chemical Society 131(19): 6650–6651.

Bi, Fei, Xiangting Dong, Jinxian Wang and Guixia Liu. 2015. Tuned magnetism–luminescence bifunctionality simultaneously assembled into flexible Janus nanofiber. RSC Advances 5(17): 12571–12577.

Bognitzki, Michael, Wolfgang Czado, Thomas Frese et al. 2001. Nanostructured fibers via electrospinning. Advanced Materials 13(1): 70–72.

Cao, Yang, Yu-Jie Dong, Hao-Lin Feng, Hong-Yan Chen and Dai-Bin Kuang. 2016. Electrospun TiO_2 nanofiber based hierarchical photoanode for efficient dye-sensitized solar cells. Electrochimica Acta 189: 259–264.

Capone, G. J. 1995. Wet-spinning technology. Acrylic Fiber Technology and Applications 69–103.

Caruso, Rachel, A., Jan H. Schattka and Andreas Greiner. 2001. Titanium dioxide tubes from sol–gel coating of electrospun polymer fibers. Advanced Materials 13(20): 1577–1579.

Casper, Cheryl L., Jean S. Stephens, Nancy G. Tassi, D. Bruce Chase and John F. Rabolt. 2004. Controlling surface morphology of electrospun polystyrene fibers: effect of humidity and molecular weight in the electrospinning process. Macromolecules 37(2): 573–578.

Chen, Gaoyun, Ying Xu, Deng-Guang Yu, Dao-Fang Zhang, Nicholas P. Chatterton and Kenneth N. White. 2015. Structure-tunable Janus fibers fabricated using spinnerets with varying port angles. Chemical Communications 51(22): 4623–4626.

Chen, Hongyan, Yong Zhao, Yanlin Song and Lei Jiang. 2008. One-step multicomponent encapsulation by compound-fluidic electrospray. Journal of the American Chemical Society 130(25): 7800–7801.

Chen, Hongyan, Nü Wang, Jiancheng Di, Yong Zhao, Yanlin Song and Lei Jiang. 2010. Nanowire-in-microtube structured core/shell fibers via multifluidic coaxial electrospinning. Langmuir 26(13): 11291–11296.

Chen, Jiun-Tai, Wan-Ling Chen, Ping-Wen Fan and Yao, I. 2014. Effect of thermal annealing on the surface properties of electrospun polymer fibers. Macromolecular Rapid Communications 35(3): 360–366.

Chen, Li-Feng, Xu-Dong Zhang, Hai-Wei Liang, Kong Mingguang, Qing-Fang Guan, Ping Chen et al. 2012. Synthesis of nitrogen-doped porous carbon nanofibers as an efficient electrode material for supercapacitors. ACS Nano 6(8): 7092–7102.

Chronakis, Ioannis S. 2005. Novel nanocomposites and nanoceramics based on polymer nanofibers using electrospinning process—a review. Journal of Materials Processing Technology 167(2): 283–293.

Cumberbirch, R. J. E., Ford, J. E. and Gee, R. E. 1961. 26—The effect of spinning conditions on the structure of viscose rayon filaments. Journal of the Textile Institute Transactions 52(7): T330–T350.

Dai, Hongqin, Jian Gong, Hakyong Kim and Doukrae Lee. 2002. A novel method for preparing ultra-fine alumina-borate oxide fibres via an electrospinning technique. Nanotechnology 13(5): 674.

Deitzel, Joseph, M., James Kleinmeyer, Harris, D. E. A. and Beck Tan, N. C. 2001. The effect of processing variables on the morphology of electrospun nanofibers and textiles. Polymer 42(1): 261–272.

Dersch, R., Taiqi Liu, Schaper, A. K., Greiner, A. and Wendorff, J. H. 2003. Electrospun nanofibers: Internal structure and intrinsic orientation. Journal of Polymer Science Part A: Polymer Chemistry 41(4): 545–553.

Ding, Bin, Tasuku Ogawa, Jinho Kim, Kouji Fujimoto and Seimei Shiratori. 2008. Fabrication of a super-hydrophobic nanofibrous zinc oxide film surface by electrospinning. Thin Solid Films 516(9): 2495–2501.

Ding, B., Li, C., Wang, D. and Shiratori, S. 2010. Fabrication and application of novel two-dimensional nanowebs via electrospinning. Nanotechnology: Nanofabrication, Patterning, and Self Assembly. New York: Nova Science Publishers, Inc. 51–69.

Ding, Bin, Yang Si, Xianfeng Wang, Jianyong Yu, Li Feng and Gang Sun. 2011a. Label-free ultrasensitive colorimetric detection of copper (II) ions utilizing polyaniline/polyamide-6 nano-fiber/net sensor strips. Journal of Materials Chemistry 21(35): 13345–13353.

Ding, Bin, Xianfeng Wang, Jianyong Yu and Moran Wang. 2011b. Polyamide 6 composite nano-fiber/net functionalized by polyethyleneimine on quartz crystal microbalance for highly sensitive formaldehyde sensors. Journal of Materials Chemistry 21(34): 12784–12792.

Ding, Bin and Jianyong Yu. 2014. Electrospun Nanofibers for Energy and Environmental Applications: Springer Science & Business Media.

Doshi, Jayesh and Darrell H. Reneker. 1995. Electrospinning process and applications of electrospun fibers. Journal of Electrostatics 35(2-3): 151–160.

Eda, Goki and Satya Shivkumar. 2006. Bead structure variations during electrospinning of polystyrene. Journal of Materials Science 41(17): 5704–5708.

Eswaramoorthy, M., Rahul Sen and Rao, C. N. R. 1999. A study of micropores in single-walled carbon nanotubes by the adsorption of gases and vapors. Chemical Physics Letters 304(3): 207–210.

Faccini, M., Vaquero, C. and Amantia, D. 2012. Development of protective clothing against nanoparticle based on electrospun nanofibers. Journal of Nanomaterials 2012: 18.

Fang, Jian, HaiTao Niu, Tong Lin and XunGai Wang. 2008. Applications of electrospun nanofibers. Chinese Science Bulletin 53(15): 2265.

Fang, Jian, Xungai Wang and Tong Lin. 2011. Functional applications of electrospun nanofibers. In Nanofibers-production, Properties and Functional Applications: InTech.

Feng, Xinliang, Yanyu Liang, Linjie Zhi, Arne Thomas, Dongqing Wu, Ingo Lieberwirth et al. 2009. Synthesis of microporous carbon nanofibers and nanotubes from conjugated polymer network and evaluation in electrochemical capacitor. Advanced Functional Materials 19(13): 2125–2129.

Fong, H., Chun, I. and Reneker, D. H. 1999. Beaded nanofibers formed during electrospinning. Polymer 40(16): 4585–4592.

Gopal, Renuga, Satinderpal Kaur, Zuwei Ma, Casey Chan, Seeram Ramakrishna and Takeshi Matsuura. 2006. Electrospun nanofibrous filtration membrane. Journal of Membrane Science 281(1): 581–586.

Graeser, Martin, Eckhard Pippel, Andreas Greiner and Joachim H. Wendorff. 2007. Polymer core-shell fibers with metal nanoparticles as nanoreactor for catalysis. Macromolecules 40(17): 6032–6039.

Greiner, Andreas and Joachim H. Wendorff. 2007. Electrospinning: a fascinating method for the preparation of ultrathin fibers. Angewandte Chemie International Edition 46(30): 5670–5703.

Gu, Yuanxiang, Dairong Chen, Xiuling Jiao and Fangfang Liu. 2007. $LiCoO_2$–MgO coaxial fibers: co-electrospun fabrication, characterization and electrochemical properties. Journal of Materials Chemistry 17(18): 1769–1776.

Gupta, Pankaj and Garth L. Wilkes. 2003. Some investigations on the fiber formation by utilizing a side-by-side bicomponent electrospinning approach. Polymer 44(20): 6353–6359.

Han, Tao, Darrell H. Reneker and Alexander L. Yarin. 2007. Buckling of jets in electrospinning. Polymer 48(20): 6064–6076.

He, Ji-Huan, Yu-Qin Wan and Lan Xu. 2007. Nano-effects, quantum-like properties in electrospun nanofibers. Chaos, Solitons & Fractals 33(1): 26–37.

Hou, Haoqing and Darrell H. Reneker. 2004. Carbon nanotubes on carbon nanofibers: a novel structure based on electrospun polymer nanofibers. Advanced Materials 16(1): 69–73.

Hu, Juanping, Xianfeng Wang, Bin Ding, Jinyou Lin, Jianyong Yu and Gang Sun. 2011. One-step electro-spinning/netting technique for controllably preparing polyurethane nano-fiber/net. Macromolecular Rapid Communications 32(21): 1729–1734.

Hua, Dawei, Zhongche Liu, Fang Wang et al. 2016. pH Responsive polyurethane (core) and cellulose acetate phthalate (shell) electrospun fibers for intravaginal drug delivery. Carbohydrate Polymers 151: 1240–1244.

Huang, Chaobo, Shuiliang Chen, Chuilin Lai et al. 2006. Electrospun polymer nanofibres with small diameters. Nanotechnology 17(6): 1558.

Huang, Zheng-Ming, Zhang, Y. -Z., Kotaki, M. and Ramakrishna, S. 2003. A review on polymer nanofibers by electrospinning and their applications in nanocomposites. Composites Science and Technology 63(15): 2223–2253.

Hwang, Tae Hoon, Yong Min Lee, Byung-Seon Kong, Jin-Seok Seo and Jang Wook Choi. 2012. Electrospun core–shell fibers for robust silicon nanoparticle-based lithium ion battery anodes. Nano Letters 12(2): 802–807.

Jang, Jyongsik and Hwanseok Park. 2002. Formation and structure of polyacrylamide–silica nanocomposites by sol–gel process. Journal of Applied Polymer Science 83(8): 1817–1823.

Jiang, Jia-Xing, Fabing Su, Hongjun Niu, Colin D Wood, Neil J Campbell, Yaroslav Z Khimyak et al. 2008a. Conjugated microporous poly (phenylene butadiynylene) s. Chemical Communications (4): 486–488.

Jiang, Jia-Xing, Fabing Su, Abbie Trewin, Colin D Wood, Hongjun Niu, James Thomas Anthony Jones et al. 2008b. Synthetic control of the pore dimension and surface area in conjugated microporous polymer and copolymer networks. Journal of the American Chemical Society 130(24): 7710–7720.

Jiang, Jia-Xing, Chao Wang, Andrea Laybourn et al. 2011. Metal–organic conjugated microporous polymers. Angewandte Chemie International Edition 50(5): 1072–1075.

Joo, Yong Lak. 2006. Incorporation of vanadium oxide in silica nanofiber mats via electrospinning and sol-gel synthesis. Journal of Nanomaterials 2006(1): 14–14.

Kang, Haigang, Yihua Zhu, Yujia Jing, Xiaoling Yang and Chungzhong Li. 2010. Fabrication and electrochemical property of Ag-doped SiO_2 nanostructured ribbons. Colloids and Surfaces A: Physicochemical and Engineering Aspects 356(1): 120–125.

Katta, P., Alessandro, M., Ramsier, R. D. and Chase, G. G. 2004. Continuous electrospinning of aligned polymer nanofibers onto a wire drum collector. Nano Letters 4(11): 2215–2218.

Kayaci, Fatma, Sesha Vempati, Cagla Ozgit-Akgun, Inci Donmez, Necmi Biyikli and Tamer Uyar. 2014. Selective isolation of the electron or hole in photocatalysis: ZnO–TiO_2 and TiO_2–ZnO core–shell structured heterojunction nanofibers via electrospinning and atomic layer deposition. Nanoscale 6(11): 5735–5745.

Kessick, Royal and Gary Tepper. 2004. Microscale polymeric helical structures produced by electrospinning. Applied Physics Letters 84(23): 4807–4809.

Khan, W. S., Asmatulu, R., Ceylan, M. and Jabbarnia, A. 2013. Recent progress on conventional and non-conventional electrospinning processes. Fibers & Polymers 14(8): 1235–1247.

Khil, Myung Seob, Hak Yong Kim, Min Sub Kim, Seong Yoon Park and Douk-Rae Lee. 2004. Nanofibrous mats of poly (trimethylene terephthalate) via electrospinning. Polymer 45(1): 295–301.

Ko, Frank, Yury Gogotsi, Ashraf A. Ali, N. Naguib, HH Ye, G.L. Yang et al. 2003. Electrospinning of continuous carbon nanotube-filled nanofiber yarns. Advanced Materials 15(14): 1161–1165.

Koombhongse, Sureeporn, Wenxia Liu and Darrell H. Reneker. 2001. Flat polymer ribbons and other shapes by electrospinning. Journal of Polymer Science Part B: Polymer Physics 39(21): 2598–2606.

Lai, Chulin, Qiaohui Guo, Xiang-Fa Wu, Darrell H. Reneker and Haoqing Hou. 2008. Growth of carbon nanostructures on carbonized electrospun nanofibers with palladium nanoparticles. Nanotechnology 19(19): 195303.

Larsen, Gustavo, Raffet Velarde-Ortiz, Kevin Minchow, Antonio Barrero and Ignacio G. Loscertales. 2003. A method for making inorganic and hybrid (organic/inorganic) fibers and vesicles with diameters in the submicrometer and micrometer range via sol-gel chemistry and electrically forced liquid jets. Journal of the American Chemical Society 125(5): 1154–1155.

Larsen, Gustavo, Rubén Spretz and Raffet Velarde-Ortiz. 2004. Use of coaxial gas jackets to stabilize Taylor cones of volatile solutions and to induce particle-to-fiber transitions. Advanced Materials 16(2): 166–169.

Lee, Chang Hun, Ho Joon Shin, In Hee Cho, Young-Mi Kang, In Ae Kim, Ki Dong Park et al. 2005. Nanofiber alignment and direction of mechanical strain affect the ECM production of human ACL fibroblast. Biomaterials 26(11): 1261–1270.

Lee, K. H., Kim, H. Y., Khil, M. S., Ra, Y. M. and Lee, D. R. 2003. Characterization of nano-structured poly (ε-caprolactone) nonwoven mats via electrospinning. Polymer 44(4): 1287–1294.

Li, Dan, Thurston Herricks and Younan Xia. 2003a. Magnetic nanofibers of nickel ferrite prepared by electrospinning. Applied Physics Letters 83(22): 4586–4588.

Li, Dan, Yuliang Wang and Younan Xia. 2003b. Electrospinning of polymeric and ceramic nanofibers as uniaxially aligned arrays. Nano Letters 3(8): 1167–1171.

Li, Dan and Younan Xia. 2004a. Direct fabrication of composite and ceramic hollow nanofibers by electrospinning. Nano Letters 4(5): 933–938.

Li, Dan and Younan Xia. 2004b. Electrospinning of nanofibers: reinventing the wheel? Advanced Materials 16(14): 1151–1170.

Li, Dan, Amit Babel, Samson A. Jenekhe and Younan Xia. 2004. Nanofibers of conjugated polymers prepared by electrospinning with a two-capillary spinneret. Advanced Materials 16(22): 2062–2066.

Li, Dan, Jesse T. McCann and Younan Xia. 2005a. Use of electrospinning to directly fabricate hollow nanofibers with functionalized inner and outer surfaces. Small 1(1): 83–86.

Li, Dan, Gong Ouyang, Jesse T. McCann and Younan Xia. 2005b. Collecting electrospun nanofibers with patterned electrodes. Nano Letters 5(5): 913–916.

Li, Wan-Ju, Cato T. Laurencin, Edward J. Caterson, Rocky S. Tuan and Frank K. Ko. 2002. Electrospun nanofibrous structure: a novel scaffold for tissue engineering. Journal of Biomedical Materials Research Part A 60(4): 613–621.

Lim, Jong-Min, Jun Hyuk Moon, Gi-Ra Yi, Chul-Joon Heo and Seung-Man Yang. 2006. Fabrication of one-dimensional colloidal assemblies from electrospun nanofibers. Langmuir 22(8): 3445–3449.

Lin, Tong, Hongxia Wang and Xungai Wang. 2005. Self-crimping bicomponent nanofibers electrospun from polyacrylonitrile and elastomeric polyurethane. Advanced Materials 17(22): 2699–2703.

Liu, P. S., Du, H. Y. and Xia, F. J. 2013. Fabrication and characterization of a porous titanium dioxide film with sub-micropores. Materials & Design 51: 193–198.

Liu, Yong, Ji-Huan He, Jian-yong Yu and Hong-mei Zeng. 2008. Controlling numbers and sizes of beads in electrospun nanofibers. Polymer International 57(4): 632–636.

Lu, Ping and Bin Ding. 2008. Applications of electrospun fibers. Recent Patents on Nanotechnology 2(3): 169–182.

Lu, Xiaofeng, Linlin Li, Wanjin Zhang and Ce Wang. 2005a. Preparation and characterization of Ag2S nanoparticles embedded in polymer fibre matrices by electrospinning. Nanotechnology 16(10): 2233.

Lu, Xiaofeng, Yiyang Zhao, Ce Wang and Yen Wei. 2005b. Fabrication of CdS nanorods in PVP fiber matrices by electrospinning. Macromolecular Rapid Communications 26(16): 1325–1329.

Lu, Xiaofeng, Ce Wang and Yen Wei. 2009. One-dimensional composite nanomaterials: Synthesis by electrospinning and their applications. Small 5(21): 2349–70.

Ma, Qianli, Jinxian Wang, Xiangting Dong, Wensheng Yu and Guixia Liu. 2015. Flexible Janus nanoribbons array: a new strategy to achieve excellent electrically conductive anisotropy, magnetism, and photoluminescence. Advanced Functional Materials 25(16): 2436–2443.

Ma, Wenjing, Qilu Zhang, Sangram Keshari Samal, Fang Wang, Buhong Gao, Hui Pan et al. 2016. Core–sheath structured electrospun nanofibrous membranes for oil–water separation. Rsc Advances 6(48): 41861–41870.

Ma, Wenjing, Zhongfu Guo, Juntao Zhao, Qian Yu, Fang Wang, Jingquan Han et al. 2017. Polyimide/cellulose acetate core/shell electrospun fibrous membranes for oil-water separation. Separation and Purification Technology 177: 71–85.

MacDiarmid, A. G., W. E. Jones, I. D. Norris, Junbo Gao, Johnson Charlie, N.J. Pinto et al. 2001. Electrostatically-generated nanofibers of electronic polymers. Synthetic Metals 119(1-3): 27–30.

Mauter, Meagan Stumpe. 2011. Implications and Applications of Nanomaterials for Membrane-Based Water Treatment: Yale University.

McCann, Jesse T., Dan Li and Younan Xia. 2005. Electrospinning of nanofibers with core-sheath, hollow, or porous structures. Journal of Materials Chemistry 15(7): 735–738.

McCann, Jesse T., Manuel Marquez and Younan Xia. 2006. Highly porous fibers by electrospinning into a cryogenic liquid. Journal of the American Chemical Society 128(5): 1436–1437.

Megelski, Silke, Jean S. Stephens, D. Bruce Chase and John F. Rabolt. 2002. Micro- and nanostructured surface morphology on electrospun polymer fibers. Macromolecules 35(22): 8456–8466.

Miyoshi, Takanori, Kiyotsuna Toyohara and Hiroyoshi Minematsu. 2005. Preparation of ultrafine fibrous zein membranes via electrospinning. Polymer International 54(8): 1187–1190.

Mou, Fangzhi, Jian-guo Guan, Weidong Shi, Zhigang Sun and Shuanhu Wang. 2010. Oriented contraction: a facile nonequilibrium heat-treatment approach for fabrication of maghemite fiber-in-tube and tube-in-tube nanostructures. Langmuir 26(19): 15580–15585.

Niessen, W. M. A. and Tinke, A. P. 1995. Liquid chromatography-mass spectrometry general principles and instrumentation. Journal of Chromatography A 703(1-2): 37–57.

Nijkamp, M. G., Raaymakers, J. E. M. J., Van Dillen, A. J. and De Jong, K. P. 2001. Hydrogen storage using physisorption–materials demands. Applied Physics A: Materials Science & Processing 72(5): 619–623.

Pan, Chao, Yong-Hao Han, Li Dong, Jing Wang and Zhong-Ze Gu. 2008. Electrospinning of continuous, large area, latticework fiber onto two-dimensional pin-array collectors. Journal of Macromolecular Science, Part B: Physics 47(4): 735–742.

Peng, Qing, Xiao-Yu Sun, Joseph C. Spagnola, Carl D. Saquing, Saad Arman Khan, Richard J. Spontak et al. 2009. Bi-directional Kirkendall effect in coaxial microtube nanolaminate assemblies fabricated by atomic layer deposition. Acs Nano 3(3): 546–554.

PPennella, Francesco, Cerino Giulia., Massai Diana, Gallo Diego, Falvo D'Urso Labate, Schiavi Alessandro et al. 2013. A survey of methods for the evaluation of tissue engineering scaffold permeability. Annals of Biomedical Engineering 41(10): 2027–2041.

Qin, Yong, Lifeng Liu, Renbin Yang, Ulrich Gösele and Mato Knez. 2008. General assembly method for linear metal nanoparticle chains embedded in nanotubes. Nano Letters 8(10): 3221–3225.

Ramakrishna, Seeram. 2005. An Introduction to Electrospinning and Nanofibers: World Scientific.
Reneker, Darrell H. and Iksoo Chun. 1996. Nanometre diameter fibres of polymer, produced by electrospinning. Nanotechnology 7(3): 216.
Reneker, Darrell H., Alexander L. Yarin, Hao Fong and Sureeporn Koombhongse. 2000. Bending instability of electrically charged liquid jets of polymer solutions in electrospinning. Journal of Applied Physics 87(9): 4531–4547.
Shin, Min Kyoon, Sun I. Kim and Seon Jeong Kim. 2006. Controlled assembly of polymer nanofibers: From helical springs to fully extended. Applied Physics Letters 88(22): 223109.
Somvipart, Siraporn, Sorada Kanokpanont, Rattapol Rangkupan, Juthamas Ratanavaraporn and Siriporn Damrongsakkul. 2013. Development of electrospun beaded fibers from Thai silk fibroin and gelatin for controlled release application. International Journal of Biological Macromolecules 55: 176–184.
Sridhar, Radhakrishnan, Rajamani Lakshminarayanan, Kalaipriya Madhaiyan, Veluchamy Amutha Barathi, Keith Hsiu Chin Lim and Seeram Ramakrishna. 2015. Electrosprayed nanoparticles and electrospun nanofibers based on natural materials: applications in tissue regeneration, drug delivery and pharmaceuticals. Chemical Society Reviews 44(3): 790–814.
Srinivasan, Gokul and Darrell H. Reneker. 1995. Structure and morphology of small diameter electrospun aramid fibers. Polymer International 36(2): 195–201.
Starr, Justin D. and Jennifer S. Andrew. 2013. A route to synthesize multifunctional tri-phasic nanofibers. Journal of Materials Chemistry C 1(14): 2529–2533.
Subbiah, Thandavamoorthy, Bhat, G. S., Tock, R. W., Parameswaran, S. and Ramkumar, S. S. 2005. Electrospinning of nanofibers. Journal of Applied Polymer Science 96(2): 557–569.
Sun, Yugang and Younan Xia. 2004. Multiple-walled nanotubes made of metals. Advanced Materials 16(3): 264–268.
Sun, Zaicheng, Eyal Zussman, Alexander L. Yarin, Joachim H. Wendorff and Andreas Greiner. 2003. Compound core–shell polymer nanofibers by co-electrospinning. Advanced Materials 15(22): 1929–1932.
Teo, Wee Eong and Seeram Ramakrishna. 2009. Electrospun nanofibers as a platform for multifunctional, hierarchically organized nanocomposite. Composites Science & Technology 69(11): 1804–1817.
Theron, A., Zussman, E. and Yarin, A. L. 2001. Electrostatic field-assisted alignment of electrospun nanofibres. Nanotechnology 12(3): 384.
Tomczak, Nikodem, Niek F. van Hulst and G. Julius Vancso. 2005. Beaded electrospun fibers for photonic applications. Macromolecules 38(18): 7863–7866.
Verreck, Geert, Iksoo Chun, Joel Rosenblatt, Jef Peeters, Alex Van Dijck, Jurgen Mensch et al. 2003. Incorporation of drugs in an amorphous state into electrospun nanofibers composed of a water-insoluble, nonbiodegradable polymer. Journal of Controlled Release 92(3): 349–360.
Wan, Huigao, Na Wang, Jianmao Yang, Yinsong Si, Kun Chen, Bin Ding et al. 2014. Hierarchically structured polysulfone/titania fibrous membranes with enhanced air filtration performance. Journal of Colloid and Interface Science 417: 18–26.
Wang, Liyan, Xiangting Dong, Guangqing Gai, Li Zhao, Shuzhi Xu and Xinfu Xiao. 2015. One-pot facile electrospinning construct of flexible Janus nanofibers with tunable and enhanced magnetism–photoluminescence bifunctionality. Journal of Nanoparticle Research 17(2): 91.
Wang, M., Singh, H., Hatton, T. A. and Rutledge, G. C. 2004. Field-responsive superparamagnetic composite nanofibers by electrospinning. Polymer 45(16): 5505–5514.
Wang, Na, Xianfeng Wang, Yongtang Jia, Xiaoqi Li, Jianyong Yu and Bin Ding. 2014. Electrospun nanofibrous chitosan membranes modified with polyethyleneimine for formaldehyde detection. Carbohydrate Polymers 108: 192–199.
Wang, Nuanxia, Chenghua Sun, Yong Zhao, Shuyun Zhou, Ping Chen and Lei Jiang. 2008. Fabrication of three-dimensional ZnO/TiO_2 heteroarchitectures via a solution process. Journal of Materials Chemistry 18(33): 3909–3911.
Wang, Xianfeng, Bin Ding, Jianyong Yu, Moran Wang and Fukui Pan. 2009. A highly sensitive humidity sensor based on a nanofibrous membrane coated quartz crystal microbalance. Nanotechnology 21(5): 055502.
Wang, Xianfeng, Bin Ding, Jianyong Yu, Yang Si, Shangbin Yang and Gang Sun. 2011a. Electro-netting: Fabrication of two-dimensional nano-nets for highly sensitive trimethylamine sensing. Nanoscale 3(3): 911–915.

Wang, Xianfeng, Bin Ding, Jianyong Yu and Jianmao Yang. 2011b. Large-scale fabrication of two-dimensional spider-web-like gelatin nano-nets via electro-netting. Colloids and Surfaces B: Biointerfaces 86(2): 345–352.

Wang, Xianfeng, Bin Ding, Gang Sun, Moran Wang and Jianyong Yu. 2013. Electro-spinning/netting: a strategy for the fabrication of three-dimensional polymer nano-fiber/nets. Progress in Materials Science 58(8): 1173–1243.

Weber, Jens and Arne Thomas. 2008. Toward stable interfaces in conjugated polymers: microporous poly (p-phenylene) and poly (phenyleneethynylene) based on a spirobifluorene building block. Journal of the American Chemical Society 130(20): 6334–6335.

Xie, Jingwei, Xiaoran Li and Younan Xia. 2008. Putting electrospun nanofibers to work for biomedical research. Macromolecular Rapid Communications 29(22): 1775–1792.

Xin, Y., Huang, Z.H., Yan, E. Y., Zhang, W. and Zhao, Q. 2006. Controlling poly (p-phenylene vinylene)/poly (vinyl pyrrolidone) composite nanofibers in different morphologies by electrospinning. Applied Physics Letters 89(5): 053101.

Xue, Jiajia, Jingwei Xie, Wenying Liu and Younan Xia. 2017. Electrospun nanofibers: New concepts, materials, and applications. Accounts of Chemical Research.

Yang, Shangbin, Xianfeng Wang, Bin Ding, Jianyong Yu, Jingfang Qian and Gang Sun. 2011. Controllable fabrication of soap-bubble-like structured polyacrylic acid nano-nets via electro-netting. Nanoscale 3(2): 564–568.

Yarin, A. L. 2011. Coaxial electrospinning and emulsion electrospinning of core–shell fibers. Polymers for Advanced Technologies 22(3): 310–317.

Yin, Duanduan, Qianli Ma, Xiangting Dong et al. 2015. Tunable and enhanced simultaneous photoluminescence–electricity–magnetism trifunctionality successfully realized in flexible Janus nanofiber. Journal of Materials Science: Materials in Electronics 26(4): 2658–2667.

Yoon, Young Il, Hyun Sik Moon, Won Seok Lyoo, Taek Seung Lee and Won Ho Park. 2008. Superhydrophobicity of PHBV fibrous surface with bead-on-string structure. Journal of Colloid and Interface Science 320(1): 91–95.

Yu, Deng-Guang, Jiao-Jiao Li, Man Zhang and Gareth R. Williams. 2017. High-quality Janus nanofibers prepared using three-fluid electrospinning. Chemical Communications 53(33): 4542–4545.

Yu, Jie, Yejun Qiu, Xiaoxiong Zha, Min Yu, Jiliang Yu, Javed Rafique et al. 2008. Production of aligned helical polymer nanofibers by electrospinning. European Polymer Journal 44(9): 2838–2844.

Yuan, Xiaoyan, Yuanyuan Zhang, Cunhai Dong and Jing Sheng. 2004. Morphology of ultrafine polysulfone fibers prepared by electrospinning. Polymer International 53(11): 1704–1710.

Zander, Nicole E. 2013. Hierarchically structured electrospun fibers. Polymers 5(1): 19–44.

Zhang, Chengcheng, Xiang Li, Tian Zheng, Yang Yang, Yongxin Li, Ye Li et al. 2012. Beaded $ZnTiO_3$ fibers prepared by electrospinning and their photocatalytic properties. Desalination and Water Treatment 45(1-3): 324–330.

Zhang, Daming and Jiang Chang. 2007. Patterning of electrospun fibers using electroconductive templates. Advanced Materials 19(21): 3664–3667.

Zhang, Daming and Jiang Chang. 2008. Electrospinning of three-dimensional nanofibrous tubes with controllable architectures. Nano Letters 8(10): 3283–3287.

Zhang, Jin, Sun-Woo Choi and Sang Sub Kim. 2011. Micro- and nano-scale hollow TiO_2 fibers by coaxial electrospinning: Preparation and gas sensing. Journal of Solid State Chemistry 184(11): 3008–3013.

Zhang, Yanzhong, Zheng-Ming Huang, Xiaojing Xu, Chwee Teck Lim and Seeram Ramakrishna. 2004. Preparation of core-shell structured PCL-r-gelatin bi-component nanofibers by coaxial electrospinning. Chemistry of Materials 16(18): 3406–3409.

Zhao, Yong, Xinyu Cao and Lei Jiang. 2007. Bio-mimic multichannel microtubes by a facile method. Journal of the American Chemical Society 129(4): 764–765.

Zhao, Yong and Lei Jiang. 2009. Hollow micro/nanomaterials with multilevel interior structures. Advanced Materials 21(36): 3621–3638.

Zheng, Jianfen, Aihua He, Junxing Li, Jian Xu and Charles C. Han. 2006. Studies on the controlled morphology and wettability of polystyrene surfaces by electrospinning or electrospraying. Polymer 47(20): 7095–7102.

Zheng, Jianfen, Haiyuan Zhang, Zhiguo Zhao and Charles C. Han. 2012. Construction of hierarchical structures by electrospinning or electrospraying. Polymer 53(2): 546–554.

Zhou, Huajun. 2006. Electrospun fibers from both solution and melt: Processing, structure and property: the American Chemical Society.

Zhou, Shaolian and Matthias Hamburger. 1996. Application of liquid chromatography-atmospheric pressure ionization mass spectrometry in natural product analysis evaluation and optimization of electrospray and heated nebulizer interfaces. Journal of Chromatography A 755(2): 189–204.

Zhou, Xuejiao, Qianli Ma, Wensheng Yu, Tingting Wang, Xiangting Dong, Jinxian Wang et al. 2015. Magnetism and white-light-emission bifunctionality simultaneously assembled into flexible Janus nanofiber via electrospinning. Journal of Materials Science 50(24): 7884–7895.

Zhu, Miaomiao, Jingquan Han, Fang Wang, Shao Wei, Ranhua Xiong, Qilu Zhang et al. 2016. Electrospun nanofibers membranes for effective air filtration. Macromolecular Materials and Engineering.

Zong, Xinhua Steven, Harold Bien, Chiung-Yin Chung, Lihong Yin, Dufei Fang, Benjamin S. Hsiao et al. 2005. Electrospun fine-textured scaffolds for heart tissue constructs. Biomaterials 26(26): 5330–5338.

Chapter 7

Improvement on Mechanical Property of Electrospun Nanofibers, Their Yarns, and Materials

Shaohua Jiang,[1,*] *Xiaojian Liao*[2] and *Haoqing Hou*[2,*]

Introduction

Electrospinning is a versatile and fascinating technology to produce ultra-fine fibers with diameter from several micrometers to a few nanometers. So far, huge number of materials such as polymers, composite ceramics, metals, carbon nanotubes, even bacteria and virus can be fabricated/incorporated into micro- and nano-fibers by directly electrospinning or through post-spinning process. Many publications including reviews and research papers were focused on the process, properties, and applications of electrospinning or electrospun nanofibers (Afifi et al. 2010, Agarwal et al. 2013, Agarwal et al. 2008, Greiner and Wendorff 2007, Huang et al. 2003, Kanani and Bahrami 2010, Li and Xia 2004, Sill and von Recum 2008, Wu et al. 2013, Zucchelli et al. 2011). The electrospun nanofibers and nanofibrous materials have unique properties, such as ultra-fine diameter, huge surface area to volume ratio (about a thousand times higher than that of human hair), large porosity, high aspect ratio of length to diameter, excellent mechanical properties due to the high molecular orientation along fiber axis, and easy-tailed nanofiber products. However, for the practical applications, electrospun nanofibers with proper mechanical properties are highly required. In this chapter, several approaches to achieve improved mechanical properties of electrospun fibers will be presented.

[1] College of Materials Science and Engineering, Nanjing Forestry University, Nanjing 210037, China.
[2] Department of Chemistry and Chemical Engineering, Jiangxi Normal University, Nanchang, 330022, China.
* Corresponding authors: shaohua.jiang@njfu.edu.cn; hhq2001911@126.com

Mechanical Properties of Single Electrospun Nanofibers

Mechanical properties of electrospun nanofibers are usually affected by the deposited formations of nanofibers, such as random or aligned assembling electrospun nanofiber mats. However, both kinds of fibrous mat contain a high porosity, which significantly influence the intrinsic mechanical properties of electrospun fiber mats. Therefore, in order to characterize the intrinsic mechanical properties of electrospun fibers, many researchers devoted their efforts to determine the mechanical properties of single electrospun nanofibers. Table 7.1 summarized the mechanical properties including tensile strength, modulus, elongation at break, and toughness of some typical single electrospun nanofiber. Hou's group comprehensively studied the mechanical properties of single electrospun nanofiber of polyimide (PI) and their precursors, polyamic acids (PAAs) (Chen et al. 2008, Chen et al. 2015a, Chen et al. 2012). They found that the copolymerization of monomers with soft and rigid moiety can greatly improve the mechanical properties. In their reports, the PAA single fibers with only soft moiety (PAA/BPDA-ODA) and rigid moiety (PAA/BPDA-PRM) showed tensile strength of 237 and 209 MPa, respectively (Chen et al. 2015a). However, their copolymer single fiber (PAA/BPDA-PRM-ODA with 5:5 molar ratio of PRM to ODA) presented the highest strength of 586 MPa, which is 2.71 and 2.39 times of the values of PAA/BPDA-ODA and PAA/BPDA-PRM single electrospun fiber (Chen et al. 2015). In their reports, they ascribed the effect of copolymerization of soft and rigid moiety to the uniform micro-blocks in the nanofibers in the level of super-molecular structure. Jiang and Hou et al. also found that the thermal imidization could significantly improve the tensile strength of the single electrospun fiber. For example, after thermal imidization from PAA/BPDA-PRM-ODA, the corresponding PI single fiber (PI/BPDA-PRM-ODA) even showed superior tensile strength of 1770 MPa, which is 211% higher than that of PAA/BPDA-PRM-ODA single fiber (Chen et al. 2015a). In another report, PI/BPDA-PPA single electrospun nanofiber showed tensile strength of 1708 MPa and modulus of 76 GPa, which are 2.2 and 5.8 times of those of PAA/BPDA-PPA single nanofiber (Chen et al. 2008). These greatly improved mechanical properties of PI could be because of the thermal induced higher molecular orientation along the single electrospun fiber. In another recent report, Jiang and Hou et al. produced PI microfibers by mixing green solvents of water and ethanol (wt/wt, 30/70) by electrospinning, and characterized the mechanical properties of the corresponding single fiber (Xu et al. 2017). The single PI microfiber with diameter around 2.32 μm showed a breaking load of 4872 μN, tensile strength of 1064 MPa, modulus of 5.6 GPa and elongation at break of 15.7%, which enable it to even hold a 0.5 g metal ring (Xu et al. 2017). Recently, Jiang and Hou et al. produced two kinds of carbide/carbon nanofibers, titanium carbide/carbon nanofiber (TiC/CNF) (Zhou et al. 2016), and tantalum carbide/carbon nanofiber (TaC/CNF) (Zhou et al. 2017), by electrospinning, followed with calcination. In these studies, Young's modulus of the single fiber was characterized by scanning probe microscopy (SPM) with Peak Force Quantitative Nanomechanics (QNM) mode and analysed by Derjaguin-Muller-Toporov (DMT) model. The results indicated that the TiC/CNF with diameter of 284 nm and TaC/CNF with diameter of 540 nm possessed super high modulus of 446 GPa and 502 GPa, respectively (Zhou et al. 2016, Zhou et al. 2017).

Table 7.1: Mechanical properties of single electrospun nanofibers.

Material	Diameter	Mechanical properties σ: tensile strength (MPa); E: E modulus (GPa); ε: elongation at break; T: Toughness	Ref.
PAA/BPDA-ODA	~ 560 nm	σ: 237	(Chen et al. 2015)
PAA/BPDA-PRM	~ 560 nm	σ: 209	(Chen et al. 2015)
PAA/BPDA-PRM-ODA	~ 560 nm	σ: 568	(Chen et al. 2015)
PI/BPDA-PRM-ODA	~ 560 nm	σ: 1770; E: 59	(Chen et al. 2015)
PAA/BPDA-PPA	234 nm	σ: 766; E: 13; ε: 43.3%	(Chen et al. 2008)
PI/BPDA-PPA	256 nm	σ: 1708; E: 76; ε: 2.8%	(Chen et al. 2008)
PI/BPDA-BPA-ODA	251 nm	σ: 1666; E: 37; ε: 7.5%	(Chen et al. 2012)
Nylon-6	A) 60 nm B) 100 nm C) 170 nm	A) σ: 364; ε: 44% B) σ: 125; ε: 85% C) σ: 94; ε: 130%	(Hwang et al. 2010)
Nylon-6,6	A) 570 nm B) 500 nm	A) σ: 110; E: 0.453; ε: 66% B) σ: 150; E: 0.950; ε: 61%	(Zussman et al. 2006)
PAN	A) 297 nm B) 357 nm C) 305 nm D) 179 nm E) 260 nm	A) E: 6.88 B) E: 3.79 C) E: 14.07 D) E: 47.49 E) E: 28.87	(Gu et al. 2005)
PAN	A) 2800 nm B) 100 nm	A) σ: 15; E: 0.36; T: 0.25 B) σ: 1750; E: 48; T: 605	(Papkov et al. 2013)
PLA	A) 890 nm B) 610 nm	A) σ: 89; E: 1.0; ε: 1.54% B) σ: 183; E: 2.9; ε: 0.45%	(Ryuji et al. 2005)
PLA	A) 1150 nm B) 920 nm C) 800 nm	A) σ: 48.39; E: 1.82 B) σ: 160.0; E: 3.31 C) σ: 195.2; E: 5.04	(Li et al. 2015)
PCL	A) 200–300 nm B) 5000 nm	A) σ: 220; E: 3200 B) σ: 20; E: 300	(Chew et al. 2006)
TiC/CNF	284 nm	E: 446	(Zhou et al. 2016)
TaC/CNF	540 nm	E: 502	(Zhou et al. 2017)
S-EPIMF	2.32 μm	σ: 1064	(Xu et al. 2017)

Note: PAA: polyamic acid; PI: polyimide; BPDA: 3,3',4,4'-biphenyltetracarboxylic dianhydride; ODA: 4,4'-Oxydianiline; PRM: 2,5-Bis(4-aminophenyl)pyrimidine; PPA: p-Phenylenediamine; BPA: 4,4'-Biphenyldiamine; PAN: polyacrylonitrile; PLA: polylactide; PCL: poly(ε-caprolactone); TaC/CNF: tantalum carbide-carbon electrospun nanofiber; TiC/CNF: titanium carbide–carbon nanofiber; and S-EPIMF: single electrospun polyimide microfiber.

Size Effect on the Improvement of Mechanical Properties of Single Electrospun Nanofiber

Besides PAA and PI single electrospun nanofiber, other polymer single electrospun nanofiber were also studied regarding their mechanical properties, such as nylon (Hwang et al. 2010, Zussman et al. 2006), polyacrylonitrile (PAN) (Gu et al. 2005,

Papkov et al. 2013), polylactide (PLA) (Li et al. 2015, Ryuji et al. 2005), and poly(ε-caprolactone) (PCL) (Chew et al. 2006). In these reports, the mechanical properties are significantly increased when the fiber diameter decreased. This phenomenon was called size effect on mechanical properties of electrospun nanofibers. Chew et al. found that when the diameter of the PCL fibers decreased from 5 μm down to the nanometer regime (200–300 nm), a dramatic increase of Young's modulus from 300 MPa to 3200 MPa and tensile strength from 20 MPa to 220 MPa was observed (Chew et al. 2006). The similar results for PCL fibers were also proved by Sun et al. (Sun et al. 2008). Recently, Papkov et al. found the size effect of polyacrylonitrile (PAN) fibers on the elastic modulus and toughness (Fig. 7.1) (Papkov et al. 2013). A linear increase in the strength, modulus, and toughness was observed, as the diameter of the electrospun nanofiber was smaller than 500 nm. The same conclusion about the size effect of fibers on the mechanical performance was also observed on the electrospun polyamide (nylon-6) fibers, polyimide (PI) fibers, carbon fibers, and so on (Bazbouz et al. 2010, Chen et al. 2008, Hwang et al. 2010, Lin et al. 2012). This effect could be explained by three aspects (Hwang et al. 2010, Zucchelli et al. 2011). First, electric force-induced molecular orientation along the fiber axis happened during the electrospinning. Second, the smaller fiber diameter leads to a higher crystallinity. Third, less surface defects per unit fiber length appeared on the smaller fiber, which result in better mechanical properties.

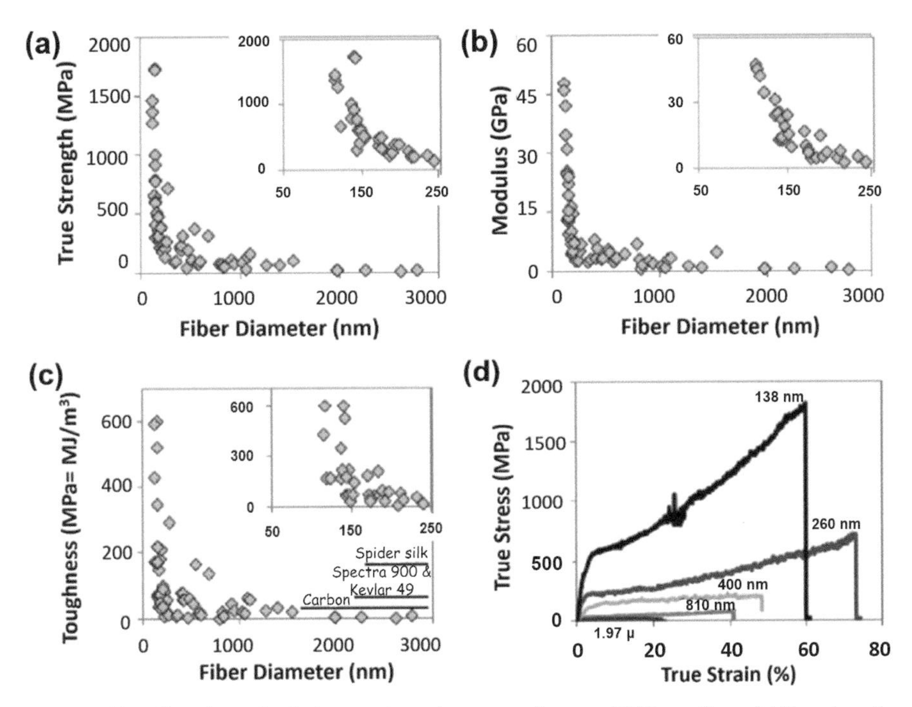

Fig. 7.1: Size effects in mechanical properties and structure of as-spun PAN nanofibers. (a) True strength; (b) modulus; (c) toughness (lines indicate comparison values for several high-performance fibers and spider silk); (d) typical stress/strain behavior. Adapted with permission from (Papkov et al. 2013) Copyright (2013) American Chemical Society.

Blending of Polymer Components

Mechanical properties of electrospun fibers could be improved by blending different polymer components with different mechanical properties. This blending can be realized via different kinds of ways, such as blending of the electrospinning solution, blending of electrospun fibers from multi-jets, and blending by coaxial/triaxial electrospinning.

Blending Solutions

Blending different polymers in one electrospinning solution is a straightforward way to produce composite electrospun fibers with modified mechanical properties. In this way, polymers with better mechanical strength, higher modulus or larger toughness were usually used to improve the mechanical properties of other polymeric fibers with weak mechanical properties. The blending solutions can be prepared by directly mixing different polymers into homogeneous electrospinning solutions or mixing polymeric crystals into the solution for electrospinning. Jiang and Hou et al. prepared highly strong and highly tough electrospun PI/PI composite nanofibers from a binary blend of PAA (He et al. 2014). Without blending, the PI/BPDA-ODA and PI/BPDA-BPA electrospun nanofiber belts showed tensile strength of 415 and 458 MPa, respectively. After blending, the corresponding PI/PI composite fiber belt showed tunable tensile strength and the highest value of 1299 MPa was obtained when the blend ratio of BPA/ODA was 40/60 (He et al. 2014). They believed that the improvement of the tensile strength could be because of the co-existence of microblock structures, interlamellar and interfibrillar segregations of the rigid and flexible moiety in the blend composite nanofibers (Fig. 7.2).

Polyvinyl alcohol (PVA) is a very useful polymer with very good electrospinnality and is often used as a blend component to improve the fiber formation of other polymers. Islam and Karim incorporated PVA into alginate for electrospinning and the obtained PVA/alginate blend nanofibers presented enhanced tensile strength as increasing the amount of alginate (Islam and Karim 2010). In another report, the addition of PVA improved the elongation at break of the PVA/chitosan blend nanofiber membranes (Duan et al. 2006). In addition, other polymers, such as gelatin (Linh et al. 2012), collagen (Shields et al. 2004), and poly(ethylene terephthalate) (PET) (Li et al. 2013), were also electrospun into fibrous scaffolds with the addition of PVA, and the resultant composite membranes showed improved mechanical properties. Kim et al. blended poly (vinyl chloride) (PVC) and polyurethane (PU) for electrospinning and systematically investigated the effects of different blend ratios on the mechanical properties of the electrospun composite nonwovens (Lee and Lee 2003). The largest modulus of 11.80 MPa and tensile strength of 3.73 MPa were obtained from PVC/PU (50/50, wt/wt) nonwoven. These values were 214% and 314% higher than those of PVC electrospun nonwoven. In another report, PU was blended with polysulfone (PSF) for electrospinning (Cha et al. 2006). The increasing amount of PU in the composites showed larger tensile strength and modulus (Fig. 7.3). For example, the PSF/PU with 80/20 weight ratio possessed 2.5 MPa tensile strength and 3.1 MPa modulus, which were 6.2 and 2.2 times of the values from sample with 90/10 weight ratio.

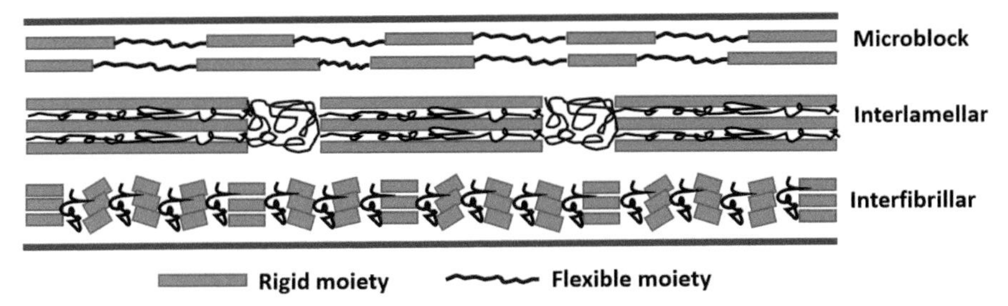

Fig. 7.2: Scheme of microstructures of the blend PI nanofiber with rigid and flexible components. Adapted with permission from (He et al. 2014). Copyright (2014) Royal Society of Chemistry.

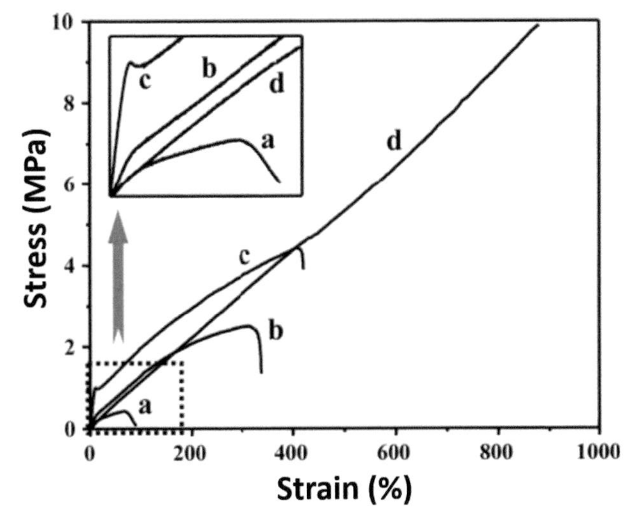

Fig. 7.3: Mechanical behaviors of polyblend electrospun nonwovens as a function of the blend composition (PSF/PU, w/w): (a) 90/10, (b) 80/20, (c) 20/80, and (d) 0/100. Adapted with permission from (Cha et al. 2006) Copyright (2006) The Polymer Society of Korea and Springer.

Cellulose is a biocompatible polymer and its nanocrystals (CNCs) can be produced from an easily accessible and abundant natural source—wood pulp. CNCs possess unique properties, such as nanoscale dimension, excellent mechanical properties, ease of chemical modifications, inherent biodegradability, and biocompatibility (Azizi Samir et al. 2005, Habibi et al. 2010). Therefore, CNCs are reported to add into other polymer solutions for electrospinning to prepare CNC reinforced polymer fibrous composites. Wu et al. prepared electrospun fibrous scaffolds reinforced with CNCs by using maleic anhydride grafted PLA (MPLA) (Zhou et al. 2013). As shown in Table 7.2, only very small amount addition (1–5 wt%) of CNCs could lead to superior enhancement on mechanical properties. Similar studies were also performed to produce reinforced electrospun composite membranes, such as PCL/CNCs (Zoppe et al. 2009), poly(vinyl acetate)/CNC (Ansari et al. 2015), PU/CNC (Yao et al. 2014), lignin/CNC (Ago et al. 2013, Ago et al. 2012), polyethylene oxide (PEO)/CNC (Zhou et al. 2011), polystyrene (PS)/CNC (Rojas et al. 2009), PAN/CNC (Cao et al. 2013), and so on.

Table 7.2: Mechanical properties of MPLA/CNCs electrospun composite scaffolds (Zhou et al. 2013).

Samples	Tensile strength (MPa)	Modulus (MPa)	Elongation at break (%)
MPLA	1.6	7.8	51.8
MPLA/CNC (1 wt%)	4.8	77.2	46.9
MPLA/CNC (2 wt%)	4.9	87.4	45.8
MPLA/CNC (5 wt%)	10.8	135.1	44.0

Blending by Multi-jet Electrospinning

Electrospinning for blend solutions requires a good solvent for both components. However, it is difficult to find proper solvents. To solve this problem, multi-jet electrospinning was applied to produce composite fibrous membranes. In this way, different polymer solutions can be fed in different syringes for electrospinning (Fig. 7.4a). The components of the resultant composite fibrous membranes can be controlled by changing the number of syringes for each polymers. Ding et al. used this method to produce a biodegradable blend fibrous nonwoven from PVA and cellulose acetate (CA) (Ding et al. 2004). By changing the number of jets for each component, the composite nonwoven presented tunable mechanical properties (Fig. 7.4b). A similar method was also applied for the preparation of PS/PAN composite nonwoven from the same group (Sun et al. 2010). The fibrous PS nonwoven possessed a tensile strength of 5.56 kPa and modulus of 0.16 MPa. By adding the PAN component, a great enhancement on mechanical properties was observed. The tensile strength and modulus of the composite nonwoven increased from 0.45 to 0.68 MPa, and from 10.21 to 16.75 kPa, when increasing the number ratio of jets of PS/PAN from 1/3 to 3/1, respectively (Sun et al. 2010). Recently, the same strategy was also used by Yoon et al. to produce polyamide 6 (PA6) reinforced PS composite nonwoven (Yoon et al. 2017). The fiber structure, the properties of each component, and the interaction between each polymer fiber are the possible reasons for the improvement of the mechanical properties by multi-jet electrospinning.

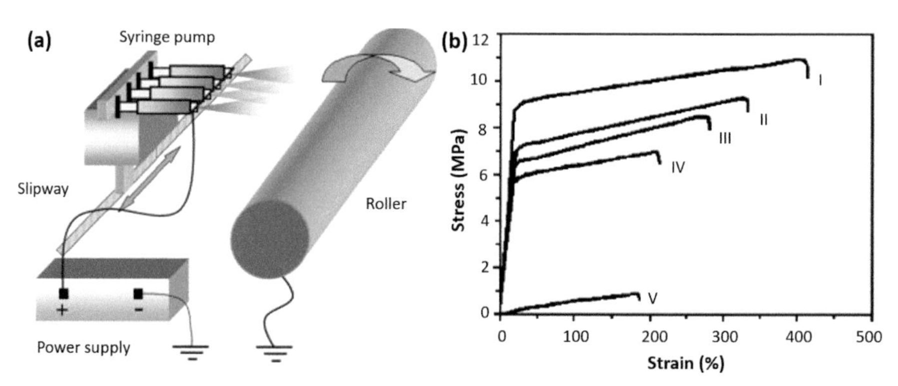

Fig. 7.4: (a) The schematic of the multi-syringe electrospinning process (Adapted with permission from (Sun et al. 2010) Copyright (2010) Elsevier). (b) stress–strain curves of nanofibrous mats with different number ratio of jets of PVA/CA: (I) 4/0; (II) 3/1; (III) 2/2; (IV) 1/3; and (V) 0/4. Adapted with permission from (Ding et al. 2004) Copyright (2004) Elsevier.

Blending by Coaxial and Triaxial Electrospinning

Modification on electrospinning nozzles, such as coaxial nozzle, triaxial nozzle, and side-by-side nozzle, makes it possible to prepare composite fibrous materials with improved mechanical properties. Coaxial electrospinning has been broadly used to combine two different materials into one electrospun nanofiber (Qu et al. 2013). Many studies applied this coaxial electrospinning to make reinforced composite electrospun fibers. Huang et al. fabricated electrospun polycarbonate (PC) (shell)/PU (core) composite fibers by coaxial electrospinning (Han et al. 2006). Compared to the shell polymer of PC, the PC/PU composite fibers showed significant improvement on tensile strength (13 MPa vs 1.04 MPa) and modulus (0.82 MPa vs 0.04 MPa) (Han et al. 2006). With this coaxial electrospinning technique, the coaxial fibers, such as PCL/Teflon (Han and Steckl 2009), PLA/chitosan (Nguyen et al. 2011), and PMMA/PAN (Zander et al. 2011), were prepared and showed enhanced mechanical properties when comparing the single component.

Triaxial electrospinning could be used to produce nano/micro-fibers with concentric three-layer morphology. It can be performed by a triaxial nozzle by which three electrospinning solutions can be electrospun into fibers with tri-layer structure (Fig. 7.5a, b). Jiang et al. prepared PS/thermoplastic polyurethane (TPU)/PS fiber by triaxial electrospinning and compared the mechanical properties of its single fiber with pure PS, TPU, and TPU/PS coaxial single fiber (Fig. 7.5c) (Jiang et al. 2014). The triaxial fiber with TPU/PS weight ratio of 30/70 presented the best mechanical properties with tensile strength > 60 MPa, toughness > 270 J/g and elongation at break > 870% in comparison to the mechanical properties of PS single fiber with tensile strength < 20 MPa, toughness < 0.5 J/g and elongation at break < 5% (Jiang et al. 2014). This great improvement on mechanical properties by triaxial electrospinning could be attributed to the formation of the strong interface, which guarantee the interfacial force transfer and providing great improvement in the ductility and toughness (Jiang et al. 2014).

Another special electrospinning nozzle, named side-by-side nozzle could also combine two different polymers into one fiber. With this method, Peng et al. prepared

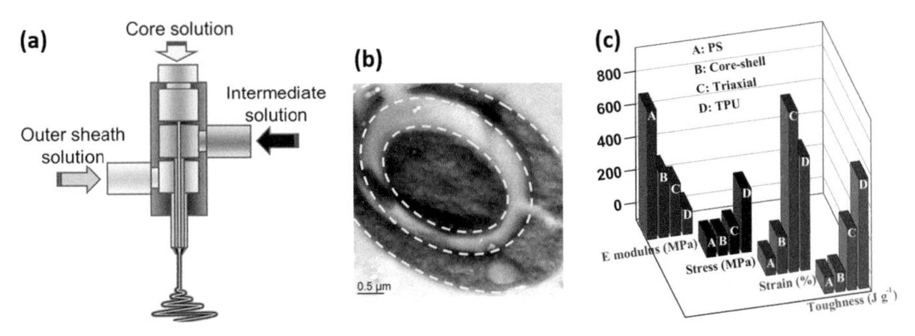

Fig. 7.5: (a) Schematic triaxial nozzle. Adapted with permission from (Han and Steckl 2013). Copyright (2013) American Chemical Society. (b) TEM image of cross-section of PS/TPU/PS triaxial single fiber, and (c) comparison of mechanical properties of pure PS, TPU, coaxial TPU/PS, and triaxial PS/TPU/PS triaxial single fiber. Adapted with permission from (Jiang et al. 2014). Copyright (2014) American Chemical Society.

silk/PLA side-by-side fibrous mat (Peng et al. 2015). The aligned side-by-side fiber mat showed higher tensile strength of 14 MPa and modulus of 299 MPa than those of silk fibers (tensile strength of 7.19 MPa and modulus of 170 MPa), which could be due to the stress-induced crystallinity in PLA fibers during alignment (Peng et al. 2015). In another study, Chen et al. presented an interesting work with side-by-side electrospinning to produce nanofibers with and without springs and compared their mechanical properties (Fig. 7.6) (Chen et al. 2009). Although the tensile strength was slightly decreased to 151 MPa, the Nomex/TPU side-by-side aligned spring composite nanofibers possessed much better elongation of 97% and toughness of 102 MPa, which were 2.93 and 1.73 times of the properties of aligned side-by-side fibers, but without springs (Chen et al. 2009).

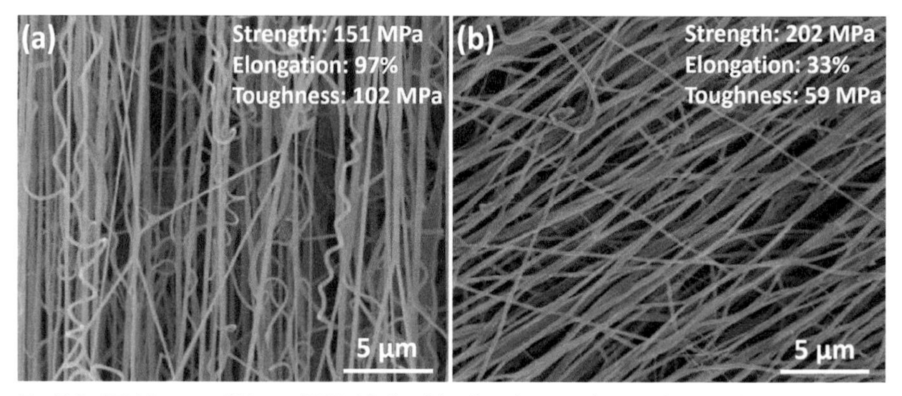

Fig. 7.6: SEM images of Nomex/TPU side-by-side aligned composite nanofibers with (a) and without (b) springs. Adapted with permission from (Chen et al. 2009). Copyright (2009) John Wiley and Sons.

Incorporation of Inorganic Fillers

Inorganic fillers often have superior mechanical properties. Many researches focus on the fabrication of electrospun composite nanofibers reinforced by inorganic fillers, such as carbon nanotubes (CNTs), graphene, TiO_2, etc.

Incorporation of Carbon Nanotubes

Carbon nanotubes (CNTs) have one dimensional cylindrical nanostructures with high aspect length-to-diameter ratio, and superior properties, especially mechanical properties (tensile strength: 11–63 GPa; modulus: 270–950 GPa) (Yu et al. 2000). So CNTs are good candidates as reinforcements to produce composites. CNTs can be dispersed into the polymer solutions. After electrospinning, CNTs reinforced electrospun polymeric nanofibers can be produced. Sen et al. compared the effect of as-prepared CNTs (AP-CNT) and ester functionalized CNTs (E-CNT) on the mechanical properties of their PU fibrous composite membranes (Sen et al. 2004). Both CNTs with only 1 wt% could lead to significant improvement on tensile strength and modulus. When compared to the PU membranes, E-CNT showed larger enhancement than AP-CNT (Table 7.3) (Sen et al. 2004). In another report, CNTs

Table 7.3: Summary of CNTs reinforced electrospun polymer nanofibers.

Type of CNTs	Composite nanofibers (amount of CNT)	Improvement of mechanical properties	Ref.
Ester functionalized	PU (1 wt%)	σ: 104%; E: 250%	(Sen et al. 2004)
As-prepared	PU (1 wt%)	σ: 46%; E: 215%	(Sen et al. 2004)
High-pressure disproportionation	Silk (1 wt%)	E: 460%	(Ayutsede et al. 2006)
As-prepared	PAN (A) 5 wt% (B) 20 wt%	A) σ: 75%; E: 72% B) σ: –19%; E: 144%	(Hou et al. 2005)
–COOH functionalized	Nylon-6,6 (1 wt%)	σ: 25%; E: 70%	(Baji et al. 2010)
–COOH functionalized	Cellulose (A) 0.11 wt% (B) 0.55 wt%	A) σ: 35%; E: 69% B) σ: 86%; E: 107%	(Lu and Hsieh 2010)
Surface functionalized	PMMA (5 wt%)	σ: 157%; E: 267%	(Liu et al. 2007)
–COOH functionalized	PLA (3 wt%)	Random (σ: 274%; E: 422%) Aligned (σ: 136%; E: 259%)	(Shao et al. 2011)
–COOH functionalized	PI (A) 3.5 wt% (B) 10 wt%	A) σ: 28%; E: 4% B) σ: –10%; E: 26%	(Chen et al. 2009)

were used to reinforce silk electrospun nanofibers, and an increase in modulus up to 460% was observed (Ayutsede et al. 2006). Hou et al. studied the effect of high amount of CNTs on the mechanical properties of PAN/CNT composite nanofibers (Hou et al. 2005). 5 wt% and 20 wt% of CNTs could lead to an improvement of 75% tensile strength and 144% modulus, respectively. Baji et al. found that only 1 wt% addition of CNTs could lead to 70% and 25% improvement on modulus and tensile strength, respectively, when compared to the neat nylon-6, 6 electrospun fibers (Baji et al. 2010). Lu and Hsieh studied the CNT reinforced cellulose electrospun composite fibers and found that very few amount of CNTs could result in a significant improvement on mechanical properties (Lu and Hsieh 2010). When the amounts of CNTs were 0.11 wt% and 0.55 wt%, the modulus and tensile strength could improve 69%/107% and 35%/86%, respectively. In another report, Wong et al. systematically studied the effect of the aspect ratio of CNTs on the reinforcement of PVA electrospun nanofibers (Wong et al. 2009). They found that significant enhancement in the modulus of the composite fibers was achieved when the aspect ratio was above 36. Many more examples by using CNTs to reinforce polymer electrospun fibers can be found in Table 7.3. In general, for the CNT reinforced polymer electrospun composite fibers, the mechanical improvement is usually affected by the following factors: (a) aspect ratio of CNTs; (b) amount of CNTs; (c) surface functionalization on CNTs to improve the dispersion in polymer matrix; and (d) hydrogen bonding between the polar groups of CNTs and polymer fiber matrix.

Incorporation of Graphene

Graphene is another kind of important carbon material. It is a two-dimensional nanomaterial with superior tensile strength of 130 GPa and modulus of 1 TPa (Lee et al. 2008). Therefore, graphene can be good additive to produce electrospun polymer nanofibers with improved mechanical properties. Mack et al. reported that 1–4 wt% addition of graphene could lead to an enhancement on modulus of electrospun PAN nanofibers due to the efficient stress transfer from the high aspect ratio of graphene (Mack et al. 2005). Das et al. applied pyrrolidone-stablized graphene to reinforce PVA electrospun nanofibers, and a 205% improvement of modulus was observed (Das et al. 2013). Wan and Chen incorporated only 0.3 wt% graphene oxide in electrospun PCL nanofibers, and the composite nanofibers showed an improvement of 95%, 66%, and 416% for tensile strength, modulus, and energy at break, respectively (Wan and Chen 2011). Gopiraman et al. modified graphene with –COOH group (graphene-COOH) to improve the dispersion in cellulose acetate (CA) solution and compared the effect of modified and unmodified graphene on the mechanical properties of CA/graphene composite nanofibers (Gopiraman et al. 2013). The summary of modulus of CA/graphene composite nanofibers was shown in Table 7.4. Due to the better distribution in the CA fibrous matrix, graphene-COOH showed much better mechanical enhancement than the unmodified graphene (Gopiraman et al. 2013). The composite nanofiber with 4 wt% graphene-COOH presented the highest modulus of 911 MPa, which was 269% of the pure CA nanofibers (246 MPa). This difference of reinforcement could be due to the hydrogen-bonding between –COOH groups in the graphene-COOH and –C=O groups in the CA molecules, as proved by the FT-IR, XRD, and Raman analysis (Gopiraman et al. 2013).

Table 7.4: Modulus of CA/graphene composite electrospun nanofibers reinforced by –COOH modified and unmodified graphene (Gopiraman et al. 2013).

Amount of graphene (wt%)	Modulus of CA/graphene (MPa)	Modulus of CA/graphene-COOH (MPa)
0	246	246
0.5	351	390
1.0	370	421
2.0	414	677
3.0	420	806
4.0	740	911
5.0	704	806

Incorporation of Other Inorganic Nanoparticles

Inorganic nanoparticles attract a lot of attention because of their novel physical properties when their sizes are in the range of nanometer scale. Numerous researchers devoted their efforts to reinforce electrospun nanofibers by addition of inorganic nanoparticles (C.-L. Zhang and Yu 2014). Two different approaches are reported to fabricate inorganic nanoparticle/polymer composite nanofibers. One is preparing nanofibers after electrospinning, combined with post-treatment. Another is directly preparing composite nanofibers during the electrospinning process. In the first method, precursor materials are usually blended together with polymer solutions for electrospinning. Then the nanoparticles would be formed from the precursors on the surfaces or inside the nanofibers by post-treatments, such as *in situ* reduction, hydrothermal-assisted method, and calcination. For the latter method, inorganic nanoparticles are often directly blended into the polymer matrix solutions for electrospinning. This method often meets the problem of aggregation of nanoparticles inside the electrospun nanofibers. Via these two methods, inorganic nanoparticles of metal, metal oxides, ceramic, and carbon can be incorporated into electrospun nanofibers, and enhance their mechanical properties.

Ji et al. blended ZnO nanoparticles with average particle size of 80 nm into electrospun PLA nanofibers, and studied the effect of particle amounts on the mechanical properties of PLA/ZnO composite electrospun fibers (Ji et al. 2013). When the amounts of ZnO were 0.4 wt% and 1.0 wt%, the tensile strength of PLA/ZnO composite nanofibers increased to 0.35 and 0.45 MPa, which were much larger than the pure PLA nanofibers (tensile strength of 0.3 MPa). However, when the amount of ZnO nanoparticles increased to 2 wt%, the mechanical properties including tensile strength and modulus were dramatically decreased, which could be attributed to the aggregation of ZnO nanoparticles (Ji et al. 2013). In another report, ZnO nanoparticles with an average size of 60 nm were used to reinforce PCL nanofibers, which showed antibacterial and cell adhesion properties (Augustine et al. 2014). As shown in Fig. 7.7, the tensile strength first showed an increase and then decrease, when increasing the amount of ZnO. This phenomenon could be explained by the following reasons: (a) better dispersion of nanoparticles in polymer matrix at low amount of nanoparticles, which increase the interfacial area for stress transfer from matrix to fillers; and (b) larger amount of ZnO nanoparticles led to the decreased crystallinity of matrix polymer, which showed better mechanical properties (Augustine et al. 2014).

Besides ZnO nanoparticles, other kinds of metal oxides nanoparticles were also reported to enhance electrospun polymer nanofibers, such as TiO_2 reinforced nylon-6, ZrO_2 and Al_2O_3 reinforced PVA nanofibers (Lamastra et al. 2008, Pant et al.

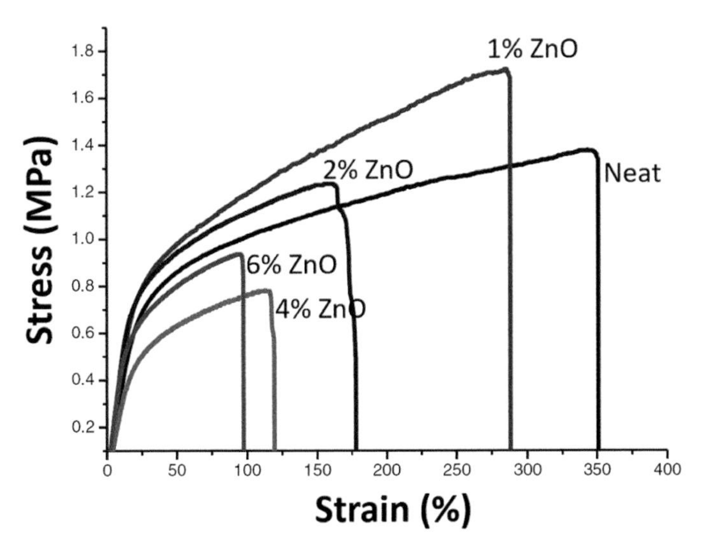

Fig. 7.7: Typical stress-strain curves of PCL/ZnO composite electrospun fibers with different amounts of ZnO nanoparticles. Adapted with permission from (Augustine et al. 2014). Copyright (2014) Springer.

2011). 1 wt% of TiO_2 nanoparticles with an average size of 21 nm was found to be the best for enhancement of nylon-6 due to the good dispersion of nanoparticles. An additional energy-dissipating mechanism was used to explain the reinforcement of TiO_2 nanoparticles to the nylon-6 nanofibers, which could result in the mobility of the nanoparticles. During the tensile test, the mobility would lead to the nanoparticles aligning under the stress, and then create temporary cross-links to enhance the mechanical properties (Pant et al. 2011). In another communication, Lamastra et al. investigate the effect of ZrO_2 and Al_2O_3 nanoparticles with average particle sizes of 20 nm, and their amounts on the mechanical properties of PVA composite nanofibers (Lamastra et al. 2008). As shown in Table 7.5, both ZrO_2 and Al_2O_3 nanoparticles led to a great enhancement of tensile strength and modulus, but accompanied with the decease of the elongation at break.

Silver nanoparticles (AgNPs) are attractive materials due to their well-known antibacterial activity. Francis et al. blended 3 wt% of AgNPs into nylon-6 solution for electrospinning, and about 3 times increase in tensile strength was observed when compared to the neat nylon-6 fibrous mat (Francis et al. 2010). In another report, An et al. prepared uniform chitosan (CS)/poly(ethylene oxide) (PEO) electrospun nanofibers containing AgNPs (An et al. 2009). In order to improve the dispersion of AgNPs in the matrix polymer, CS/AgNP colloids were first prepared by an *in situ* chemical reduction. After electrospinning, the AgNPs could be uniformly distributed in the composite fibers without any aggregation, as shown in TEM images in Fig. 7.8 (An et al. 2009). In addition, they also investigated the mechanical properties of CS/PEO/AgNP composite nanofibers in dry and wet states. As shown in the typical stress-strain curves in Fig. 7.8, both in dry and wet state, the CS/PEO/AgNP composite nanofiber membranes showed obvious enhancement on tensile strength and modulus, which could be attributed to the good dispersion of AgNPs in the polymer fiber matrix, the strong adhesion between CS and AgNPs, and the strong hydrogen-bonding between CS and PEO (An et al. 2009).

Besides the above metal and metal oxide nanoparticles reinforced electrospun nanofiber composites, electrospun nanofibers could also be enhanced by some other inorganic particles, such as hydroxyapatite nanoparticle reinforced silk fibroin nanofibers (Kim et al. 2014), $CaCO_3$ nanoparticles reinforced chitosan/poly (vinyl alcohol) nanofibers (Sambudi et al. 2015), carbon black nanoparticles reinforced poly (vinyl alcohol) nanofiber mat (Chuangchote et al. 2007), SiO_2 reinforced nylon-6 (Ding et al. 2009), and so on.

Table 7.5: Summary of mechanical properties of PVA nanofiber mat, PVA/ZrO_2 nanofiber mat at low amount (3 wt%, PVA/ZrO_2-L) and high amount (10 wt%, PVA/ZrO_2-H) and, PVA/Al_2O_3 nanofiber mat at low amount (3 wt%, PVA/Al_2O_3-L) and high amount (10 wt%, PVA/Al_2O_3-H) (Lamastra et al. 2008).

Samples	Tensile strength (MPa)	Modulus (MPa)	Elongation at break (%)
PVA	3.2 ± 0.2	90 ± 20	90 ± 20
PVA/ZrO_2-L	5.4 ± 0.2	120 ± 45	70 ± 5
PVA/ZrO_2-H	6.4 ± 0.6	190 ± 55	70 ± 6
PVA/Al_2O_3-L	5.4 ± 0.1	280 ± 80	50 ± 10
PVA/Al_2O_3-H	6.0 ± 2.0	350 ± 40	60 ± 20

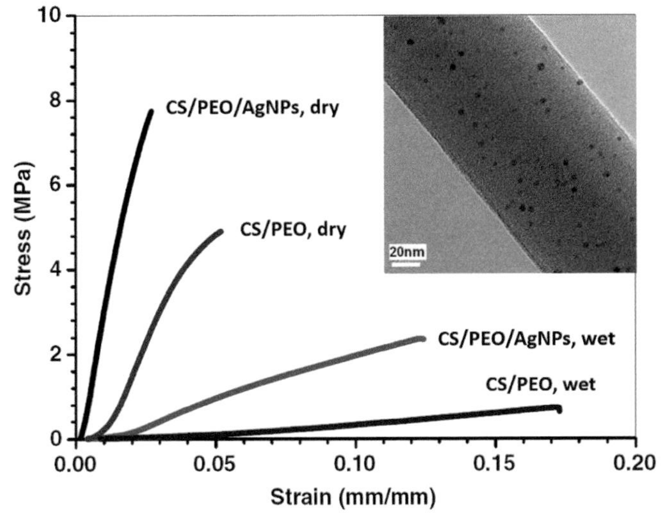

Fig. 7.8: Typical stress-strain curves of CS/PEO/AgNPs composite and CS/PEO electrospun fibrous membranes in dry and wet states. Adapted with permission from (An et al. 2009). Copyright (2009) Springer.

Cross-linking of Polymeric Nanofibers

Cross-linking is linking one polymer chain to another by a bone-like covalent bond and noncovalent bond (ionic bond or hydrogen bond). As the polymer chains are linked together, the individual polymer chains lose certain ability to move. As a result, the physical properties of polymer can change a lot by the use of cross-links, such as viscosity, T_g, melting point, density, hardness, solubility, and mechanical properties, which depend strongly on the cross-linking density. Cross-linking has been proved to be an effective technique to improve the mechanical properties of polymeric nanofibers (Agarwal et al. 2010).

To modify the electrospinning nanofibers with high tensile strength, cross-linking can be induced in the electrospinning nanofibers by forming chemical reactions with cross-linking regents of an unpolymerized or partially polymerized resin with specific chemicals. The inter- and intra-molecular covalent bonds and the bonding between the fiber junctions are responsible for tremendous improvement in the mechanical properties of nanofibers. Besides, formation of point-bonded structures favors the structural integrity of electrospun fibers and the enhanced intermolecular interaction during the cross-linking are ascribed to improved mechanical properties (Lee et al. 2003, Zhang et al. 2006).

As shown in the Table 7.6, the glutaraldehyde (GTA), genipin, hexamethylene diisocyanate and 1-ethyl-3-(dimethyl-aminopropyl) carbodiimide hydrochloride and N-hydroxyl succinimide are usually used to form a cross-linking in the nanofibers. GTA has been used to cross-link hydroxyl-containing nanofibers (e.g., PVA, PEO, starch, chitosan, silk, and gelatin) through a vapor phase cross-linking reaction with high efficiency, short reaction time, and low cost (Ramires and Milella 2002, Wang et al. 2016, Zhang et al. 2006). To conquer the issue of the electrospun starch nanofiber

Table 7.6: Mechanical properties of electrospun nanofiber materials before and after cross-linking.

Polymer	**Cross-linker**	**Mechanical properties before cross-linking**	**Mechanical properties after cross-linking**	**Ref.**
Chitosan–PEO	Glutaraldehyde	σ: 9.46	σ: 141.2	(Vondran et al. 2008)
Gelatin	Glutaraldehyde	σ: 1.28; E: 46.5; ε: 32.4%	σ: 12.62; E: 424.7; ε: 48.8%	(Zhang et al. 2006)
Starch	Glutaraldehyde	σ: 1.2	σ: 12.2	(Wang et al. 2016)
Polybutadiene	UV		σ: 1.19; E: 0.52; ε: 90%	(Choi et al. 2006)
PEI/PVA	Glutaraldehyde	σ: 8.31; E: 29.18; ε: 133.6%	σ: 9.84; E: 162.48; ε: 35.6%	(Fang et al. 2011)
Gelatin	1-ethyl-3-(dimethyl-aminopropyl) carbodiimide hydrochloride and N-hydroxyl succinimide	σ: 1.11; E: 9.5; ε: 37%	σ: 2.44; E: 156; ε: 1.3%	(Zhang et al. 2009)
PCL/gelatin	Genipin	σ: 36.2; E: 0.9; ε: 105%	σ: 78.3; E: 2.9; ε: 181.3%	(Kim et al. 2010)
Zein	Hexamethylene diisocyanate	σ: 1.696; E: 184.5; ε: 7.934%	σ: 4.24; E: 95.15; ε: 15.838%	(Yao et al. 2007)
Chitosan	Genipin	σ: 2.1; E: 44.2	σ: 3.5; E: 147.4	(Frohbergh et al. 2012)
PEO	UV	σ: 2.49; E: 13.7; ε: 245%	σ: 14.22; E: 36.3; ε: 167%	(Zhou et al. 2012)
PBz/PBI	Thermally cross-linker	σ: 85; E: 2270; ε: 6.4%	σ: 115; E: 6570; ε: 10%	(Li and Liu 2013)
silk fibroin/ chitosan	Genipin, glutaraldehyde	σ: 2.68; ε: 4.7%	σ: 12.66; ε: 4.33%	(Zhang et al. 2010)
chitosan/PVA	Gluteraldehyde	σ: 5.65; E: 34.56; ε: 17.18%	σ: 8.35; E: 156.65; ε: 3.68%	(Liao et al. 2011)

with water solubility and low mechanical properties, GTA vapor phase was used to cross-link the starch nanofibers (Wang et al. 2016). Compared to non-cross-linked starch nanofibers, the cross-linked nanofibers by cross-linking treatment for 24 hours were increased by nearly 10 times in tensile strength and 9 times in modulus, as shown in Fig. 7.9.

In addition, water soluble (PVA, PEO, PVP) nanofibers are biocompatible nanomaterials, which can be electrospun from a relatively cheap and environmentally friendly solvent like water. However, the poor mechanical strength limit the reuse of these nanofibrous materials used for practical applications. Based on a similar concept, Fang et al. recently also demonstrated the improved mechanical properties of PEI/PVA nanofibers after TGA vapour cross-linking (Fig. 7.10) (Fang et al. 2011).

Ultraviolet (UV) induced cross-linking is also an effective way to form cross-links in the nanofibers (Choi et al. 2006, Tsanov et al. 1995, Zhou et al. 2012). Zhou et al. reported UV-initiated cross-linking of electrospun poly(ethylene oxide) (PEO)

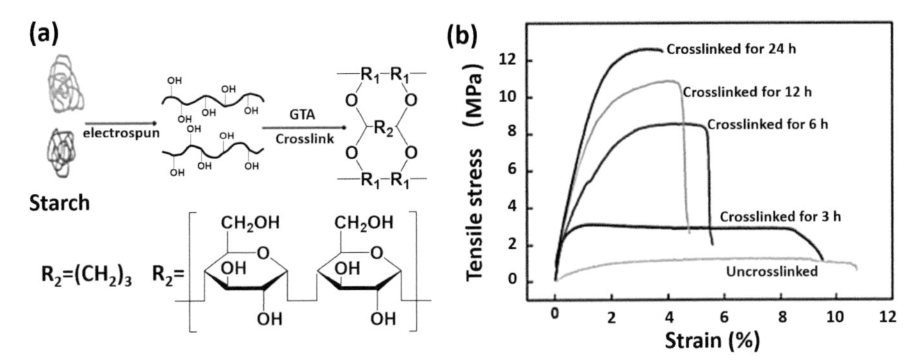

Fig. 7.9: (a) Effect of the electrospinning and on glutaraldehyde treatment on starch molecular structure. (b) strain–stress curves of electrospun starch fibers before and after cross-linking for different periods of time. Adapted with permission from (Wang et al. 2016). Copyright (2016) Elsevier.

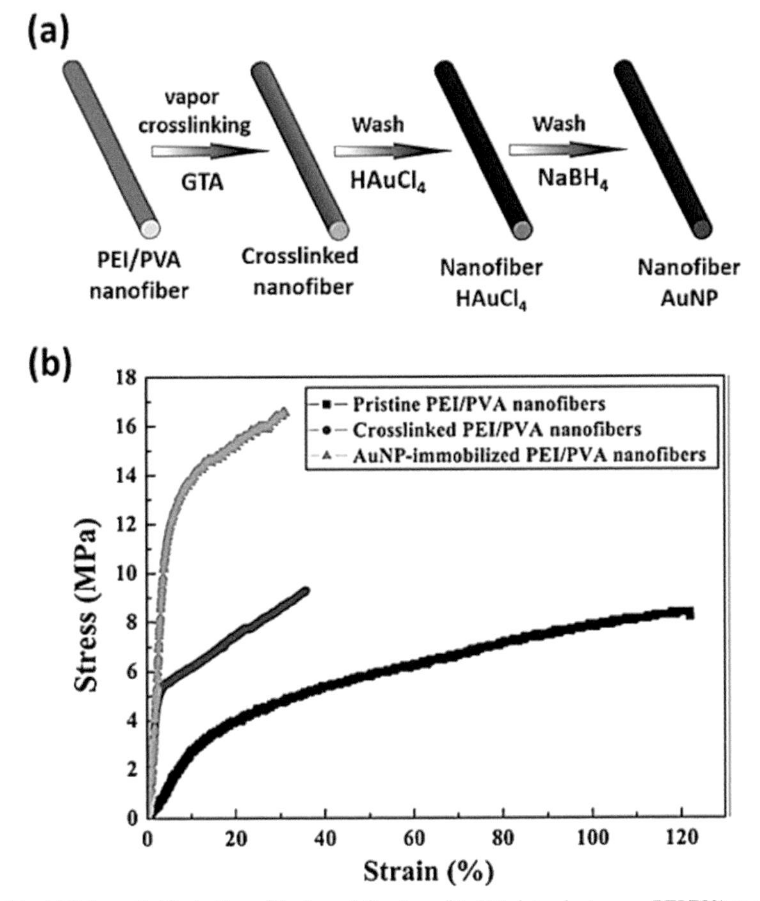

Fig. 7.10: (a) Schematic illustration of the immobilization of AuNPs into electrospun PEI/PVA nanofibers. (b) Stress–strain curves of the pristine non-cross-linked electrospun PEI/PVA nanofibrous mats, the GA vapour-cross-linked PEI/PVA nanofibers, and the AuNP-immobilized PEI/PVA nanofibers. Adapted with permission from (Fang et al. 2011). Copyright (2011) The Royal Society of Chemistry.

nanofibers (Zhou et al. 2012). The cellulose nanocrystals (CNCs) and pentaerythritol triacrylate (PETA) were performed as both photo-initiator and cross-linker. The experimental results showed that the cross-linked PEO/CNCs composite nanofibers exhibited a significant improvement in mechanical properties. At the CNCs content of 10 wt%, the maximum tensile stress and Young's modulus of the cross-linked PEO/CNCs composite fibrous mats increased by 377.5 and 190.5% than the un-cross-linked PEO mats. Choi et al. prepared cross-linked electrospun polybutadiene (BR) nanofibers by electrospinning and UV curing method, as shown in the Fig. 7.11 (Choi et al. 2006). Tensile strength, modulus, and elongation at break of the electrospun BR fiber mats increased with increase of the cross-linker content.

Polybenzoxazine (PBz) modified polybenzimidazole (PBI) nanofibers have been reported by Li and Liu (Li and Liu 2013). The nanofibers have been cross-linked through the ring-opening addition reaction of the benzoxazine groups of PBz. A significant enhancement in mechanical strength can be found in the cross-linked PBI nanofibers (Fig. 7.12). And the relatively high amplitude of increase in the

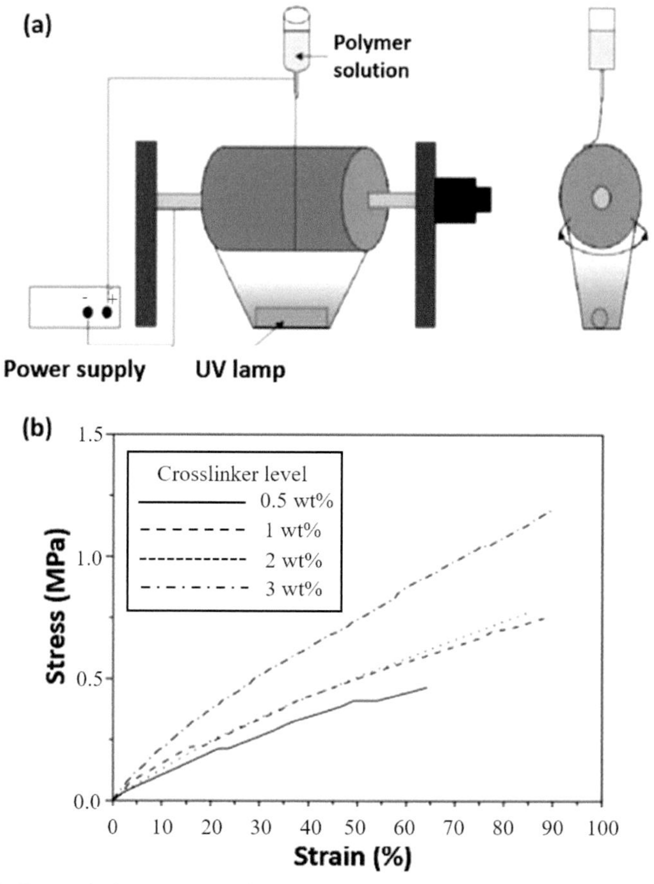

Fig. 7.11: (a) Electrospinning apparatus with *in situ* UV irradiation system. (b) Stress-strain curves for UV-cured electrospun BR fiber mats. Adapted with permission from (Choi et al. 2006). Copyright (2006) John Wiley and Sons.

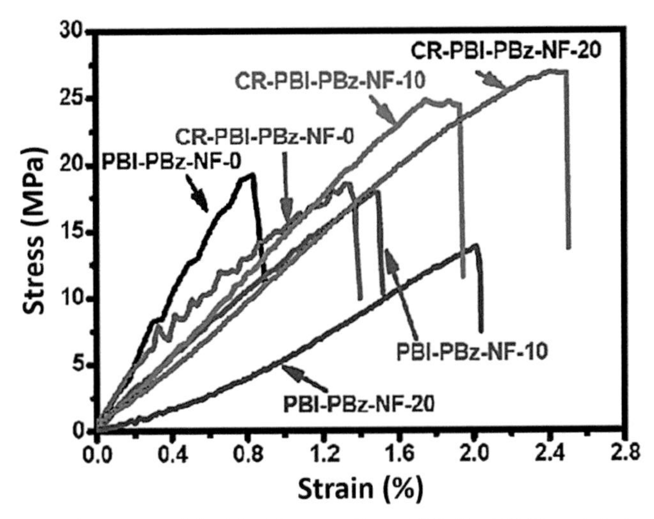

Fig. 7.12: Stress-strain curves of the PBI-PBz nanofibers (PBI-PBz-NF-X) and their corresponding cross-linked samples (CR-PBI-PBz-NF-X). Adapted with permission from (Li and Liu 2013). Copyright (2013) John Wiley and Sons.

mechanical strength could be attributed to the relatively high cross-linking density of the benzoxazine group.

Thermal Treatment

Thermal treatment on electrospun nanofibers can improve the crystallinity, create junctions between nanofibers, and increase the molecular orientation along nanofibers, which results in the mechanical improvement of electrospun nanofibrous membranes. In the following, thermal treatments including thermal calendaring, annealing, and stretching on the electrospun nanofibers to improve their mechanical properties will be introduced.

Thermal Calendaring

Thermal calendaring is a process to apply pressure and temperature to the materials. This process on electrospun nanofibers can create bonding point between nanofibers and combine different polymer fibers together. This approach can effectively decrease the thickness and improve the mechanical properties of electrospun nanofiber membranes.

Many researchers applied this method to prepare composite nanofiber membranes for battery separator applications. Chen et al. prepared poly(vinylidene fluoride-co-hexafluoropropylene) (PVdF-HFP)/PI composite nanofibers by cross-electrospinning followed with thermal calendaring for use as battery separator (Chen et al. 2014). Because of the fusion of PVdF-HFP during thermal calendaring, many bonding points were formed between PVdF-HFP and PI nanofibers (Fig. 7.13a), which led to the more condensed composite membrane than that before thermal calendering (Chen et al. 2014). As a result, the tensile strength of the composite membrane after thermal calendaring increased to 7.5 MPa, which was 275% higher than that before thermal calendaring (Fig. 7.13b). In the following research, they added 2 wt% of functionalized TiO_2 (f-TiO_2) into the above composite nanofibers (f-T/P-H/PI) (Chen et al. 2015b). With further thermal calendaring, the f-T/P-H/PI composite nanofiber membrane showed a strength of 11 MPa, which was 4.4 times the sample before thermal calendaring (Chen et al. 2015).

Yildiz et al. reported a novel hybrid fabrics containing electrospun PEO nanofibers and CNTs (Yildiz et al. 2015). By applying thermal calendaring at 70°C, the hybrid fabrics showed obvious improvement in mechanical properties. The typical stress-strain curves before and after thermal calendaring are shown in Fig. 7.14. Before calendaring, the composite fabrics only showed tensile strength and modulus in the range of 0.15–2.23 MPa and 15.69–38.79 MPa, respectively.

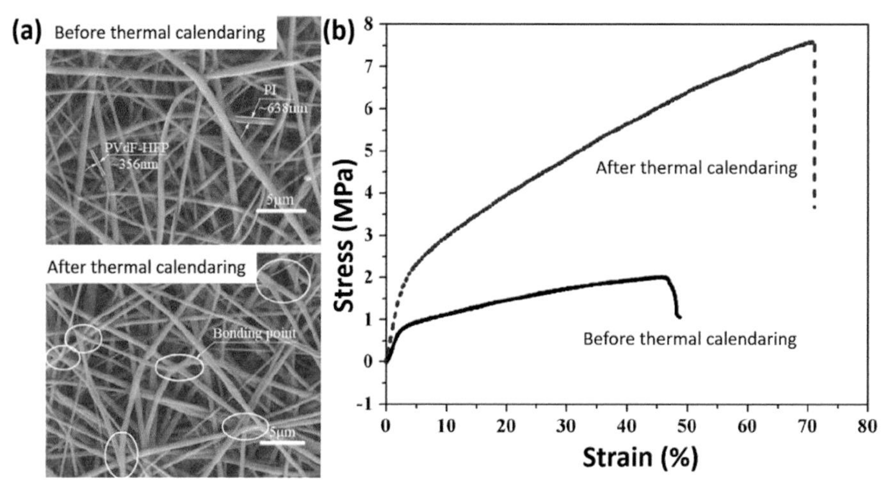

Fig. 7.13: SEM images of PVdF/PI composite nanofibers before and after thermal calendering (a) and (b) the corresponding typical stress-strain curves. Adapted with permission from (Chen et al. 2014). Copyright (2014) Elsevier.

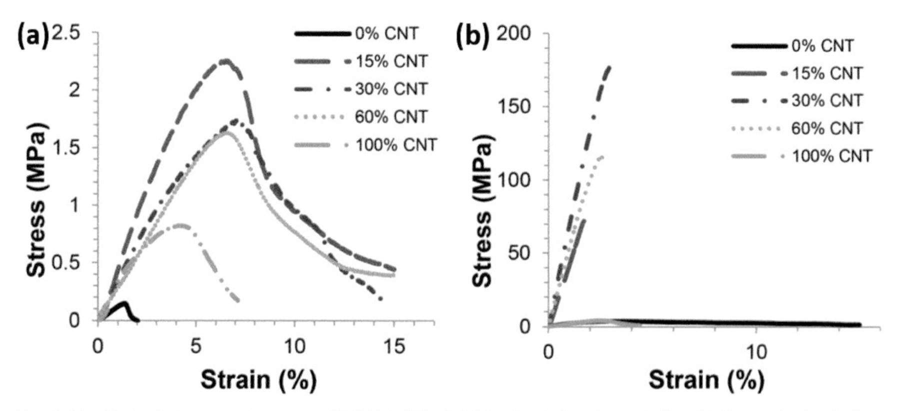

Fig. 7.14: Typical stress-strain curves of PEO/CNT hybrid fabrics before (a) and after (b) thermal calendering. Adapted with permission from (Yildiz et al. 2015). Copyright (2015) Royal Society of Chemistry.

After thermal calendaring, dramatic increase on tensile strength and modulus were achieved. The composite with 15 wt% CNTs possessed a strength of 75 MPa and modulus of 5053 MPa, while the composites with 30 wt% CNTs showed a further higher strength of 172 MPa, and higher modulus of 9996 MPa, respectively (Yildiz et al. 2015). This great improvement on mechanical properties could be attributed to the much stronger adhesion between PEO polymer nanofibers and CNTs after thermal calendaring.

Thermal Annealing

Thermal annealing is to heat materials above their recrystallization temperature. This process can improve the mechanical properties of electrospun nanofibers. Tan and Lim found that thermal annealing on electrospun PLA nanofibers could increase the modulus due to the increased crystallinity and the changed microstructures from a fibrillar structure to a mixture of fibrillary and nano-granular structure with enhanced interfibrillar bonding (Tan and Lim 2006). Similar results on the effect of annealing on electrospun PLA nanofibers were also reported for PLA/CNT composite nanofibers and PLA/silk composite nanofibers (Peng et al. 2015, Ramaswamy et al. 2011).

In another report, Mannarino and Rutledge systematically investigated the effect of thermal annealing temperature on the mechanical properties of electrospun poly(trimethyl hexamethylene terephthalamide) (PA 6(3)T) nanofiber nonwovens (Mannarino and Rutledge 2012). By increasing the annealing temperatures, the tensile strength and modulus were increased, as shown in Fig. 7.15a. The annealing samples showed increased toughness when the temperatures were < 160°C. Further increasing temperature to 170°C led to a significant decrease on the tensile strain. The possible explanation on the above results could be the changeable fibrous morphology of the nonwovens. When the treating temperatures were approaching the glass transition temperature (T_g) of 153°C, the molecules of PA 6(3)T had enough energy to orient along the fiber axis. Further increasing the temperature to 170°C (above T_g) made the nonwoven welding together (Mannarino and Rutledge 2012). On the one hand, the welding created many junction points, which led to the increase

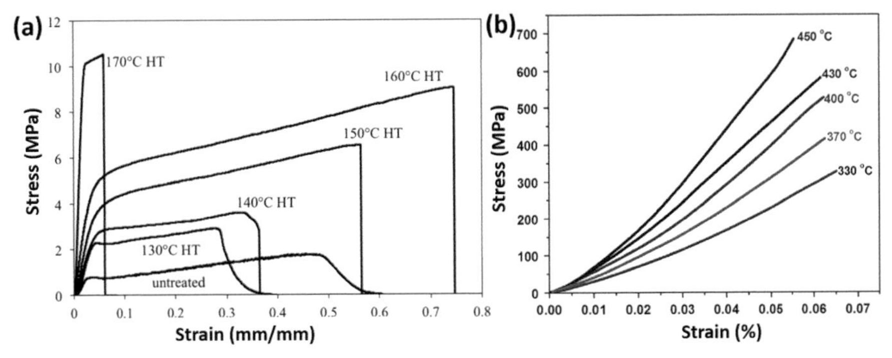

Fig. 7.15: (a) Effect of thermal annealing temperatures on the mechanical properties of electrospun PA 6(3)T nanofibrous nonwovens. Adapted with permission from (Mannarino and Rutledge 2012). Copyright (2012) Elsevier. (b) Effect of thermal annealing on the mechanical properties of electrospun aligned PI nanofibers. Adapted with permission from (Jiang et al. 2015). Copyright (2015) Elsevier.

of tensile strength and modulus. On the other hand, the porous film-like morphology could act as stress concentrators to initiate the crack formation, leading to the great decrease on the tensile strain (Mannarino and Rutledge 2012).

Jiang et al. systematically studied the effect of annealing temperature on electrospun aligned PI nanofiber mats (A-PI-NFB, Fig. 7.15b) (Jiang et al. 2015). The A-PI-NFB annealing at 330°C possessed the smallest tensile strength of 326 MPa and modulus of 5.0 GPa. As increasing the annealing temperatures gradually, an increase in tensile strength and modulus were also observed. The highest strength of 689 MPa and modulus of 13.2 GPa were obtained when annealing the fibers at 450°C. This enhanced mechanical properties of PI aligned fiber mat could be attributed to the temperature-induced higher crystallinity and larger molecular orientation along the fiber axis (Jiang et al. 2015).

Thermal Stretching

Thermal stretching is a technique to apply forces to stretch materials. With this process, the crystallinity and molecular orientation along the stretching direction increase, which lead to the increase of tensile strength and modulus. This method is often used to improve the mechanical properties of aligned electrospun nanofibers.

Lai et al. stretched the electrospun aligned PAN bundles in steam at about 100°C for 2, 3, and 4 times of the original lengths (Lai et al. 2011). With these stretching ratios, the fiber diameter decreased to about 220 nm, as compared to the diameter of as-spun fibers (~ 300 nm). Further characterizations by XRD, polarized FT-IR, and DSC indicated that the stretching led to the increase of crystallinity, molecular orientation, and therefore improve the mechanical properties (Fig. 7.16). The as-spun

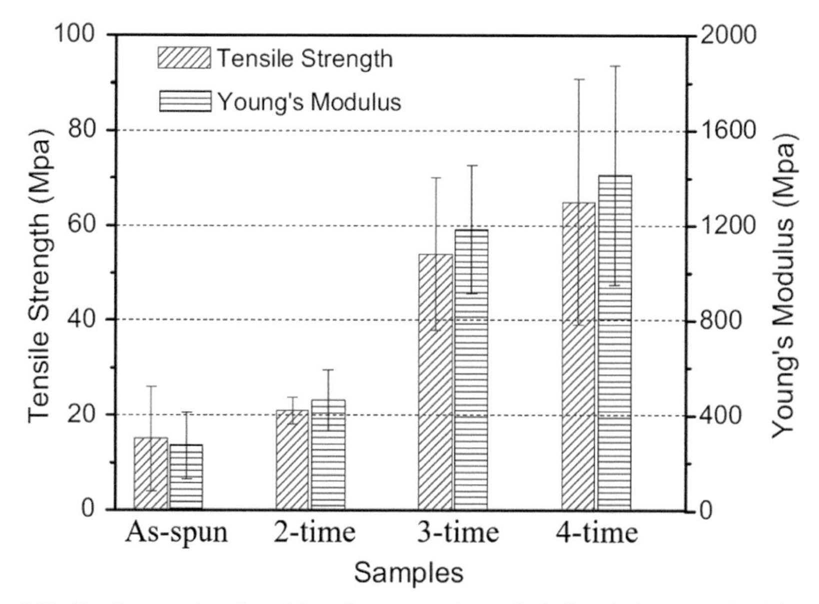

Fig. 7.16: Tensile strength and modulus of as-spun and stretched aligned electrospun PAN bundles. Adapted with permission from (Lai et al. 2011). Copyright (2011) Elsevier.

PAN nanofibers showed a tensile strength of 15 MPa and modulus of 275 MPa. After stretching for 2, 3, and 4 times, the values of tensile strength/modulus increased to 22/430 MPa, 53/1190 MPa, and 65/1412 MPa, respectively (Lai et al. 2011). These exciting results provide promising opportunities to fabricate electrospun carbon nanofibers with superior mechanical properties.

Besides the aligned electrospun PAN bundles, many other researchers focus on the study of electrospun PAN nanofiber yarns by thermal stretching. Xie et al. applied dry-drawing at 140°C to the electrospun PAN yarns, and up to 6 times stretching ratio could be achieved (Xie et al. 2015). After stretching, the diameter of the yarn decreased dramatically from 279 μm for the as-spun yarn to 151, 84, and 64 μm for the yarns with 2, 4, and 6 times stretching ratio, respectively. The yarn with 5 times stretching ratio presented the highest tensile strength of 362 MPa and modulus of 9.2 GPa, which were 800% and 1800% higher than those of as-spun PAN yarns. Similar work regarding the thermal stretching to improve the mechanical performance of electrospun PAN nanofiber yarns was also reported by Zhu's group (Bin et al. 2016, Wang et al. 2008). These kinds of electrospun PAN nanofiber yarns would on the one hand open a new gate for the preparation of yarns with excellent mechanical performance, and on the other hand provide opportunities to form next generation of precursor yarns for the high performance carbon nanofiber yarns.

In addition, Zong et al. reported an interesting work to thermally stretch the electrospun poly(glycolide-co-lactide) (GA/LA: 90:10, PLA10GA90) nonwovens (Zong et al. 2003). After stretching at 90°C for 20 minutes under a constant deformation strain of 450%, the nanofibers were highly aligned along the stretching direction, compared to the random deposition of nanofibers before stretching (Fig. 7.17a). The highly aligned PLA10GA90 nanofibers also showed great mechanical improvement, as shown in Fig. 7.17b. The as-prepared nonwoven only

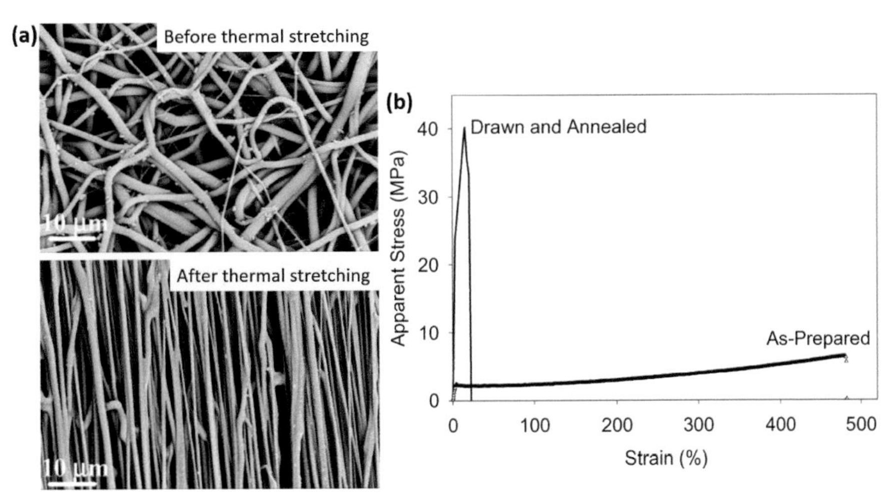

Fig. 7.17: SEM images of the as-spun PLA10GA90 nonwoven before and after thermal stretching at 90°C for 20 minutes under a constant deformation strain of 450% (a), and the corresponding typical stress-strain curves for the as-prepared and post-drawn samples (b). Adapted with permission from (Zong et al. 2003). Copyright (2003) Elsevier.

has a tensile strength of 5 MPa, but high elongation at break of 450% (Zong et al. 2003). After stretching, the membrane became much stronger with the improved tensile strength of 40 MPa, but also brittle with much smaller tensile strain. These results could be because of the increase of the crystallinity and orientation after the thermal stretching (Zong et al. 2003).

Twisting of Nanofiber Yarns

As is well-known, the single electrospun nanofiber is hardly to be applied to the real applications due to the nano-size and small breaking force. Normally, the products of electrospinning are obtained in the form of nonwoven mats. However, these nonwovens with randomly aligned electrospun nanofibers often showed poor mechanical properties, which severely limited the application of nanofibers. Recently, in order to improve the desirable properties of nanofibers and expand their applications, many researchers are focusing on the preparation of 1D nanofiber yarns (Abbasipour and Khajavi 2013). Furthermore, twists were usually applied to fabricate a fiber yarn with the purpose of improving the mechanical properties in the conventional fiber and textile areas. By compressing the yarn, increasing the friction and adhesion between fibers, and improving the alignment of fibers in the yarn, the mechanical properties can be improved efficiently. Several works about twisting of nanofiber yarns have been reported through a post-twisting process or during electrospinning (Shuakat and Lin 2014).

Some twisted nanofiber yarns were prepared by cutting electrospun membrane into narrow strips and twisting them into nanofiber yarns (Baniasadi et al. 2015, Fennessey and Farris 2004, Moon and Farris 2009, Zhou et al. 2011). Baniasadi et al. used a high speed rotated collector drum to obtain aligned PVDF-TrFE nanofiber ribbons (Baniasadi et al. 2015). Then these nanofiber ribbons were utilized to prepare nanofiber yarns by twisting. As shown in the Fig. 7.18, during the twist process, a DC motor was induced to twist the ribbon into a uniform nanofiber yarn. The results of tensile test of nanofiber ribbons and nanofiber yarns showed an initial increase in tensile strength after twisting the ribbon into yarn.

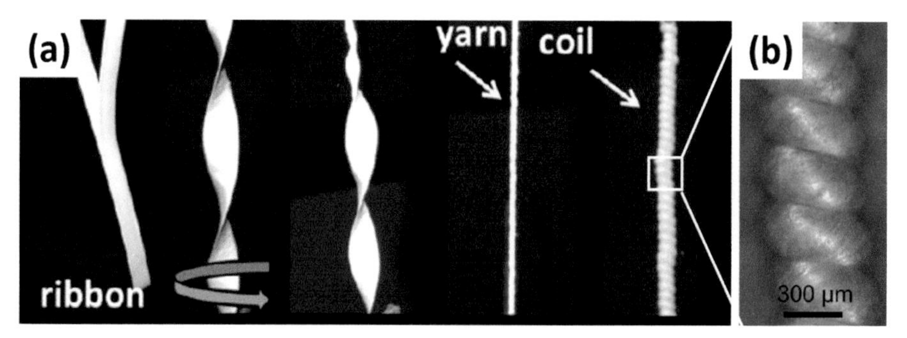

Fig. 7.18: Fabrication of yarns and coils from electrospun ribbons by twisting process (a) and optical microscope image of a coil (b). Adapted with permission from (Baniasadi et al. 2015). Copyright (2015) American Chemical Society.

Based on the same fabrication, Fennessey et al. observed an initial increase in the ultimate strength and modulus with increasing the twist angle (Fennessey and Farris 2004). However, the mechanical properties decreased with further increase in twist. Twisted short nanofiber yarns made from PVDF-HFP nanofiber strips were investigated by Zhou et al. (Zhou et al. 2011). The effects of nanofiber orientation, diameter, morphology, and twist level on the mechanical properties of yarns were studied. They found that: (a) diameter and orientation played important roles in mechanical properties; (b) finer uniform nanofiber yarns exhibited better mechanical properties than randomly oriented uniform nanofiber yarns; (c) within the twist level range of 1000–8000 tpm, the tensile strength increased with the increase in the twist level; and (d) the short nanofiber yarns showed the highest tensile strength and modulus when the nanofibers were oriented along the nanofiber length direction (Zhou et al. 2011).

Besides, several researchers studied the formation of yarns by twisting during the electrospinning process (Dabirian et al. 2011, He et al. 2014, He et al. 2013, Nakashima et al. 2013, Yan et al. 2011, Yousefzadeh et al. 2011). Yan et al. described twisted nanofiber yarn by using a homemade set-up of single needle nozzle and three rotating tubes (Yan et al. 2011). Two tubes rotated in reverse direction were used to twist the nanofibers into yarn, and then a third roller was used to collect yarn (Fig. 7.19a). The twist angles were controlled by the speed of these three rotating tubes during the electrospinning process. The relationship of modulus and twist angle of the yarn followed the Hearle's equation. When the twist angle was about 35°, the tensile stress reached a maximum value of 190 MPa (Fig. 7.19b).

Recently, a lot of works about conjugated electrospinning nanofiber yarn were described (Ali et al. 2016, Fakhrali et al. 2016, He et al. 2014, Li et al. 2008). For a typical conjugated electrospinning, opposite needles, a rotated funnel and winding device were used (Fig. 7.20a). During the electrospun process, the nanofiber cone formed at the end of funnel could be twisted into yarn due to the rotating effect of funnel and drawing force induced by the winding device (He et al. 2014). The twist angle of nanofibers in the yarn could be controlled by modifying the speed of funnel, yarn winding speed, electrospun voltage, and distance between needles and funnel. The results of tensile tests showed the twist angle directly affects the mechanical properties of yarn. An increase in the twist angle led to an improvement in the mechanical properties of yarn, which was shown in Fig. 7.20b.

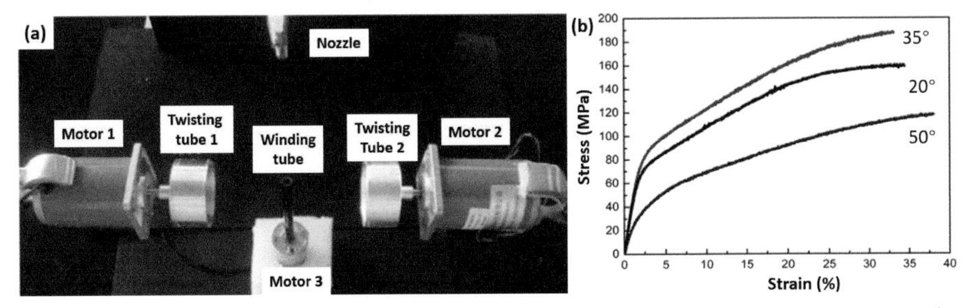

Fig. 7.19: (a) Top view of the modified electrospinning apparatus and (b) typical stress–strain curves for PAN nanofiber staple yarns with different twist angles. Adapted with permission from (Yan et al. 2011). Copyright (2011) American Chemical Society.

Chang et al. introduced a direct twisting electrospinning method (Chang et al. 2016). As shown in the Fig. 7.21a, the method differed greatly from the typical electrospinning set-ups, which used a high speed motor induced in the spinneret tip and a short collecting distance during the electrospinning process. On the one hand, due to tractive effort of the high rotational speed of the nozzle, the jet could be twisted into a fibrous rope with helical form. On the another hand, the proper collecting distance provided the ejected fibers enough time to achieve self-bundling and twisting process, but not ready for deposition into the collector similar to normal electrospinning process. By this direct twisted electrospinning method, the poly(ether sulfones) (PES) nanofiber yarns exhibited significantly higher tensile strength of 30 MPa than that of PES nonwoven membrane fabricated by typical electrospinning process (Fig. 7.21b) (Chang et al. 2016).

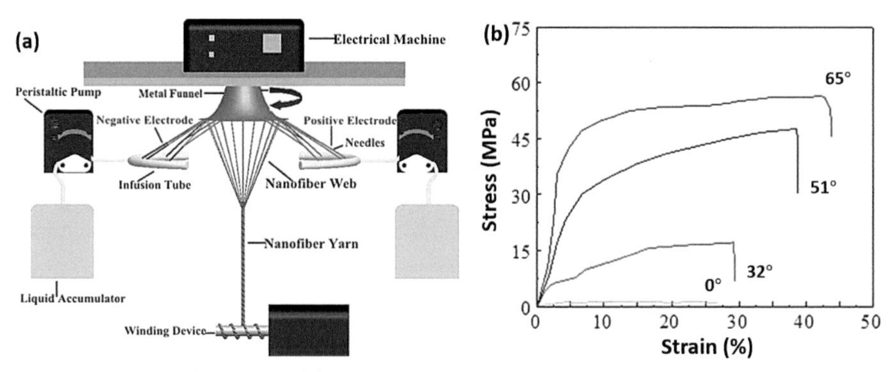

Fig 7.20: (a) Schematic diagram of the multiple conjugate electrospinning apparatus, and (b) stress-strain curves for PAN nanofiber yarns with different twist angles. Adapted with permission from (He et al. 2014). Copyright (2014) John Wiley and Sons.

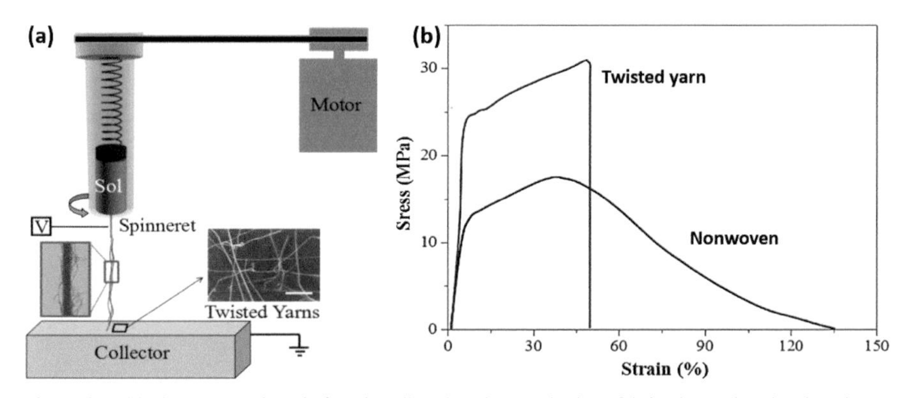

Fig. 7.21: (a) Set-up employed for the directly electrospinning fabrication of twisted polymer microfiber/nanofiber yarns. (b) Typical stress–strain curves of electrospun nonwoven membrane and the corresponding twisted fibrous yarn. Adapted with permission from (Chang et al. 2016). Copyright (2016) American Chemical Society.

Acknowledgements

The authors wish to acknowledge grant from the National Natural Science Foundation of China (Grants No. 21564008 and 51803093) and Priority Academic Program Development of Jiangsu Higher Education Institutions (PAPD).

References

Abbasipour, M. and Khajavi, R. 2013. Nanofiber bundles and yarns production by electrospinning: A review. Advanced Polymer Technology 32(3): 21363.

Afifi, A. M., Nakano, S., Yamane, H. and Kimura, Y. 2010. Electrospinning of continuous aligning yarns with a 'funnel' target. Macromolecular Materials Engineering 295(7): 660–665.

Agarwal, S., Wendorff, J. H. and Greiner, A. 2008. Use of electrospinning technique for biomedical applications. Polymer 49(26): 5603–5621.

Agarwal, S., Wendorff, J. H. and Greiner, A. 2010. Chemistry on electrospun polymeric nanofibers: merely routine chemistry or a real challenge? Macromolecular Rapid Communications 31(15): 1317–1331.

Agarwal, S., Greiner, A. and Wendorff, J. H. 2013. Functional materials by electrospinning of polymers. Progress Polymer Science 38(6): 963–991.

Ago, M., Okajima, K., Jakes, J. E., Park, S. and Rojas, O. J. 2012. Lignin-based electrospun nanofibers reinforced with cellulose nanocrystals. Biomacromolecules 13(3): 918–926.

Ago, M., Jakes, J. E. and Rojas, O. J. 2013. Thermomechanical properties of lignin-based electrospun nanofibers and films reinforced with cellulose nanocrystals: A dynamic mechanical and nanoindentation study. ACS Applied Materials Interfaces 5(22): 11768–11776.

Ali, U., Niu, H., Abbas, A., Shao, H. and Lin, T. 2016. Online stretching of directly electrospun nanofiber yarns. RSC Advances 6(36): 30564–30569.

An, J., Zhang, H., Zhang, J., Zhao, Y. and Yuan, X. 2009. Preparation and antibacterial activity of electrospun chitosan/poly(ethylene oxide) membranes containing silver nanoparticles. Colloid. Polymer Science 287(12): 1425–1434.

Ansari, F., Salajková, M., Zhou, Q. and Berglund, L. A. 2015. Strong surface treatment effects on reinforcement efficiency in biocomposites based on cellulose nanocrystals in Poly(vinyl acetate) Matrix. Biomacromolecules 16(12): 3916–3924.

Augustine, R., Malik, H. N., Singhal, D. K., Mukherjee, A., Malakar, D., Kalarikkal, N. et al. 2014. Electrospun polycaprolactone/ZnO nanocomposite membranes as biomaterials with antibacterial and cell adhesion properties. Journal Polymer Research 21(3): 347.

Ayutsede, J., Gandhi, M., Sukigara, S., Ye, H., Hsu, C.-M., Gogotsi, Y. et al. 2006. Carbon nanotube reinforced bombyx mori silk nanofibers by the electrospinning process. Biomacromolecules 7(1): 208–214.

Azizi Samir, M. A. S., Alloin, F. and Dufresne, A. 2005. Review of recent research into cellulosic whiskers, their properties and their application in nanocomposite field. Biomacromolecules 6(2): 612–626.

Baji, A., Mai, Y. -W., Wong, S. -C., Abtahi, M. and Du, X. 2010. Mechanical behavior of self-assembled carbon nanotube reinforced nylon 6,6 fibers. Composites Science Technology 70(9): 1401–1409.

Baniasadi, M., Huang, J., Xu, Z., Moreno, S., Yang, X., Chang, J. et al. 2015. High-performance coils and yarns of polymeric piezoelectric nanofibers. ACS Applied Materials Interfaces 7(9): 5358–5366.

Bazbouz, M. B. and Stylios, G. K. 2010. The tensile properties of electrospun nylon 6 single nanofibers. Journal Polymer Science, Part B: Polymer Physics 48(15): 1719–1731.

Bin, Y., Hao, Y., Meifang, Z. and Hongzhi, W. 2016. Continuous high-aligned polyacrylonitrile electrospun nanofibers yarns via circular deposition on water bath. Journal Nanoscience Nanotechnology 16(6): 5633–5638.

Cao, X., Huang, M., Ding, B., Yu, J. and Sun, G. 2013. Robust polyacrylonitrile nanofibrous membrane reinforced with jute cellulose nanowhiskers for water purification. Desalination 316: 120–126.

Cha, D. I., Kim, K. W., Chu, G. H., Kim, H. Y., Lee, K. H. and Bhattarai, N. 2006. Mechanical behaviors and characterization of electrospun polysulfone/polyurethane blend nonwovens. Macromolecular Research 14(3): 331–337.

Chang, G., Li, A., Xu, X., Wang, X. and Xue, G. 2016. Twisted polymer microfiber/nanofiber yarns prepared via direct fabrication. Industrial Engineering Chemistry Research 55(25): 7048–7051.

Chen, D., Liu, T., Zhou, X., Tjiu, W. C. and Hou, H. 2009. Electrospinning fabrication of high strength and toughness polyimide nanofiber membranes containing multiwalled carbon nanotubes. Journal Physical Chemistry B 113(29): 9741–9748.

Chen, F., Peng, X., Li, T., Chen, S., Wu, X. F., Reneker, D. H. et al. 2008. Mechanical characterization of single high-strength electrospun polyimide nanofibres. Journal Physicals D: Applied Physics 41: 025308.

Chen, L., Jiang, S., Chen, J., Chen, F., He, Y., Zhu, Y. et al. 2015a. Single electrospun nanofiber and aligned nanofiber belts from copolyimide containing pyrimidine units. New Journal Chemistry 39(11): 8956–8963.

Chen, S., Hou, H., Hu, P., Wendorff, J. H., Greiner, A. and Agarwal, S. 2009. Effect of different bicomponent electrospinning techniques on the formation of polymeric nanosprings. Macromolecular Materials Engineering 294(11): 781–786.

Chen, W., Liu, Y., Ma, Y., Liu, J. and Liu, X. 2014. Improved performance of PVdF-HFP/PI nanofiber membrane for lithium ion battery separator prepared by a bicomponent cross-electrospinning method. Materials Letters 133: 67–70.

Chen, W., Liu, Y., Ma, Y. and Yang, W. 2015b. Improved performance of lithium ion battery separator enabled by co-electrospinning polyimide/poly(vinylidene fluoride-co-hexafluoropropylene) and the incorporation of TiO_2-(2-hydroxyethyl methacrylate). Journal Power Sources 273: 1127–1135.

Chen, Y., Han, D., Ouyang, W., Chen, S., Hou, H., Zhao, Y. et al. 2012. Fabrication and evaluation of polyamide 6 composites with electrospun polyimide nanofibers as skeletal framework. Composites Part B-Engineering 43(5): 2382–2388.

Chew, S. Y., Hufnagel, T. C., Lim, C. T. and Leong, K. W. 2006. Mechanical properties of single electrospun drug-encapsulated nanofibres. Nanotechnology 17(15): 3880.

Choi, S. -S., Hong, J. -P., Seo, Y. S., Chung, S. M. and Nah, C. 2006. Fabrication and characterization of electrospun polybutadiene fibers crosslinked by UV irradiation. Journal Applied Polymer Science 101(4): 2333–2337.

Chuangchote, S., Sirivat, A. and Supaphol, P. 2007. Mechanical and electro-rheological properties of electrospun poly (vinyl alcohol) nanofibre mats filled with carbon black nanoparticles. Nanotechnology 18(14): 145705.

Dabirian, F., Ravandi, S. A. H., Sanatgar, R. H. and Hinestroza, J. P. 2011. Manufacturing of twisted continuous PAN nanofiber yarn by electrospinning process. Fibers Polymers 12(5): 610.

Das, S., Wajid, A. S., Bhattacharia, S. K., Wilting, M. D., Rivero, I. V. and Green, M. J. 2013. Electrospinning of polymer nanofibers loaded with noncovalently functionalized graphene. Journal Applied Polymer Science 128(6): 4040–4046.

Ding, B., Kimura, E., Sato, T., Fujita, S. and Shiratori, S. 2004. Fabrication of blend biodegradable nanofibrous nonwoven mats via multi-jet electrospinning. Polymer 45(6): 1895–1902.

Ding, Y., Zhang, P., Jiang, Y., Xu, F., Yin, J. and Zuo, Y. 2009. Mechanical properties of nylon-6/SiO_2 nanofibers prepared by electrospinning. Materials Letters 63(1): 34–36.

Duan, B., Yuan, X., Zhu, Y., Zhang, Y., Li, X., Zhang, Y. et al. 2006. A nanofibrous composite membrane of PLGA–chitosan/PVA prepared by electrospinning. European Polymer Journal 42(9): 2013–2022.

Fakhrali, A., Ebadi, S. V., Gharehaghaji, A. A., Latifi, M. and Moghassem, A. 2016. Analysis of twist level and take-up speed impact on the tensile properties of PVA/PA6 hybrid nanofiber yarns. E-Polymers 16(2): 125–135.

Fang, X., Ma, H., Xiao, S., Shen, M., Guo, R., Cao, X. et al. 2011. Facile immobilization of gold nanoparticles into electrospun polyethyleneimine/polyvinyl alcohol nanofibers for catalytic applications. Journal Materials Chemistry 21(12): 4493–4501.

Fennessey, S. F. and Farris, R. J. 2004. Fabrication of aligned and molecularly oriented electrospun polyacrylonitrile nanofibers and the mechanical behavior of their twisted yarns. Polymer 45(12): 4217–4225.

Francis, L., Giunco, F., Balakrishnan, A. and Marsano, E. 2010. Synthesis, characterization and mechanical properties of nylon–silver composite nanofibers prepared by electrospinning. Current Applied Physics 10(4): 1005–1008.

Frohbergh, M. E., Katsman, A., Botta, G. P., Lazarovici, P., Schauer, C. L., Wegst, U. G. et al. 2012. Electrospun hydroxyapatite-containing chitosan nanofibers crosslinked with genipin for bone tissue engineering. Biomaterials 33(36): 9167–9178.

Gopiraman, M., Fujimori, K., Zeeshan, K., Kim, B. and Kim, I. 2013. Structural and mechanical properties of cellulose acetate/graphene hybrid nanofibers: spectroscopic investigations. Express Polymer Letters 7(6).

Greiner, A. and Wendorff, J. H. 2007. Electrospinning: a fascinating method for the preparation of ultrathin fibers. Angewandte Chemie International Edition 46(30): 5670–5703.

Gu, S. -Y., Wu, Q. -L., Ren, J. and Vancso, G. J. 2005. Mechanical properties of a single electrospun fiber and its structures. Macromolecular Rapid Communications 26(9): 716–720.

Habibi, Y., Lucia, L. A. and Rojas, O. J. 2010. Cellulose nanocrystals: Chemistry, self-assembly, and applications. Chemical Reviews 110(6): 3479–3500.

Han, D. and Steckl, A. J. 2009. Superhydrophobic and oleophobic fibers by coaxial electrospinning. Langmuir 25(16): 9454–9462.

Han, D. and Steckl, A. J. 2013. Triaxial electrospun nanofiber membranes for controlled dual release of functional molecules. ACS Applied Materials Interfaces 5(16): 8241–8245.

Han, X. -J., Huang, Z. -M., He, C. -L., Liu, L. and Wu, Q. -S. 2006. Coaxial electrospinning of PC(shell)/PU(core) composite nanofibers for textile application. Polymer Composites 27(4): 381–387.

He, J., Zhou, Y., Qi, K., Wang, L., Li, P. and Cui, S. 2013. Continuous twisted nanofiber yarns fabricated by double conjugate electrospinning. Fibers Polymers 14(11): 1857–1863.

He, J., Qi, K., Zhou, Y. and Cui, S. 2014. Multiple conjugate electrospinning method for the preparation of continuous polyacrylonitrile nanofiber yarn. Journal Applied Polymer Science 131(8): 40137.

He, Y., Han, D., Chen, J., Ding, Y., Jiang, S., Hu, C. et al. 2014. Highly strong and highly tough electrospun polyimide/polyimide composite nanofibers from binary blend of polyamic acids. RSC Advances 4(104): 59936–59942.

Hou, H., Ge, J. J., Zeng, J., Li, Q., Reneker, D. H., Greiner, A. et al. 2005. Electrospun polyacrylonitrile nanofibers containing a high concentration of well-aligned multiwall carbon nanotubes. Chemistry Materials 17(5): 967–973.

Huang, Z. -M., Zhang, Y. Z., Kotaki, M. and Ramakrishna, S. 2003. A review on polymer nanofibers by electrospinning and their applications in nanocomposites. Composites Science Technology 63(15): 2223–2253.

Hwang, K. Y., Kim, S. -D., Kim, Y. -W. and Yu, W. -R. 2010. Mechanical characterization of nanofibers using a nanomanipulator and atomic force microscope cantilever in a scanning electron microscope. Polymer Testing 29(3): 375–380.

Islam, M. S. and Karim, M. R. 2010. Fabrication and characterization of poly(vinyl alcohol)/alginate blend nanofibers by electrospinning method. Colloids Surfaces Physicochemical Engineering Aspects 366(1): 135–140.

Ji, X., Wang, T., Guo, L., Xiao, J., Li, Z., Zhang, L. et al. 2013. Effect of nanoscale-ZnO on the mechanical property and biocompatibility of electrospun poly (L-lactide) acid/nanoscale-ZnO mats. Journal Biomedical Nanotechnology 9(3): 417–423.

Jiang, S., Duan, G., Zussman, E., Greiner, A. and Agarwal, S. 2014. Highly flexible and tough concentric triaxial polystyrene fibers. ACS Applied Materials Interfaces 6(8): 5918–5923.

Jiang, S., Duan, G., Chen, L., Hu, X. and Hou, H. 2015. Mechanical performance of aligned electrospun polyimide nanofiber belt at high temperature. Materials Letters 140: 12–15.

Kanani, A. G. and Bahrami, S. H. 2010. Review on electrospun nanofibers scaffold and biomedical applications. Trends Biomaterials Artificial Organs 24(2): 93–115.

Keun Hyung Lee, Hak Yong Kim, Young Jun Ryu, Kwan Woo Kim and Choi, S. W. 2003. Mechanical behavior of electrospun fiber mats of poly(vinyl chloride)/polyurethane polyblends. Journal Polymer Science, Part B: Polymer Physics 41(11): 1256–1262.

Kim, H., Che, L., Ha, Y. and Ryu, W. 2014. Mechanically-reinforced electrospun composite silk fibroin nanofibers containing hydroxyapatite nanoparticles. Materials Science Engineering C-Materials 40: 324–335.

Kim, M. S., Jun, I., Shin, Y. M., Jang, W., Kim, S. I. and Shin, H. 2010. The development of genipin-crosslinked poly(caprolactone) (PCL)/gelatin nanofibers for tissue engineering applications. Macromolecular Bioscience 10(1): 91–100.

Lai, C., Zhong, G., Yue, Z., Chen, G., Zhang, L., Vakili, A. et al. 2011. Investigation of post-spinning stretching process on morphological, structural, and mechanical properties of electrospun polyacrylonitrile copolymer nanofibers. Polymer 52(2): 519–528.

Lamastra, F., Bianco, A., Meriggi, A., Montesperelli, G., Nanni, F. and Gusmano, G. 2008. Nanohybrid PVA/ZrO_2 and PVA/Al_2O_3 electrospun mats. Chemical Engineering Journal 145(1): 169–175.

Lee, C., Wei, X., Kysar, J. W. and Hone, J. 2008. Measurement of the elastic properties and intrinsic strength of monolayer graphene. Science 321(5887): 385–388.

Lee, K. H., Kim, H. Y., Ryu, Y. J., Kim, K. W. and Choi, S. W. 2003. Mechanical behavior of electrospun fiber mats of poly(vinyl chloride)/polyurethane polyblends. Journal Polymer Science, Part B: Polymer Physics 41(11): 1256–1262.

Li, D. and Xia, Y. 2004. Electrospinning of nanofibers: reinventing the wheel? Advanced Materials 16(14): 1151–1170.

Li, G., Zhao, Y., Lv, M., Shi, Y. and Cao, D. 2013. Super hydrophilic poly(ethylene terephthalate) (PET)/ poly(vinyl alcohol) (PVA) composite fibrous mats with improved mechanical properties prepared via electrospinning process. Colloids Surfaces Physicochemical Engineering Aspects 436: 417–424.

Li, H. -Y. and Liu, Y. -L. 2013. Polyelectrolyte composite membranes of polybenzimidazole and crosslinked polybenzimidazole-polybenzoxazine electrospun nanofibers for proton exchange membrane fuel cells. Journal Materials Chemistry A 1(4): 1171–1178.

Li, X., Yao, C., Sun, F., Song, T., Li, Y. and Pu, Y. 2008. Conjugate electrospinning of continuous nanofiber yarn of poly(L-lactide)/nanotricalcium phosphate nanocomposite. Journal Applied Polymer Science 107(6): 3756–3764.

Li, Y., Lim, C. T. and Kotaki, M. 2015. Study on structural and mechanical properties of porous PLA nanofibers electrospun by channel-based electrospinning system. Polymer 56(0): 572–580.

Liao, H., Qi, R., Shen, M., Cao, X., Guo, R., Zhang, Y. et al. 2011. Improved cellular response on multiwalled carbon nanotube-incorporated electrospun polyvinyl alcohol/chitosan nanofibrous scaffolds. Colloids Surfaces B Biointerfaces 84(2): 528–535.

Lin, Y., Clark, D. M., Yu, X., Zhong, Z., Liu, K. and Reneker, D. H. 2012. Mechanical properties of polymer nanofibers revealed by interaction with streams of air. Polymer 53(3): 782–790.

Linh, N. T. B. and Lee, B. -T. 2012. Electrospinning of polyvinyl alcohol/gelatin nanofiber composites and cross-linking for bone tissue engineering application. Journal Biomaterials Applications 27(3): 255–266.

Liu, L. Q., Tasis, D., Prato, M. and Wagner, H. D. 2007. Tensile mechanics of electrospun multiwalled nanotube/poly(methyl methacrylate) nanofibers. Advanced Materials 19(9): 1228–1233.

Lu, P. and Hsieh, Y. -L. 2010. Multiwalled carbon nanotube (MWCNT) reinforced cellulose fibers by electrospinning. ACS Applied Materials Interfaces 2(8): 2413–2420.

Mack, J. J., Viculis, L. M., Ali, A., Luoh, R., Yang, G., Hahn, H. T. et al. 2005. Graphite nanoplatelet reinforcement of electrospun polyacrylonitrile nanofibers. Advanced Materials 17(1): 77–80.

Mannarino, M. M. and Rutledge, G. C. 2012. Mechanical and tribological properties of electrospun PA 6(3)T fiber mats. Polymer 53(14): 3017–3025.

Moon, S. and Farris, R. J. 2009. Strong electrospun nanometer-diameter polyacrylonitrile carbon fiber yarns. Carbon 47(12): 2829–2839.

Nakashima, R., Watanabe, K., Lee, Y., Kim, B. -S. and Kim, I. -S. 2013. Mechanical properties of poly(vinylidene fluoride) nanofiber filaments prepared by electrospinning and twisting. Advances Polymer Technology 32(S1): E44–E52.

Nguyen, T. T. T., Chung, O. H. and Park, J. S. 2011. Coaxial electrospun poly(lactic acid)/chitosan (core/shell) composite nanofibers and their antibacterial activity. Carbohydrate Polymers 86(4): 1799–1806.

Pant, H. R., Bajgai, M. P., Nam, K. T., Seo, Y. A., Pandeya, D. R., Hong, S. T. et al. 2011. Electrospun nylon-6 spider-net like nanofiber mat containing TiO_2 nanoparticles: a multifunctional nanocomposite textile material. Journal Hazardous Materials 185(1): 124–130.

Papkov, D., Zou, Y., Andalib, M. N., Goponenko, A., Cheng, S. Z. D. and Dzenis, Y. A. 2013. Simultaneously strong and tough ultrafine continuous nanofibers. ACS Nano 7(4): 3324–3331.

Peng, L., Jiang, S., Seuß, M., Fery, A., Lang, G., Scheibel, T. et al. 2015. Two-in-one composite fibers with side-by-side arrangement of silk fibroin and poly(l-lactide) by electrospinning. Macromolecular Materials Engineering 48–55. Doi: 10.1002/mame.201500217.

Qu, H., Wei, S. and Guo, Z. 2013. Coaxial electrospun nanostructures and their applications. Journal Materials Chemistry A 1(38): 11513–11528.

Ramaswamy, S., Clarke, L. I. and Gorga, R. E. 2011. Morphological, mechanical, and electrical properties as a function of thermal bonding in electrospun nanocomposites. Polymer 52(14): 3183–3189.

Ramires, P. A. and Milella, E. 2002. Biocompatibility of poly(vinyl alcohol)-hyaluronic acid and poly(vinyl alcohol)-gellan membranes crosslinked by glutaraldehyde vapors. Journal Materials Science Materials Medicine 13(1): 119–123.

Rojas, O. J., Montero, G. A. and Habibi, Y. 2009. Electrospun nanocomposites from polystyrene loaded with cellulose nanowhiskers. Journal Applied Polymer Science 113(2): 927–935.

Ryuji, I., Masaya, K. and Seeram, R. 2005. Structure and properties of electrospun PLLA single nanofibres. Nanotechnology 16(2): 208.

Sambudi, N. S., Sathyamurthy, M., Lee, G. M. and Park, S. B. 2015. Electrospun chitosan/poly (vinyl alcohol) reinforced with $CaCO_3$ nanoparticles with enhanced mechanical properties and biocompatibility for cartilage tissue engineering. Composites Science Technology 106: 76–84.

Sen, R., Zhao, B., Perea, D., Itkis, M. E., Hu, H., Love, J. et al. 2004. Preparation of single-walled carbon nanotube reinforced polystyrene and polyurethane nanofibers and membranes by electrospinning. Nano Letters 4(3): 459–464.

Shao, S., Zhou, S., Li, L., Li, J., Luo, C., Wang, J. et al. 2011. Osteoblast function on electrically conductive electrospun PLA/MWCNTs nanofibers. Biomaterials 32(11): 2821–2833.

Shields, K. J., Beckman, M. J., Bowlin, G. L. and Wayne, J. S. 2004. Mechanical properties and cellular proliferation of electrospun collagen type II. Tissue Engineering 10(9-10): 1510–1517.

Shuakat, M. N. and Lin, T. 2014. Recent developments in electrospinning of nanofiber yarns. Journal Nanoscience Nanotechnology 14(2): 1389–1408.

Sill, T. J. and von Recum, H. A. 2008. Electrospinning: applications in drug delivery and tissue engineering. Biomaterials 29(13): 1989.

Sun, L., Ray, P. S. H., Jun, W. and Lim, C. T. 2008. Modeling the size-dependent elastic properties of polymeric nanofibers. Nanotechnology 19(45): 455706.

Sun, M., Li, X., Ding, B., Yu, J. and Sun, G. 2010. Mechanical and wettable behavior of polyacrylonitrile reinforced fibrous polystyrene mats. Journal Colloid Interface Science 347(1): 147–152.

Tan, E. P. and Lim, C. 2006. Effects of annealing on the structural and mechanical properties of electrospun polymeric nanofibres. Nanotechnology 17(10): 2649.

Tsanov, T., Vassilev, K., Stamenova, R. and Tsvetanov, C. 1995. Crosslinked poly(ethylene oxide) modified with tetraalkylammonium salts as phase transfer catalyst. Journal Polymer Science, Part A: Polymer Chemistry 33(15): 2623–2628.

Vondran, J. L., Sun, W. and Schauer, C. L. 2008. Crosslinked, electrospun chitosan–poly(ethylene oxide) nanofiber mats. Journal Applied Polymer Science 109(2): 968–975.

Wan, C. and Chen, B. 2011. Poly (ε-caprolactone)/graphene oxide biocomposites: mechanical properties and bioactivity. Biomedical Materials 6(5): 055010.

Wang, W., Jin, X., Zhu, Y., Zhu, C., Yang, J., Wang, H. et al. 2016. Effect of vapor-phase glutaraldehyde crosslinking on electrospun starch fibers. Carbohydrate Polymers 140: 356–361.

Wang, X., Zhang, K., Zhu, M., Hsiao, B. S. and Chu, B. 2008. Enhanced mechanical performance of self-bundled electrospun fiber yarns via post-treatments. Macromolecular Rapid Communications 29(10): 826–831.

Wong, K. K. H., Zinke-Allmang, M., Hutter, J. L., Hrapovic, S., Luong, J. H. T. and Wan, W. 2009. The effect of carbon nanotube aspect ratio and loading on the elastic modulus of electrospun poly(vinyl alcohol)-carbon nanotube hybrid fibers. Carbon 47(11): 2571–2578.

Wu, J., Wang, N., Zhao, Y. and Jiang, L. 2013. Electrospinning of multilevel structured functional micro-/nanofibers and their applications. Journal Materials Chemistry A 1(25): 7290–7305.

Xie, Z., Niu, H. and Lin, T. 2015. Continuous polyacrylonitrile nanofiber yarns: preparation and dry-drawing treatment for carbon nanofiber production. RSC Advances 5(20): 15147–15153.

Xu, H., Jiang, S., Ding, C., Zhu, Y., Li, J. and Hou, H. 2017. High strength and high breaking load of single electrospun polyimide microfiber from water soluble precursor. Materials Letters 201: 82–84.

Yan, H., Liu, L. and Zhang, Z. 2011. Continually fabricating staple yarns with aligned electrospun polyacrylonitrile nanofibers. Materials Letters 65(15): 2419–2421.

Yao, C., Li, X. and Song, T. 2007. Electrospinning and crosslinking of zein nanofiber mats. Journal Applied Polymer Science 103(1): 380–385.

Yao, X., Qi, X., He, Y., Tan, D., Chen, F. and Fu, Q. 2014. Simultaneous reinforcing and toughening of polyurethane via grafting on the surface of microfibrillated cellulose. ACS Applied Materials Interfaces 6(4): 2497–2507.

Yildiz, O., Stano, K., Faraji, S., Stone, C., Willis, C., Zhang, X. et al. 2015. High performance carbon nanotube—polymer nanofiber hybrid fabrics. Nanoscale 7(40): 16744–16754.

Yoon, J. W., Park, Y., Kim, J. and Park, C. H. 2017. Multi-jet electrospinning of polystyrene/polyamide 6 blend: thermal and mechanical properties. Fashion and Textiles 4(1): 9.

Yousefzadeh, M., Latifi, M., Teo, W. -E., Amani-Tehran, M. and Ramakrishna, S. 2011. Producing continuous twisted yarn from well-aligned nanofibers by water vortex. Polymer Engineering Science 51(2): 323–329.

Yu, M. -F., Lourie, O., Dyer, M. J., Moloni, K., Kelly, T. F. and Ruoff, R. S. 2000. Strength and breaking mechanism of multiwalled carbon nanotubes under tensile load. Science 287(5453): 637–640.

Zander, N. E., Strawhecker, K. E., Orlicki, J. A., Rawlett, A. M. and Beebe, T. P. 2011. Coaxial electrospun poly(methyl methacrylate)–polyacrylonitrile nanofibers: atomic force microscopy and compositional characterization. Journal Physical Chemistry B 115(43): 12441–12447.

Zhang, C. -L. and Yu, S. -H. 2014. Nanoparticles meet electrospinning: recent advances and future prospects. Chemical Society Reviews 43(13): 4423–4448.

Zhang, K., Qian, Y., Wang, H., Fan, L., Huang, C., Yin, A. et al. 2010. Genipin-crosslinked silk fibroin/hydroxybutyl chitosan nanofibrous scaffolds for tissue-engineering application. Journal Biomedical Materials Research A 95(3): 870–881.

Zhang, S., Huang, Y., Yang, X., Mei, F., Ma, Q., Chen, G. et al. 2009. Gelatin nanofibrous membrane fabricated by electrospinning of aqueous gelatin solution for guided tissue regeneration. Journal Biomedical Materials Research A 90(3): 671–679.

Zhang, Y. Z., Venugopal, J., Huang, Z. M., Lim, C. T. and Ramakrishna, S. 2006. Crosslinking of the electrospun gelatin nanofibers. Polymer 47(8): 2911–2917.

Zhou, C., Chu, R., Wu, R. and Wu, Q. 2011. Electrospun polyethylene oxide/cellulose nanocrystal composite nanofibrous mats with homogeneous and heterogeneous microstructures. Biomacromolecules 12(7): 2617–2625.

Zhou, C., Wang, Q. and Wu, Q. 2012. UV-initiated crosslinking of electrospun poly(ethylene oxide) nanofibers with pentaerythritol triacrylate: Effect of irradiation time and incorporated cellulose nanocrystals. Carbohydrate Polymers 87(2): 1779–1786.

Zhou, C., Shi, Q., Guo, W., Terrell, L., Qureshi, A. T., Hayes, D. J. et al. 2013. Electrospun bio-nanocomposite scaffolds for bone tissue engineering by cellulose nanocrystals reinforcing maleic anhydride grafted PLA. ACS Applied Materials Interfaces 5(9): 3847–3854.

Zhou, G., Xiong, T., Jiang, S., Jian, S., Zhou, Z. and Hou, H. 2016. Flexible titanium carbide–carbon nanofibers with high modulus and high conductivity by electrospinning. Materials Letters 165: 91–94.

Zhou, S., Zhou, G., Jiang, S., Fan, P. and Hou, H. 2017. Flexible and refractory tantalum carbide-carbon electrospun nanofibers with high modulus and electric conductivity. Materials Letters 200: 97–100.

Zhou, Y., Fang, J., Wang, X. and Lin, T. 2011. Strip twisted electrospun nanofiber yarns: Structural effects on tensile properties. Journal Materials Research 27(03): 537–544.

Zong, X., Ran, S., Fang, D., Hsiao, B. S. and Chu, B. 2003. Control of structure, morphology and property in electrospun poly(glycolide-co-lactide) non-woven membranes via post-draw treatments. Polymer 44(17): 4959–4967.

Zoppe, J. O., Peresin, M. S., Habibi, Y., Venditti, R. A. and Rojas, O. J. 2009. Reinforcing poly(ε-caprolactone) nanofibers with cellulose nanocrystals. ACS Applied Materials Interfaces 1(9): 1996–2004.

Zucchelli, A., Focarete, M. L., Gualandi, C. and Ramakrishna, S. 2011. Electrospun nanofibers for enhancing structural performance of composite materials. Polymers Advanced Technologies 22(3): 339–349.

Zussman, E., Burman, M., Yarin, A. L., Khalfin, R. and Cohen, Y. 2006. Tensile deformation of electrospun nylon-6,6 nanofibers. Journal Polymer Science, Part B: Polymer Physics 44(10): 1482–1489.

Chapter 8

Functionalization of Electrospun Nanofibrous Materials

Wei Wang,[1,*] *Zhigao Zhu,*[1] *Qiao Wang*[1] and *Ruisha Shi*[2]

Introduction

Electrospinning technology has been applied in various fields and exhibits lots of advantages due to the intrinsic properties of the ultrafine-fiber products. In order to make this technology more competent for applications, researchers are paying ever-growing research attention to the functionalization of electrospun fibrous products. Different kinds of electrospun products were designed and fabricated for diverse applications. Thus, in this chapter, we will review the recent progress of functionalization strategies for nanofibers.

Direct-dispersed Electrospinning

In the earliest stage, electrospinning technology was developed to fabricate polymeric products. As the demand in various applications increased, certain functional components were required to be combined with the polymeric fibrous matrices. The most general and straightforward strategy is the direct-dispersed electrospinning, which generally involves the mixing of inorganic nanoparticles or organic components into the fiber matrices by a direct blending process followed by electrospinning.

Inorganic Nanocomponent Directly Combining in Electrospun Fiber Matrix

Metallic Nanoparticles Dispersed Directly in Fibers. In order to combine inorganic nanocomponent into a polymer fibrous matrix by a direct-dispersed route, the

[1] School of Environment, Harbin Institute of Technology, Huanghe Road 73, Harbin, 150090, R. P. China.
[2] Library of Harbin Institute of Technology, Huanghe Road 73, Harbin, 150090, R. P. China.
* Corresponding author: wangweirs@hit.edu.cn

functional components were first fabricated and then dispersed in the polymer solution, followed by electrospinning. Wang's group (Yang et al. 2003) did one of the earliest efforts to demonstrate this strategy and introduced as-synthesized silver nanoparticles into PAN nanofibers. The nanoparticles were easily aggregated without effective stabilization by PAN, exhibiting the smallest diameter of about 20 nm and the largest size of around 200 nm. Thus, in order to make the inorganic nanoparticles more effectively disperse in polymer nanofibers, sometimes surfactants or stabilizers were required to stabilize the NPs (Zhang et al. 2012a, He et al. 2009). For example, Khan and co-workers (Saquing et al. 2009) described an *in situ* fabrication of Ag NP–PEO composite nanofibers via direct-dispersed electrospinning at ambient conditions. The high molecular weight PEO acted as both the reducing agent for $AgNO_3$ and protecting agent for the resulting Ag NPs, since pseudo-crown ethers with high reducing property were formed in the high molecular weight PEO chains. The NF sample fabricated from 2 wt% 2000 kDa PEO and 0.27 wt% $AgNO_3$ mixed in H_2O were characterized by TEM. Ag NPs were observed as spherical spots well dispersed on the surface of the nanofibrous matrix, with an average diameter of 8.1 ± 1.3 nm. Additionally, it was found, growing with the Ag^+ content, Ag NPs tended to arrange in a line (Fig. 8.1). The reason was proposed to be the combined effect of applied electrical field on the polymer and the differences between the electrical conductivity and polarizability of the polymer and metal NPs.

Indeed, in earlier work, researchers had paid their attention to this *in situ* direct-dispersed electrospinning for fabrication of polymer/nanocomponent composite

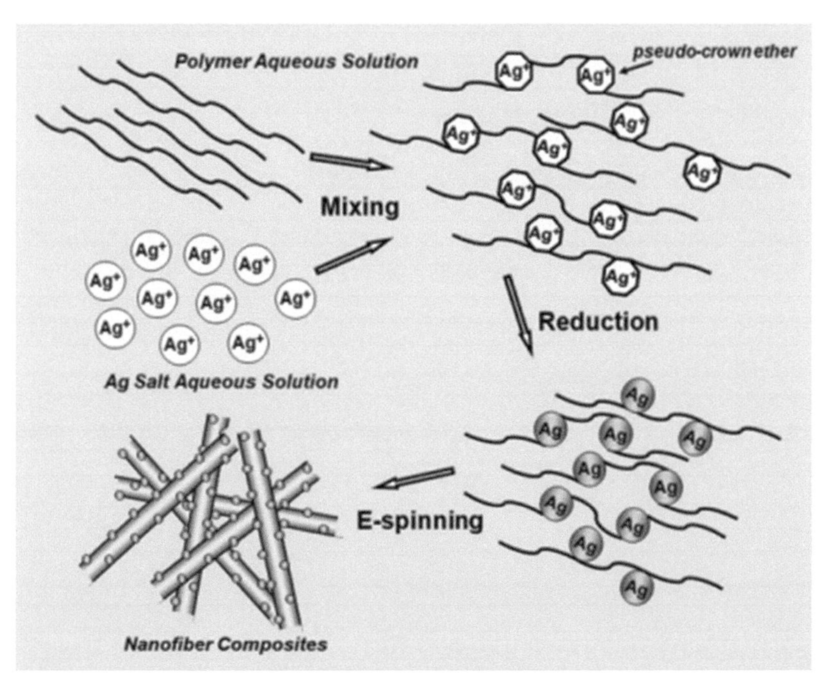

Fig. 8.1: Schematic of the *in situ* direct-dispersed electrospinning for fabricating Ag NP–PEO NF composites (Saquing et al. 2009).

nanofibers. For example, Li et al. successively dissolved $CuCl_2$, $NaHSO_3$, polymer PVA, surfactant AOT, and hydrazine hydrate into distilled water. Hydrazine hydrate acted as the reductant to reduce Cu^{2+} in to Cu(0) NPs at pH = 10, and $NaHSO_3$ served as the oxygen scavenger to prohibit the oxidation of Cu(0). Cu(0)/(PVA) nanocables were successfully obtained by electrospinning the above PVA protected Cu(0) NPs solution (molar ratio of Cu ions to VA repeating units is 1:40). The average diameter of the Cu(0) cores and PVA shells was around 100 and 400 nm, respectively. When molar ratio of Cu ions to VA was lower, both Cu nanoparticles and Cu nanowires were found in the PVA nanofibers. The author proposed that when the solution was polarized in an electric field, the PVA protected Cu NPs were positively charged on one side and negatively charged on the other, which then led to static interaction among the polarized copper nanoparticles in the electrospinning solution. Therefore the copper NPs tended to form a line-like structure oriented along the electric field, similar to electrorheological fluids (Li et al. 2006a).

Semiconductors Dispersed Directly in Fibers. Semiconductor nanoparticles, due to their attractive photonic, electronic, magnetic, and chemical properties, are also of interest in functionalizing electrospun polymer products. In order to improve the dispersion of semiconductor nanoparticles, stabilizers based on organic small molecules or polymers were often applied. Wang et al. inserted CdTe quantum dots (QDs) in PVP nanofibrous matrix by the direct dispersed electrospinning with cetyltrimethylammonium bromide (CTAB) as the stabilizer (Wang et al. 2007). The CdTe QDs could be well dispersed in the fibrous matrix after adding a small amount of CTAB, but in contrast aggregated in the absence of CTAB. Direct-dispersed electrospinning was also effective to disperse aqueous synthesized CdTe QDs in water-soluble PVA nanofibers. A uniform and separate distribution of the CdTe QDs was clearly observed in the PVA fibers matrices when the CdTe QD concentration was lower than 4 wt% (Li et al. 2007), as demonstrated in Yang's work. During the synthesis, a little dioctyl sulfosuccinate sodium salt (AOT) was added into the electrospun polymer solution in order to lower the surface tension and viscosity of the PVA solution. We think that both PVA and AOT may make some contribution to the uniformity and well-dispersed nature of CdTe QDs in polymer fibers. Lu and Wang successfully prepared Ag_2S/PVP composite nanofibers using an *in situ* direct-dispersed electrospinning. First, they prepared Ag_2S sol using an *in situ* reaction with $AgNO_3$ as precursor and CS_2 as the sulfide source in the presence of PVP at room temperature, and then the sample solution was electrospun into composite nanofibers (Lu et al. 2005a). As the Ag_2S sol was compatible with the polymer of PVP, the obtained Ag_2S nanoparticles could be well dispersed into solid PVP fibers after electrospinning, with an average size of about 15 nm.

Till now many metal or semiconductor nanocomponents were successfully incorporated into polymeric fibers by direct-dispersed electrospinning for obtaining functional materials, including both *in situ* and *ex situ* processes (Wang et al. 2006, Jin et al. 2005). In some literature, the *in situ* direct-dispersed process was also named as sol-gel method.

Carbon Nanomaterials Dispersed Directly in Fibers. Carbon nanomaterials were often incorporated into electrospun polymer fiber matrix (PAN, PEO, PVA,

PLA, PC, PS, PU, PMMA, PVDF, and so on) (Lu et al. 2009), besides metal and semiconductor nanocomponents. Carbon nanotubes (CNTs), possessing the appropriate size, have attracted the most attention. Especially, the functionalization by CNTs was expected to tailor the mechanical properties of polymer fibers for the application in tissue engineering for the fabrication of cell-growth scaffolds and vascular grafts. CNTs were inserted in electrospun polymer fibers after being well dispersed in the electrospinning solutions first. For instance, single-walled carbon nanotubes (SWNTs) were incorporated into PAN fibrous matrices by being dispersed in dimethylformamide (DMF) solution of PAN, followed by electrospinning. SWNTs kept parallel to the axis direction of the PAN fiber, meanwhile maintaining their straight shape. The electrospun PAN/SWNTs composite nanofibers exhibited a significant reinforcement effect at less than 3% volume SWNTs (Ko et al. 2003). SWNTs were also embedded in natural polymer electrospun fibers, *Bombyx mori* silk nanofibers through sonication dispersion in formic acid followed by electrospinning. The final products exhibited up to 3.6 times higher Young's modulus in comparison to the un-reinforced aligned fiber (Ayutsede et al. 2006). Researchers studied the behavior of CNT fillers in electrospinning process and in the fiber products. A theoretical model proposed by Cohen and co-workers indicated that rod-like structured CNT were randomly oriented initially but gradually oriented along the jet line for the sink like flow in a wedge (Dror et al. 2003). The CNTs reinforce the polymer fibers by hindering crazing extension, reducing stress concentration, and dissipating energy by pullout.

Sometimes, CNTs are functionalized before being incorporated in fibers, by means of small molecules (sodium dodecyl sulfate) (SDS), highly branched polymers, or some other surface chemical functionalization methods. For instance, polymers/SWNTs composite nanofibers were prepared by electrospinning a mixture of polymer and ester (EST)-functionalized SWNTs (Sen et al. 2004). The tensile strength of EST-SWNT-PU mat was strengthened by 104% in comparison to pure PU electrospun fibrous mats, in contrast SWNTs-PU fiber mats only exhibited a 46% increase in mechanical properties. Reneker and co-workers (Jason et al. 2004, Haoqing et al. 2005) fabricated surface-oxidized MWNTs, and then dispersed them together with PAN in DMF. They obtained PAN/MWNTs composite nanofibers via electrospinning. MWNTs exhibited a high degree of orientation in the PAN/MWNTs nanofibers, as observed by TEM and 2D wide-angle X-ray diffraction (WAXD).

In addition to the enhancement of mechanical properties, direct dispersion of CNTs can also improve the electrical conductivity of the fiber products. Seoul et al. (Chang et al. 2003) reported that the CNTs/PVDF electrospun fiber mat derived from a PVDF/DMF solution (with 0.1 wt% CNTs) presented a conductivity of 7×10^{-6} S.cm^{-1}. Jin and co-workers (Sung et al. 2004) indicated that the conductivity of PMMA composite fibrous films combined with well-dispersed MWNTs ranged from 10^{-4} and 10^{-2} S.cm^{-1}, when the MWNT ratio was 1–5 wt% in the fibrous matrices.

Compared to CNTs, graphene gained much less attention, maybe for their bigger size. Polystyrene (PS)/graphene nanoplatelets (GNP) nanofibers were produced via electrospinning of DMF stabilized GNP and PS solutions. Morphological analysis confirmed uniform fiber formation and good GNP dispersion/distribution within the

PS matrix. GNP modified PS nanofibers showed a 6-fold increase in the thermal conductivity and an increase of 7–8 orders of magnitude in electrical conductivity of the nanofibers at 10 wt.% GNP loading (Li et al. 2016a).

Organic Components Directly Combining in Electrospun Fiber Matrix

Electrospinning provides an effective and facile way to encapsulate functional organic components in fiber products. The composite nanofibers exhibit important potential in biomedical application, biocides for water treatment, enzyme immobilization, electronics, and so on. The principle for preparing this kind nanofibrous products was the dissolution or dispersion of the functional small molecular components within the electrospinning solution. Generally, the types of the electrospinning solutions include solution system, emulsion system, and sol-gel system.

Solution Electrospinning for Organic Components Incorporation. By forming homogeneous solutions, the functional components can combine well with the matrices. Alan G. MacDiarmid and his co-workers (Pinto et al. 2003) have fabricated polyaniline/polyethylene oxide nanofibers for field-effect transistor. In their work, 100 mg of emeraldine base PANi was doped with 129 mg of camphorsulfonic acid and dissolved in 10 ml $CHCl_3$. The solution was filtered and then used to dissolve 20 mg of PEO (molecular weight 900,000). The PANi/PEO composite fibers was assembled on FET substrates, it exhibited a hole mobility in the depletion regime of 1.4×10^{-4} cm^2/Vs, while the one-dimensional charge density (at zero gate bias) is calculated to be approximately 1 hole per 50 two-ring repeat units of polyaniline, consistent with the rather high channel conductivity ($10 \times^{-3}$ S/cm). The mobility was low, which should be due to the charge barrier effect of the insulated PEO.

This solution method was also used to combine biocides in fibrous mats for water disinfection. For example, poly [(dimethylimino)(2-hydroxy-1,3-propanedily) chloride] (WSCP), a kind of quaternary ammonium compound with excellent antibacterial properties on both Gram-positive and Gram-negative bacteria, was applied as the functional component to be incorporated in PA nanofibers. Due to the positive surface charge of WSCP, PA nanofibers with 5 wt.% WSCP showed much higher removal of bacteria compared to pure PA nanofibers. Short term experiments demonstrated that a 5.2 log10 CFU/100 ml removal is possible with hospital waste water. This removal is maintained after filtration with 650 ml of water inoculated with *S. aureus* and after rinsing the water with 25 L of demineralized water. Additionally, it was found, the direct dispersed method is more effective than the post impregnation method in organic biocides immobilization. The biocide appeared within the matrix of the nanofiber material in case the biocide is added to the spinning solution (Fig. 8.2a), and presented on the surfaces of the nanofibers (Fig. 8.2b) by post impregnation method (Daels et al. 2011).

Emulsion Electrospinning for Organic Components Incorporation. Sometimes, emulsion electrospinning was applied to prepare organic component functionalized nanofibers. For example, for controlled drug delivery, the medicines are required to not release too fast or too slow. If a hydrophobic drug was capsulated into hydrophobic fiber matrices, normally a very slow release kinetics occurred (Zeng et al. 2003). If water-soluble drugs were electrospun into water-soluble polymers,

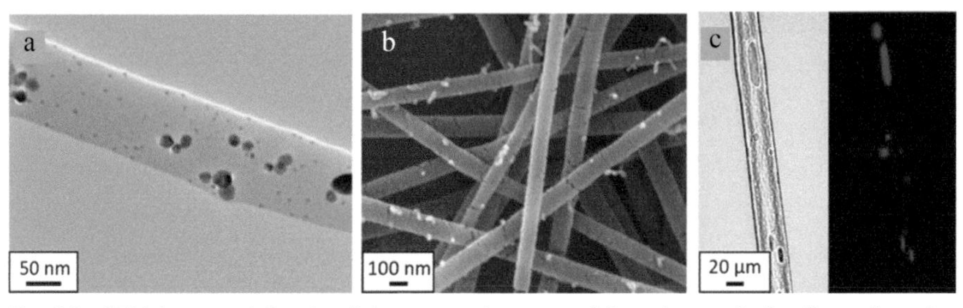

Fig. 8.2: SEM images of functionalizing agents incorporated into the matrix by direct dispersion electrospinning (a) and presented on the surface of post functionalized nanofibers (b) (Daels et al. 2011). (c) Visualization of fluorescently labeled protein encapsulated in reservoirs using visible (left) and ultraviolet (right) light (Sanders et al. 2003).

the composite fibers would dissolve quickly in body fluid and fail in drug delivery (Zeng et al. 2005). In this context, emulsion electrospinning was proposed to be a promising strategy for drug/polymer system. In this method, hydrophilic drugs could be distributed inside fibers, which avoided the burst release. Wnek and co-workers successfully encapsulated aqueous bovine serum albumin (BSA) in the poly(ethylene-co-vinyl acetate) fibers (Sanders et al. 2003). In a typical electrospinning process, a poly(ethylene-*co*-vinyl acetate) mother solution in CH_2Cl_2 was mixed with bovine serum albumin (BSA) dissolved in phosphate-buffered saline (PBS). Thus, two-phase system of EVA-CH_2Cl_2/BSA-water was formed. Then the cloudy suspension containing dispersed water droplets was electrospun into composite fibers. Observed by fluorescence microscope, small and larger domains within the fibers appeared, and BSA resided indeed in the large domains (Fig. 8.2c). This method was also used to fabricate durg-Dox/poly(ethylene glycol)-poly(L-lactic acid) (PEG-PLLA) diblock copolymer composites (Xu et al. 2005). The aqueous drug solution was dropped into the amphiphilic polymer solution under stirring with the assistant by SDS surfactant, followed by electrospinning. Water-soluble drugs could be well encapsulated and distributed into the nanofibrous matrices. The drug release experiment from Dox/PEG-PLLA was conducted *in vitro*. The diffusion was predominant at the early stage and the enzymatic degradation mechanism became predominant a certain time later. Moreover, a loss of cytotoxicity was not observed during the release of Dox from the fibers.

Sol-gel Electrospinning for Organic Components Incorporation. Sol-gel electrospinning was used to fabricate nanofibrous products with the organic component combined. Yen and his co-workers demonstrated an interesting work by this method for enzyme immobilization for biosensing. In that work, tetramethyl orthosilicate (TMOS) was used as a silica precursor, poly(vinyl alcohol) (PVA, Mn 30,000) was used as an extender that assists in easy spinning of silica fibers, and glucose not only functions as a non-surfactant template for controlling mesoporosity of the resultant sol-gel silica fibers, but also helps in the fiber spinning by increasing viscosity. The enzyme immobilized inside the mesoporous silica fibers was horseradish peroxidase (HRP). A typical procedure for preparing the composites included two steps. The first was the preparation of the spinning solution. TMOS was pre-hydrolyzed using

H_2O under the HCl catalysis and then mixed with glucose, PVA, and commercially available HRP in phosphate buffer solution. The second was electrospinning, by which white, flexible, dry fiber mats containing approximately HRP were obtained. The fibers with mesoporosity showed fourfold greater activity than the conventional non-templated silica samples and threefold greater activity than HRP immobilized silica powders (Patel et al. 2006).

In addition, nonwovens composed of polymer nanofibers and living cells, e.g., bacterial and virus cells, can also be prepared directly by blending electrospinning without total loss of biological functionality (Gensheimer et al. 2010, Lee and Belcher 2009).

Calcination Transformation

Inorganic materials often exhibit outstanding performances in various field, such as catalysis, electronics, magnetics, and so on. Promoting by the super large aspect ratio and high surface area, functional inorganic nanofibers were fabricated by direct blending electrospinning, followed by appropriate calcination processes. By this route, pure oxide nanofibers, inorganic composite nanofibers, and carbon nanofibers were produced.

Pure Oxide Nanofiber Fabricated by Calcination

For oxide nanofiber fabrication, the oxide component can form before electrospinning, after electrospinning but before calcination, during heat treatment.

First of all, SiO_2 nanofibers were successfully prepared in 2002 by electrospinning and subsequent calcination, reported by Shao et al. (Shao et al. 2002). In the process on SiO_2 fiber formation, a silica gel, with the molar composition TEOS:H_3PO_4:H_2O = 1:0.01:11, was first prepared by hydrolysis and polycondensation by the drop-wise addition of aqueous H_3PO_4 as the catalyst to TEOS with vigorous stirring at room temperature. After the above mixture had reacted for several hours, a certain amount of aqueous PVA solution was dropped slowly into the obtained silica gels. Thus, a viscous gel of PVA/silica composites was obtained for electrospinning. After calcination at required temperature, PVA/silica composite nanofibers were transferred into pure SiO_2 nanofibers.

Sometimes, oxide component is formed in the precursor nanofibers after electrospinning but before calcination. Li and Xia reported the most famous work on calcination transformation method in 2003. In their work, pure titania ceramic nanofibers were fabricated by combination of electrospinning with sol-gel method and subsequent calcination. Composite nanofibers composed by poly(vinyl pyrrolidone) (PVP) and amorphous TiO_2 were first obtained by electrospinning an solution containing PVP, tetraisopropoxide (or tetrabutyltitanate), ethanol, and acetic acid, then followed by hydrolysis of tetraisopropoxide into amorphous TiO_2 in air. Subsequently, proper calcination (500°C) in air was conducted to selectively remove PVP, and meanwhile convert the composite nanofibers into anatase. The average diameter of these ceramic nanofibers could be controlled in the range from 20 to 200 nm by varying a number of electrospinning parameters. Observed by TEM, nanofiber was found to be formed through the sintering of TiO_2 nanoparticles that were ~ 10 nm in diameter, and voids existed between adjacent nanoparticles. In the above procedure, it is critical to suppress the hydrolysis of titania precursors in electrospinning solution to form excellent morphology (Li and Xia 2003). Figure 8.3 provides the photograph, SEM, and TEM images.

For most other oxide nanofibers, the oxide components were formed during the calcination, when at the same time polymer template components were selectively removed. In this process, polymer/metal salt composite nanofibers were first produced by electrospinning the solution of polymers and metal salts (inorganic or organometallic compounds). Then calcination was carried on to remove polymeric substrates, and in the meantime convert metal salt precursors into metal oxides. During the above procedure, the most important issue is the calcination process. Unsuited calcination may destroy the nanofibrillar structure and lead to the disintegration of the ceramic nanofibers. For example, a slow heating rate between 300 and 500°C was selected during the fabrication of WO_3 nanofibers (Wang et al. 2011a). However, for the formation of ZnO, CuO, and Fe_2O_3 nanofibers, a very slow heating rate was suggested to be adopted at the range of 150 to 350°C. These slow heating procedures were aimed at slowing down the removal of the continuous polymeric phase in the fiber precursors, thus preventing the disintegration of the nanofibrous structure.

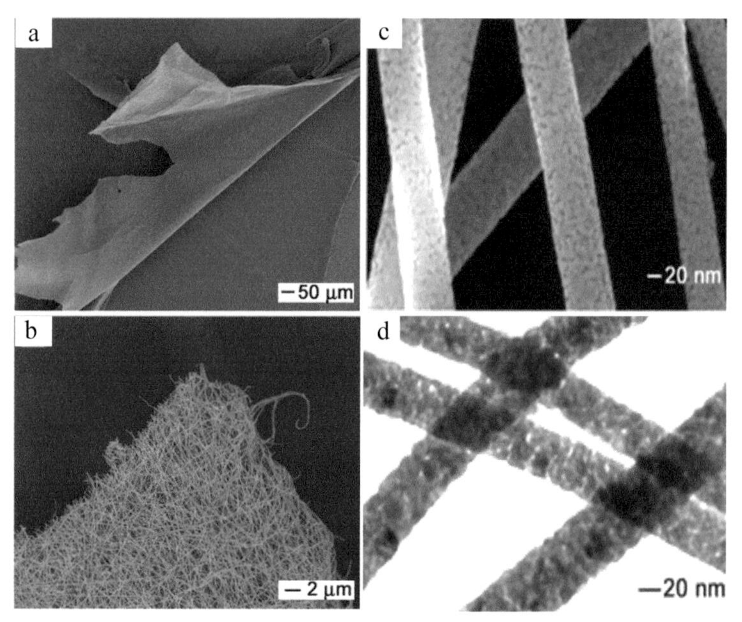

Fig. 8.3: Photograph (a), SEM in low-magnification (b), high-magnification (c) and TEM image of TiO_2 nanofibers (d) (Li and Xia 2003).

Oxide Composite Fibers Fabricated by Calcination

The last decade also witnessed the development of inorganic composite fibers, into which a series of functional components was briefly incorporated via electrospinning and subsequent calcination. These functional components include metals, carbon materials, pore forming materials, or flexibilizers.

Much research attention was paid on compound-doping composite nanofibers for their facile fabrication process and excellent performance in various areas. For example, Wang's group reported the first highly sensitive humidity sensor based on LiCl-doped TiO_2 nanofibers. In a typical fabrication process, a solution of tetrabutyltitanate, LiCl, and poly(vinyl pyrrolidone) (PVP) in acetic acid and ethanol was electrospun into nanofibers, followed by calcination to remove PVP and to afford LiCl-doped TiO_2 nanofibers. The composite nanofiber sensors with 30.0% LiCl showed the best sensing performance than others, with excellent stability, high sensitivity, outstanding linearity in broad humidity range, ultra-fast response, and recovery behavior. Such a good performance could be explained by the structures of 1D TiO_2 nanofibers. The large surface of the nanofiber makes the absorption of water molecules on the surface of the sensors easy. The 1D structure of the fibers can facilitate fast mass transfer of the water molecules to and from the interaction region, as well as improve the rate for charge carriers to transverse the barriers induced by molecular recognition along the fibers. On the other hand, Li ions in the fiber matrices are also a good water adsorbent, which can dissolve in the adsorbed water layers and then transfer freely, thus resulting in a sharp decrease of device impendence (Li et al. 2008).

Oxide components were more frequently integrated into fiber matrices. In a typical process, polymer templates, precursors of the all components were dissolved in a solvent to form electrospinning solution. After electrospinning and subsequent calcination, oxide-doped composite nanofibers were then obtained. For example, PdO_2 nanoparticles were combined into CuO fibers for the first high performance non-enzyme glucose electrochemical sensors (Wang et al. 2009); NiO nanoparticles were dispersed into SnO_2 nanofibers to detect hydrogen gas (Wang et al. 2010). Sometimes, by this direct calcination route, hierarchitectures could also be obtained. Xia et al. reported that V_2O_5 nanorods could directly grow on electrospun V_2O_5/TiO_2 composite nanofibers by calcinating $PVP/Ti(OiPr)_4/VO(OiPr)_3$ precursor nanofibers. The shape and size of V_2O_5 nanorods could be controlled by the content of titania and calcination conditions. At the nucleation stage of V_2O_5 nanorods, the presence of titania networks in the composite fibers or low temperature favored the formation of small nuclei, leading to the formation of thin and long nanorods. The spatial confinement of vanadia by titania might also force some of the V_2O_5 nanocrystals to grow toward the outside. Figure 8.4 presents the SEM images V_2O_5-TiO_2 nanofibers that were electrospun from a 2-propanol solution containing 20% $VO(OiPr)_3$, 20% $Ti(OiPr)_4$, 4% PVP, and 1% HTAB, followed by calcination at 475°C for different periods of time: (a) 2.5 minutes; (b) 10 minutes; and (c) 60 minutes. The ratio of $VO(OiPr)_3$ to $Ti(OiPr)_4$ (r) was 1:1. Figure 8.4d is the EDS microanalysis of the composite nanofibers (Ostermann et al. 2006).

Sometimes, people desired to replace some lattice oxygen in the oxide nanofibers with some heteroatoms, e.g., nitrogen, to form N-doped oxide nanofibers and then enhance the photoelectric properties. Thus, nitrogen sources, e.g., urea, cyanamide, etc. were added into the as-prepared electrospinning solutions followed by calcination treatment (Hassen et al. 2016, Babu et al. 2012).

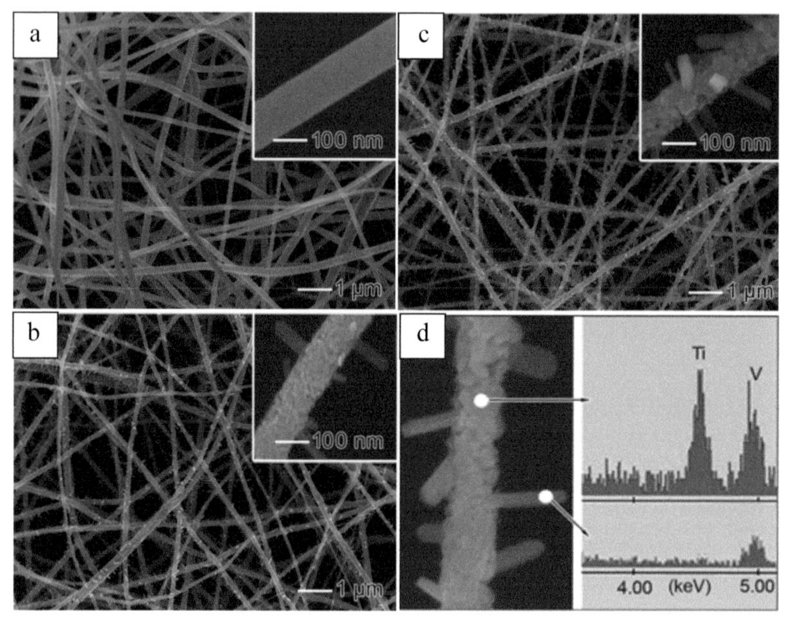

Fig. 8.4: SEM images V_2O_5-TiO_2 nanofibers fabricated by calcination at 475°C for different periods of time: (a) 2.5 minutes; (b) 10 minutes; and (c) 60 minutes. (d) is the EDS microanalysis of the composite nanofibers (Ostermann et al. 2006).

Noble metal nanoparticles have attracted much research attention for their broad applications in various fields. They could be derived from metal salts by appropriate heat treatment. Thus, metal nanoparticles (including Au, Ag, etc.) modified inorganic composite nanofibers were successfully fabricated by blending electrospinning followed by calcination. Porous silica nanofibers containing catalytic silver nanoparticles have been synthesized by a method that combines sol-gel chemistry and electrospinning technique. Tetraethyl orthosilicate (TEOS), poly-[3-(trimethoxysily) propyl methacrylate] (PMCM), and silver nitrate ($AgNO_3$) were used as precursors for the production of silica-PMCM hybrid fibers containing $AgNO_3$. Calcination of the hybrid fibers at high temperatures results in Ag doped flexible silica fibers because of thermal decomposition of PMCM polymer and conversion of $AgNO_3$ to silver nanoparticles. The corresponding digital picture and TEM micrographs of the flexible electrospun porous silica fibers containing silver nanoparticles are shown in Fig. 8.5a–c. The size and density of the silver particles in the silica fibers could be tuned by varying the size of the fibers, amount of $AgNO_3$ introduced, and the thermal treatment conditions. The final product exhibited good catalytic performances for MB dye reduction (Patel et al. 2007). Besides, metallic nanoparticles were also incorporated into other oxide nanofibers by calcination transformation to fabricate composite nanofibers, e.g., Ag in ZnO nanofibers and Au in TiO_2 nanofibers (Lin et al. 2009).

Recent years also witnessed the combination of carbon nano-materials and pore forming agents into oxide nanofibrous matrices. For carbon nano-materials functionalization, carbon nanotubes were usually employed and dispersed in

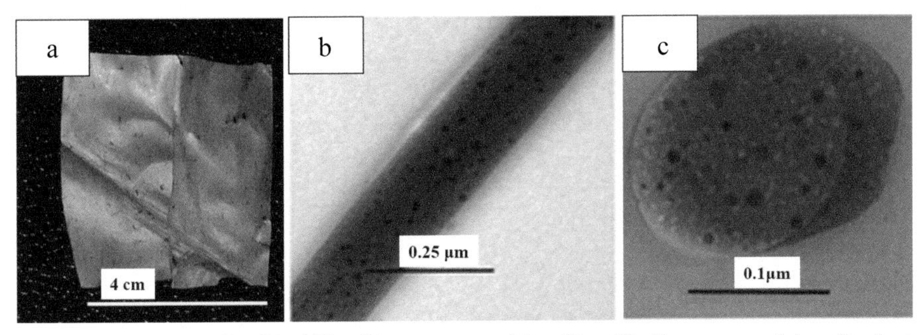

Fig. 8.5: (a) Photograph of Ag/silica fiber mats containing 10 wt% silver nanoparticles after heat treatment at 600°C. (b and c) TEM cross section micrographs of electrospun silica fiber containing Ag nanoparticles (Patel et al. 2007).

electrospinning solution. After electrospinning and subsequent calcination, the final products were obtained. If the annealing temperature was lower than 500ºC, the calcination could be conducted in the atmosphere. If the annealing temperature was higher, the calcination should be performed under the protection of nitrogen or argon (Yang et al. 2007, Toprakci et al. 2012).

The aim of loading pore forming agents into nanofibrous matrix was to fabricate mesoporous electrospun nanofibers to enhance the performance of the fibers during some reaction. For example, Wang et al. (Song et al. 2009) fabricated mesoporous ZnO–SnO_2 nanofibers by adding some surfactant-directed agent $H(C_2H_5O)_{20}(C_3H_7O)_{70}(C_2H_5O)_{20}OH$ (pluronic, P123) into the $SnCl_2/ZnCl_2$ electrospinning solution followed by electrospinning and calcination. The resulting mesoporous ZnO–SnO_2 nanofibers exhibited much higher ethanol sensing performance than common ZnO–SnO_2 nanofibers. Cubic mesoporous silica nanofibers were successfully prepared by Goutam De et al. (Saha and De 2013) using a F127–PVA–SiO_2 tri-constituent assembly approach by the electrospinning technique. PVA was used to protect the F127 directed cubic micelles which usually deform during electrospinning. The highly ordered cubic mesoporous electrospun SiO_2 nanofibers were obtained after calcination under 500ºC for 2 hours. The authors expected this mesoporous silica nanofiber to be applied in drug delivery tissue engineering and enzyme immobilization.

Carbon Nanofibers Fabricated by Calcination

Calcination treatment can also produce carbon nano/sub-micro fibers. For this purpose, many carbon precursor polymers were selected, such as polyacrylonitrile (PAN), pitches, poly(vinyl alcohol) (PVA), polyimides (PIs), polybenzimidazol (PBI), poly(vinylidene fluoride) (PVDF), phenolic resin, and so on (Li et al. 2016b, Yan et al. 2014a, Nagamine et al. 2016, Chung et al. 2005, Kim et al. 2004, Hong et al. 2014, Si et al. 2012). Among them, PAN was the most frequently employed precursor because of its high molecular weight, high degree of molecular orientations, as well as high carbon yield (Ko et al. 1992, Sawai et al. 2006, Liu and Zhang 2005).

To fabricate carbon fibers, polymer fibers were first fabricated via electrospinning and then stabilized by heat treatment at around 250ºC ~ 300ºC in oxygen-containing atmosphere under tension (Yusof and Ismail 2012). At this step, the oxidative stabilization of fibrous structure was aimed to cross-link PAN chains with a stabilized structure preventing fibers from melting, fusing, and avoiding excessive volatilization of carbon elemental during carbonization process at high temperature. The chemistry of the stabilization process is complex, but the cyclization of the nitrile groups and cross-linking of the chain molecules in the form of –C=N-C=N- are considered as the main reactions in the oxidative stabilization process (Zhao et al. 1992, Ko 2010). Moreover, the external tension applied to the as-prepared PAN fiber during the oxidation process is of great importance to prevent PAN chains from relaxing and losing their orientation, which becomes locked-in through cross-linking (Wu et al. 2005a). The pyrolysis temperature, atmosphere, heating, and airflow rate in the furnace together determine and vary the optimum stretch ratio for PAN fiber to obtain high-quality carbon fibers. The carbonization under much higher temperature (generally around 700–2000°C) was performed in nitrogen, argon, or the mixture of Ar/H_2 to produce carbon fibers with certain content of graphite phase (Qie et al. 2012, Wu et al. 2013a, Streckova et al. 2016). During the carbonization process, controlling the fiber diameters, stretching the fibrous mats, or increasing the annealing temperature can all dramatically affect the electric properties and mechanical properties of the final carbon electrospun fibers (Inagaki et al. 2012, Gupta et al. 2005, Zhou et al. 2009). In some cases, activated carbon fibers with high specific surface area were in demand because of their attractive performances in catalytic reactions or in energy storage devices. Thus, activation process of carbon fibers by CO_2 or concentrated alkaline solutions were usually applied to prepare activated carbon fibers (Tavanai et al. 2009, Yoon et al. 2004). Alternatively, researcher also loaded dissoluble components into the carbon fibrous matrices during the fabrication, and then washed to release the pores (Teng et al. 2012).

Uniform metallic nanoparticles were usually required to be loaded in the electrospun carbon nanofibers (CNFs) to enhance the performance of the fibers for energy storage, catalysis, etc. (Fan et al. 2011, Xie et al. 2016). Metallic nanoparticles or metal precursors were first mixed with the carbon precursor fibers (polymers), and then followed by stabilization in air, calcination in inactive gases, and metallic nanoparticles are embedded and immobilized into the resultant carbon nanofibers. For example, Wang et al. demonstrated a facile strategy to directly fabricate the nano-Fe/CNF catalyst, comprised of a hierarchical structure and active α-Fe NPs with uniform distribution (35 ± 5 nm) and strong magnetism, through co-electrospinning and pyrolysis self-reduction. The resultant composite Fe/CNFs exhibited excellent activity in Fenton reaction, recyclability, and easy recovery in the oxidative-degradation of dyes and phenol (Zhu et al. 2017a). Du et al. (Zhu et al. 2017b) designed and constructed three-dimensional architectures for superior electrocatalysis through the integration of one-dimensional (1D) electrospun carbon nanofibers, 1D carbon nanotubes, and 0D oxygen-deficient Mn_3Co_7-$Co_2Mn_3O_8$ nanoparticles. The 3D CoMnO@CNTs/CNFs architectures exhibit superior electrocatalytic activity and stability for the oxygen reduction, oxygen evolution, and hydrogen evolution reactions. Up to the present, the carbon nanofibers fabricated from electrospun polymers were mainly concentrated on robust physical and chemical properties with large specific surface area, high porosity, great electrical conductivity, and excellent biocompatibility rather than mechanical properties and flexibility in membrane applications due to the brittle nature of the carbon matrix. Therefore, it is an urgent demand to fabricate flexible functional carbon nanofibrous membrane to further promote their applications in membrane. Recently, the in suit synthesized single or multi-component metallic nanoparticles like SiO_2, SnO_2, Fe_3O_4, etc. distributed uniformly in the interior of the carbon fiber matrix with a seamless boundary with carbon phase (Li et al. 2016b, Tai et al. 2014, Ge et al. 2016). On one hand, the bending stress loading on the composite carbon membrane would finally result in a deformation of a single fiber due to the ultra-high length to diameter ratio of CNFs and their interpenetrating networks in the carbon nanofibrous membranes. On the other hand, the functional nanoparticles doped in carbon matrix not only effectively transmit and scatter the stress derived from the outside of fiber, but also finally avoid the generation of cracks. In addition, the microporous structures and graphitic carbon layers could also absorb the energy through the deformation of pores and slips of graphitic carbon nanosheets. A plausible mechanism for the enhanced mechanical flexibility of SnO_2/CNF membranes is shown in Fig. 8.6 (Ge et al. 2016).

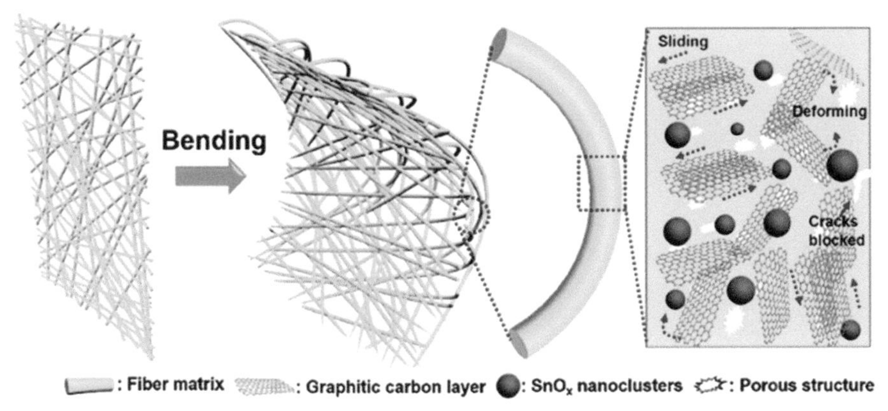

Fig. 8.6: A plausible mechanism for the enhanced mechanical flexibility of SnO_2/CNF membranes (Ge et al. 2016).

In situ Two-phase Reaction

The definition of *in situ* two-phase reaction here is that one or more components in nanofibers (solid phase) react with the components in gas or solution phase (fluid phase), and then form new materials combined in the nanofiber matrices. It mainly included gas-solid reaction and liquid-solid reaction post-treatment processes. Sometimes, solid-solid reaction was also applied to fabricated composite nanofibers. Wang's group from Jilin University of China for the first time developed a gas–solid reaction combining with electrospinning technique to incorporate semiconductor nanostructures into polymeric nanofibers. Till now, gas–solid reaction or liquid-solid reaction method has been widely applied in the production of composite nanofibers, including sulfides, chlorides, nitrides, and metals containing nanofibers. This method exhibits many advantages, such as facile operation, universality, resistance to the aggregation of nanocomponents in fiber matrix, and so on. For the broad application of the nano-components, the as-prepared composite nanofibrous products by two-phase reaction have been applied in solar cells, supercapacitors, catalysis, adsorption, water purification, batteries, sensors, and so on.

Two-phase Sulfurization

It is well known metal sulfides are mostly semiconductors and exhibited broad application in various fields for their excellent performances. Hence, the fabrication of electrospun nanofiber products containing sulfides is of ever-growing research interest. Sulfurization is the most popular strategy to synthesize sulfide semiconductors.

The gas-solid reaction was at first developed for the fabrication of composite nanofibers incorporated with sulfide semiconductor nanoparticles. The preparation process generally involved three steps. (1) The semiconductor precursor (metal salt) was in the first place co-dissolved with polymer into an appropriate solvent to form a homogeneous solution. (2) The solution was electrospun to fabricate metal salt/polymer composite nanofibers. (3) The precursor nanofibers were exposed in some gas at room temperature, then the metal salts in the fiber matrix react with the reactants in gas phase. As an example, Lu and Wang (Lu et al. 2010) prepared PdS nanoparticles with well dispersion in polyvinylpyrrolidone (PVP) nanofibers by this method. In their work, the Pd^{2+} in the precursor fibers reacted with H_2S and then *in situ* formed PdS nanoparticles. The as-prepared PbS nanoparticles were dispersed both inside and on the fibers. They did not aggregate and were spherical in shape with an average diameter of 5 nm (Fig. 8.7a and b). The effect of the kind of metal salt on the shape of the nano-semiconductors in the fibrous matrix fabricated by gas-solid reaction was also studied by Lu et al. It is found that when cadmium acetate was employed as the semiconductor precursor to fabricate PVP/CdS composite nanofibers, the resulting CdS presented as nanorod in shape instead of dense spherical nanoparticles (Fig. 8.7c and d) (Lu et al. 2005b). It was proposed to be ascribed to interaction differences between the polymer molecules and various metal ions.

The semiconductor components can not only be incorporated into polymeric fibers, but are also able to be introduced into carbon or ceramic fibers by similar gas-solid reaction process. For example, Zhang et al. (Chen et al. 2017) fabricated polyacrylonitrile-polymethyl methacrylate solution dissolved in DMF with SnO_2/Sb_2O_5 dispersed in it at first. Then electrospinning and carbonization process were carried on to fabricate SnSb/carbon nanofibers. Afterwards, the as-prepared SnSb/carbon nanofiber mat was loaded into a ceramic boat, and another ceramic boat containing thiourea was put at the upflow of the furnace. The SnSb/PCNFs were sulfurated at 350ºC under the protection of nitrogen gas flow, respectively. During the thermal treatment, the thiourea decomposed, generating hydrogen sulfide gas. The SnSb/PCNFs reacted with the produced hydrogen sulfide and transformed into SnSbSx/carbon nanofibers. The as-prepared SnSbSx/carbon nanofibers exhibit a unique two-dimensional nano-sheet morphology. As a result, the SnSbSx/carbon nanofibers can deliver a high reversible capacity of 566.7 $mAhg^{-1}$ after 80 cycles and

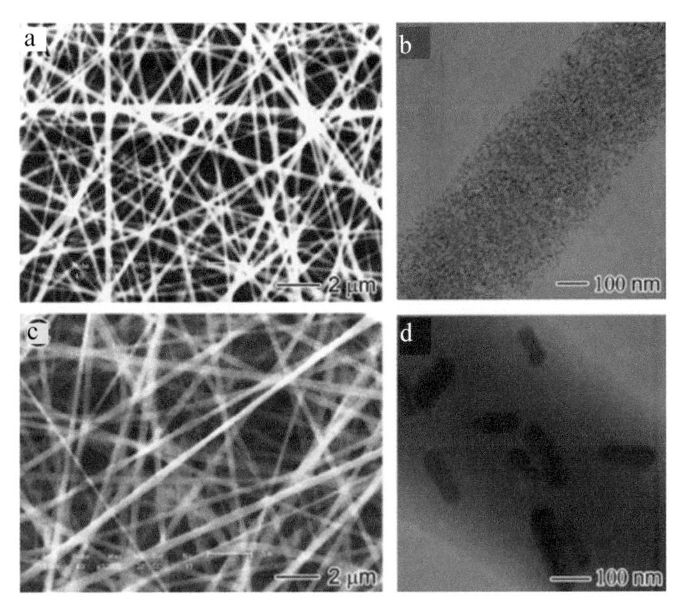

Fig. 8.7: Electrospun PVP nanofibers incorporated with semiconductor nanoparticles synthesized by electrospinning combined with gas–solid reaction. (a) SEM image of PVP/PbS composite fibers. (b) TEM image of spherical PbS nanoparticles incorporated in PVP fibers, showing that the PbS nanoparticles with a diameter of about 5 nm are well dispersed in PVP fibers. (c) SEM image of PVP/CdS composite fibers. (d) TEM image of CdS nanorods formed in PVP fibers after reaction with H_2S gas. CdS nanorods are also well dispersed in PVP fibers (Lu et al. 2010, Lu et al. 2005b).

achieve good cycling stability and rate capability when used as anode materials for sodium-ion batteries.

In addition to gas-solid sulfurization, liquid-solid sulfurization was also frequently adopted. In this strategy, nanofibers containing semiconductor precursors (the metallic ions) were first fabricated by direct-dispersed electrospinning. Then the above fibers were exposed to sulfur source in the liquid phased to generate metallic sulfides. Due to the protection of the fibrous matrix, the particles sizes are generally in the range of several nano-meters. For example, Nie et al. soaked PAN/$CuAc_2 \cdot H_2O$ fibrous mats into 1.0 M thioacetamide solution in Teflon autoclave, and the sulfurization reaction proceeded at 150°C for 1 hour. They obtained PAN/CuS composite nanofibers with monodispersed CuS nanoparticles uniformly distributed on the PAN substrate nanofibers. The as-fabricated products exhibited an excellent catalytic activity and reusability for the removal of dyes in the presence of H_2O_2 during Fenton-like reaction for water purification (Nie et al. 2013).

Sometimes, solid-solid two-phase reaction can also achieve sulfurization to fabricated sulfide nano-semiconductors. For instance, Zhang et al. first fabricated polymer/$Fe(NO_3)_3$ nanofibers by directly-dispersion electrospinning, and then obtained Fe_2O_3 nanofibers by calcination treatment in air. To synthesize iron sulfide nanofibers, the Fe_2O_3 nanofibers were mixed with sulfur powder with a weight ratio of 1:1, and treated at 800°C in an argon atmosphere for 4 hours. The final product was applied in counter electrodes for dye-sensitized solar cells. It was found that upon the conversion of n-type semiconductor a-Fe_2O_3 into p-type semiconductor FeS through the sulfurization process, the photoenergy conversion efficiency was increased from 3.79% to 6.47%, which is comparable to Pt. This increment might be related to the higher amount of electron-hole pairs and higher electrical conductivity, as well as the mixed valence of Fe element in the iron sulfide which can facilitate the transfer of electrons to the electrolyte (Zhang et al. 2017).

Two-phase Chlorination, Nitridation and Hydroxylation

Besides sulfide semiconductors, chloride and nitride nanostructures are also capable of being introduced into electrospun fibers by combination of this two-phase reaction with electrospinning method.

Chlorination. For example, Yang et al. (Bai et al. 2007) produced $AgNO_3$/PAN composite nanofibers first, and then exposed them to HCl gas to *in situ* synthesize AgCl nanoparticles in the PAN matrix by gas-solid chlorination reaction. The AgCl nanoparticles displayed a face-centered cubic structure and exhibited a controllable and uniform size (around 19.4 nm in diameter). Ce Wang and Wei Wang (Han et al. 2012) combined liquid-solid chlorination reaction with electrospinning to fabricate Ag/AgCl/PAN composite nanofiber films for visible-light photocatalysts. They first used Electrospinning to fabricate PAN/Ag composite nanofibers, and then an *in situ* oxidation and chlorination between Ag nanoparticles in fibers and $FeCl_3$ solution was conducted to prepare the composite nanofiber films. The as-prepared materials can be used as high-performance photocatalysts, taking the advantage of the visible-light activity, flexibility, and high photocatalytic kinetics.

Nitridation. Nitridation was also employed to synthesize nitride containing nanofibrous structures. Pan and Wu (Wu et al. 2009) reported the synthesis of nitride nanofibers by electrospinning for the first time. They obtained GaN nanofibers by electrospinning technology combined with gas-solid nitridation reaction. The strategy to obtain GaN nanofibers was divided into three steps: (1) fabrication of precursor composite nanofibers that comprise polymer and gallium nitrate by electrospinning, (2) calcination of the composite nanofibers in air for 4 hours at 500ºC to thoroughly remove polymeric components and transform gallium nitrate into gallium oxide nanofibers, (3) *in situ* conversion of the as-prepared oxide nanofibers into GaN nanofibers at 850ºC for 2 hours in an ammonia atmosphere. The GaN nanofibers appeared yellowish in color and exhibited polycrystalline character, with a diameter of 40 nm and a length of over a centimeter (Fig. 8.8a–c). The electrical characterization of a single GaN nanofiber shows an intrinsic n-type semiconductive nature. These polycrystalline GaN nanofibers were applied in UV-light sensors and exhibited superior performance in sensitivity, response speed, and reversibility.

Indeed, two-phase nitridation strategy was also applied for the generation of nitrogen doped oxide nanofibers in other works. For example, researchers produced oxide nanofibers by electrospinning and subsequent calcination. Then for N-doping, the obtained oxide nanofibers were dispersed in urea solution and then reacted with urea in a Teflon container at 180°C for 5 hours. Based on the above process,

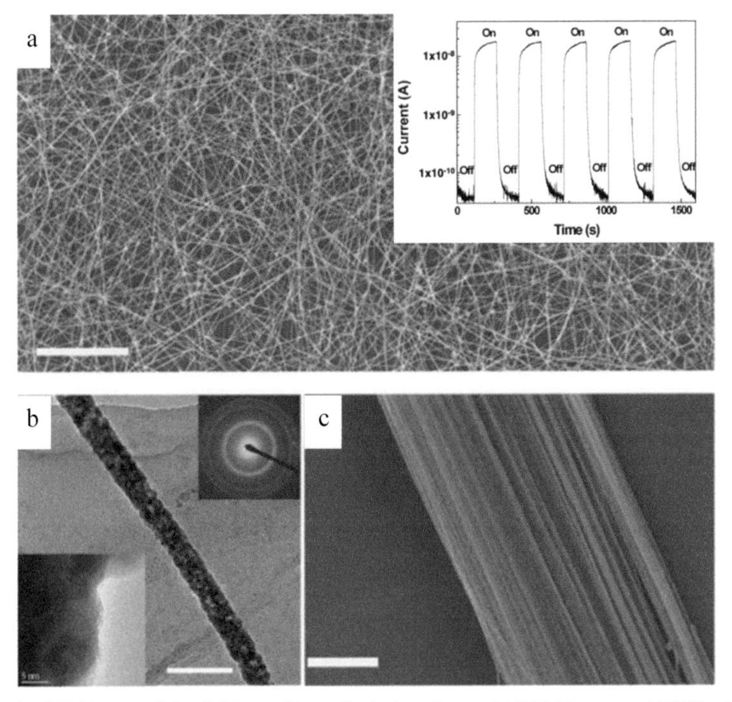

Fig. 8.8: (a) SEM image of the GaN nanofibers. Scale bar: 5 mm. (b) TEM image and EDX pattern of a single GaN nanofiber. Scale bar: 100 nm. (c) SEM image of a bundle of oriented GaN nanofibers. Scale bar: 1 mm. The inset of (a) Conductance response of a GaN nanofiber upon pulsed illumination from a 254 nm wavelength UV light with a power density of 3 mW cm^{-2} (Wu et al. 2009).

N-doped oxide nanofibers were obtained (Mohamed et al. 2017). The obtained oxide nanofibers can also react with NH_3 at around 350ºC to realize gas-solid nitridation (Li et al. 2012). The above N-doped oxides nanofibers exhibited better performance than their oxide counterparts for dye-sensitized solar cells or photocatalysis.

Metallic Hydroxides Fabrication. In some cases, synthesis of metallic hydroxides containing nanofibers is more attractive, because without heteroatom they will lead to less secondary pollution during application than sulfides and nitrides. Wang's group developed liquid-solid two-phase reaction for the *in situ* synthesis of lanthanum hydroxide, which was incorporated in the PAN nanofibrous matrix. In their work, PAN and certain amounts of $La(NO_3)_3 \cdot 6H_2O$ were first dissolved in DMF. Then electrospinning of the above solution was conducted to fabricate PAN/ $La(NO_3)_3$ precursor nanofiber mat. Subsequently, liquid-solid reaction was carried on in 0.1 M NaOH bath between the La^{3+} ions in the fibers and OH^- in the solution to generate $La(OH)_3$. The time-dependent growth process of the $La(OH)_3$ in the PAN nanofibers during two-phase precipitation was studied in their work and shown in Fig. 8.9. When the samples were treated by NaOH for a short time (30 minutes), a few dark nanodots emerged (Fig. 8.9b) and grew in size with increasing time (1 hour, Fig. 8.9c). As the immersion time in NaOH further increased, many nanodots agglomerated (2 hours, Fig. 8.9d) along a 1D direction and subsequently formed

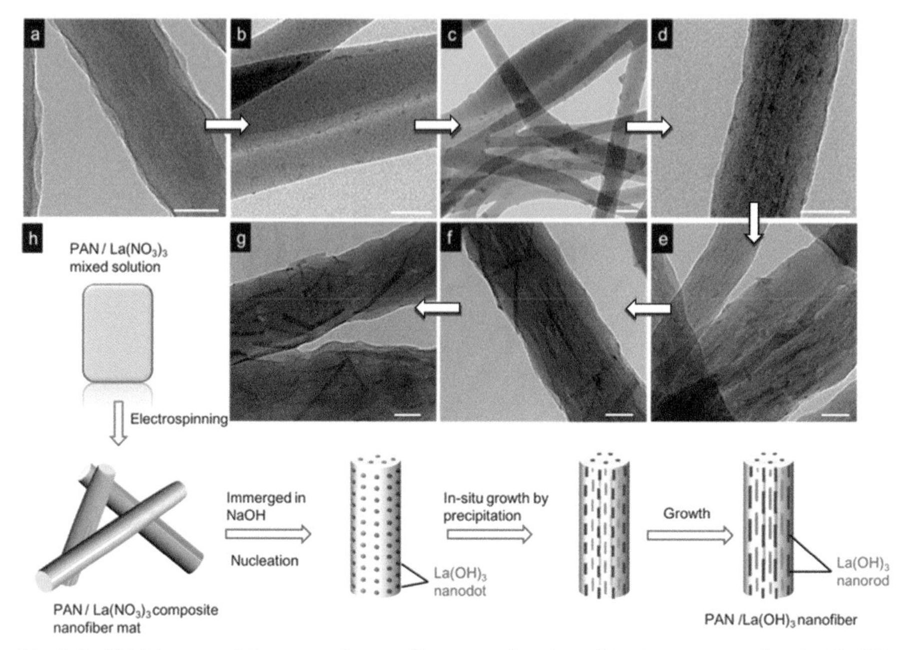

Fig. 8.9: TEM images of the composite nanofibers as a function of *in situ* treatment time by NaOH: (a) 0 minute (PAN/La$(OH)_3$), (b) 30 minute, (c) 1 hour, (d) 2 hours, (e) 6 hours, (f) 10 hours, (g) 12 hours. (scale bar: 50 nm), (h) Scheme for the formation of dispersed-La$(OH)_3$ nanorods in PAN nanofibers by electrospinning and the subsequent *in situ* precipitation (He et al. 2015).

short nanorods (6 hours, Fig. 8.9e). Finally, after being soaked in NaOH solution for more than 10 hours, the *in situ* La$(OH)_3$ nanorods with a definite 1D configuration in PAN nanofibers were formed (Fig. 8.9f and g). A possible formation process for the La$(OH)_3$ nanorods was proposed and is schematically displayed in Fig. 8.9h. La$(OH)_3$ fabricated by precipitation in alkaline condition tended to form anisotropic structures due to the intrinsical anisotropy of the hexagonal La$(OH)_3$. Additionally, PAN fibers acted as a nanoreactor, which could not only provide a place for the reaction between La^{3+} and alkali solution, but also impact the nucleation and growth process of the assemblies. By the growth-controlling and agglomeration-inhibiting functions of the PAN fibers, as well as the intrinsical anisotropy of the hexagonal structure La$(OH)_3$, the aeolotropic structure was finally formed. The final product exhibited a fast kinetics, high-removal rate, and easy separation for phosphate adsorption, even toward low concentration targets. This benefited from the increased active sites, which were provided by the well-dispersed La$(OH)_3$ nanorods to chemically bind with phosphate. Finally, an effective nutrient-starvation antibacterial strategy was successfully demonstrated through scavenging free bioavailable phosphate by the composite fibers, and endowed drinking water with long-term biostability (He et al. 2015). The above PAN/La$(OH)_3$ electrospun nanofibers were also carbonized to fabricate carbon nanofiber matrix with embedded $LaCO_3OH$, which could synchronously capture phosphate and organic carbon to achieve more effective starvation strategy to kill bacteria (Zhang et al. 2016a).

Two-phase Reduction

Two-phase Reduction in Polymer Fibers. Two-phase reactions, involving gas-solid reaction and liquid-solid reaction were also applied for the reduction of metal ions in the fibrous matrix and production of metallic nanostructures. At the early stage, this strategy was developed to reduce the noble metal ions in polymeric fibers to form noble metal nanoparticles embedded nanofibers. Taking Ag nanoparticle as an example, one can first fabricate polymer/$AgNO_3$ precursor nanofibers, and then soak the products in a solution containing proper reductants (Wang et al. 2005). The reductants could be hydrazine hydrate, citrate, ascorbic acid, sodium borohydride, ethylene glycol, and so on. Notably, if ethylene glycol was used as reducer for reduction of Ag^+, the reaction should be operated in alkaline condition under irradiation by a microwave to accelerate the formation of Ag(0) nanoparticles in fibers (Yan et al. 2014b). Sometimes, if the polymer nanofibers were water-soluble or the aim was to reduce the loss of metal ions during the reduction process, the vapor of hydrazine hydrate can be employed to carry on gas-solid reduction (Dong et al. 2012).

Metal nanoparticle embedded nanofibers attracted broad attention for application in catalysis, antibacteria, electronics, etc. For example, Jeong et al. (Park et al. 2012) demonstrated a first highly stretchable electric circuit based on a composite material of silver nanoparticles and elastomeric fibers, which was fabricated by electrospinning and subsequent liquid-solid reaction. Figure 8.10a presents the schematic illustration of the overall fabrication process. Poly (styrene-block-butadiene-block-styrene) (SBS) rubber fibers were obtained by electrospinning SBS solution in solvent mixture of THF/DMF (3:1). Then the SBS fibers were dipped in a silver precursor solution ($AgCF_3COO$ in ethanol). The $AgCF_3COO$ and the ethanol were absorbed by the fibers, such that the fiber mat becomes swollen. After drying, the precursor is reduced by a solution of hydrazine hydrate, generating silver nanoparticles inside the fibers and silver shells at the surfaces of the fibers. The conductive composite elastic fiber mats exhibited excellent stretchability. Even though at large strains the silver shell broke into small pieces of debris, the electrical conductance was still maintained by percolation of the silver nanoparticles inside the fibers (Fig. 8.10b and c), as well as by inter-fiber bridges formed from the pieces of silver shell. It was found the bulk conductivity was as high as 2,200 S cm^{-1} at 100% strain for a 150-mm-thick mat. This material was expected to be applied in various electronic devices.

Two-phase Reduction in Inorganic Fibers. Metal nanostructures are often demanded to be incorporated in electrospun inorganic fibers to enhance the performance of the composite materials in chemical sensors, batteries, catalysis, etc. Inorganic fibers, especially oxide nanofibers, are generally derived from the calcination of the polymer precursor fiber products in air atmosphere. However, during the calcination process, many metal ions, except Ag, Au, Pt, tend to converse to oxides. Therefore, additional reduction steps are required to synthesize metal nanoparticles containing inorganic nanofibers. For example, in order to obtain ultrasensitive hydrogen sensor, Pd(0) loaded SnO_2 nanofibers were fabricate by reduction of the electrospun Pd(II)/SnO_2 nanofibers in excess hydrazine hydrate solution for 30 minutes via liquid-solid reaction. The resulting Pd(0) loaded SnO_2 nanofiber was used for H_2

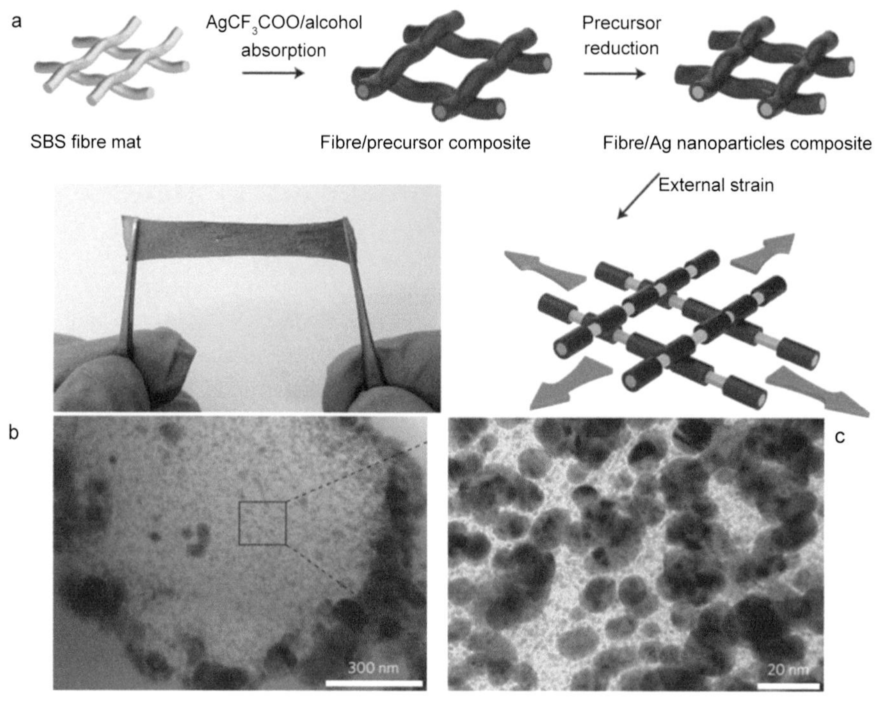

Fig. 8.10: (a) Schematic illustration of the overall fabrication process of SBS/Ag. The composite fiber mat is highly stretchable, without losing its conductivity at large strains. (b) and (c) Cross-sectional TEM image of an SBS/Ag composite fiber. The silver nanoparticles within each SBS fiber are interconnected, providing the main contribution to electrical conductivity (Park et al. 2012).

detection, which displayed an ultralow limit of detection (20 ppb), high response, fast response and recovery, and selectivity at room temperature (Wang et al. 2013). Excessive consumption of fertilizer and poor disposal of animal waste render nitrate contamination a big challenge in drinking water safety and public healthcare. Pd-Cu bimetallic catalyst has been widely accepted as one of the most promising combinations for catalytic reduction of nitrate to nitrogen. Wang and co-workers demonstrated a highly-dispersed Pd-Cu bimetallic catalyst supported by titanium dioxide nanofibers, which was *in situ* fabricated by electrospinning and subsequent calcination and liquid-solid chemical reduction. They produced CuO/PdO loaded TiO_2 nanofibers first by electrospinning and subsequent calcination in air. Then liquid-solid reduction was operated in sodium borohydride solution (0.1 M) to reduce Cu(II) and Pd(II) into Cu(0) and Pd(0), respectively. For catalytic removal of nitrate, the catalyst exhibited high nitrate removal efficiency, selectivity and fast kinetics, due to the well-dispersed and closely connected Pd-Cu bimetallic nanoparticles in TiO_2 nanofiber matrix (Wang et al. 2017).

All-metallic Nanofibers Fabrication by Two-phase Reduction. All-metallic nanofibers have attracted special attention the last years, for their important application for transparency electrodes and electronic devices. Several methods can be used for metal nanofiber preparation, including direct calcination transformation, two-phase reaction and post-growth by template method. The direct calcination transformation has been depicted in detail in the part of "calcination transformation". Differently, for metal nanofiber fabrication, the precursor, metal salt/polymer nanofibers should be calcined at high temperature in inert gas atmosphere (e.g., argon) instead of in air (Barakat et al. 2009). The two-phase reaction will be discussed here and the post-growth method later in "surface functionalization" section.

The first work on metal nanofibers was reported by Andreas Greiner et al. (Bognitzki et al. 2006) in 2006. They communicated the preparation of long copper nanofibers by the electrospinning of copper nitrate–polyvinylbutyral (PVB) solutions to corresponding composite fibers followed by thermal treatment in air at 450°C for 2 hours (conversion of copper nitrate to copper oxide and degradation of PVB). Afterwards, gas-solid reaction treatment was carried on in a hydrogen atmosphere at 300°C for 60 minutes to convert CuO to Cu. Finally, copper fibers with an average diameter of 270 nm were obtained. Inspired by this work, Cui and co-workers demonstrated a high-performance transparent electrode with copper nanofiber networks. The Cu nanofibers had ultra-high aspect ratios of up to 100,000 and fused crossing points with ultralow junction resistances, which resulted in high transmittance at low sheet resistance, e.g., 90% at 50 Ω/sq. Organic solar cells using the Cu nanofiber networks as transparent electrodes presented a power efficiency of 3.0%, comparable to devices made with ITO electrodes. In addition, due to their large aspect ratios, nanoscale diameters, and metallic bonding natures of Cu nanofibers, transparent Cu nanofiber electrodes on poly(dimethylsiloxane) (PDMS) substrates exhibited great flexibility and stretchability. They were fabricated by simply transferring free-standing CuO nanofiber networks to PDMS, followed by gas-solid reaction in an H_2 atmosphere at 300°C to transform into Cu (Wu et al. 2010). Besides Cu nanofibers, Pt, Au, Ag, Ni, and some alloy electrospun nanofibers can also be prepared by the above strategy.

Two-phase Crosslinking

Water-soluble polymeric nanofiber mats are a kind of important products fabricated by electrospinning technology, including polyvinyl akohol (PVA) nanofibers, gelatin nanofibers, collagen nanofibers, zein nanofibers, chitosan nanofibers, and so forth. Due to the numerous merits such as biological origin, nonimmunogenicity, biodegradability, biocompatibility, and commercial availability at relatively low cost, water-soluble polymeric nanofibers are widely applied in the pharmaceutical and medical fields, such as sealants for vascular prostheses, carriers for drug delivery, dressings for wound healing, and so on. Many water-soluble polymers with good biocompatibility generally possess a large amount of hydroxyl groups on their molecular chains. Therefore, their electrospun products often suffer from the drawbacks of being weak mechanically and fragility in aqueous media. Hence, in order to achieve stable and long-term application, these electrospun products must be crosslinked in advance to improve both water-resistant ability and mechanical performances.

There are many routes to realizing crosslinking for the crosslinkable polymers, e.g., thermal *in situ* treatment (Mani and Dupuy BSonenshein 2002), UV irradiation

(Mahanta and Valiyaveettil 2012), and chemical crosslinking method. Amongst these, chemical crosslinking by two-phase reaction is the most effective one, which generally occurs between the hydroxyl groups of the nanofibers with aldehydes (or amines) vapor or solution by condensation reaction. Many chemicals such as formaldehyde, glyoxal, glutaraldehyde, Hexamethylene diisocyanate, and dextran dialdehyde (Yao et al. 2010) have been employed for chemically curing the water-soluble polymers. Thereinto, glutaraldehyde is by far the most broadly used chemical, due to its high efficiency. A glutaraldehyde molecule can form four linkages to PVA fibers, hence should serve as a tetra-functional crosslinking agent to stabilize the PVA fibers. Glutaraldehyde has been applied for curing PVA nanofiber, gelatin nanofibers, collagen nanofibers, chitosan nanofibers, or their composite samples (Chen and Wang 2010, Wu et al. 2005b, Zhang et al. 2006, Zhu et al. 2016). Generally, the as-prepared crosslinkable nanofibers were placed in the GTA vapor to achieve water-stabilization. It is also reported that chemical crosslinking can reduce the crystallinity and perfection of polymers, thus Differential Scanning Calorimetry (DSC) can be used for investigating the chemical curing reaction.

Physical cross-linking by liquid-solid reaction was also developed for curing PVA nanofibers. Wnek et al. (Yao et al. 2003) produced fully hydrolyzed PVA fibers via electrospinning and then soaked them in methanol for at least 12 hours. The crosslinking should occur by the removal of residual water within the PVA fibers by the methanol, allowing PVA-water hydrogen bonding to be replaced by intermolecular polymer hydrogen bonding, resulting in additional crystallization. Thus, the methanol treatment served to increase the degree of crystallinity, and hence the number of physical cross-links in the electrospun PVA fibers, which was proved by DSC experiments. The methanol-treated PVA fiber mats exhibited much better water-resistance property. The crosslinking reactions of the electrospun products are discussed in detail in Chapter 7.

Surface Functionalization

Most of the functionalization methods mentioned above generally occur both inside and outside of the fibers. Another case it that functionalization is mainly expected to be conducted on the outside surface of the fibers, with the electrospun fibers as templates. By far, functional groups, functional polymers, inorganic semiconductors, metals, and so forth have been modified on electrospun fibers, and found important applications in biomedicine, environmental treatment, nano-electronics, catalysis, etc.

Surface Modification with Organic Components

Surface Grafting. Surface grafting can functionalize the electrospun fibers with functional groups or functional polymers. Researchers often functionalized polymeric electrospun fibers with multiple amino or hydroxyl groups to bind heavy metal ions, then remove them from water. For example, Zhao and co-workers reported a phosphorylated PAN nanofiber mat by electrospinning process and a subsequent grafting modification. In that work, the as-spun PAN nanofibers were first cross-linked by hydrazine hydrate aqueous during a reflux reaction. Afterwards, the cross-linked PAN was filtered, washed, and then soaked in diethylenetriamine aqueous solution to conduct amination by another reflux reaction at 90°C. Finally, paraformaldehyde and phosphorous acid reacted with the above PAN to generate phosphorylated PAN fibers. Due to the chelating ligands from the multiple hydroxyl groups on the functional nanofibers, Cu^{2+}, Pb^{2+}, Cd^{2+}, and Ag^{+} could be effectively adsorbed (Zhao et al. 2015). In the above procedure, the aim of the crosslinking step is to avoid the destruction of the functionalized fibers in water, since functional groups of amino and hydroxyl groups tend to increase the solubleness of the fibers. By far, researchers have successfully synthesized polyethylenimine grafted nanofibers (Zhao et al. 2016), thioamide-group grafted nanofibers (Li et al. 2013), amidoxime-modified nanofibers (Saeed et al. 2008), and amine groups functionalized nanofibers (Neghlani et al. 2011) by surface grafting strategy for water treatment.

Surface grafting method is also effective for the functionalization of inorganic fibers. Electrospinning products have been widely applied in biomedical fields, because they can provide excellent biocompatibility, 3D porous structure, appropriate, geometry on the surface, etc. To enhance the properties of inorganic electrospun nanofiber matrix for biomedical application, surface grafting was used. Taking an example, Zhang and co-workers, introduced a new detection assay for circulating tumor cells (CTC) based on an electrospun nanofibers deposited substrate, which was grafted with cell capture agent (i.e., anti-EpCAM) (Fig. 8.11). They at first fabricated densely packed TiO_2 nanofibers (TiNF) with diameters of ~ 100–300 nm on silicon wafers by calcinating titanium n-butoxide (TBT)/polyvinyl pyrrolidone (PVP) composited nanofibers. After preparing the TiNF substrates, silane chemistry was applied to modify the TiNF substrates with 3-mercaptopropyl trimethoxysilane (MPTMS). Afterwards, they used the coupling agent N-maleimidobutyryloxy succinimide ester (GMBS) to treat with the substrate, resulting in a layer of GBMS coating onto the TiNF substrate. Subsequently, they utilized N-hydroxysuccinimide (NHS)/maleimide chemistry to introduce streptavidin (SA) onto the surfaces of the TiNF substrate. Finally, biotinylated anti-EpCAM was freshly conjugated onto the streptavidin-coated substrates prior to the cell-capture experiments. Considering the geometric orientation of nanostructures embedded in the ECM scaffolds, electrospun TiO_2 nanofibers can better mimic these horizontally oriented nanostructures that could lead to improved cell/substrate affinity (Zhang et al. 2012b).

The above surface grafting procedures were all operated in solution. Also, in recent years, plasma treatments in some gas atmospheres are frequently demonstrated for surface grafting. The high energy from the plasma drive the gas molecules to react with electrospun fibrous surface, and then modify the surface with functional groups. For instance, in NH_3 atmosphere, plasma treatment could graft-NR chemical groups on polyvinyl chloride nanofibers and then endow the surface with positive charges. The positively charged composites are capable of binding with negative charged materials to achieve various applications (Feng et al. 2015). In order to constructed a super hydrophobic surface on electrospun fibers, Yoon et al. have demonstrated that the hydrophobicity of polymer fibrous membranes can be further improved by treatment of a CF_4 plasma (Yoon et al. 2008). In fact, plasma treatment on electrospun nanomaterials is a complex process and is just in the beginning phase. However, due to its universality on surface grafting of solid surfaces, plasma treatment will attract

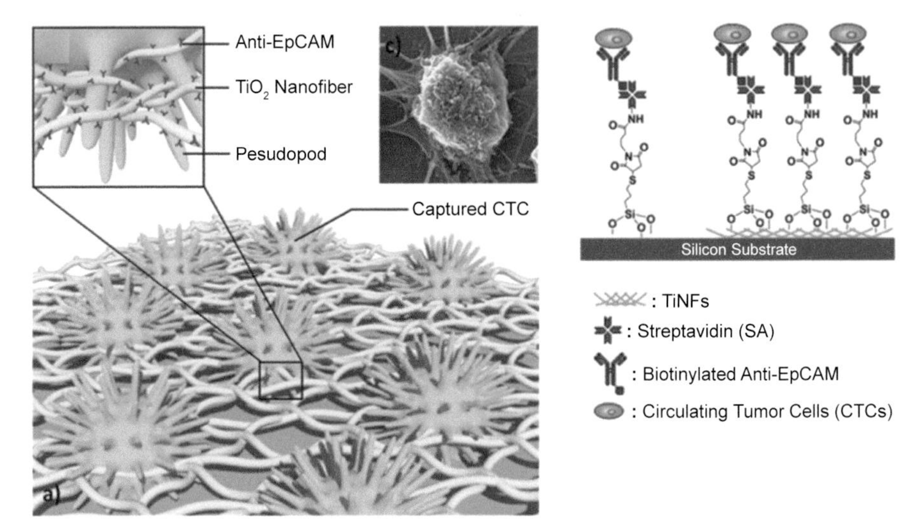

Fig. 8.11: Schematic of the horizontally packed TiNFs for improved CTC capture through combining cell-capture-agent (i.e., Anti-EpCAM) and cancer cell-preferred nano-scale topography. Biotinylated epithelial-cell adhesion-molecule antibody (Anti-EpCAM) grafted flat Si/TiNFs alternately patterned substrate (Zhang et al. 2012b).

more and more attention. Indeed, plasma treatment in gas atmospheres is a kind of chemical vapor deposition.

Gas-phase Polymerization. In recent years, conductive polymers (CPs) have been applied in various fields, including electronics, catalysis, energy storage, and so on. Researchers often desire to incorporate these CPs in the electrospun products, promoted by both the super-long 1D nano-structures and the electric activity of CPs. At an earlier stage, direct electrospinning was employed, which was seriously hindered by the low solubility and poor spinnability of CPs. Although blended electrospinning was demonstrated in many works, in the case of blend fibers, the CP components are not continuous. The carrier polymers (e.g., poly(ethylene oxide), poly($\mathcal{E}$-caprolactone), and so on) block the charge transport, and thus result in much lower charge mobility of below 10^{-3} cm^2 V^{-1} s^{-1} (Lee et al. 2009a).

In that case, researchers have developed a series of post-polymerization methods to fabricate high-performance CP nanofibers, including by gas-phase polymerization and solution-phase polymerization. In those methods, a continuous layer of CPs can successfully modify on the surface of electrospun fibers, with the fibers as one-dimensional templates. Taking gas-phase polymerization as an example, the typical procedure is that an appropriate oxidative initiator is incorporated in the electrospun fibrous matrix followed by subsequent gas-phase polymerization in vapor atmosphere of monomer of CPs. Wang et al. reported a facile route for the fabrication of Au-doped polyacrylonitrile (PAN)–polyaniline (PANi) core–shell nanofibers, which involves electrospinning of the core and subsequent gas-phase polymerization of the shell. In detail, core nanofibers comprising PAN and $HAuCl_4$ were first electrospun from a dimethylformamide (DMF) solution. The mass ration of polymer to $HAuCl_4.4H_2O$ ranged from 2~5. Then gas-phase polymerization of a

PANi shell onto the nanofibers was subsequently conducted at 60°C for 5 hours in aniline vapor, using $HAuCl_4$ in the as-spun nanofibers as the oxidant. During this process, not only was a shell of PANi obtained on the surface of PAN nanofibers, but also evenly dispersed Au nanoparticles were also formed in the nanofibers (Fig. 8.12a and b). The as-prepared Au-doped PAN–PANi core–shell nanofibers displayed a very high field-effect mobility of up to 11.6 $cm^2 V^{-1} s^{-1}$. This could be due to the continuous PANi shell promoted charge transfer and reduction of the grain-boundary effect. Additionally, the doping of Au nanoparticles could serve as "conducting bridges" between the PANi semiconducting domains to enhance the electrical properties (Wang et al. 2011b).

For polypyrrole (PPy) gas-phase polymerization, the reaction condition was more moderate, because PPy is easier to be triggered than PANi. If $HAuCl_4$ served as the oxidative initiator, a reaction temperature close to room temperature is enough. Fe^{3+} is also strong enough to trigger the gas-phase polymerization of PPy. For example,

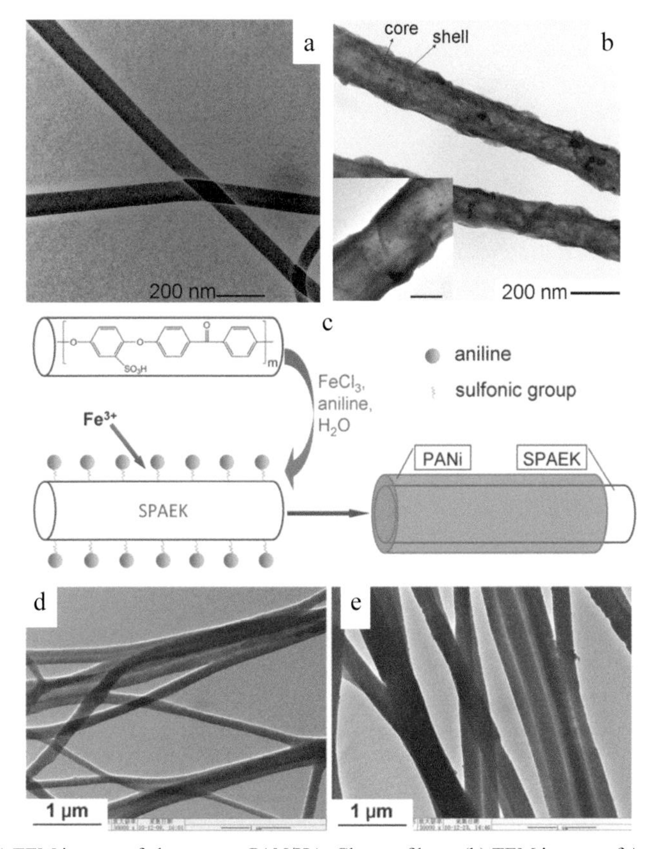

Fig. 8.12: (a) TEM images of electrospun PAN/$HAuCl_4$ nanofibers. (b) TEM images of Au-doped PAN–PANi core–shell nanofibers fabricated by gas-polymerization. Wang et al. (2011b) with a lower and higher magnification (inset). The scale bar in the inset is 100 nm. (c) Scheme of the synthesis of SPAEK/PANi nanofibers by solution-phase selective polymerization combined with an electrospinning technique. TEM images of (d) SPAEK and (e) SPAEK/PANi fibers prepared by electrospinning and solution-phase selective polymerization (Wang et al. 2011c).

Jiang and Wang reported a gas-polymerization for construction of PPy coated electrospun nanofibers. In their work, PVP, equimolar $SnCl_2 \cdot 2H_2O$ and $FeCl_3 \cdot 9H_2O$ were dissolved in 1:1 weight ratio of DMF and ethanol to prepare electrospinning solution. Subsequently, after electrospinning and calcination, SnO_2/Fe_2O_3 composite nanofiber were formed. Then SnO_2/Fe_2O_3 nanofibers were exposed to HCl to activate the Fe^{3+} followed by the gas-polymerization in saturated pyrrole vapor in a vacuum drier at room temperature. Finally, SnO_2/PPy core-shell nanofibers were obtained (Jiang et al. 2013). For various applications, SnO_2 can be replaced by other fibrous components (Lu et al. 2012). Other CPs, like polythiophene (PTH) (Wang et al. 2011d), poly(3,4-ethylenedioxythiophene) (PEDOT) (Kwon et al. 2010) can also generate on electrospun fibrous substrates by the similar process.

In addition to CPs, the surface functionalization by gas-phase polymerization is suitable for other functional polymers as well, such as hydrophobic fluoropolymers (Guo et al. 2015), poly(p-xylylene) (PPX) (Bognitzki et al. 2000), and so forth. Impressively, Greiner and co-workers (Mitschang et al. 2014) demonstrated a tea-bag-like catalyst based on gold-containing polymer nanotubes with both high catalytic activity and resistance to leaching. The catalyst was fabricated by electrospinning of poly(l-lactide)-Au followed by coating of poly(*p*-xylylene) (PPX) and subsequent removal of core nanofibers. The coating of PPX was operated in paracyclophane vapor below 30°C by gas-polymerization route.

Solution-phase Polymerization. Solution procedure attracts more research attention for preparation of composite materials than gas-phase procedure, due to the environmental-friendliness, cost-efficiency, easier-fabrication, and higher yield. Solution-phase polymerization has been broadly applied for surface coating of nanofiber matrix by CPs. The most facile method is direct deposition polymerization. Taking PANi as an example, in a typical process, core nanofibers were at first fabricated by electrospinning at the template. Then the template nanofibers were immersed in an aqueous solution of aniline (aniline in HCl solution). After adding ammonium persulfate solution as oxidant ($(NH_4)_2S_2O_8$), the PANi layer gradually formed on the surfaces of the template nanofibers. If the desired coating was PPy, the generally used oxidative initiator in solution polymerization was Fe^{3+}, Au^{3+}, Pd^{2+}, etc. (Wang et al. 2012a, Zhang et al. 2016b).

However, for this direct deposition process, the polymerization not only occurs on fiber surface, but also takes place in solution. The low selectivity renders both a waste of raw material and a rough coating surface with low controllability. Therefore, a selective solution polymerization was developed for CPs coating on nanofibers. Lu reported the first work on a surfactant directed solution polymerization with selectivity to prepare PPy/TiO_2 coaxial nanocables in aqueous solution. In their system, surfactant molecule, sodium dodecyl sulphate were absorbed on the surface of TiO_2 nanofibers. A columnar microregion containing a hard core (TiO_2) and a soft interface (surfactant) was formed. After adding the pyrrole monomers and the $FeCl_3$ oxidant, polymerization occurred between the surfactant layer and the TiO_2 nanofibers. PPy was selectively deposited on the surface of the TiO_2 nanofibers and PPy/TiO_2 nanocables are thus formed (Lu et al. 2007).

Selective coating by solution procedure is meaningful for construction of nano-electronics. For example, Wang et al. demonstrated a fabrication of an organic field-effect transistors (OFETs) by electrospinning and subsequent solution-phase selective polymerization. First, aligned nanofibers consisting of sulfonated poly(arylene ether ketone) (SPAEK) were collected between two parallel aluminum plates via electrospinning and transferred onto SiO_2 wafers. Secondly, the SiO_2 wafers with SPAEK nanofiber attached were immersed in aniline aqueous solution to conduct the polymerization with $FeCl_3$ as oxidant. During the process of solution-phase selective polymerization, the sulfonic groups on the SPAEK nanofibers can chemically combine aniline monomer, resulting in an aniline rich region along the SPAEK nanofibers. It is important to note that the concentration of aniline in the aqueous medium was kept at a low level to ensure the PANi selectively coated on the SPAEK nanofibers, and not on SiO_2 wafers, as illustrated in Fig. 8.12c. The increase of the nanofiber diameters after polymerization indicated the successful growth of PANi on SPAEK (Fig. 8.12d and e). Benefiting from the selective polymerization, a one-dimensional active channel was constructed for OFETs, which exhibited reduced grain boundary effects. In addition, the SPAEK core nanofibers did not only provide 1D structure, it also could provide an internal modulation for charge transfer from the nano-interface of the 1D core/shell nanostructure under gate bias. Based on the SPAEK/PANi nanofibers, a high mobility of $\approx 3\ cm^2\ V^{-1}\ s^{-1}$ is obtained with the current on/off ratio of exceeding 10^4. The mobility was much higher than that of previously reported PANi-based FETs (Wang et al. 2011c). Wang employed this solution-phase selective polymerization as well to fabricate sulfonated polymer/PPy core–shell nanofibers for ultrasensitive ammonia detector. The PPy shell coating was uniform and controllable. The adsorption effect of the HSO_4^- in core fibers toward NH_3 facilitated the mass diffusion of ammonia through the PPy layers, resulting in the enhanced sensing signals. The sensors exhibit large gas responses, even when exposed to 20 ppb ammonia at room temperature (Wang et al. 2012b).

In addition to the above process, solution-phase selective polymerization can also be achieved by coating electrospun nanofibers with polydopamine (PDA) and its composites, which should be due to their high adhesive force of PDA towards solid surface (Zhu et al. 2017c). Moreover, electrochemical polymerization has been applied for functional polymer coating, whereas the potential differences act as the polymerization initiators (Abidian and Martin 2009).

Surface Modification with Inorganic Component

A series of measures are being developed to deposit inorganic components on the surface of electrospun fibers to functionalize the composites. These methods can be roughly divided into two groups according to the procedure, the gas (vapor) phase deposition and the solution phase deposition. The vapor-phase routes generally include chemical vapor deposition, physical vapor deposition, and atomic layer deposition; while the solution-phase one contains hydrothermal or solvothermal reaction, electroless-plating, sol-gel coating, self-assembly method, and so forth.

Chemical Vapor Deposition and Physical Vapor Deposition. Different form PVD, there must be some chemical reaction happening during the chemical vapor deposition process. CVD has emerged as a powerful strategy to synthesize both polymeric structures and inorganic structures. For polymer polymerization, CVD is also named as gas-phase polymerization, which has been discussed in the above text. CVD enhanced by plasma was also briefed in the above text for surface grafting by functional groups. Here, we will briefly introduce the CVD process for the fabrication of inorganic nanostructures on electrospun fibrous substrates. The first CVD process combined with electrospinning was demonstrated by Hou et al. in 2004. In that work, they obtained hierarchical carbon nanotubes modified CNFs. Fe nanoparticles embedded CNFs was first fabricated by electrospinning, calcinations, and reduction process (Fig. 8.13a). Then the above product was placed in CVD furnace, where hexane vapor was employed as the source of carbon to form carbon nanotubes on CNFs (Fig. 8.13b and c). The hexane molecules could be decomposed on the surface of the Fe nanoparticles by the catalytic action of the Fe at 700ºC. The carbon atoms were adsorbed on and dissolved in the metal, transported to the interface between the Fe particle and the growing end of the nanotube, and finally incorporated into the tubular structure (Hou and Reneker 2004). Recently, a quite similar work on electrospun CNFs modified with carbon web of hairy fibers was also reported. The products were synthesized by CVD using a mixture of C_2H_6 and H_2, and the Ni nanoparticles embedded in the electrospun CFs as the growth catalyst (Liu et al. 2016).

In some cases, CVD process was conducted without pre-loaded growth catalysts in fibrous matrix. By this process, coating of semiconductor structure, Si containing structure, MoS_2, etc. can be generated (Qiao et al. 2013, Chen et al. 2016). For example, Chen et al. manufactured a composite comprising of MoS_2@electrospun carbon nanofibers by electrospinning, carbonization and subsequent CVD. MoS_2 were deposited onto CNFs by CVD in a tube furnace under ambient-pressure at 550ºC. Molybdenum chloride powder and excessive sulfur powder acted as the MoS_2 precursors. During the synthesis, the reaction between vapor phase precursors occurred to produce gaseous sheet-structured MoS_2. Simultaneously, the generated gaseous MoS_2 species were precipitated onto the CNF substrate to form MoS_2@CNFs (Chen et al. 2016).

Physical deposition, such as physical vapor deposition (PVD) and dip-coating, has also applied to functionalize electrospun fibers or fabricate tubular structures combined with sacrificial template methods. Dip-coating was reported to fabricate polymer coating in earlier work, but for the quite low controllability, it is scarcely employed right now (Bognitzki et al. 2000). PVD was frequently employed to prepare inorganic coatings (e.g., oxide and metallic coating) with electrospun fibers serving as templates (Choi et al. 2009a, Wu et al. 2013b). For example, Wu et al. coated free-standing polymer nanofiber networks with a thin layer of conducting material using PVD techniques, such as thermal evaporation, electron-beam evaporation, or magnetron sputtering (Fig. 8.14a). Then they transferred the composite fibers to a desired substrate and removed the template fiber for preparation of transparent electrodes. They have successfully manufactured various continuous and conductive nanotrough networks by the PVD procedure, including silicon, indium tin oxide (ITO), silver, copper, platinum, aluminium, chromium, nickel, and their alloys. The nanotroughs were firmly attached to the substrates with 90% transmission, high conductivity (sheet resistance of ~ 2 Ω) and flexibility, which should be due to the super long 1D nanotrough structures provided by electrospun fibers (Wu et al. 2013b).

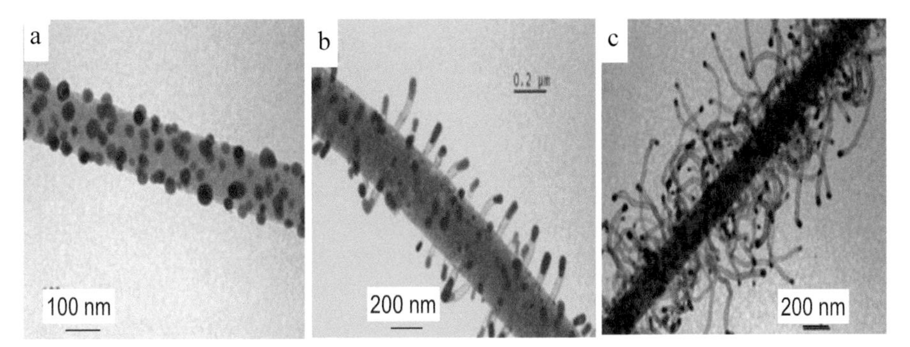

Fig. 8.13: (a) TEM image of carbon nanofibers containing Fe NPs (carbon/Fe). (b) and (c) The growth of carbon nanotubes on the carbon/Fe nanofibers via CVD process in hexane vapor precursor for 3 minutes and 5 minutes, respectively (Hou and Reneker 2004).

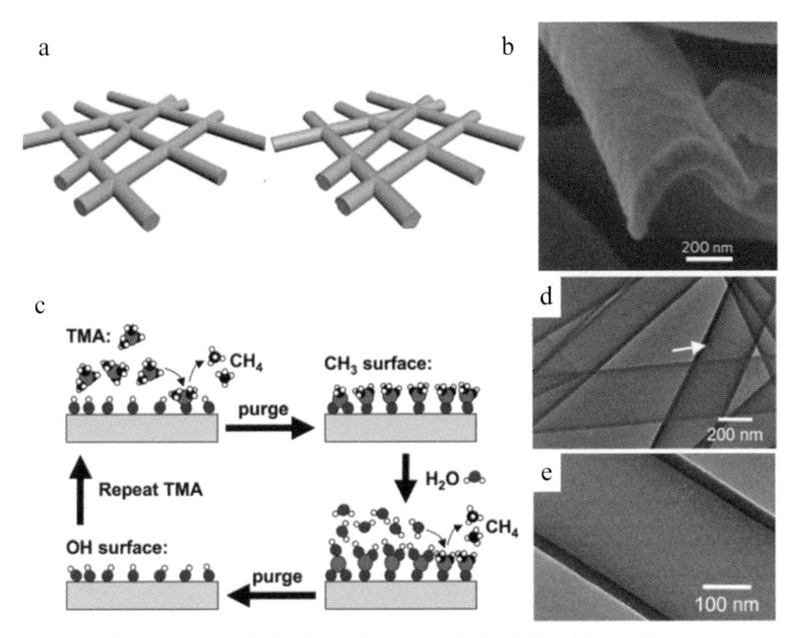

Fig. 8.14: (a) Schematic of the fabrication of nanotroughs by PVD with nanofibers as the template. (b) SEM image of the cross-section of a single gold nanotrough, revealing its concave shape (Wu et al. 2013b). (c) Schematic diagram of formation Al_2O_3 layer on nanofibers by ALD with TMA and H_2O as precursors. (d and e) TEM images of Al_2O_3 microtubes formed when ALD was performed on electrospun PVA fibers for 300 cycles at 45°C (Peng et al. 2007).

In a word, physical deposition is an efficient route to prepare composite electrospun fibers, however, it is a line-of-sight deposition technique, so it is challenged to permit conformal deposition on fibers throughout the matrix. It can be proved by Fig. 8.14b, where a concave shaped 1D structure was obtained, not conformal cylindrical structure.

Atomic Layer Deposition. ALD is considered one deposition method with great potential for producing very thin, conformal layers with control of the thickness at the atomic level. ALD has been combined with electrospinning to produce long nano-/microshells or tubes with smooth outer surfaces and uniform walls of controlled thickness. During ALD process, a layer is grown on a substrate by exposing its surface to alternate gaseous species (typically referred to as precursors). In contrast to CVD, the precursors are never present simultaneously in the reactor, but they are inserted as a series of sequential, non-overlapping pulses. In each of these pulses the precursor molecules react with the surface in a self-limiting way, so that the reaction terminates once all the reactive sites on the surface are consumed. Consequently, by varying the number of cycles it is possible to grow materials uniformly with high controllability. Peng and co-workers demonstrated the first ALD method combining with electrospinning to fabricate functional materials. In their work, $Al(CH_3)_3$ (TMA) and H_2O, were alternately introduced into the reactor as ambient-temperature precursor vapors to generate Al_2O_3 layer. By controlling the deposition cycles, a uniform Al_2O_3 shell with precisely controlled thickness was obtained on PVA electrospun nanofibers. After the removal of PVA by calcination, Al_2O_3 long microtubes were produced (Fig. 8.14c, d, and e) (Peng et al. 2007). Up to now, ALD has be extended to construct other functional coating on electrospun fibers (Kayaci et al. 2014, Choi et al. 2009b, Kayaci et al. 2012).

Hydrothermal or Solvothermal Reaction. Hydrothermal or solvothermal reaction has been established as an effective solution strategy for preparing functional composites with various morphologies. Combined with electrospinning technology, it can functionalize the electrospun fibers with versatile hierarchical structures for different applications. Nanostructures based on metal, semiconductors, etc. can effectively grow on the outside surface of both inorganic and polymeric fibers. Notably, it is much easier to bind the inorganic hierarchical structures onto the oxide nanofibers, which should be due to the fact that oxide nanofiber itself can provide active seed sites for guiding growth of inorganic hierarchical structures. For example, SnO_2 hierarchical nanostructures with versatile nano-topologies, hexagonal SnS_2, isomeric TiO_2 nanorods, and so forth have been incorporated onto electrospun TiO_2 nanofibers by hydrothermal reaction after electrospinning. These products exhibited potential application in enhanced photocatalysis, nano-electronics, sensors, and so on (Zhang et al. 2013, Dong et al. 2013). If the substrate fibers are composed of polymers, active seed are generally required to load on fibrous matrix at first to obtain hierarchical structures. Chang (Chang 2011) reported a facile fabrication of firecracker-shaped ZnO/PI hybrid nanofibers composed of ZnO nanorods grown on electrospun PI nanofibers by combining electrospinning technique with hydrothermal process. First, PI nanofibers were coated with ZnO nanoparticles as seeds by a dip-coating technique. Subsequently, the PI nanofibers covered with ZnO seeds were immersed into an aqueous solution of zinc nitrate hexahydrate and hexamethylenetetramine to conduct the hydrothermal process. The ZnO seed can guide the anisotropic growth of ZnO nanorods, thus promoting the formation of the hierarchical nanostructure. This similar route was also demonstrated in other report (Kayaci et al. 2014). If there were no pre-loaded seeds, nanoparticles with low roughness, instead of hierarchical structure, would be generally modified on the fiber surface via hydrothermal reaction (Pant et al. 2013). Functional components can be modified on carbon electrospun fibers by hydrothermal process as well, and the principle was similar to that on polymeric fibers. Zhang et al. successfully assembled Ag nanoparticles on electrospun carbon nanofibers (CNFs) by two steps consisting of the preparation of the CNFs by electrospinning and the hydrothermal growth of the AgNPs. They also modified ZnO nanoparticles on CNFs to construct a heterojunction for photocatalysis (Zhang et al. 2011, Mu et al. 2011).

Layer-by-Layer Self-assembly. The LBL assembly process involves the sequential adsorption of oppositely charged materials to construct conformal ultrathin coatings on proper substrates with controlled film thickness and chemical properties. The first work on LBL assembly combined with electrospinning should be reported by Li et al., where electrospun hollow fibers decorated with colloidal gold particles were reported (Li et al. 2005). Since then, LBL has been widely applied for construction of functional fibrous composites for various application. Ma et al. (Ma et al. 2007) first introduced the LBL technique to construct superhydrophobic surfaces on the electrospun fibers. They alternately coated electrospun nylon mats with positively charged poly(allylamine hydrochloride) and negatively charged 50 nm silica NPs by LBL self-assembly technique. The final products exhibited a big roughness and displayed an average large contact angle of 168°, after the superficial silica NPs reacted with (tridecafluoro-1,1,2,2-tetrahydrooctyl)-1-trichlorosilane. Besides superhydrophobic materials, Lee et al. reported a high-performance photocatalyst constructed by electrospinning and subsequent LBL assembly (Lee et al. 2009b). Electrospun fibers were first prepared and modified with negative charge by plasma treatment. Then positively charged polyhedral oligosilsesquioxane (POSS) molecules were pre-adsorbed on negatively charged electrospun fibers, then to capture negatively charged colloidal TiO_2 nanoparticles onto the fibrous surface and hence form an ultrathin conformal coating. After multiple LBL cycles, a photocatalyst with synergetic effect between POSS and TiO_2 were obtained (Fig. 8.15a, b, and c). Recently, graphene was successfully assembled onto electrospun fibers to construct soft graphene nanofibers by LBL technique (Feng et al. 2015). Positively charged poly(vinyl chloride) (PVC) nanofibers was prepared at first by electrospinning combining with subsequent NH_3 plasma treatment. Afterwards, the self-assembly was driven by the electrostatic interaction between positively charged PVC fibers and negatively charged GO. The GO nanosheets entered the interior of the electrospun nanofibers film and adsorbed onto the entire surface of the nanofibers, forming a shell with typical nano-winkle-like topography (Fig. 8.15d). After reduction of GO, the graphene nanofibers served as a soft cell modulation scaffold for cellular electrical stimulation, and lead to an unprecedented accelerated growth and development of primary motor neurons. The conformable 1D coating surfaces constructed by LBL assembly combined with electrospinning would have important applications as antifouling, self-cleaning, water resistant coatings and water purification, biomedical application, etc.

Electroless-plating. Electroless-plating technique is surface functionalization method that is often effective for metal coating. It is an auto-catalytic chemical deposition technique that can grow a layer of metal coating on proper solid substrates. Four important factors are generally involved in an electroless-plating process, i.e., pre-load catalyst seeds, a reducing agent, a metal precursor solution and a complexing

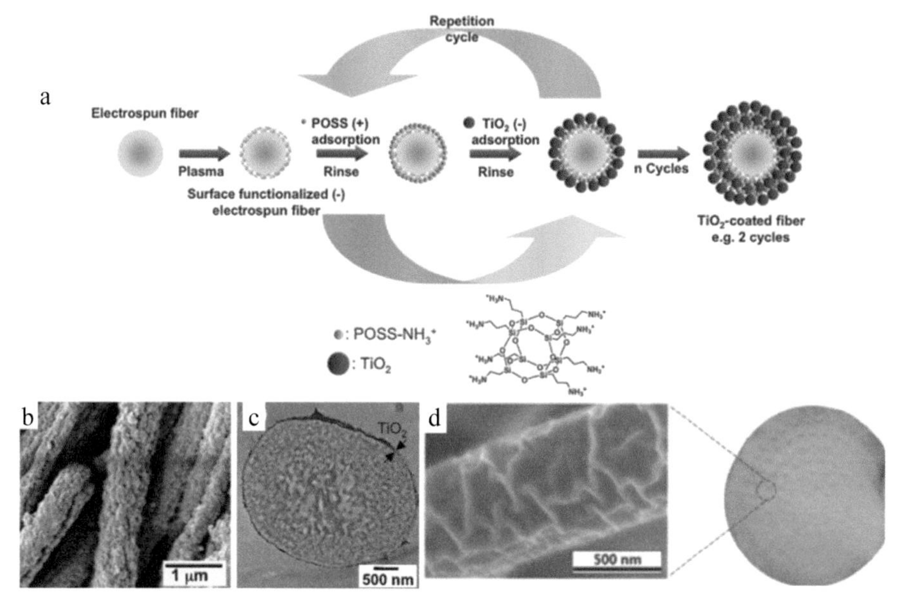

Fig. 8.15: (a) Schematic illustration of the preparation of TiO_2-coated electrospun polymer fibers using LBL deposition method. (b and c) SEM image and cross-view of TiO_2 coated electrospun fiber (Lee et al. 2009). (d) Conformal GO layer modified on PVC fibers by LBL (Feng et al. 2015).

agent that can control the reaction rate. Wang et al. first combined electroless-plating with electrospinning to fabricate Ag and Cu shells coated on PAN electrospun fibers (Song et al. 2008). Additionally, they transferred the Ag nano-shell into Ag/AgCl composite nano-shell by solution chlorination process, then obtained a high-performance visible-light photocatalyst. Taking Ag coating fabricated by electroless-plating as an example, electrospun fibers incorporated with Ag nanoparticles were prepared at first. Then the sample was immersed into glucose aqueous solution, which is the reducing agent. Subsequently, $Ag(NH_3)^{2+}$ solution fabricated by the coordination between Ag precursor and NH_4OH solution was added drop-wise into the vigorously stirred glucose solution until Ag nanoshells reached desired thickness (Lei et al. 2011). Cu deposition can also be achieved by the catalytic function of Ag nano-seeds in Cu precursor solution with suitable reducing agents and complexants. Hsu et al. (Hsu et al. 2014) has employed this electroless-plating combined with electrospinning to synthesize interconnected, ultralong, high-performance metal nanofibers for transparent electrode. The as-prepared products possessed fused junction and ultralong 1D nanostructure, which are quite meaningful for high-performance transparent electrodes. In principle, electroless-plating technique have the advantages of low-cost, room temperature operation, and selective coating along the catalytic substrates.

There are also some other functionalization methods reported in literature, e.g., photo-reduction of metal nanoparticles on TiO_2 fiber surfaces, sol-gel coating combined with template method to prepare TiO_2 nanotubes, and so forth. The reader can refer to corresponding publications (Li et al. 2006b, Caruso et al. 2001).

Acknowledgements

The authors gratefully acknowledge the National Natural Science Foundation of China (51573034) and Scientific Research Foundation for Returned Scholars, Heilongjiang of China (LC2017023).

References

Abidian, M. R. and Martin, D. C. 2009. Multifunctional nanobiomaterials for neural interfaces. Advanced Functional Materials 19: 573–585.

Ayutsede, J., Gandhi, M., Sukigara, S., Ye, H., Hsu, C. M., Gogotsi, Y. et al. 2006. Carbon nanotube reinforced *Bombyx mori* silk nanofibers by the electrospinning process. Biomacromolecules 7(1): 208–214.

Babu, V. J., Kumar, M. K., Nair, A. S., Tan, L. K., Allakhverdiev, S. I. and Ramakrishna, S. 2012. Visible light photocatalytic water splitting for hydrogen production from N-TiO_2 rice grain shaped electrospun nanostructures. International Journal of Hydrogen Energy 37(10): 8897–8904.

Bai, J., Li, Y., Yang, S., Du, J., Wang, S., Zhang, C. et al. 2007. Synthesis of AgCl/PAN composite nanofibres using an electrospinning method. Nanotechnology 18(30): 410–415.

Barakat, N. A. M., Kim, B. and Kim, H. Y. 2009. Production of smooth and pure nickel metal nanofibers by the electrospinning technique: nanofibers possess splendid magnetic properties. Journal of Physical Chemistry C 113(2): 531–536.

Bognitzki, M., Hou, H., Ishaque, M., Frese, T., Hellwig, M., Schwarte, C. et al. 2000. Polymer, metal, and hybrid nano- and mesotubes by coating degradable polymer template fibers (TUFT process). Advanced Materials 12: 637–640.

Bognitzki, M., Becker, M., Graeser, M., Massa, W., Wendorff, J. H., Schaper, A. et al. 2006. Preparation of sub-micrometer copper fibers via electrospinning. Advanced Materials 18(18): 2384–2386.

Caruso, R. A., Schattka, J. H. and Greiner, A. 2001. Titanium dioxide tubes from sol-gel coating of electrospun polymer fibers. Advanced Materials 13: 1577–1579.

Chang, S., Kim, Y. T. and Baek, C. K. 2003. Electrospinning of poly(vinylidene fluoride)/dimethylformamide solutions with carbon nanotubes. Journal of Polymer Science Part B: Polymer Physics 41(13): 1572–1577.

Chang, Z. 2011. "Firecracker-shaped" ZnO/polyimide hybrid nanofibers via electrospinning and hydrothermal process. Chemical Communications 47(15): 4427–4429.

Chen, C., Li, G., Lu, Y., Zhu, J., Jiang, M., Hu, Y. et al. 2016. Chemical vapor deposited MoS_2/electrospun carbon nanofiber composite as anode material for high-performance sodium-ion batteries. Electrochimica Acta 222: 1751–1760.

Chen, C., Li, G., Zhu, J., Lu, Y., Jiang, M., Hu, Y. et al. 2017. *In situ* formation of tin-antimony sulfide in nitrogen-sulfur Co-doped carbon nanofibers as high performance anode materials for sodium-ion batteries. Carbon 120: 380–391.

Chen, Z. G. and Wang, P. 2010. Electrospun collagen-chitosan nanofiber: a biomimetic extracellular matrix for endothelial cell and smooth muscle cell.Acta Biomaterialia 6(2): 372–382.

Choi, S. H., Ankonina, G., Youn, D. Y., Oh, S. G., Hong, J. M., Rothschild, A. et al. 2009a. Hollow ZnO nanofibers fabricated using electrospun polymer templates and their electronic transport properties. ACS Nano 3(9): 2623–2631.

Choi, S. W., Park, J. Y. and Kim, S. S. 2009b. Synthesis of SnO_2-ZnO core-shell nanofibers via a novel two-step process and their gas sensing properties. Nanotechnology 20(46): 465603.

Chung, G. S., Jo, S. M. and Kim, B. C. 2005. Properties of carbon nanofibers prepared from electrospun polyimide. Journal of Applied Polymer Science 97(1): 165–170.

Daels, N., Vrieze, S. D., Sampers, I., Decostere, B., Westbroek, P., Dumoulin, A. et al. 2011. Potential of a functionalised nanofibre microfiltration membrane as an antibacterial water filter. Desalination 275(1): 285–290.

Dong, B., Lu, X., Li, Z., Zhang, H., Wang, Z., Wang, C. et al. 2012. Effect of the thickness of PAN/Au modified electrode on biosensor sensitivity. Functional Materials Letters 5(03): 1250035.

Dong, R., Tian, B., Zeng, C., Li, T., Wang, T. and Zhang, J. 2013. Ecofriendly synthesis and photocatalytic activity of uniform cubic Ag@AgCl plasmonic photocatalyst. The Journal of Physical Chemistry C 117(1): 213–220.

Dror, Y., Salalha, W., Khalfin, R. L., Cohen, Y., Yarin, A. L. and Zussman, E. 2003. Carbon nanotubes embedded in oriented polymer nanofibers by electrospinning. Langmuir 19: 7012–7020.

Fan, Z., Yan, J., Wei, T., Zhi, L., Ning, G., Li, T. et al. 2011. Asymmetric supercapacitors based on graphene/MnO_2 and activated carbon nanofiber electrodes with high power and energy density. Advanced Functional Materials 21(12): 2366–2375.

Feng, Z. Q., Wang, T., Zhao, B., Li, J. and Jin, L. 2015. Soft graphene nanofibers designed for the acceleration of nerve growth and development. Advanced Materials 27(41): 6462–6468.

Ge, J., Qu, Y., Cao, L., Wang, F., Dou, L., Yu, J. et al. 2016. Polybenzoxazine-based highly porous carbon nanofibrous membranes hybridized by tin oxide nanoclusters: durable mechanical elasticity and capacitive performance. Journal of Materials Chemistry A 4(20): 7795–7804.

Gensheimer, M., Becker, M., Bris-Heep, A., Wendorff, J. H., Thauer, R. K. and Greiner, A. 2010. Novel biohybrid materials by electrospinning: nanofibers of poly(ethylene oxide) and living bacteria. Advanced Materials 19(18): 2480–2482.

Guo, F., Servi, A., Liu, A., Gleason, K. K. and Rutledge, G. C. 2015. Desalination by membrane distillation using electrospun polyamide fiber membranes with surface fluorination by chemical vapor deposition. ACS Applied Materials & Interfaces 7(15): 8225–8232.

Gupta, P., Elkins, C., Long, T. E. and Wilkes, G. L. 2005. Electrospinning of linear homopolymers of poly(methyl methacrylate): exploring relationships between fiber formation, viscosity, molecular weight and concentration in a good solvent. Polymer 46(13): 4799–4810.

Han, Y. Y., Wang, W., Song, M. X., Zhen-Yu, L. I., Wang, C. and Sun, J. H. 2012. Ag/AgCl composite nanoparticles/polyacrylonitrile nanofiber films for visible-light photocatalysis. Chemical Journal of Chinese Universities 33(3): 604–607.

Haoqing Hou, Jason J. Ge, Jun Zeng, Qing Li, Darrell H. Reneker, Andreas Greiner et al. 2005. Electrospun polyacrylonitrile nanofibers containing a high concentration of well-aligned multiwall carbon nanotubes. Chemistry of Materials 17(5): 967–973.

Hassen, D., Shenashen, M. A., El-Safty, S. A., Selim, M. M., Isago, H., Elmarakbi, A. et al. 2016. Nitrogen-doped carbon-embedded TiO_2 nanofibers as promising oxygen reduction reaction electrocatalysts. Journal of Power Sources 330: 292–303.

He, D., Hu, B., Yao, Q. F., Wang, K. and Yu, S. H. 2009. Large-scale synthesis of flexible free-standing SERS substrates with high sensitivity: electrospun PVA nanofibers embedded with controlled alignment of silver nanoparticles. ACS Nano 3(12): 3993–4002.

He, J., Wang, W., Sun, F., Shi, W., Qi, D., Wang, K. et al. 2015. Highly efficient phosphate scavenger based on well-dispersed $La(OH)_3$ nanorods in polyacrylonitrile nanofibers for nutrient-starvation antibacteria. ACS Nano 9(9): 9292–9302.

Hong, S. -M., Kim, S. H., Jeong, B. G., Jo, S. M. and Lee, K. B. 2014. Development of porous carbon nanofibers from electrospun polyvinylidene fluoride for CO_2 capture. RSC Advances 4(103): 58956–58963.

Hou, H. and Reneker D. H. 2004. Carbon nanotubes on carbon nanofibers: a novel structure based on electrospun polymer nanofibers. Advanced Materials 16(1): 69–73.

Hsu, P. C., Kong, D., Wang, S., Wang, H., Welch, A. J., Wu, H. et al. 2014. Electrolessly deposited electrospun metal nanowire transparent electrodes. Journal of the American Chemical Society 136(30): 10593–10596.

Inagaki, M., Yang, Y. and Kang, F. 2012. Carbon nanofibers prepared via electrospinning. Advanced Materials24(19): 2547–2566.

Jason, J. Ge, Haoqing Hou, Qing Li, Matthew J. Graham, Andreas Greiner, Darrell H. Reneker et al. 2004. Assembly of well-aligned multiwalled carbon nanotubes in confined polyacrylonitrile environments: electrospun composite nanofiber sheets. Journal of the American Chemical Society 126(48): 15754–15761.

Jiang, T., Wang, Z., Li, Z., Wang, W., Xu, X., Liu, X. et al. 2013. Synergic effect within n-type inorganic–p-type organic nano-hybrids in gas sensors. Journal of Materials Chemistry C 1(17): 3017–3025.

Jin, W. -J., Lee, H. K., Jeong, E. H., Park, W. H. and Youk, J. H. 2005. Preparation of polymer nanofibers containing silver nanoparticles by using poly(N-vinylpyrrolidone). Macromolecular Rapid Communications 26(24): 1903–1907.

Kayaci, F., Ozgitakgun, C., Donmez, I., Biyikli, N. and Uyar, T. 2012. Polymer-inorganic core-shell nanofibers by electrospinning and atomic layer deposition: flexible nylon-ZnO core-shell nanofiber mats and their photocatalytic activity. ACS Applied Materials & Interfaces 4(11): 6185–6194.

Kayaci, F., Vempati, S., Ozgit-Akgun, C., Biyikli, N. and Uyar, T. 2014. Enhanced photocatalytic activity of homoassembled ZnO nanostructures on electrospun polymeric nanofibers: A combination of atomic layer deposition and hydrothermal growth. Applied Catalysis B: Environmental 156-157(9): 173–183.

Kim, C., Park, S. -H., Lee, W. -J. and Yang, K. -S. 2004. Characteristics of supercapacitor electrodes of PBI-based carbon nanofiber web prepared by electrospinning. Electrochimica Acta 50(2-3): 877–881.

Ko, F., Gogotsi, Y., Ali, A., Naguib, N., Ye, H., Yang, G. et al. 2003. Electrospinning of continuous carbon nanotube-filled nanofiber yarns. Advanced Materials 15(14): 1161–1162.

Ko, T. H., Liau, S. C. and Lin, M. F. 1992. Preparation of graphite fibres from a modified PAN precursor. Journal of Materials Science 27(22): 6071–6078.

Ko, T. H. 2010. Influence of continuous stabilization on the physical properties and microstructure of PAN-based carbon fibers. Journal of Applied Polymer Science 42(7): 1949–1957.

Kwon, O. S., Park, E., Kweon, O. Y., Park, S. J. and Jang, J. 2010. Novel flexible chemical gas sensor based on poly(3,4-ethylenedioxythiophene) nanotube membrane. Talanta 82(4): 1338–1343.

Lee, S., Moon, G. D. and Jeong, U. 2009a. Continuous production of uniform poly(3-hexylthiophene) (P3HT) nanofibers by electrospinning and their electrical properties. Journal of Materials Chemistry 19(6): 743–748.

Lee, J. A., Krogman, K. C., Ma, M., Hill, R. M., Hammond, P. T. and Rutledge, G. C. 2009b. Highly reactive multilayer-assembled TiO_2 coating on electrospun polymer nanofibers. Advanced Materials 21(12): 1252–1256.

Lee, S. -W. and Belcher, A. M. 2009. Virus-based fabrication of micro- and nanofibers using electrospinning. Nano Letters 4(3): 387–390.

Lei, J., Wang, W., Song, M., Dong, B., Li, Z., Wang, C. et al. 2011. Ag/AgCl coated polyacrylonitrile nanofiber membranes: synthesis and photocatalytic properties. Reactive and Functional Polymers 71(11): 1071–1076.

Li, D. and Xia, Y. 2003. Fabrication of titania nanofibers by electrospinning. Nano Letters 3(4): 555–560.

Li, D., Mccann, J. T. and Xia, Y. 2005. Use of electrospinning to directly fabricate hollow nanofibers with functionalized inner and outer surfaces. Small 1(1): 83–86.

Li, G., Zhu, Z., Qi, B., Liu, G., Wu, P., Zeng, G. et al. 2016b. Rapid capture of Ponceau S via a hierarchical organic–inorganic hybrid nanofibrous membrane. Journal of Materials Chemistry A 4(15): 5423–5427.

Li, H., Zhang, W., Huang, S. and Pan, W. 2012. Enhanced visible-light-driven photocatalysis of surface nitrided electrospun TiO_2 nanofibers. Nanoscale 4(3): 801–806.

Li, M., Zhang, J., Zhang, H., Liu, Y., Wang, C., Xu, X. et al. 2007. Electrospinning: a facile method to disperse fluorescent quantum dots in nanofibers without Förster resonance energy transfer. Advanced Functional Materials 17(17): 3650–3656.

Li, X., Zhang, C., Zhao, R., Lu, X., Xu, X., Jia, X. et al. 2013. Efficient adsorption of gold ions from aqueous systems with thioamide-group chelating nanofiber membranes. Chemical Engineering Journal 229(4): 420–428.

Li, Y., Huang, Z., Huang, Z., Zhang, H., Bilotti, E. and Peijs, T. 2016a. Enhanced thermal and electrical properties of polystyrene-graphene nanofibers via electrospinning. Journal of Nanomaterials 2016(4): 18.

Li, Z., Huang, H. and Wang, C. 2006a. Electrostatic forces induce poly(vinyl alcohol)-protected copper nanoparticles to form copper/poly(vinyl alcohol) nanocables via electrospinning. Macromolecular Rapid Communications 27(2): 152–155.

Li, Z., Huang, H., Shang, T., Yang, F., Zheng, W., Wang, C. et al. 2006b. Facile synthesis of single-crystal and controllable sized silver nanoparticles on the surfaces of polyacrylonitrile nanofibres. Nanotechnology 17(3): 917.

Li, Z., Zhang, H., Zheng, W., Wang, W., Huang, H., Wang, C. et al. 2008. Highly sensitive and stable humidity nanosensors based on LiCl doped TiO_2 electrospun nanofibers. Journal of the American Chemical Society 130(15): 5036–5037.

Lin, D., Wu, H., Zhang, R. and Pan, W. 2009. Enhanced photocatalysis of electrospun Ag–ZnO heterostructured nanofibers. Chemical of Materials 21(15): 3479–3484.

Liu, J. and Zhang, W. 2005. Structural changes during the thermal stabilization of modified and original polyacrylonitrile precursors. Journal of Applied Polymer Science 97(5): 2047–2053.

Liu, Y., Luo, J., Helleu, C., Behr, M., Ba, H., Romero, T. et al. 2016. Hierarchical porous carbon fibers/ carbon nanofibers monolith from electrospinning/CVD processes as a high effective surface area support platform. Journal of Materials Chemistry A 5(5): 2151–2162.

Lu, X., Li, L., Zhang, W. and Wang, C. 2005a. Preparation and characterization of Ag_2S nanoparticles embedded in polymer fibre matrices by electrospinning. Nanotechnology 16(10): 2233–2237.

Lu, X., Zhao, Y., Wang, C. and Wei, Y. 2005b. Fabrication of CdS nanorods in PVP fiber matrices by electrospinning. Macromolecular Rapid Communications 26(16): 1325–1329.

Lu, X., Mao, H. and Zhang, W. 2007. Surfactant directed synthesis of polypyrrole/TiO_2 coaxial nanocables with a controllable sheath size. Nanotechnology 18(2): 025604.

Lu, X., Wang, C. and Wei, Y. 2009. One-dimensional composite nanomaterials: synthesis by electrospinning and their applications. Small 5(21): 2349–2370.

Lu, X., Zhao, Y. and Wang, C. 2010. Fabrication of PbS nanoparticles in polymer-fiber matrices by electrospinning. Advanced Materials 17(20): 2485–2488.

Lu, X., Bian, X., Nie, G., Zhang, C., Wang, C. and Wei, Y. 2012. Encapsulating conducting polypyrrole into electrospun TiO_2 nanofibers: a new kind of nanoreactor for *in situ* loading Pd nanocatalysts towards p-nitrophenol hydrogenation. Journal of Materials Chemistry 22(25): 12723–12730.

Ma, M., Gupta, M., Li, Z., Zhai, L., Gleason, K.K., Cohen, R. E. et al. 2007. Decorated electrospun fibers exhibiting superhydrophobicity. Advanced Materials 19(2): 255–259.

Mahanta, N. and Valiyaveettil, S. 2012. *In situ* preparation of silver nanoparticles on biocompatible methacrylated poly(vinyl alcohol) and cellulose based polymeric nanofibers. RSC Advances 2: 11389–11396.

Mani, N. and Dupuy BSonenshein, A. L. 2002. Preparation and characterization of nanoscaled poly(vinyl alcohol) fibers via electrospinning. Fiber and Polymers 3(2): 73–79.

Mitschang, F., Schmalz, H., Agarwal, S. and Greiner, A. 2014. Tea-bag-like polymer nanoreactors filled with gold nanoparticles. Angewandte Chemie International Edition 53(19): 4972–4975.

Mohamed, I. M., Dao, V. -D., Yasin, A. S., Mousa, H. M., Yassin, M. A., Khan, M. Y. et al. 2017. Physicochemical and photo-electrochemical characterization of novel N-doped nanocomposite ZrO_2/TiO_2 photoanode towards technology of dye-sensitized solar cells. Materials Characterization 127: 357–364.

Mu, J., Shao, C., Guo, Z., Zhang, Z., Zhang, M., Zhang, P. et al. 2011. High photocatalytic activity of ZnO-carbon nanofiber heteroarchitectures. ACS Applied Materials & Interfaces 3(2): 590–596.

Nagamine, S., Matsumoto, T., Hikima, Y. and Ohshima, M. 2016. Fabrication of porous carbon nanofibers by phosphate-assisted carbonization of electrospun poly(vinyl alcohol) nanofibers. Materials Research Bulletin 79: 8–13.

Neghlani, P. K., Rafizadeh, M. and Taromi, F. A. 2011. Preparation of aminated-polyacrylonitrile nanofiber membranes for the adsorption of metal ions: Comparison with microfibers. Journal of Hazardous Materials 186(1): 182–189.

Nie, G., Li, Z., Lu, X., Lei, J., Zhang, C. and Wang, C. 2013. Fabrication of polyacrylonitrile/CuS composite nanofibers and their recycled application in catalysis for dye degradation. Apply Surface Science 284(11): 595–600.

Ostermann, R., Li, D., Yin, Y., Mccann, J. T. and Xia, Y. 2006. V_2O_5 nanorods on TiO_2 nanofibers: a new class of hierarchical nanostructures enabled by electrospinning and calcination. Nano Letters 6(6): 1297–1302.

Pant, H. R., Pant, B., Pokharel, P., Han, J. K., Tijing, L. D., Chan, H. P. et al. 2013. Photocatalytic TiO_2–RGO/nylon-6 spider-wave-like nano-nets via electrospinning and hydrothermal treatment. Journal of Membrane Science 429(4): 225–234.

Park, M., Im, J., Shin, M., Min, Y., Park, J., Cho, H. et al. 2012. Highly stretchable electric circuits from a composite material of silver nanoparticles and elastomeric fibres. Nature Nanotechnology 7(12): 803–809.

Patel, A. C., Li, S., Jianmin Yuan, A. and Wei, Y. 2006. *In situ* encapsulation of horseradish peroxidase in electrospun porous silica fibers for potential biosensor applications. Nano Letters 6(6): 1042–1046.

Patel, A. C., Li, S., Wang, C., Zhang, W. and Wei, Y. 2007. Electrospinning of porous silica nanofibers containing silver nanoparticles for catalytic applications. Chemistry of Materials 19(6): 1231–1238.

Peng, Q., Sun, X. Y., Spagnola, J. C., Hyde, G. K., Spontak, R. J. and Parsons, G. N. 2007. Atomic layer deposition on electrospun polymer fibers as a direct route to Al_2O_3 microtubes with precise wall thickness control. Nano Lettes 7(3): 719–722.

Pinto, N. J., Johnson, A. T., Macdiarmid, A. G., Mueller, C. H., Theofylaktos, N., Robinson, D. C. et al. 2003. Electrospun polyaniline/polyethylene oxide nanofiber field effect transistor. Applied Physics Letters 83(20): 4244–4246.

Qiao, L., Sun, X., Yang, Z., Wang, X., Wang, Q. and He, D. 2013. Network structures of fullerene-like carbon core/nano-crystalline silicon shell nanofibers as anode material for lithium-ion batteries. Carbon 54(2): 29–35.

Qie, L., Chen, W. M., Wang, Z. H., Shao, Q. G., Li, X., Yuan, L. X. et al. 2012. Nitrogen-doped porous carbon nanofiber webs as anodes for lithium ion batteries with a superhigh capacity and rate capability. Advanced Materials 24(15): 2047–2050.

Saeed, K., Haider, S., Oh, T. J. and Park, S. Y. 2008. Preparation of amidoxime-modified polyacrylonitrile (PAN-oxime) nanofibers and their applications to metal ions adsorption. Journal of Membrane Science 322(2): 400–405.

Saha, J. and De, G. 2013. Highly ordered cubic mesoporous electrospun SiO_2 nanofibers. Chemical Communications 49(56): 6322–6324.

Sanders, E. H., Kloefkorn, R., Bowlin, G. L., Simpson, D. G. and Wnek, G. E. 2003. Two-phase electrospinning from a single electrified jet: microencapsulation of aqueous reservoirs in poly (ethylene-co-vinyl acetate) fibers. Macromolecules 36: 3803–3805.

Saquing, C. D., Manasco, J. L. and Khan, S. A. 2009. Electrospun nanoparticle-nanofiber composites via a one-step synthesis. Small 5(8): 944–951.

Sawai, D., Fujii, Y. and Kanamoto, T. 2006. Development of oriented morphology and tensile properties upon superdawing of solution-spun fibers of ultra-high molecular weight poly(acrylonitrile). Polymer 47(12): 4445–4453.

Sen, R., Zhao, B., Perea, D., Itkis, M. E., Hu, H., Love, J. et al. 2004. Preparation of single-walled carbon nanotube reinforced polystyrene and polyurethane nanofibers and membranes by electrospinning. Nano Letters 4(3): 459–464.

Shao, C. L., Kim, H., Gong, J. and Lee, D. 2002. A novel method for making silica nanofibres by using electrospun fibres of polyvinylalcohol/silica composite as precursor. Nanotechnology 13(5): 635–637.

Si, Y., Ren, T., Li, Y., Ding, B. and Yu, J. 2012. Fabrication of magnetic polybenzoxazine-based carbon nanofibers with Fe_3O_4 inclusions with a hierarchical porous structure for water treatment. Carbon 50(14): 5176–5185.

Song, X., Lei, J., Li, Z., Li, S. and Wang, C. 2008. Synthesis of polyacrylonitrile/Ag core–shell nanowire by an improved electroless plating method. Materials Letters 62(17): 2681–2684.

Song, X., Wang, Z., Liu, Y., Wang, C. and Li, L. 2009. A highly sensitive ethanol sensor based on mesoporous ZnO–SnO_2 nanofibers. Nanotechnology 20(7): 75501.

Streckova, M., Mudra, E., Orinakova, R., Markusova-Buckova, L., Sebek, M., Kovalcikova, A. et al. 2016. Nickel and nickel phosphide nanoparticles embedded in electrospun carbon fibers as favourable electrocatalysts for hydrogen evolution. Chemical Engineering Journal 303: 167–181.

Sung, J. H., Kim, H. S., Jin, H. J., And, H. J. C. and Chin, I. J. 2004. Nanofibrous membranes prepared by multiwalled carbon nanotube/poly(methyl methacrylate) composites. Macromolecules 37(26): 9899–9902.

Tai, M. H., Gao, P., Tan, B. Y. L., Sun, D. D. and Leckie, J. O. 2014. Highly efficient and flexible electrospun carbon–silica nanofibrous membrane for ultrafast gravity-driven oil–water separation. ACS Applied Materials & Interfaces 6(12): 9393–9401.

Tavanai, H., Jalili, R. and Morshed, M. 2009. Effects of fiber diameter and CO_2 activation temperature on the pore characteristics of polyacrylonitrile based activated carbon nanofibers. Surface and Interface Analysis 41(10): 814–819.

Teng, M., Qiao, J., Li, F. and Bera, P. K. 2012. Electrospun mesoporous carbon nanofibers produced from phenolic resin and their use in the adsorption of large dye molecules. Carbon 50(8): 2877–2886.

Toprakci, O., Toprakci, H. A., Ji, L., Xu, G., Lin, Z. and Zhang, X. 2012. Carbon nanotube-loaded electrospun $LiFePO_4$/carbon composite nanofibers as stable and binder-free cathodes for rechargeable lithium-ion batteries. ACS Applied Materials & Interfaces 4(3): 1273–1280.

Wang, G., Ji, Y., Huang, X., Yang, X., Gouma, P. I. and Dudley, M. 2011a. Fabrication and characterization of polycrystalline WO_3 nanofibers and their application for ammonia sensing. Journal of Physical Chemistry B 110(47): 23777–23782.

Wang, L., Hu, J., Zhang, H. and Zhang, T. 2011d. Au-impregnated polyacrylonitrile (PAN)/polythiophene (PTH) core-shell nanofibers with high-performance semiconducting properties. Chemical Communications 47(24): 6837–6839.

Wang, Q., Wang, W., Yan, B., Shi, W., Cui, F. and Wang, C. 2017. Well-dispersed Pd-Cu bimetals in TiO_2 nanofiber matrix with enhanced activity and selectivity for nitrate catalytic reduction. Chemical Engineering Journal 326: 182–191.

Wang, S., Li, Y., Wang, Y., Yang, Q. and Wei, Y. 2007. Introducing CTAB into CdTe/PVP nanofibers enhances the photoluminescence intensity of CdTe nanoparticles. Materials Letters 61(25): 4674–4678.

Wang, W., Li, Z., Zheng, W., Yang, J., Zhang, H. and Wang, C. 2009. Electrospun palladium (IV)-doped copper oxide composite nanofibers for non-enzymatic glucose sensors. Electrochemistry Communications 11(9): 1811–1814.

Wang, W., Li, Z., Xu, X., Dong, B., Zhang, H., Wang, Z. et al. 2011b. Au-doped polyacrylonitrile-polyaniline core-shell electrospun nanofibers having high field-effect mobilities. Small 7(5): 597–600.

Wang, W., Lu, X., Li, Z., Lei, J., Liu, X., Wang, Z. et al. 2011c. One-dimensional polyelectrolyte/polymeric semiconductor core/shell structure: sulfonated poly(arylene ether ketone)/polyaniline nanofibers for organic field-effect transistors. Advanced Materials 23(43): 5109–5112.

Wang, W., Lu, X., Li, Z., Li, X., Xu, X., Lei, J. et al. 2012a. Weak-acceptor-polyacrylonitrile/donor-polyaniline core–shell nanofibers: A novel 1D polymeric heterojunction with high photoconductive properties. Organic Electronics 13(11): 2319–2325.

Wang, W., Li, Z., Jiang, T., Zhao, Z., Li, Y., Wang, Z. et al. 2012b. Sulfonated poly(ether ether ketone)/polypyrrole core-shell nanofibers: a novel polymeric adsorbent/conducting polymer nanostructures for ultrasensitive gas sensors. ACS Applied Materials & Interfaces 4(11): 6080–6084.

Wang, Y., Yang, Q., Shan, G., Wang, C., Du, J., Wang, S. et al. 2005. Preparation of silver nanoparticles dispersed in polyacrylonitrile nanofiber film spun by electrospinning. Materials Letters 59(24): 3046–3049.

Wang, Y., Li, Y., Yang, S., Zhang, G., An, D., Wang, C. et al. 2006. A convenient route to polyvinyl pyrrolidone/silver nanocomposite by electrospinning. Nanotechnology 17(13): 3304–3307.

Wang, Z., Li, Z., Sun, J., Zhang, H., Wang, W., Zheng, W. et al. 2010. Improved hydrogen monitoring properties based on p-NiO/n-SnO_2 heterojunction composite nanofibers. Journal of Physical Chemistry C 114(13): 6100–6105.

Wang, Z., Li, Z., Jiang, T., Xu, X. and Wang, C. 2013. Ultrasensitive hydrogen sensor based on Pd^0-loaded SnO_2 electrospun nanofibers at room temperature. ACS Applied Materials & Interfaces 5(6): 2013–2021.

Wu, G., Lu, C., Ling, L., Hao, A. and He, F. 2005a. Influence of tension on the oxidative stabilization process of polyacrylonitrile fibers. Journal of Applied Polymer Science 96(4): 1029–1034.

Wu, H., Sun, Y., Lin, D., Zhang, R., Zhang, C. and Pan, W. 2009. GaN nanofibers based on electrospinning: facile synthesis, controlled assembly, precise doping, and application as high performance UV photodetector. Advanced Materials 21(2): 227–231.

Wu, H., Hu, L., Rowell, M. W., Kong, D., Cha, J. J., Mcdonough, J. R. et al. 2010. Electrospun metal nanofiber webs as high-performance transparent electrode. Nano Letters 10(10): 4242–4248.

Wu, H., Kong, D., Ruan, Z., Hsu, P. C., Wang, S., Yu, Z. et al. 2013b. A transparent electrode based on a metal nanotrough network. Nature Nanotechnology 8(6): 421–425.

Wu, L., Yuan, X. and Sheng, J. 2005b. Immobilization of cellulase in nanofibrous PVA membranes by electrospinning. Journal of Membrane Science 250(1): 167–173.

Wu, Z. Y., Li, C., Liang, H. W., Chen, J. F. and Yu, S. H. 2013a. Ultralight, flexible, and fire-resistant carbon nanofiber aerogels from bacterial cellulose. Angewandte Chemie International Edition 52(10): 2925–2929.

Xie, J., Torres Galvis, H. M., Koeken, A. C., Kirilin, A., Dugulan, A. I., Ruitenbeek, M. et al. 2016. Size and promoter effects on stability of carbon-nanofiber-supported iron-based Fischer-Tropsch catalysts. ACS Catalysis 6(6): 4017–4024.

Xu, X., Yang, L., Xu, X., Wang, X., Chen, X., Liang, Q. et al. 2005. Ultrafine medicated fibers electrospun from W/O emulsions. Journal of Controlled Release 108(1): 33–42.

Yan, H., Mahanta, N. K., Wang, B., Wang, S., Abramson, A. R. and Cakmak, M. 2014a. Structural evolution in graphitization of nanofibers and mats from electrospun polyimide–mesophase pitch blends. Carbon 71: 303–318.

Yan, T., Lu, X., Sun, W., Nie, G., Liu, Y. and Wang, C. 2014b. Electrospun polyacrylonitrile nanofibers supported Ag/Pd nanoparticles for hydrogen generation from the hydrolysis of ammonia borane. Journal of Power Sources 261(3): 221–226.

Yang, A., Tao, X., Wang, R., Lee, S. and Surya, C. 2007. Room temperature gas sensing properties of SnO_2/multiwall-carbon-nanotube composite nanofibers. Applied Physics Letters 91: 133110.

Yang, Q. B., Li, D. M., Hong, Y. L., Li, Z. Y., Wang, C., Qiu, S. L. et al. 2003. Preparation and characterization of a pan nanofibre containing Ag nanoparticles via electrospinning. Synthetic Metals 137(1-3): 973–974.

Yao, C., Li, X. and Song, T. 2010. Electrospinning and crosslinking of zein nanofiber mats. Journal of Applied Polymer Science 103(1): 380–385.

Yao, L., Thomas W. Haas, Anthony Guiseppielie, Gary L. Bowlin, David. G. Simpson and Gary E. Wnek. 2003. Electrospinning and stabilization of fully hydrolyzed poly(vinyl alcohol) fibers. Chemistry of Materials 15(9): 1860–1864.

Yoon, S. -H., Lim, S., Song, Y., Ota, Y., Qiao, W., Tanaka, A. et al. 2004. KOH activation of carbon nanofibers. Carbon 42(8-9): 1723–1729.

Yoon, Y. I., Moon, H. S., Lyoo, W. S., Lee, T. S. and Park, W. H. 2008. Superhydrophobicity of PHBV fibrous surface with bead-on-string structure. Journal of Colloid and Interface Science 320(1): 91–95.

Yusof, N. and Ismail, A. F. 2012. Post spinning and pyrolysis processes of polyacrylonitrile (PAN)-based carbon fiber and activated carbon fiber: A review. Journal of Analytical and Applied Pyrolysis 93: 1–13.

Zeng, J., Xu, X., Chen, X., Liang, Q., Bian, X., Yang, L. et al. 2003. Biodegradable electrospun fibers for drug delivery. Journal of Controlled Release 92(3): 227–231.

Zeng, J., Yang, L., Liang, Q., Zhang, X., Guan, H., Xu, X. et al. 2005. Influence of the drug compatibility with polymer solution on the release kinetics of electrospun fiber formulation. Journal of Controlled Release 105(1): 43–51.

Zhang, C., Deng, L., Zhang, P., Ren, X., Li, Y. and He, T. 2017. Electrospun FeS nanorods with enhanced stability as counter electrodes for dye-sensitized solar cells. Electrochimica Acta 229: 229–238.

Zhang, C. L., Lv, K. P., Cong, H. P. and Yu, S. H. 2012a. Controlled assemblies of gold nanorods in PVA nanofiber matrix as flexible free-standing SERS substrates by electrospinning. Small 8(5): 647–653.

Zhang, J., Chen, H., Chen, Z., He, J., Shi, W., Liu, D. et al. 2016b. Microstructured macroporous adsorbent composed of polypyrrole modified natural corncob-core sponge for Cr(VI) removal. RSC Advances 6: 59292–59298.

Zhang, N., Deng, Y., Tai, Q., Cheng, B., Zhao, L., Shen, Q. et al. 2012b. Electrospun TiO_2 nanofiber-based cell capture assay for detecting circulating tumor cells from colorectal and gastric cancer patients. Advanced Materials 24(20): 2756–2760.

Zhang, P., Shao, C., Zhang, Z., Zhang, M., Mu, J., Guo, Z. et al. 2011. *In situ* assembly of well-dispersed Ag nanoparticles (AgNPs) on electrospun carbon nanofibers (CNFs) for catalytic reduction of 4-nitrophenol. Nanoscale 3(8): 3357–3363.

Zhang, X., Wang, W., Shi, W., He, J., Feng, H., Xu, Y. et al. 2016a. Carbon nanofiber matrix with embedded $LaCO_3OH$ synchronously captures phosphate and organic carbon to starve bacteria. Journal of Materials Chemistry A 4(33): 12799–12806.

Zhang, Y. Z., Venugopal, J., Huang, Z. M., Lim, C. T. and Ramakrishna, S. 2006. Crosslinking of the electrospun gelatin nanofibers. Polymer 47(8): 2911–2917.

Zhang, Z., Shao, C., Li, X., Sun, Y., Zhang, M., Mu, J. et al. 2013. Hierarchical assembly of ultrathin hexagonal SnS_2 nanosheets onto electrospun TiO_2 nanofibers: enhanced photocatalytic activity based on photoinduced interfacial charge transfer. Nanoscale 5(2): 606–618.

Zhao, G. X., Chen, B. J. and Qian, S. A. 1992. Kinetics of the -C≡N bond transformation into the conjugated C=N- bond in acrylonitrile copolymer using *in situ* Fourier transform infrared spectroscopy. Journal of Analytical and Applied Pyrolysis 23(1): 87–97.

Zhao, R., Li, X., Sun, B., Shen, M., Tan, X., Ding, Y. et al. 2015. Preparation of phosphorylated polyacrylonitrile-based nanofiber mat and its application for heavy metal ion removal. Chemical Engineering Journal 268: 290–299.

Zhao, R., Li, X., Sun, B., Li, Y., Li, Y., Yang, R. et al. 2016. Branched polyethylenimine grafted electrospun polyacrylonitrile fiber membrane: a novel and effective adsorbent for Cr(VI) remediation in wastewater. Journal of Materials Chemistry A 5: 1133–1144.

Zhou, Z., Lai, C., Zhang, L., Qian, Y., Hou, H., Reneker, D. H. et al. 2009. Development of carbon nanofibers from aligned electrospun polyacrylonitrile nanofiber bundles and characterization of their microstructural, electrical, and mechanical properties. Polymer 50(13): 2999–3006.

Zhu, H., Gu, L., Yu, D., Sun, Y., Wan, M., Zhang, M. et al. 2017b. The marriage and integration of nanostructures with different dimensions for synergistic electrocatalysis. Energy & Environmental Science 10(1): 321–330.

Zhu, Z., Wu, P., Liu, G., He, X., Qi, B., Zeng, G. et al. 2016. Ultrahigh adsorption capacity of anionic dyes with sharp selectivity through the cationic charged hybrid nanofibrous membranes. Chemical Engineering Journal 957–966.

Zhu, Z., Xu, Y., Qi, B., Zeng, G., Wu, P., Liu, G. et al. 2017a. Adsorption-intensified degradation of organic pollutants over bifunctional α-Fe@carbon nanofibres. Environmental Science Nano 4(2): 302–306.

Zhu, Z., Wu, P., Liu, G., He, X., Qi, B., Zeng, G. et al. 2017c. Ultrahigh adsorption capacity of anionic dyes with sharp selectivity through the cationic charged hybrid nanofibrous membranes. Chemical Engineering Journal 313: 957–966.

Chapter 9

Applications of Electrospun Nanofibers

Yun-Ze Long,[1,*] *Xiao-Xiong Wang,*[1] *Jun Zhang,*[1] *Xu Yan*[2] and *Hong-Di Zhang*[2]

Introduction

Electrospinning (E-spinning) is a versatile and effective method for the preparation of nanofibers with advantages of fiber continuity, large specific surface area, and wide material sources. With these advantages, electrospun (E-spun) nanofibers are widely used in different fields, which is the topic of this chapter. The filtration efficiency of the fibrous materials increases with the decrease in fiber diameter, which makes the E-spun nanofibers perform well in gas filtration, liquid filtration, and separation, and these applications are discussed first. Taking advantage of its high specific surface area, the interaction region of the E-spun nanofibers and the detectable objects or light is large, thereby enhancing the sensitivity of nanodevices, and these applications are summarized then. In the field of biomedicine, E-spun nanofibers can be used as cell scaffolds for drug release or for trauma repair, and these applications are discussed in the following section. Then the interesting applications of E-spun fibers in fuel cell electrode catalytic carriers, photocatalysts, biosensor catalysts, and gas sensor catalytic carriers are summarized. Higher surface area of the E-spun nanofibers, as electrodes, provide batteries better performance, and the good compatibility with fabrics also offers the relevant energy conversion system a good application prospect. Related applications are presented in the last section.

[1] Collaborative Innovation Center for Nanomaterials & Devices, College of Physics, Qingdao University, Qingdao 266071, China.
[2] Industrial Research Institute of Nonwovens & Technical Textiles, Qingdao University, Qingdao 266071, China.
* Corresponding author: yunze.long@163.com or yunze.long@qdu.edu.cn

Filtration, Separation, and Insulation

E-spun Nanofibers for Gas Filtration

Particle pollutants have been a concomitant of industry, which aroused increasing concerns for its health hazard, causing diseases like lung cancer, chronic obstructive pulmonary disease, or asthma (Mannucci et al. 2015). Fine particle (diameter < 2.5 μm, $PM_{2.5}$) pollutants are more dangerous than other particle pollutants, because they cannot be blocked by nasal cavity or expelled by trachea cilia. Particles with smaller diameters can even penetrate the pulmonary alveoli to enter the circulatory system and cause more serious health problems. Traditional filtration materials can deal with nanoparticles with larger diameters, but they are inefficient filtrating submicron particles, which can be well settled by E-spun nanofibrous membranes. Such membranes are mainly applicable in four forms. (1) Individual protection as a form of facial masks for protection of $PM_{2.5}$, which can integrate sensing ability to be smart masks as well. (2) Vehicle air filters and cabin air filters, which work better in filtrating smaller particles. (3) Indoor air filters or anti-smog screen window. (4) Industrial dust collector, which takes advantage of the self-cleaning feature of the membranes.

There are several filter mechanisms (Yang 2012). They are gravity, clogging, inertia impaction-deposition, interception, electrostatic deposition, and Brownian diffusion, arranged by filtration preference varying from larger to smaller particles. The gravity effect can be ignored if the particles are no larger than 0.5 μm, and the clogging effect can only happen when the particles are larger than the vacant diameter. The flow line of air is tortuous, caused by the arrangement of the fibers. The tiny particles move independently of the air flow line and impact or deposit onto the fibers due to the inertia. This mechanism becomes more dominant for larger particles and at higher gas flow velocities. On the contrary, when the particles follow the air flow line, interception effect will happen when the particle contacts the surface of fibers due to the van der Waals forces. Efficiency of such effect will increase with increasing particle size. It's a daily phenomenon that polymers charge easily. Such static electricity will attract particles carrying opposite charge or those which are electrically neutral. That's helpful for the fibers to attract the particles keeping their vent ability. When the particles are small enough, the random Brownian motion will also cause particles to collide with the fibers, leading to deposition. The practical filtration process may be a combination of such effects, leading to a complex dependence on particle size, gas velocity, vacant diameter, and fiber diameter. By controlling such parameters well, E-spun nanofibers can be optimized to fit different environments or applications. For example, in an enclosed space polluted by $PM_{2.5}$, most effects by flowing air will vanish, making it difficult to clean it. Zhang et al. have invented an *in situ* E-spinning apparatus (Zhang et al. 2017), which can generate movable nanofibers continuously in order to capture $PM_{2.5}$. The fibers capture $PM_{2.5}$ mainly taking advantage of electrostatic attraction, as they are charged in the E-spinning process. As shown in Fig. 9.1, a removal rate as high as 3.7 μg m^{-3} s^{-1} can be obtained using this technology with an efficiency of more than 95% $PM_{2.5}$ capture.

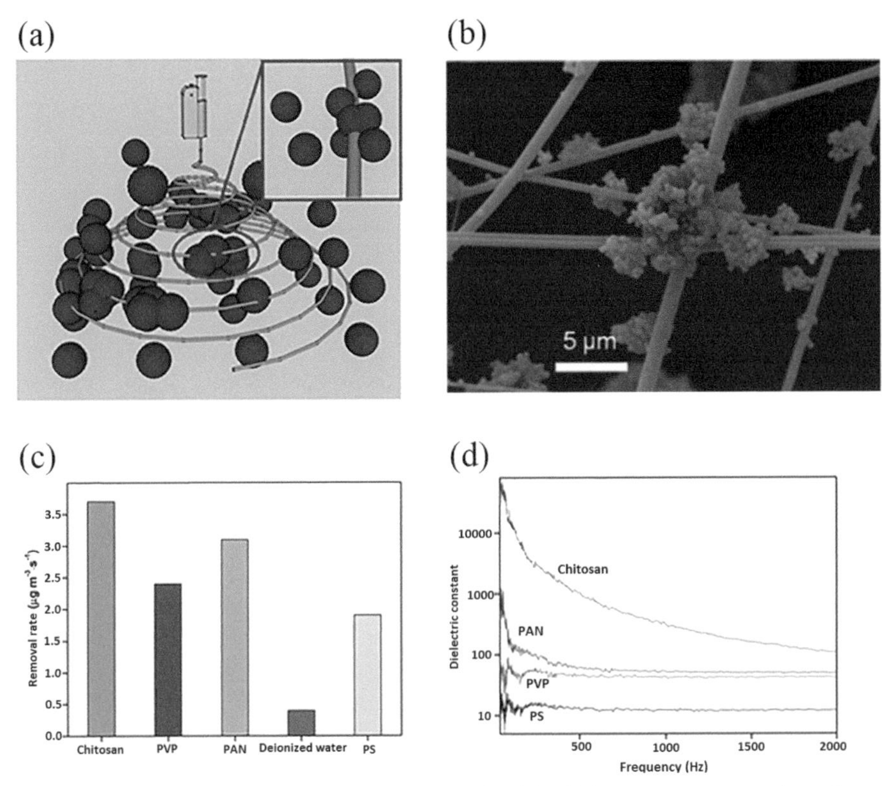

Fig. 9.1: (a) Schematic illustration of the mechanism for high-efficiency $PM_{2.5}$ capture using polymer nanofibers via *in situ* E-spinning. (b) SEM image of the E-spun nanofibers in the polluted air. (c) Removal efficiency of different polymers. (d) Dielectric constants of the polymers as a variation of frequency (Zhang et al. 2017).

Four parameters of the membranes will influence the filtration process (Leung et al. 2010). They are (1) fibrous diameter and diameter distribution, (2) surface area and pore size distribution, (3) fiber basis weight/thickness, and (4) packing density. Reduction of fibrous diameter will increase available surface area, which increases filtration efficiency, decreasing air resistance. Porous structures also increase the surface area, increasing the filtration efficiency, but such structures are more easily clogged. Increase in the membrane thickness will surely increase the filtration efficiency, but the air resistance will also increase. Increasing the packing density increases the basis weight of the fibrous membranes, causing increase of the air resistance. The filtration can also be affected by (1) face velocity, (2) particle size, (3) temperature and humidity. These make the filtration efficiency complex but optimizable parameter.

In recent years, researchers are focusing on a few types of membranes for different purposes. Composite air filter membranes are ones to optimize the membranes properties maintaining their filtration performance by different membranes composition. Wang et al. developed polyvinyl chloride/polyurethane polymer (PVC/

PU) two-tier composites, achieving tensile strength enhancement up to 9.9 MPa in relative to PVC membranes (Wang et al. 2013). The high filtration efficiency (99.5%) and low pressure drop (144 Pa) were kept. Such composites also showed good abrasion resistance performance ($\approx$ 134 cycles). They also fabricated nylon 6/polyacrylonitrile (PA-6/PAN) 3D binary-structured composites (Wang et al. 2015), which showed high filtration efficiency (99.99%) and a great quality factor in the low basis weight region. Additionally, PAN/PU composites with superamphiphobic character were also fabricated (Wang et al. 2014), showing superior antifouling properties. Multilevel structured air filter membranes with nanostructures on E-spun nanofibers can also improve the filtration performance. For example, Wang et al. reported a superhydrophobic polysulfone (PSU) fibers with TiO_2 nanoparticles, achieving outstanding filtration efficiency up to 99.997% with pressure dropping down to 45.3 Pa (Wan et al. 2014). Charges, superhydrophobicity, multipore, and cavity structures can also be induced by such multilevel structures, which increases filtration performance of such membranes. Air filter membranes with thermal stability property is an interesting topic due to the recent development of flexible inorganic fibrous media, which can stand high temperature without melting down or burning. Wang et al. fabricated flexible alumina fibrous (Wang et al. 2014). The membranes will stand calcination up to 700ºC, keeping the high filtration efficiency (99.848%) and low pressure drop (239.12 Pa). Such membranes will be applicable for industrial and vehicle filtration, where the flexibility is less required. Additives with antibacterial properties can also be induced into the membranes. Reproducibility and self-cleaning are required for the membranes to serve longer life. Xiaoqi et al. added hydrophobic SiO_2 nanoparticles onto E-spun PA-56 filtration membranes (Xiaoqi et al. 2015). Water drops rolling off the membranes will pick up dirt along the way, achieving self-cleaning. Table 9.1 summarizes the filtering capabilities of some reported E-spun nanofibrous membranes.

E-spun Nanofibers for Liquid Filtration

Liquid filtration has played a pivotal role in the defense, industry, agriculture, medical, and other fields (Fig. 9.2). People have been committed to the liquid filter materials research and development over the years. Membrane filtration technology, as the most promising liquid filtration technology in 21st century, has aroused more and more attention. Microfiltration (MF), ultrafiltration (UF), nanofiltration (NF), and reverse osmosis (RO) are four most commonly used membrane technologies. As a new technology for the preparation of membrane materials, E-spun nanofibrous membranes have small pore size, high porosity (up to 80% or more) compared to the conventional filtration membrane, high surface roughness, and low weight. They have broad applications of liquid filtration prospects (Thavasi et al. 2008).

E-spun nanofibrous membrane used in liquid filtration has two outstanding advantages. First, the filtration efficiency is high. The E-spun membrane can improve the filtration efficiency by 70% compared to the conventional fiber membrane. The other is the water flux. Yoon et al. prepared the E-spun PAN nanofibrous membrane for water filtration (Yoon et al. 2009). The flux reaches 3353 L m^{-2} h^{-1}, while the water flux of traditional PAN ultrafiltration membrane is only 52.2 L m^{-2} h^{-1}.

Table 9.1: Filtering capabilities of several E-spun nanofibrous membranes.

Polymer	**Particle diameter (μm)**	**Filter efficiency (%)**	**Basis weight (g m^{-2})**	**Air velocity (cm s^{-1})**	**Air resistance (9.80665 Pa)**	**References**
Polylactic acid/Poly β Hydroxybutyrate (PLA/PHB)	0.3	98.5	6.1	5.3	16.315	Nicosia et al. 2015
PVDF	2.5	99.9	9.6	5.3	4.130	Zhao et al. 2017
Polyether Sulfone/Polyamide 66 (PES/PA66)	0.3	99.999	–	42	52.006	Nicosia et al. 2015
M-benzoyl phthalamide (PMIA)	0.3	99.999	0.365	–	9.381	Zhang et al. 2017
Polyamide-56 (PA-56)	2.5	99.995	49	–	11.319	Liu et al. 2015
PAN	0.3	99.989	2.04	–	11.931	Wang et al. 2014
Polyvinyl chloride/Polyurethane (PVC/PU)	0.3	99.5	21	15.41	14.684	Wang et al. 2013
PVA	0.075	99.95	–	–	–	Li 2013
Polyamide 6 (PA-6)	0.3	99.996	0.9	–	9.687	Zhang et al. 2017
Gelatin	0.3	99.3	3.43	5	20.496	Souzandeh et al. 2016
Polytetrafluoroethylene (PVDF/PTFE)	0.3	99.972	9	5.3	5.812	Wang et al. 2016
Silk fibroin (SF)	0.3	96.2	3.4	5.31	–	Wang et al. 2016
1,3,5-three amide benzene	0.2	95	32.2	25	–	Weiss et al. 2016
Polyamide-6/Polyacrylonitrile/Polyamide-6 (PA-6/PAN/PA-6)	0.3	99.9998	–	–	11.982	Zhu et al. 2017
Polyacrylonitrile/Polyacrylic acid (PAN/PAA)	0.3	99.994	–	5.3	16.315	Liu et al. 2015

The combination of E-spinning technology and plasma treatment, electrostatic layer-by-layer self-assembly (LBL), grafting technology can optimize the pore structure (such as pore size distribution, porosity, and hole connectivity) of E-spun membrane, changing the surface properties of the E-spun membrane can further improve the filtration efficiency and water flux of the nanofibrous membrane.

When used for the filtration of drinking water, the E-spun membrane can not only completely remove the micron micelles of colloidal particles, suspensions, and algae

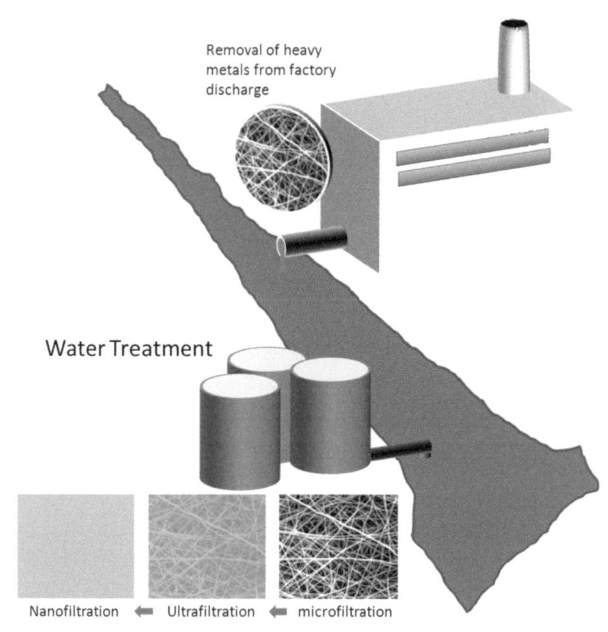

Fig. 9.2: General schematic of electrospun liquid filtration membrane [http://electrospintech.com/waterfilterintro.html#.WYLPklVOImY].

in water, but also effectively intercept the harmful bacteria, viruses, macromolecules, and other organic matter, keeping aquatic safety while retaining the trace elements needed in the human body. Gopal et al. investigated the filtration efficiency of E-spun polyvinylidene fluoride (PVDF) nanofibrous membranes (Gopal et al. 2006). The results show that the interception efficiency of E-spun PVDF nanofibrous membrane to polystyrene (PS) particles with diameter of 1 μm is more than 98%. On the basis of this, Kaur et al. further studied the porosity and the water flux of the E-spun PVDF nanofibrous membrane used as the water filter material (Kaur et al. 2007). Compared to the ordinary microfiltration membrane, it was found that the porosity of the E-spun PVDF fiber membrane is more than 80%, while the ordinary microfiltration membrane porosity is only 65%. With the same surface wettability, E-spun PVDF membrane possesses twice the water flux as the traditional microfiltration membrane.

Electrospun nanofibrous membrane can be applied not only to microfiltration, but also to removing harmful substances such as bacteria and viruses by functionalizing it. Wang et al. prepared a cross-linked PVA scaffold using an E-spinning method, coated with a layer of (PVA) hydrogel (96% degree of hydrolysis) on its surface to obtain a hydrophilic high porosity (> 80%) PVA nanofibrous membrane (Wang et al. 2006). The experimental results show that the water flux of the nanofibrous membrane can reach 130 L m^{-2} h^{-1} under the pressure of 100 Pa, which is obviously higher than that of the ordinary ultrafiltration membrane material under the same filtration efficiency (> 99.5%) [57 L m^{-2} h^{-1}]. In addition, the researchers also found that the water flux is inversely proportional to the thickness of the coating on fiber surface, so that higher filter flux can be obtained by reducing the coating thickness.

E-spun nanofibers can not only effectively filter out the particulate matter in the water by physical action, but can also kill the harmful bacteria in the water body by biochemical action after the surface modification in order to further improve the purification effect of drinking water. Bjorge et al. deposited silver nanoparticles on polyacrylic acid (PAA) E-spun nanofibers and then used them to filter and kill bacteria during drinking water filtration (Bjorge et al. 2009). The results show that the PAA fiber membrane containing silver nanoparticles can achieve efficient interception and killing of bacteria in water.

E-spun Nanofibers for Oil-water Separation

With the development of petroleum industry and increased level of people's lives, oil pollution such as spill accidents, industrial wastewater, and sanitary sewage has become one of the major problems in our modern society. During the accumulation of waste oil, it has caused long-term damaging effects and a lot of resources have been wasted in the ecological environment.

Conventional oil-water technologies including flotation, ultrasonic separation, and skimming have a certain function in the separation of oil-water mixtures, but suffer from the low efficiency of separation, complex construction technology, high energy cost, and second pollution. E-spinning technology is an effective and extensive method for nanofibers membrane with controllable structures.

Recently, E-spun nanofibrous membrane has been successfully used in the system of oil-water separation, which has been achieved in commercialization. Ge et al. have fabricated a practical superhydrophilic and underwater superoleophobic nanofibrous membrane with high porosity, large surface, and easily tunable structures, for the separation of oil-water complex, which was prepared via E-spinning and electrospraying methods, exhibited high separation efficiency, robust antifouling properties, and extremely excellent flux solely driven by gravity (Ge et al. 2017), as shown in Fig. 9.3. According to the result, the membrane including any roughness level on the surface showed quite a separation efficiency (> 99.9%) and high separation fluxes (6290 ± 50 L m^{-2} h^{-1}) for the surfactant-free emulsion. Different types of polymers produce nanofibers with different characteristics. Various polymers have been used to prepare nanofibers, including polystyrene (Min et al. 2013), polyurethane (Lin et al. 2013), polyvinyl chloride (Zhu et al. 2011), and cellulose acetate (Shang et al. 2012). Type of oil samples from the offshore should be studied optionally, such as motor oil (Zhu et al. 2011); peanut oil (Wu et al. 2012); diesel oil (Lin et al. 2013); bean oill (Lin et al. 2012); sunflower oil (Lin et al. 2012); and silicon oil (Wu et al. 2012).

Generally speaking, electrospinning technology is a facile strategy to fabricate nanofibrous membrane with highly efficient oil-water separation under the force of gravity, which was significantly higher than traditional methods.

E-spun Nanofibers for Hazardous Substances Adsorption

Over the past few years, heavy-metal pollution the factory eliminated has been badly hazardous to human health and ecological environment, which could tightly

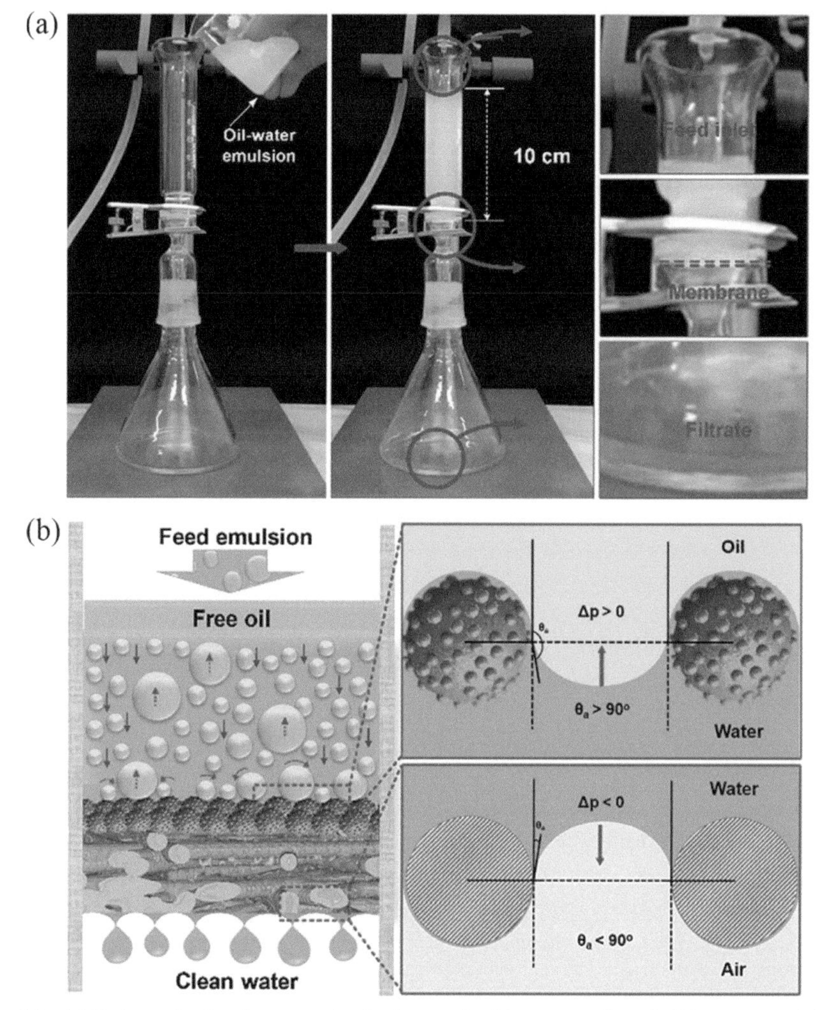

Fig. 9.3: (a) Photographs showing apparatus for the gravity-driven separation of oil-in-water emulsions with a continuous fluid infusion to maintain a stable liquid level. (b) Schematic diagram illustrating the plausible mechanism of the oil-in-water emulsion separation: oil droplets roll round and coalescence on the surface; oil was sustained by the hierarchical structured microspheres trapping the water layer at the interface; water transported into the channels formed by the nanofibers (Ge et al. 2017).

be connected together forming a cluster on the surface of food or floating in water, and cause toxicity of amplification effect with bioaccumulation. Herein, avoiding heavy pollution is a serious issue in modern life and a great number of contributions have been devoted by researchers to explore methods of hazardous substances adsorption. E-spinning with several advantages such as high specific surface area, high porosity, exhibited super adsorption property (Abdouss et al. 2014, Bai and Tsai 2014, Huang et al. 2014, Sharma et al. 2014, Tabatabaeefar et al. 2015). As an example, Byun et al. used $BaFe_{12}O_{19}$ to remove arsenic (Byun et al. 2014), as shown in Fig. 9.4. Wei et al. studied the adsorption of various heavy mental ion

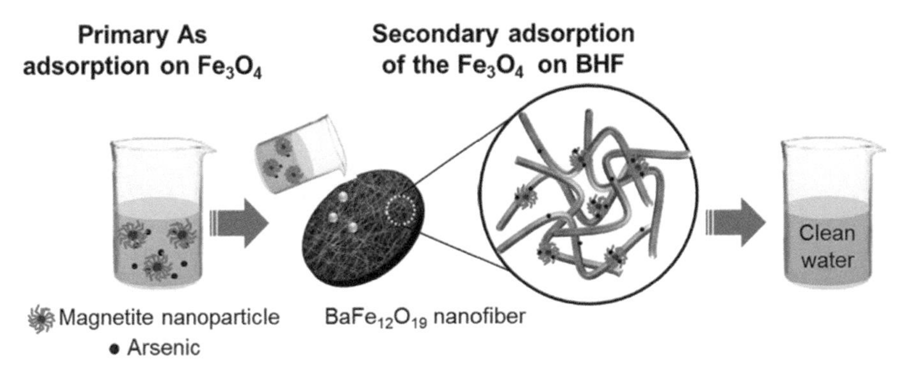

Fig. 9.4: Schematic illustration of Fe_3O_4 nanoparticles separation and arsenic removal by magnetic $BaFe_{12}O_{19}$ nanofiber (Byun et al. 2014).

in the water by manufacture of PVA nanofibers as the substance (Wei et al. 2014). PVA previously modified by epoxy chloropropane what could be further modified using rhodamine derivatives and then E-spun into nanofibrous membrane, which could be applied in removal of heavy mental ion. For the adhesion of metal ions and nanoparticles on the surface of nanofibers from an aqueous media, polymers with surface functional groups such as amino, carboxyl, pyridyl and thiol can be E-spun to form nanofibers. Dong et al. fabricated poly(4-vinylpyridine) and poly(4-vinylpyridine)/PMMA composite nanofibers to attract gold and silver ions and their nanoparticles since poly(4-vinylpyridine) contains pyridyl group (Dong et al. 2011). Incorporation of gold nanoparticles may increase the conductivity of the nanofibers, while incorporation of silver nanoparticles imparts antimicrobial properties to the nanofibers (Son et al. 2006).

The nanofibers with absorption ability of heavy-metal could selectively adsorb metal ions and not affect other components transmitting. Hence E-spinning as an emerging technology offers a promising strategy to fabricate nanofibrous membrane with tunable microstructure and surface chemical property for the effective hazardous substances adsorption. Moreover, E-spun nanofibrous membrane exhibits better property of getting rid of heavy mental ion, resulting in great application prospect to keep the body healthy.

E-spun Nanofibers for Bioengineering Separation

Protein separation and purification refers to the technology used to separate and purify the protein from the mixture, which is the core technology in the contemporary biological industry. At present, the main protein separation technology consists of precipitation, chromatography, electrophoresis, dialysis, and ultracentrifugation. These methods have some technical difficulty, and the separation process cost is high. E-spun nanofibrous membrane not only has the advantages of large specific surface area and high porosity, but also has the advantages of simple preparation process and low cost, and has been widely used in the separation and purification of protein. Researchers have found that when using a single E-spun nanofibrous membrane to separate and purify the protein, it is necessary to modify its surface to form a complex. Ma et al. prepared the cellulose acetate fiber membrane by E-spinning and used for the separation and purification of bovine serum albumin and bilirubin, respectively (Ma et al. 2005). The separation efficiencies were 13 mg g^{-1} and 4 mg g^{-1}, respectively. However, due to the low mechanical strength of the cellulose acetate membrane, researchers also carried out in-depth study of the E-spun nanofibrous membrane for protein separation and purification. Ma et al. used the E-spinning technology to prepare polysulfone (PSU) fiber membranes, followed by grafting methacrylic acid (MAA), diamino-diphenylamine (DADPA), and pigment ligands Cobalt blue F3GA (CB) to obtain polysulfone-polymethacrylic acid-diamino—diphenylamine—ciba blue film, the nanofibrous membrane not only has a good separation performance for bovine serum albumin (22 mg g^{-1}), but also good mechanical properties (Ma et al. 2006).

E-spun Nanofibers for Insulation

E-spinning provides porous structures. These pores can hold air or water, thus greatly reducing heat convection. The thermal conductivity of air and water are low, so the E-spun nanofibers can act as a good thermal insulator.

Kim et al. prepared a PU E-spun membrane containing TiO_2 and assembled it onto a wallpaper by laminating it. Compared to conventional wallpapers, nanofiber laminated paper shows better thermal insulation performance, which indicates that nanofiber layers play an effective role in insulation. This wallpaper also exhibits good moisture transmittance and antibacterial properties (Kim et al. 2011).

Recently, Si et al. prepared inorganic SiO_2 flexible electrodeless ceramics by electrospinning, which can withstand higher temperatures. This material has a very low thermal conductivity (0.0058 W m^{-1} K^{-1}), even lower than air (Si et al. 2014). Such membrane has good performance in thermal insulation, as shown in Fig. 9.5.

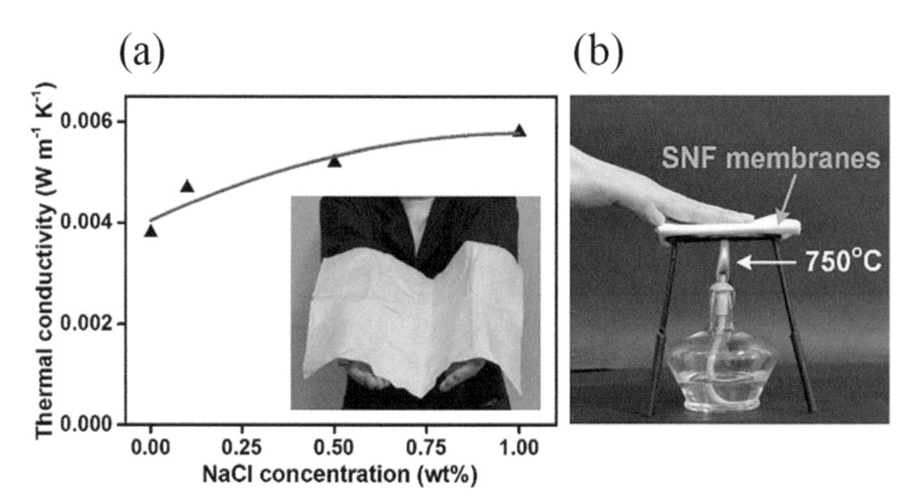

Fig. 9.5: Thermal conductivity of the silica nanofibrous membranes. Inset: (a) photograph shows the large-scale (60 cm × 60 cm) of soft SNF membranes with NaCl content of 1 wt%. (b) A designed concept test shows the robust thermal insulation performance of the relevant membranes at 750°C (Si et al. 2014).

Nanodevices

Sensor is a kind of device or equipment which can detect specified materials and transduce them into available output signals. In order to improve the sensor performance, nanomaterials with high specific surface area are introduced in the design of sensing materials. Compared to other fabrication methods, E-spinning is versatile and superior for producing and constructing ordered and complex nanofibrous materials. Moreover, the E-spun nanofibrous membrane is characterized with 3D structure, high porosity, large specific surface area, and good structure controllability, and is one kind of ideal nanomaterials for high performance sensors. At present, based on the different sensing principles, various electrospun nanofibers-based sensors have been developed, such as frequency vibration sensors, resistance sensors, photoelectric sensors, optical sensors, amperometric sensors, etc. The relative reports are summarized in Table 9.2.

Frequency Vibration Sensors

According to the working principles, frequency vibration sensors are mainly divided into surface acoustic wave (SAW) sensors and quartz crystal microbalance (QCM) sensors.

Surface Acoustic Wave Sensors. SAW sensors are based on mass deposition effect and acoustoelectric effect. Depositing sensing film that interacts with the detected materials on the travel path of sound wave, the density and elastic property of the sensing film will change. These changes will affect the propagation speed of surface acoustic wave, and then change the output frequency of the oscillator, so the relevant information can be obtained.

Recently, there has been a growing attention toward developing SAW devices for gas sensing applications. Herein, E-spun nanofibers have been introduced to the design of SAW sensors to greatly improve sensing property (He et al. 2010, Li et al. 2010). In 2010, a PVP fibers/36°$LiTaO_3$ SAW device was successfully developed by E-spinning the PVP fibers on the interdigital electrodes of the SAW transducer (He et al. 2010), the sensing property was displayed in Fig. 9.6. The concentration of the PVP solution has great effect on the morphology and the average diameter of the PVP fibers. Besides, changing the electrode-to-collector distance has no obvious impacts on the morphology and average diameter of the PVP fibers. The electrode-to-collector distance significantly affected the fiber density and the large distance tends to the smaller density. The developed SAW device coated with PVP fibers E-spun from 58% solution shows the large response to H_2 at room temperature.

Quartz Crystal Microbalance Sensors. QCM is a widely applied acoustic sensing technology. It can reflect the mass changes at nanogram level. QCM consists of sensitive water, resonance circuit, and frequency counter. The basic principle is based on the proportional relation of resonance frequency change and surface absorption mass change m, as shown in Sauerbrey Eq. 9.1.

$$\mathbf{\Delta f} = \frac{-f_0^2 \Delta m}{NS\rho} \tag{9.1}$$

Table 9.2: The application of E-spun nanofibers in sensors.

Sensor types	Materials	Fibers arrangement	Fiber diameter	Detected materials	Applicable temperature	Detection limit	References
Frequency vibration sensors	PVP/$LiTaO_3$	Random orientation	200~400 nm	H_2	Room temperature	111.6 mg m^{-3} (1250 ppm)	He et al. 2010
	Polyacrylicacid (PAA)-PVA	Random orientation	100~400 nm	NH_3	Room temperature	37.94 mg m^{-3} (50 ppm)	Ding et al. 2004
	PAA	Random orientation	1~7 μm	NH_3	Room temperature	0.0987 mg m^{-3} (130 ppb)	Ding et al. 2005
	Polyethyleneimine (PEI)-PVA	Random orientation	100~600 nm	H_2S	Room temperature	0.7589 mg m^{-3} (500 ppb)	Dirote 2006
	PAA	Random orientation	11~264 nm	H_2O	Room temperature	6%	Wang et al. 2009
	PEI-PVA	Random orientation	0.04~1.8 μm	HCHO	Room temperature	13.39 mg m^{-3} (10 ppm)	Ding et al. 2010
	PEDOT:PSS/PVP	Alignment	400~1000 nm	CO	Room temperature	5 ppm	Zhang et al. 2016a
Resistance sensors	TiO_2	Random orientation	200~500 nm	NO_2	150ºC	1.028 mg m^{-3} (500 ppb)	Ildoo Kim et al. 2006
	TiO_2	Random orientation	120~850 nm	CO	400ºC	0.0625 mg m^{-3} (50 ppb)	Landau et al. 2008
	LiCl-TiO_2	Random orientation	150~260 nm	H_2O	300ºC	11%	Wang et al. 2010
	Mg^{2+}, Na^+-TiO_2	Random orientation	~ 200 nm	H_2O	400ºC	11%	Zhang et al. 2009
	TiO_2-ZnO	Random orientation	~ 250 nm	O_2	Room temperature	0.68 Pa	Park et al. 2009
	SnO_2	Single fiber	~ 700 nm	H_2O	Room temperature	–	Wang et al. 2007
	SnO_2	Random orientation	~ 100 nm	C_2H_5OH	300ºC	0.0205 mg m^{-3} (10 ppb)	Zhang et al. 2008

	SnO_2	Random orientation	80~160 nm	Benzene	Room temperature	34.82 mg m^{-3} (10 ppm)	Qi et al. 2009
	MWCNTs/SnO_2	Random orientation	300~800 nm	CO	330°C	58.75 mg m^{-3} (47 ppm)	Yang et al. 2007
	ZnO-SnO_2	Random orientation	100~150 nm	C_2H_5OH	350°C	6.16 mg m^{-3} (3 ppm)	Song et al. 2009
	Fe-SnO_2	Random orientation	60~150 nm	C_2H_5OH	Room temperature	20.53 mg m^{-3} (10 ppm)	Song et al. 2009
	KCl-SnO_2	Random orientation	100~200 nm	H_2O	300°C	11%	Song et al. 2009
	ZnO-SnO_2	Random orientation	100~200 nm	Benzene	300°C	34.832 mg m^{-3} (10 ppm)	Yang et al. 2007
	α-Fe_2O_3	Random orientation	150~280 nm	C_2H_5OH	Room temperature	205.3 mg m^{-3} (100 ppm)	Zheng et al. 2009
	ZnO	Random orientation	80~235 nm	C_2H_5OH	200°C	20.53 mg m^{-3} (10 ppm)	Wu et al. 2009
	Ag-In_2O_3	Random orientation	60~130 nm	CH_2O	400°C	6.696 mg m^{-3} (5 ppm)	Wang et al. 2009
	Polypyrrole (PPy)-TiO_2/ZnO	Random orientation	~ 100 nm	NH_3	200°C	0.04553 mg m^{-3} (60 ppb)	Zheng et al. 2009
	Pt/In_2O_3	Random orientation	60~100 nm	H_2S	400°C	75.88 mg m^{-3} (50 ppm)	Zheng et al. 2009
	WO_3	Random orientation	20~140 nm	NH_3	220°C	37.94 mg m^{-3} (50 ppm)	Wang et al. 2006
	$SrTi_{0.8}Fe_{0.2}O_{3-\delta}$	Random orientation	~ 100 nm	CH_3OH	115°C	7.142 mg m^{-3} (5 ppm)	Sahner et al. 2007

Table 9.2 contd. ...

...Table 9.2 contd.

Sensor types	Materials	Fibers arrangement	Fiber diameter	Detected materials	Applicable temperature	Detection limit	References
	TiO_2-PEDOT	Random orientation	72~108 nm	NO_2	Room temperature	0.01437 mg/m^3 (7 ppb)	Ying et al. 2009
	TiO_2-PEDOT	Random orientation	72~108 nm	NH_3	140°C	0.5122 mg/m^3 (675 ppb)	Liu et al. 2004
	PANI-PEO	Single fiber	100~500 nm	NH_3	300°C	0.3794 mg/m^3 (500 ppb)	Bishop-Haynes and Gouma 2007
	PANI-PVP	Random orientation	1~10 μm	NO_2	350°C	2.053 mg/m^3 (1 ppm)	Manesh et al. 2007
	Poly(diphenylamine) (PDPA)-PMMA	Random orientation	~ 400 nm	NH_3	400°C	0.7589 mg/m^3 (1 ppm)	Gao et al. 2008
	PANI	Random orientation	0.3~1.5 μm	NH_3	Room temperature	75.89 mg/m^3 (100 ppm)	Ji et al. 2008
	PMMA-PANI	Random orientation	250~600 nm	$(C_2H_5)_3N$	Room temperature	90.17 mg/m^3 (20 ppm)	Pinto et al. 2008
	HCSA-PANI	Single fiber	20~150 nm	C_2H_5OH	Room temperature	–	Li et al. 2009
	PEO-PANI	Random orientation	250~500 nm	H_2O	Room temperature	22%	Yu et al. 2016
	PVDF-PANI	Patterned membrane	–	Strain	Room temperature	85%	Huang et al. 2014
	PVDF-PANI	Alignment	–	Strain	Room temperature	–	Jia et al. 2016
	PMMA-PANI	Random orientation	8.76 μm 9.25 μm	NH_3, Strain	Room temperature	0.08 ppm	Sun et al. 2013
	PEDOT:PSS/PVP	Alignment	–	Strain	Room temperature	4%	Lin et al. 2014
	PEDOT:PSS/PVP	Alignment	20 μm	Strain	Room temperature	35%	Aussawasathien et al. 2008

	HCSA-polyotoluidine (POT)/polystyrene (PS)	Random orientation	0.2~1.9 μm	H_2O	Room temperature	–	Lala et al. 2009
	MWCNTs/polyamide (PA)	Random orientation	110~140 nm	Volatile gas	Room temperature	–	Kessick and Tepper 2006
	Carbon black-polyepichlorohydrin (PECH)	Alignment	~ 3 μm	CH_3OH	Room temperature	1428 mg/m^3 (1000 ppm)	Shi et al. 2009
	Carbon black-PECH	Alignment	~ 3 μm	$C_5H_{10}Cl_2$	Room temperature	31.47 mg/m^3 (5 ppm)	Wu et al. 2009
	Carbon black-PECH	Alignment	~ 3 μm	$C_6H_5CH_3$	Room temperature	1026 mg/m^3 (250 ppm)	Xianyan Wang et al. 2002
	Carbon black-PECH	Alignment	~ 3 μm	C_2HCl_3	Room temperature	2935 mg/m^3 (500 ppm)	Luoh and Hahn 2006
Photoelectric sensors	Au/SiO_2	Random orientation	130~170 nm	Wave length	Room temperature	–	Chae et al. 2007
	Co-ZnO	Random orientation	50~400 nm	O_2	Room temperature	42.66 Pa	Ren et al. 2006
	GaN	Random orientation	32~48 nm	Wave length	Room temperature	–	Manesh et al. 2007
Optical sensors	PAA-poly-(pyrenemethanol) (PM)	Random orientation	100~400 nm	Fe^{3+}, Hg^{2+}, 2,4,6-trinitrotoluene	Room temperature	–	Liu et al. 2009
	PAN	Random orientation	50~200 nm	CO_2	Room temperature	1374 mg/m^3 (700 ppm)	Wang et al. 2009
	Polydiacetylene (PDA)	Single fiber	3 μm	α-cyclodextrin	30°C	–	Wang et al. 2009

Table 9.2 contd. ...

...Table 9.2 contd.

Sensor types	Materials	Fibers arrangement	Fiber diameter	Detected materials	Applicable temperature	Detection limit	References
Amperometric sensors	Glucoseoxidase (GO_x)/PVA	Random orientation	70~250 nm	Glucose	Room temperature	0.05 mmol/L	Ding et al. 2010
	PVDF/PAPBA*	Random orientation	~ 150 nm	Glucose	Room temperature	1 mmol/L	Sawicka et al. 2005
	Ni/carbon nanofiber (CNF)	Random orientation	200~400 nm	Glucose	Room temperature	1 μmol/L	Li et al. 2010
	CNF	Random orientation	90~140 nm	Glucose	Room temperature	190 μmol/L	Gomes et al. 2000
	CuO	Random orientation	170 nm	Glucose	Room temperature	80 μmol/L	Sasaki et al. 2002
	Hemoglobin	Random orientation	~ 2.5 μm	H_2O_2	Room temperature	0.61 μmol/L	Lin and Shih 2003
	Hemoglobin	Random orientation	~ 2.5 μm	$NaNO_2$	Room temperature	0.47 μmol/L	Guo et al. 2002
	PVDF/Urease	Random orientation	~ 100 nm	Urea	Room temperature	0.5 mmol/L	Syritski et al. 1999

Note: *Polyaminophenylboronic acid (PAPBA).

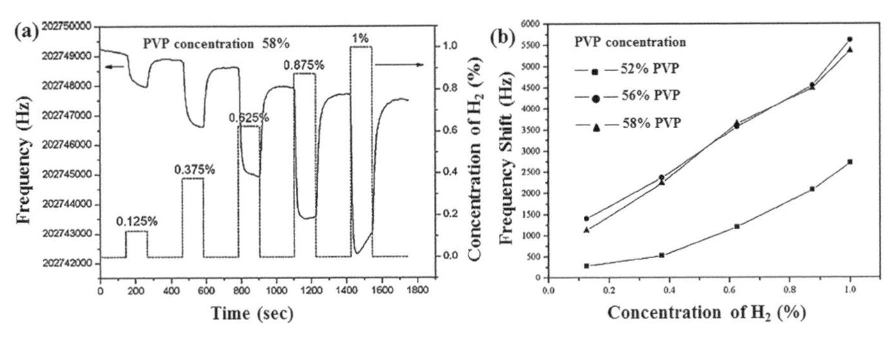

Fig. 9.6: Sensing property of the E-spun SAW transducer. (a) Dynamic response of PVP/36°$LiTaO_3$ SAW device toward H_2 at room temperature; (b) the frequency shift of PVP/36°$LiTaO_3$ SAW device upon exposure to H_2 (He et al. 2010).

where **Δf** indicates the vibration frequency change after absorbing foreign materials, f_0 is the basic frequency of quartz crystal wafer, m is the absorption mass, N is the unique constant of quartz crystal wafer, S is the electrode area, ρ is the density of quartz crystal wafer.

The sensitivity of QCM sensors has great relationship (Gomes et al. 2000) with the interaction of sensing film and target material. Based on different principles, many sensing materials such as zeolite (Sasaki et al. 2002), fullerence (Lin and Shih 2003), chiral material (Guo et al. 2002), polypyrrole (Syritski et al. 1999), graphite (Kim et al. 1999), ITO layer (Zhang et al. 2002) and oligonucleotide (Duman et al. 2003) have been successfully used in QCM detecting. To further improve sensing property, exploring new nanostructured sensing materials will be one of the feasible approaches. Due to the remarkable advantages of controllable thickness, elaborate structure, diverse materials, high specific surface area, and high porosity, electrospun nanofibers are considered an ideal QCM sensing material.

Poly(3,4-ethylenedioxythiophene):poly(styrenesulfonate)/polyvinylpyrrolidone (PEDOT:PSS/PVP) composite nanofibers were successfully fabricated via electrospinning and used as a QCM sensor for detecting CO gas (Guo et al. 2016). As shown in Fig. 9.7, the QCM sensor based on PEDOT:PSS/PVP nanofibers was sensitive to low concentration (5–50 ppm) CO. In the range of 5–50 ppm CO, the relationship between the response of PEDOT:PSS nanofibers and the CO concentration was linear. Nevertheless, when the concentration exceeded 50 ppm, the adsorption of the nanofiber membrane for CO gas reached saturation and the resonant frequency range had no change. Therefore, the results open an approach to create E-spun PEDOT:PSS/PVP for gas sensing applications.

Resistance Sensors

Semiconductor Oxide Nanofibrous Resistance Sensors. Semiconductor oxide nanofibers with nanostructure could detect the mass transfer of the detected material at a faster pace. Carrier transport along the fibers' axis is easier, herein, the resistance sensor possesses better detection performance. Based on the above advantages, these TiO_2, ZnO, WO_3, MoO_3, SnO_2, In_2O_3 semiconductor oxide nanofibers fabricated via E-spinning have been successfully used in high-performance resistance sensor, and

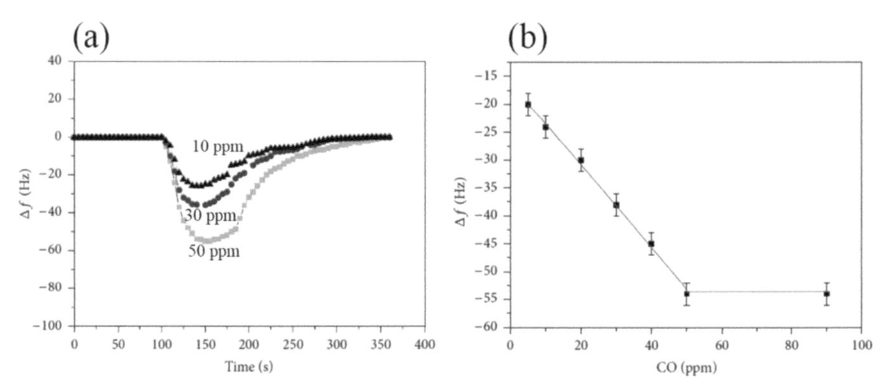

Fig. 9.7: (a) The response characteristics of PEDOT: PSS/PVP nanofiber membrane sensing to 10(black triangle), 30(red dot), and 50(green square) ppm CO; (b) relationship of frequency difference with CO concentration (Zhang et al. 2016b).

many kinds of gas detection have reached a new level, such as NH_3, H_2S, CO, NO_2, O_2, CO_2, toluene (C_7H_8), and butanol ($C_4H_{10}O$) (Qi et al. 2009, Ying et al. 2009, Zhang et al. 2009, Ding et al. 2009, Choi et al. 2009, Kim et al. 2016, Yang et al. 2015).

Humidity nanosensors based on LiCl doped TiO_2 E-spun nanofibers has been reported (Li et al. 2008). When relative humidity is 11%~95%, the resistance decreases from 10^7 Ω to 10^4 Ω, this nanosensor's response time is 3 seconds and its recovery time is 10 seconds. Resistance sensors based on TiO_2 nanofibers is also adapted to detect NO_2 (Ildoo Kim et al. 2006). In addition to TiO_2 and SnO_2, WO_3 also could be used as resistance sensors. The SnO_2 nanofibers with Pt nanoparticles have been reported because of its toluene-sensing performance (Kim et al. 2016). Au functionalized WO_3 composite nanofibers have been reported for testing different gases (methanol, ethanol, acetone, and n-butanol). In another article, a new synthetic method assisted by E-spinning technique was proposed to develop H_2S sensitive sensors using catalytic Pt functionlized macroporous WO_3 nanofibers (Yang et al. 2015). Figure 9.8 shows the E-spun fibers and the principal component. Polycrystalline WO_3 nanofibers also could be used to detect NH_3 at 350ºC.

Conducting Polymer Nanofibrous Resistance Sensors. Conducting polymer materials with nanofibrous structure not only have high specific surface area, chemical specificity, and adjustable conductivity, but are also characterized with good flexibility and processability. They are one kind of ideal materials for resistance sensors. At present, conducting polymers exploited to resistance sensors mainly include polyaniline (PANI) (Yu et al. 2016, Huang et al. 2014, Jia et al. 2016), polypyrrole (PPY) (Guo et al. 2016, Bagchi and Ghanshyam 2017), and polythiophene (Sun et al. 2013, Lin et al. 2014), among them, PANI is the most universal sensing material in conducting polymers.

Via E-spinning, various nanofibrous conductive polymeric materials with different structures have been applied in resistance sensors, including patterned PANI/PVDF nanofibrous membrane (Yu et al. 2016), nano-branched coaxial PANI/PVDF fibers (Huang et al. 2014), aligned microfibrous PEDOT:PSS-PVP arrays with curled architectures (Sun et al. 2013), and twisted microropes obtained from PEDOT:PSS-PVP fiber arrays (Lin et al. 2014), etc.

A simple and cost-effective method has been reported to fabricate highly stretchable patterned nanofibrous PANI/PVDF nanofibrous membrane (as shown in Fig. 9.9) via electrospinning and *in situ* polymerization (Yu et al. 2016). Owing to the patterned structure, the nanofibrous PANI/PVDF strain sensor can detect a strain up to 110%, for comparison, which is 2.6 times higher than the common nonwoven PANI/PVDF mat. Additionally, the patterned PANI/PVDF strain sensor can completely recover to its original electrical and mechanical values within a strain range of more than 22% and exhibits good durability over 10,000 folding-unfolding tests. Furthermore, the strain sensor also can be used to detect finger motion.

Sun et al. reported stretchable strain sensors based on aligned microfibrous arrays of PEDOT:PSS-PVP with curled architectures by a novel reciprocating-type E-spinning set-up with a spinneret in straightforward simple harmonic motion (Sun et al. 2013). Owing to the curled architectures of the as-spun fibrous polymer arrays, the sensors can be stretched reversibly with a linear elastic response to strain up to 4%, which is three times higher than that from E-spun nonwoven mats. In addition,

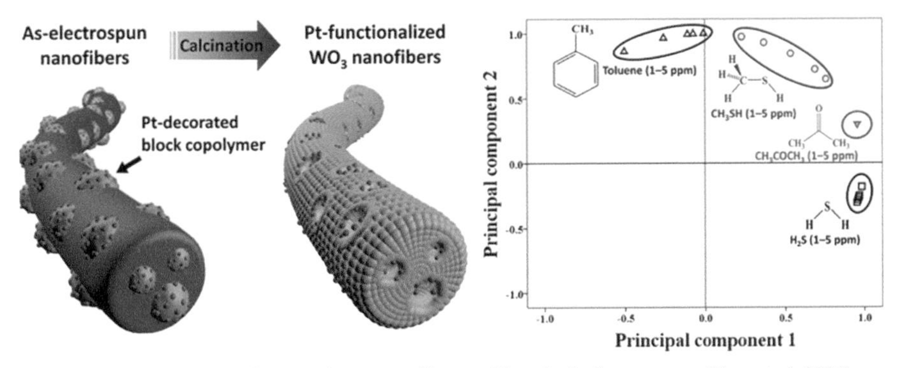

Fig. 9.8: Morphology of the as-electrospun fibers and its principal component (Yang et al. 2015).

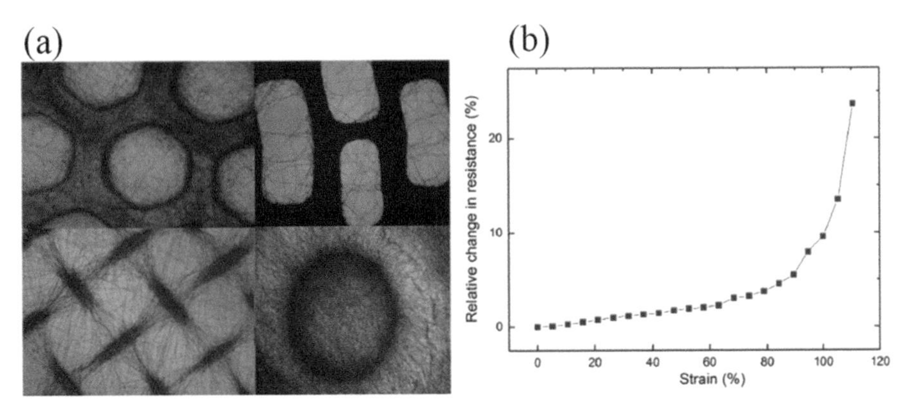

Fig. 9.9: (a) Optical images of the patterned membranes obtained from different collectors; (b) relative change in the resistance of the patterned sensor under different strains (Yu et al. 2016).

the stretchable strain sensor with a high repeatability and durability has a gauge factor of about 360.

Photoelectric Sensors

Ultraviolet (UV) photodetectors are used in military and civil applications, including missile detection, as probes to detect burning and in chemical analysis. Historically, the most common UV detectors have been photomultipliers and silicon-based UV photodiodes; however, these detectors require expensive filters and are also sensitive to visible light. ZnO is a promising candidate for use in UV photodetectors because of its excellent UV photosensitivity (Bie et al. 2011). Besides, ZnO has exceptional properties, such as a wide, direct band gap, and large photo response, which makes it an ideal candidate for use in UV photodetectors. Shinde and Rajpure reported the synthesis of a Ga-doped ZnO UV detector on an alumina substrate by spraypyrolysis (Shinde and Rajpure 2011).

Pure ZnO is an n-type semiconductor as a result of its intrinsic defects. New techniques to produce high-quality p-type ZnO thin films will facilitate their application. However, it is difficult to prepare p-type ZnO by doping, mainly due to the low solubility of the acceptor dopants. We synthesized n-type pure ZnO nanofibers and p-type Ce doped nanofibers via electrospinning and calcination (Liu et al. 2015). ZnO nanofibrous homojunctions were prepared by the same technique, as shown in Fig. 9.10. The UV-visible absorption wave length of the Ce-doped ZnO nanofibers was larger than that of pure ZnO. A device based on ZnO nanofibrous p–n homojunctions was fabricated. The I–V curves of the homojunction showed good rectifying behavior, with rectification ratios of 55.74 (254 nm) and 21.48 (365 nm). The turn-on voltage for the p–n homojunction decreased under UV illumination. These results suggest a convenient method of fabricating ZnO nanofiber p–n homojunctions for applications in optoelectronic devices.

Amperometric Sensors

For a long time, researchers in the field of biosensor have made many great achievements. Many kinds of testing technology, including chemiluminescence

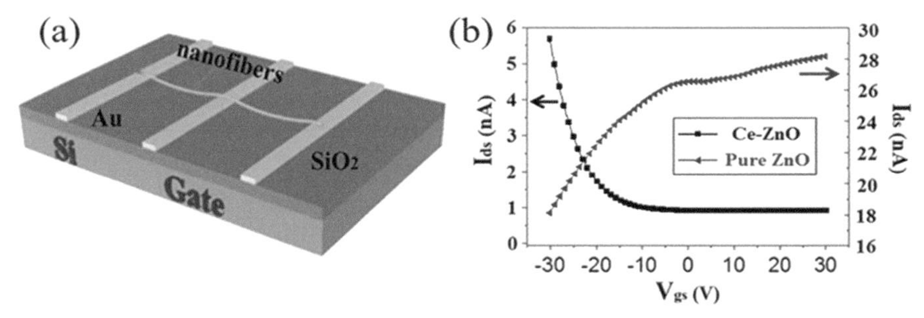

Fig. 9.10: (a) Schematic illustration of a single pure ZnO nanofiber and Ce-doped ZnO nanofiber tested using a two-metal-microprobe. (b) FET curves of an individual ZnO nanofiber prepared via E-spinning; the red (triangles) and black (squares) curves represent pure ZnO and Ce doped ZnO nanofibers, respectively (Liu et al. 2016).

method, chromatography and electrochemical processes, have been successively developed and found applications in bio-detection as biosensors. Among them, the amperometric sensor based on electrochemistry is recognized as the most effective detection method and has been widely used. Nanomaterials, especially E-spun nanofibers, have been widely used in amperometric sensor due to their remarkable specific surface area, high porosity, and good structural controllability. These make E-spun nanomaterials highly attractive to ultrasensitive sensors and of increasing importance in other nanotechnological applications. It is worth noting that most amperometric sensors based on E-spun nanofibers have attracted significant interest in detecting glucose. Fast and accurate determination of glucose level in blood is very important for humans because any deviance from the normal glucose level may induce sickness and disease. Therefore, the development of highly sensitive and selective glucose sensors has attracted much attention due to its importance in areas of biotechnology, clinical diagnostics, and the food industry (Cash and Clark 2010, Courjean and Mano 2011, Deng et al. 2012). Glucose sensors can be mainly classified into glucose oxidase (GOX)-based sensing and non-enzymatic glucose sensing (Ding et al. 2010, Wang et al. 2009). Due to the high sensitivity and selectivity to glucose, GOX has been widely used to construct various sensors for glucose detection (Ahmad et al. 2010, Tang et al. 2010, Wang et al. 2009). However, there are some disadvantages for sensors based on enzymes. For instances, enzymes are expensive and prone to losing their activity. The immobilization procedure is complicated and the enzyme-based electrodes are usually unstable. All these largely limit its wide applications (Singh et al. 2013). By contrast, non-enzymatic electro-oxidation of glucose could avoid the disadvantages of enzyme electrode (Sun et al. 2001). The majority of non-enzymatic electro-chemical glucose sensors mainly rely on the current response of glucose oxidation directly at the electrode surface (Mayorga-Martinez et al. 2012); thus, non-enzymatic glucose sensors have higher stability than enzymatic sensors. Therefore, more and more attempts have been made to construct non-enzymatic sensors for glucose detection in recent years (Guascito et al. 2012, Liu et al. 2013, Wang et al. 2012a, Yang et al. 2010, Zhao et al. 2013).

Until now, great efforts have been made to fabricate different E-spun nanofibers including Ni nanoparticle loaded carbon nanofibers (Liu et al. 2009), CuO nanofibers (Wang et al. 2009), cobalt(II) oxide nanofibers (Ding et al. 2010), and nickel(II) oxide microfibers (Gao et al. 2008) for non-enzyme glucose sensors. For example, Zhou et al. (Zhou et al. 2014) prepared ZnO-CuO hierarchical nanocomposites by E-spinning technique to construct ultrasensitive non-enzymatic glucose sensor. In such a case, 3D porous ZnO–CuO hierarchical nanocomposites (HNCs) nonenzymatic glucose electrodes with different thicknesses were fabricated by co-electrospinning and compared to 3D mixed ZnO/CuO nanowires and pure CuO nanowires electrodes. Figure 9.11 shows SEM images of the as-fabricated 3D porous ZnO–CuO HNCs. These randomly oriented hierarchical nanocomposites all have uniform and long continuous surface in a large scale. Figure 9.12 shows the I-t curve of different 3D electrodes performed at 10.7 V (vs Ag/AgCl) in 0.1 M NaOH solution by addition of different concentration of glucose. It is clear that well-defined and fast amperometric responses are observed, except the pure ZnO nanowires electrode, which also has no response to glucose in I-t curve.

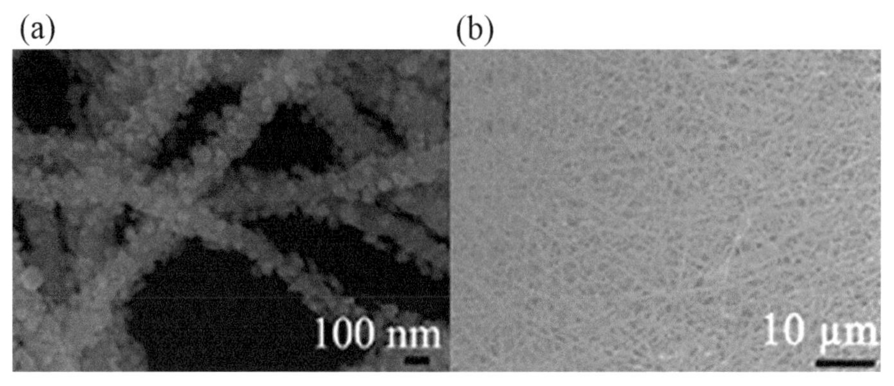

Fig. 9.11: (a) and (b) typical SEM image of 3D porous ZnO–CuO HNCs electrode (20 minutes) on the surface of FTO (Zhou et al. 2014).

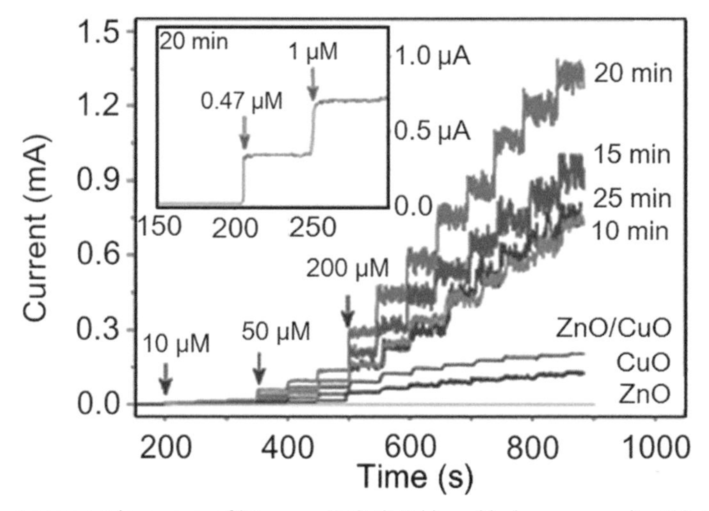

Fig. 9.12: Amperometric response of 3D porous ZnO–CuO hierarchical nanocomposites (10, 15, 20, and 25 minutes) electrodes as well as 3D mixed ZnO/CuO, 3D pure CuO and ZnO nanowires electrodes at an applied potential of 0.7 V upon successive additions of different concentration of glucose in a step of 10, 50, and 200 mM, respectively for each current step, inset is the current response of 3D porous ZnO–CuO hierarchical nanocomposites (20 minutes) to 0.47 and 1 mM glucose (Zhou et al. 2014).

Medical/Biological Applications

Wound Dressing

The nanofibrous mats used as wound dressing should possess biocompatibility, biodegradability, non-toxicity, and good mechanical properties. Many natural and synthetic materials such as gelatin, chitosan, PCL, and PLA meet these properties. The nanofibrous mats fabricated by E-spinning technique have porous structures that facilitate gas and liquid exchange in wound site. In addition, compared to traditional wound dressing, the E-spun nanofiber mats for wound dressing can keep a moist environment, which is necessary for wound healing process. The high specific surface area can also promote the absorbance of exudate. The recent application of wound dressing mainly focused on rapid hemostasis and good antibacterial properties.

Antibacterial Property. There are several antibacterial materials such as silver nanoparticles, antibiotics, triclosan, chlorhexidine, quaternary ammonium compounds (QACs), biguanides, and metal oxide nanoparticles have been used to fabricate antibacterial nanofibers. Recent works on antibacterial nanofibers fabricated by electrospinning are shown as Table 9.3. For example, it is well known that silver shows broad-spectrum antimicrobial activity against two types of common pathogens, that is, *Staphylococcus aureus* and *Escherichia coli*. Silver ions can destroy bacterial cell membranes at very low concentrations or make the enzyme activity lost by destroying cells' respiratory and nutrient transport systems that bring about cell death. It is reported that Barani (Barani 2014) fabricated antibacterial nanofibrous mats of PVA and poly-l-lactide acid (PLLA) with different concentrations of silver nanoparticles. The antibacterial efficiency of all samples was over 99.99%. Nguyen et al. (Nguyen et al. 2010) utilized methods that combine E-spinning and microwave-assisted process to fabricate silver nanoparticles loaded onto PVA mats. Magainin II (Mag II) was also demonstrated to have antibacterial property by Yüksel et al. (Yüksel and Karakeçili 2014), which was immobilized on electrospun nanofibrous mats of electron poly(lactide-co-glycolide) (PLGA) and PLGA/gelatin. The antibacterial activity of Mag II against both *Escherichia coli* and *Staphylococcus aureus* was tested through the phenomenon of the attachment and survival of bacteria inhibited in Mag II area. Calama et al. showed that E-spun PEI/silk fibroin (SF) mats could keep gram-positive and gram-negative bacterial from clinging through adding 10, 20, and 30 percent PEI into SF solutions (Çalamak et al. 2014). Cytotoxicity results demonstrated that all SF and PEI/fibroin extracts have no cytotoxicity on L929 cell lines. Dhand et al. took the advantage of the strong interfacial interaction between polyhydroxy antibiotics (changing number of single bond –OH groups) and gelatin to *in situ* crosslinking polydopamine (PDA) via ammonium carbonate diffusion method (Dhand et al. 2017). This approach that antibacterial and antifungal drugs incorporated into nanofibrous mats can be developed to produce mats with broad spectrum antimicrobial properties. Dong et al. prepared electrospun PCL nanofibers doped with silver/silica nanoparticles to have the bactericidal and anti-inflammatory effect (Dong et al. 2016). In the study, silver nanoparticles were incorporated into mesoporous silica to avoid particle aggregation and oxidation, which might lead to the loss of antibacterial activity. The obtained dressing exhibits antibacterial properties, sterilization, anti-inflammatory, promoted wound healing, and many other features.

Rapid Hemostasis. Traditional rapid hemostasis methods are usually mechanical methods that use bandage or stress, thermal devices like heater or electrodes, and chemical drugs spray or smear on wound surface to make the bleeding blood clotting. The disadvantages of these traditional methods are apparently such as the inaccurate clots formation, low adhesion to the raw wound surface, secondary damages to the wound, and poor comfort. On the contrary, the nanofibrous mats used for rapid hemostasis fabricated by E-spinning through adding some hemostatic components into fibers have attracted great of interests. This is because E-spun

Table 9.3: Recent works on antibacterial nanofibers fabricated by E-spinning.

Materials	**Antibacterial agents**	**Method of incorporation**	**Bacterial species**	**References**
PLA/PCL	Tetracycline	Blending	*Staphylococcus aureus*	Zahedi et al. 2012
PLGA	Cefoxitin	Blending	*Staphylococcus aureus*	Kim et al. 2004
PLA	Mupirocin	Blending	*Staphylococcus aureus*	Thakur et al. 2008
coPLA, coPLA/PEG	Ciprofloxacin	Blending	*Staphylococcus aureus*	Toncheva et al. 2012
PLA, PLA/Collagen	Gentamicin	Core/sheath	*Escherichia coli, Staphylococcus epidermidis, Pseudomonas aeruginosa*	Torres-Giner et al. 2012
PLLACL	Tetracycline	Core/sheath	*Escherichia coli*	Torres-Giner et al. 2012
PMMA/nylon 6	Ampicillin	Core/sheath	*Listeria innocua*	Sohrabi et al. 2013
PLGA	amoxicillin	Encapsulation	*Staphylococcus aureus*	Wang et al. 2012b
PAA	doxycycline hyclate	Blending	*Staphylococcus aureus, Streptococcus agalactiae*	Khampieng et al. 2014
Cyclodextrin inclusion complexes	triclosan	Blending	*Escherichia coli, Staphylococcus aureus*	Celebioglu et al. 2014
PCL/PLA	triclosan	Blending	*Staphylococcus epidermidis, Escherichia coli*	Valle et al. 2011
PLA	triclosan	Complexing with cyclodextrin	*Staphylococcus aureus, Escherichia coli*	Kayaci et al. 2013
PAN	N-halamine	Blending	*Staphylococcus aureus, Escherichia coli*	Ren et al. 2013
cellulose acetate (CA)	chlorhexidine	Blending	*Escherichia coli, Staphylococcus epidermidis*	Chen et al. 2008
PAN	quaternary ammonium salts	Blending	*Staphylococcus aureus, Escherichia coli*	Gli et al. 2013
CA/PEU	polyhexamethylene biguanide (PHMB)	Blending	*Escherichia coli*	Liu et al. 2012
PAN	polyhexamethylene guanidine hydrochloride	Covalent immobilization	*Staphylococcus aureus, Escherichia coli*	Mei et al. 2012
PEO/Chitosan	potassium 5-nitro-8-quinolinolate	Blending	*Staphylococcus aureus, Escherichia coli, Candida albicans*	Spasova et al. 2004
PDLLA/PEO	antimicrobial peptides	Blending	*Enterococcus faecium*	Heunis et al. 2011

PVDF	silver nanoparticles	Suspended in polymer solution	*Staphylococcus aureus, Klebsiella pneumoniae*	Liu et al. 2017b
PAN	silver nanoparticles	Synthesis in polymer solution	*Staphylococcus aureus, Escherichia coli, Bacillus subtilis*	Mahapatra et al. 2012
PLA/Chitosan	silver nanoparticles	*In situ* synthesis	*Staphylococcus aureus, Escherichia coli*	Hang et al. 2012
PEO/Chitosan	silver nanoparticles	*In situ* chemical reduction	*Escherichia coli*	An et al. 2009
PVA/Chitosan	silver nanoparticles	*In situ* synthesis	*Escherichia coli*	Hang et al. 2010
PVA/Chitosan	silver nanoparticles	*In situ* chemical reduction	*Escherichia coli*	Abdelgawad et al. 2014
PEO/carboxymethyl Chitosan	silver nanoparticles	Dispersion	*Pseudomonas aeruginosa, Escherichia coli, Staphylococcus aureus*	Fouda et al. 2013
TiO_2/nylon-6	silver nanoparticles	Photocatalytic reduction	*Escherichia coli*	Pant et al. 2011
silk fibroin	silver nanoparticles	Dispersion	*Staphylococcus aureus, Staphylococcus epidermidis, Pseudomonas aeruginosa*	Calamak et al. 2015
PVA	titanium dioxide (TiO_2)	Dispersion	*Staphylococcus aureus, Klebsiella pneumoniae*	Lee and Lee 2012
polyurethane (PU)	titanium dioxide (TiO_2)	*In situ* synthesis	*Pseudomonas aeruginosa, Staphylococcus aureus*	Yan et al. 2011
chitosan	sericin	Blending	*Escherichia coli, Bacillus subtilis*	Rui•et al. 2014
Nylon 6	ZnO	Electrospray on fiber surface	*Escherichia coli, Bacillus cereus*	Vitchuli et al. 2011
PMMA	ZnO/TiO_2	Synthesis in solution	*Escherichia coli, Staphylococcus aureus*	Hwang and Jeong 2011
PVP	Nano-Silver	Blending	*Staphylococcus aureus, Klebsiella pneumoniae and Escherichia coli*	Hwang and Jeong 2011
PAN	chitosan	Blending	*Escherichia coli, Staphylococcus aureus, Micrococcus luteus*	Kim and Lee 2014
PAA/PCL	quaternized chitosan	Core/shell	*Staphylococcus aureus, Escherichia coli*	Kalinov et al. 2014
PVA	Honey/chitosan	Blending	*Staphylococcus aureus*	Sarhan and Azzazy 2015
PEO	chitosan	Blending	*Staphylococcus aureus, Escherichia coli*	Liu et al. 2017b
PLA	Chitosan	Coaxial	*Escherichia coli*	Nguyen et al. 2011

wound dressing possesses high specific surface area and porous structures which benefit the liquid absorption and drug release, then help the wound closure and stop bleeding. Nowadays, many studies have been reported to accelerate hemostasis using E-spun wound dressing, including different hemostatic components and apparatus for E-spinning.

Among many studies on hemostatic components nanofibers, E-spun poly-(ε-caprolactone) (PCL) nanofibrous mats loaded by berberine appear to stop *in vitro* bleeding and inhibit bacteria breeding to protect the wound from infection (Bao et al. 2013); they can absorb exudates to build a comfort environment, and speed the cells grow in the healing process. The average diameter was 300 nm, which was fabricated by E-spinning the mixture of gelatin, chitosan, and polyurethane, and then treated with silver nitrate solution. It finally showed good hemostatic compared to gauze. Xia et al. have developed a biodegradable tri-layered barrier that consists of an E-spun PLGA/PLA-b-PEG layer as the interlayer between carboxymethyl chitosan layers (Xia et al. 2015). The chitosan layers aim to realize the rats' hemostasis and stop the shape of adhesion. Jiang et al. (Jiang et al. 2014) and Dong et al. (Dong et al. 2015) have designed a kind of electrostatic spraying device based on traditional E-spinning technology. And this device was successfully used for E-spinning OCA medical glue into micro or nanofibers. The nanofibers were deposited on the targeted wound surface precisely with the help of airflow to achieve a complete, effective, and reliable hemostasis. Compared to other approaches, this technique can clean up the wounds first and then very rapidly, precisely deposit nanostructured hemostats on complex, irregular, large area wounds to form some continuous, compact, flexible mats with excellent integrity, which act as a powerful barrier to stop bleeding immediately and prevent possible blood or bile seeping (Fig. 9.13). Due to the unknown harm of high voltage and other factors, traditional E-spinning technology choose to deposit materials onto a collector first, followed by transferring the material onto the wound. The whole process is very cumbersome, so Long et al. operated another E-spinning device: a handy E-spinning device (Xia et al. 2015, Dong et al. 2016), which employed ingeniously two batteries and a converter to provide operating voltages up to 10 kV, having the potential application in rapid hemostatic treatments or wound healing (Fig. 9.14).

Drug Delivery

In recent years, nanofibers as the drug storage/release systems have attracted much attention for their excellent performance in drug delivery—especially by means of E-spinning, which not only has high surface area, controllable porosity, and good biocompatibility, but can also slow degradation in the body and reduce drug burst release to ensure the sustained release of drugs in the target tissue. These nanofibers can be used in inhalation therapy and cancer treatment to reduce the pain of patient, etc. Table 9.4 summarizes some drug-loaded E-spun nanofibrous mats.

In drug delivery system, the types of drugs and the suitable biocompatible polymers are considered. E-spinning provides a way to encapsulate drug or biomolecules into polymer nanofibers. It has been reported that drugs for drug delivery systems include anticancer drugs, antibiotic, growth factor, protein, bacteria

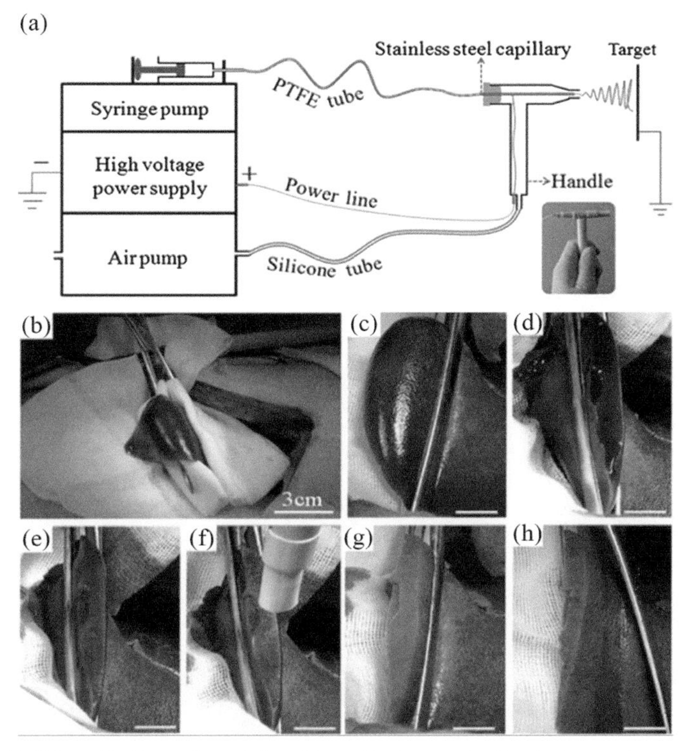

Fig. 9.13: (a) Schematic illustration of the airflow-directed *in situ* E-spinning device and (b–h) the complete process of *in vivo* rapid hemostasis in pig liver resection (Jiang et al. 2014).

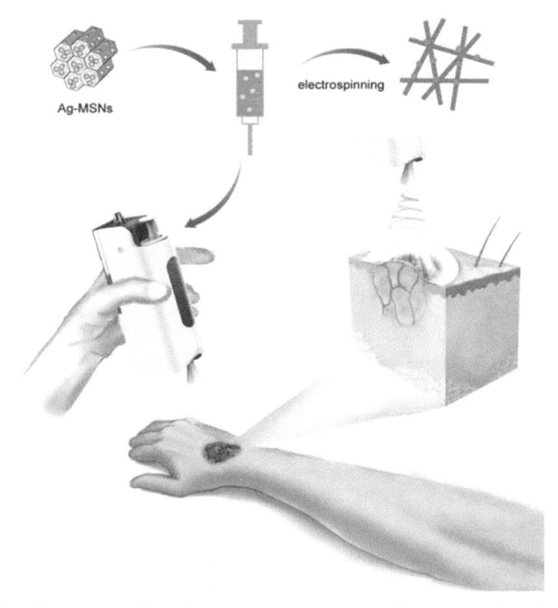

Fig. 9.14: Schematic illustration of the E-spun personalized nanofibrous dressing via a handy E-spinning device for wound healing (Dong et al. 2016).

Table 9.4: E-spinning nanofibrous coated drug species.

Drug types	Coated drug	E-spinning polymer	References
Anticancer drugs	Doxorubicin	PLGA	Zheng et al. 2013
	5-Fluorouracil (FU)	PCL	Iqbal et al. 2017
	Paclitaxel	PCL	Iqbal et al. 2017
	Oxaliplatin	PLA	Zhang et al. 2014
	Curcumin	PVP	Wang et al. 2015
Antibiotic	tetracycline hydrochloride	halloysite nanotubes/poly-(lactic-co-glycolic acid)	Qi et al. 2013
	Metronidazole	polydioxanone (PDS)	Albuquerque et al. 2015
	ciprofloxacin	PDS	Albuquerque et al. 2015
	minocydine	PDS	Albuquerque et al. 2015
	chlorhexidine	gelatin	Albuquerque et al. 2015
	Penicillin	polylactide-polyglycolide	Chao et al. 2015
Growth factor	PDGF-BB	chitosan and PEO	Xie et al. 2013
	VEGF	chitosan and PEO	Xie et al. 2013
	fibroblast growth factor 18 (FGF18)	PEO/PCL	Kang et al. 2015
	FGF2	PEO/PCL	Kang et al. 2015
	Nerve Growth Factor (NGF)	Silk fibroin-PEO	Dinis et al. 2015
	Ciliary Neurotropic Factor (CNTF)	Silk fibroin-PEO	Dinis et al. 2015
Protein	anti-CD20 antibodies	PCL	Cohn 2016
	fibronectin	PCL–gelatin	Lee et al. 2014
	osteocalcin	PCL–gelatin	Lee et al. 2014
Nano silver particles	nano silver particles	PCL	Dong et al. 2016
Gene therapy	pDNA	PCL-PEG	Saraf et al. 2010

and viruses, and silver nanoparticles. The key function of designing a drug delivery system is to transport the various drugs into the body's target in a safe manner and to regulate the release mechanism by controlling the amount of drug and the time of treatment.

Anticancer Drugs. As antitumor drugs have strong cytotoxicity, these drugs are more stringent for *in vivo* release. It needs slow and stable release to avoid the drug content in the blood being too high. In addition, the drug-loaded fibers obtained by the E-spinning method are commonly used for tumor localization therapy and adjuvant therapy after cancer surgery. The advantages of these nanofibers are sustained release, targeted drug delivery, and reduced side effects. Chen and his colleagues (Chen et al. 2015) reported multifunctional E-spun composite fibers for orthotopic cancer treatment *in vivo*. Doxorubicin (DOX) were incorporated into poly(ε-caprolactone) (PCL) and gelatin loaded with antiphlogistic drugs of indomethacin

(MC) to form nanofibrous fabrics via E-spinning process. In the research, DOX and MC are sustained release. PCL and gelatin nanofibers as drug carriers have good biocompatibility. *In vivo* experiments showed that DOX-loaded nanofibers had a good anti-cancer effect.

Antibiotics. Some antibiotics are insoluble in water or are difficult to be absorbed by the body. Therefore, encapsulating antibiotics into nanofibers which have porous and large surface area can effectively expand the surface area of the drug. These fibers containing antibiotics are promising for applications in postoperative antibacterial and wound dressing.

Torres-Giner and his coworkers encapsulated gentamicin antibiotic into PLA/collagen biodegradable fibers (Torres-Giner et al. 2012). For comparison, they prepared three different nanofiber membranes which are pure PLA fibers, gentamicin encapsulated in PLA–collagen fibers, and gentamicin encapsulated in PLA/collagen coaxial fibers, respectively. The results show that the drug-encapsulated coaxial fibers are promising drug delivery vehicles with strong and time-controllable antibacterial properties.

Growth Factors. E-spun fibers can not only be used for controlled release of drug, but also can be used for tissue engineering in the biological template, which can release growth factor to assist cell growth. Xie and his coworkers prepared multi-functional nanofibers with dual growth factor releasing for wound healing (Xie et al. 2013). In this research, Platelet Derived Growth Factor-BB (PDGF-BB) and Vascular Endothelial Growth Factor (VEGF) are encapsulated in a biomimetic nanofibrous scaffold. VEGF with designed quick releasing helped angiogenesis in an early stage of the healing process, while PDGF-BB with slow expelling improved the epithelium regeneration, collagen deposition, and functional tissue remodeling.

Proteins. A good biological activity of the proteins is required for long-term and controllable proteins/drugs delivery in E-spun fibers, and they also need to avoid the influence from the surrounding environment. Wang et al. (Li et al. 2015) prepared core-shell nanofibers with proteins encapsulation by emulsion E-spinning. They encapsulate water-soluble protein (protein-BSA as the core) into hydrophobic polystyrene (PS) polymer via emulsion E-spinning technique. By tailoring the molecular weight of PS, they can control the release rate of fibers (as shown in Fig. 9.15).

Silver Nanoparticles. Dong et al. prepared PCL nanofibers loaded with Ag-MSNs (mesoporous silica nanoparticles decorated with silver nanoparticles) by a novel handy electrospinning device (Dong et al. 2016), as shown in Fig. 9.16. Rat experiments showed that these fibrous membranes had good biocompatibility and antibacterial activity.

Other Drugs. Sarioglu et al. encapsulate Pseudomonas aeruginosa bacteria in E-spun PVA and PEO nanofibrous webs (Sarioglu et al. 2017). The viable cell 15 counting (VCC) assay and fluorescence microscopy imaging were used to confirm the bacterial cell viabilities. Such fibers can be used in removing methylene blue in water. Anita Saraf et al. prepared coaxial E-spun fiber mesh scaffolds with a non-viral gene delivery vector (r-PEI-HA) as the sheath and pDNA as core, respectively (Saraf

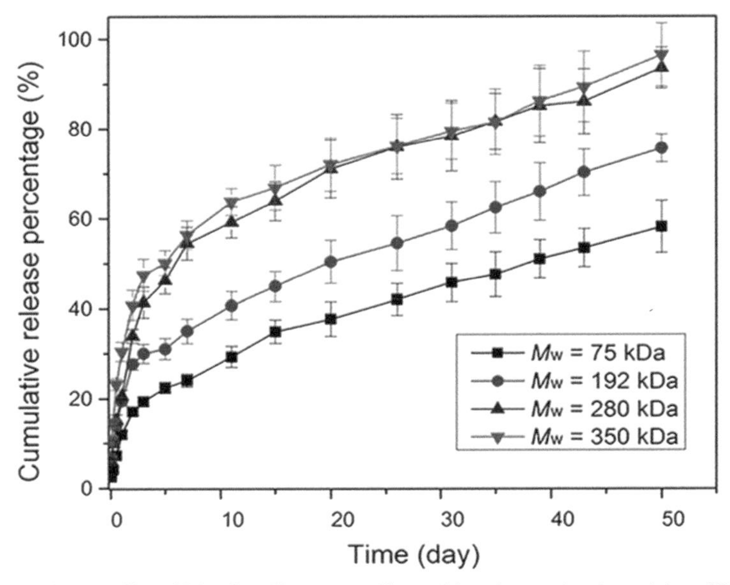

Fig. 9.15: Release profiles of BSA from E-spun nanofibers with various molecular weights of PS in PBS (pH 7.0) at room temperature (Li et al. 2015).

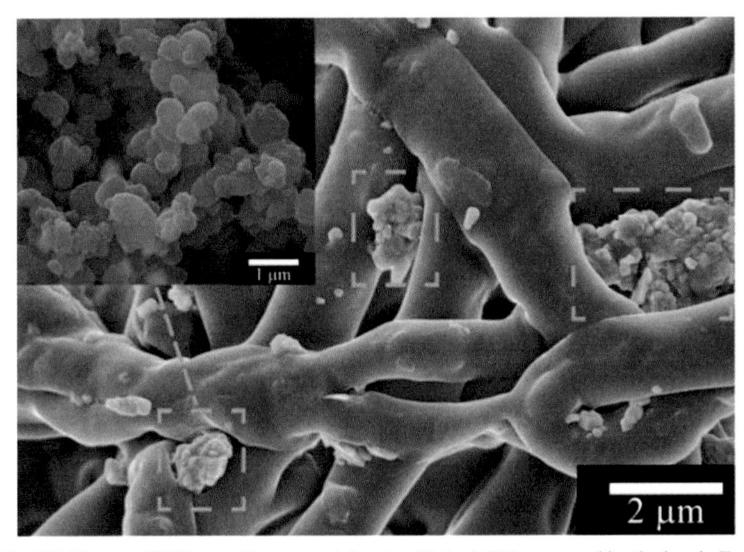

Fig. 9.16: SEM image of PCL nanofibers containing 5 wt% Ag-MSNs prepared by the handy E-spinning device, inset: SEM image of Ag-MSNs (Dong et al. 2016).

et al. 2010). The effects of different processing parameters on the fiber diameter are investigated. Such scaffolds with variable and sustained transfection properties can be used in tissue engineering and other gene delivery applications involving gene therapy.

Tissue Engineering

Organ failure and tissue damage are common problems usually arising from disease or injury, which cannot be self-repaired during a patient's lifetime, having serious impacts on normal life and health quality. Tissue engineering, also known as regenerative medicine, is a promising solution to these defects for they can generate necessary tissues to repair or replace damaged organs. In the developing of tissue engineering construct, four key elements should be taken into consideration, that is, biomaterials scaffolds, cells, extracellular matrix, and growth factors. Among these elements, the biomaterials scaffold is the most essential element, for it provides basal body for cell growth, and thus the scaffold should possess properties of biocompatibility, biodegradability, sterilizability, porosity and mechanics self-adaptability. E-spinning technique is a good method to prepare 3D porous nanofibrous membranes that imitate the natural structure and function of extracellular matrix. These nanofibrous membranes can be used as ideal biomaterials scaffolds to facilitate cell adhesion, proliferation, migration, and differentiation, and finally to duplicate and replace damaged organs. In recent years, biomaterials scaffolds prepared by E-spinning technique have been widely used in tissue engineering studies such as skin, bone, vascular, nerve, and so on.

Skin Tissue Engineering. Autograft and allograft are common methods used for skin injury. However, skin injury is difficult to treat and these traditional methods cannot satisfy the increasing demands for skin treatments. E-spun nanofibers as scaffolds are good candidates for skin treatments due to their ideal properties, such as porosity, biocompatibility, hydrophilic surface, and controllable biodegradability. Recently, much work has been done using E-spinning technique to explore new materials for scaffolds, especially natural polymers (chitosan, gelatin, collagen, fibrin, etc.) and biodegradable polymers (PGA, PCL, PLGA, PDLLA, etc.).

Chen et al. grafted ε-caprolactone oligomers onto the hydroxyl groups of chitosan and prepared cationic nanofibrous mats by e-spinning chitosan-graft-poly (ε-caprolactone) (CS-g-PCL) mixed solution (Chen et al. 2011). These nanofibrous mats show improved cellular adhesion property, and L929 cells grew well on the CS-g-PCL/PCL mats, which is a potential scaffold for skin tissue engineering. The shrinkage of E-spun mats is unfavorable for the triggering of cell adhesion. Ru et al. prepared PLGA nanofibers via E-spinning that were assembled by heat sealing on a polypropylene auxiliary supporter (Ru et al. 2015). The obtained scaffolds keep long-term integrity without dimensional shrinkage. Keratinocyte cells were seeded on the scaffold and exposed to air. The results show that human skin cells of keratinocytes can proliferate on the scaffold and infiltrate into the scaffold, which can be used in skin tissue engineering.

Bone Tissue Engineering. Bone defect is a common problem in orthopedic fields. For a big area of bone defect, it can not heal by itself, so bone grafts are usually required. An ideal synthetic bone substitute for repairing human bones should mimic the mechanical and physiological characteristics of bones, and also should possess multiple functions, such as protecting internal organs and supporting muscular contraction. Bone tissue engineering using E-spinning technique provides an emerging therapy for patients suffering from the bone defect. Different kinds of bone scaffolds have been reported for bones repairing.

For example, natural polymer cellulose derived hydroxyethyl cellulose (HEC) scaffolds were prepared by E-spinning. PVA was used as a solvent for supporting the E-spinning of HEC. The HEC/PVA scaffolds show good biocompatibility and increased cell proliferation, indicating that HEC based nanofibrous scaffolds is a promising candidate for bone tissue engineering (Chahal et al. 2016). Li et al. prepared PET fibers and nano-silica modified PET fibers via E-spinning (Li et al. 2015). Experimental results show that HOB cells have a better spreading behavior on silica-modified PET fibers, which can accelerate the repairing of bone defects. Li et al. also fabricated PLGA and PLGA/nanohydroxyapatite (nano-HA) composite scaffolds (Li et al. 2013). The nano-HA in PLGA/nano-HA composite scaffold could significantly enhance the formation of the bone-like apatites. Experimental results of cell viability, alkaline phosphatase (ALP) activity, and osteocalcin concentration demonstrated that the PLGA/nano-HA fibers could promote the proliferation of HOB, indicating that it is a promising scaffold for human bone repair.

Vascular Tissue Engineering. Cardiovascular diseases are still one of the largest causes for death, and half of deaths worldwide are due to the failure of vascular system. With the increasing number of patients diagnosed with cardiovascular diseases every year, the vascular grafts are in great demand, but in short supply. At present, the *in vitro* preparations of blood vessels with multiple functions are still full of challenges. E-spinning technique could tune the composition and mechanical property of nanofiber scaffolds, which could satisfy the demands for blood vessels. Recently, many relevant researches have been carried out.

De Valence et al. used PCL to prepare a porous and cell-friendly scaffold because of its good mechanical strength and biocompatibility (de Valence et al. 2012). Long term *in vivo* performance of PCL vascular grafts was investigated. This study showed that E-spun PCL is an excellent scaffold for shelf ready vascular graft applications. Stefani and Cooper-White studied the fabrication of slowly degradable, composite tubular polymer scaffolds made from PCL and acrylated L-lactide-co-trimethylene carbonate (aPLA-co-TMC) (Stefani and Cooper-White 2016). The core-shell fibers were used in the study, in which the core is composed of PCL and the shell is crosslinked polymers of aPLA-co-TMC. The resulting fibrous scaffolds show burst pressures and suture retention strengths, which is comparable to human arteries.

Augustine et al. prepared E-spun PCL scaffolds incorporated with zinc oxide nanoparticles (Augustine et al. 2014). A large number of matured blood vessels with highly branched capillary network were observed in the scaffolds. This is the first report on PCL membranes containing oxide nanoparticle. The results show that the fabricated membranes can induce angiogenesis, which can be used as a successful tissue engineering scaffold material.

Nerve Tissue Engineering. A human will lose his perception if the nervous system is damaged. However, the regeneration capacity of nervous system for adults is limited. The major issue for nerve repair is the non-renewal of neurites in human body. At present, the common method for nerve repair is to construct artificial structures that mimic extracellular matrix and provide axon guidance cues to stimulate the re-growth of native tissues. Among these methods, artificial nerve conduits prepared by E-spinning technique have attracted much attention. A tubular construct with a highly aligned fibrous structure can imitate the endoneurium layer around the inner axons of a nerve fascicle, which is a suitable candidate for nerve guides.

Prabhakaran et al. fabricated both random and aligned nanofibers of PHBV to evaluate their potential use as substrates for nerve regeneration (Prabhakaran et al. 2013). Nerve cells of PC12 were cultured and grown on the E-spun nanofibers. Compared to random nanofibers, they observe that aligned nanofibers can direct the growth orientation of nerve cells along the nanofibers direction, which is a promising substrate for applications in bioengineered grafts. Many natural and synthetic polymers can be used for nerve tissue engineering, such as, silk fibroin-PLGA (Liu et al. 2017a), as silk fibroin-P(LLA-CL) (Zhang et al. 2013), polypyrrole-PLLA (Jin et al. 2012), P(LLA-CL)-collagen (Kijeńska et al. 2012), polypyrrole-cellulose (Thunberg et al. 2015), PGS-PMMA (Kai et al. 2016). Among these polymers, shape memory materials for nerve tissue engineering have also attracted large attention. Hu et al. (Hu et al. 2017) reported a preparation of porous shape memory scaffolds with both biomimetic structures and electrical conductivity properties. This new kind of shape memory polyurethane polymers consists of inorganic PDMS segments with organic PCL segments. A series of conductive nanofibers were E-spun by doping with different amounts of carbon-black, and their resistivity reduced from 3.6 GΩ mm^{-1} (undoped) to 1.8 kΩ mm^{-1} (doped). Experimental results showed that carbon black did not apparently influence their shape memory properties in nanofibers. PC12 cells were cultured on these shape memory nanofibers, and these composite scaffolds showed a good biocompatibility, which could be potentially used as smart 4-dimensional scaffolds for nerve tissue regeneration.

Catalysis

In a low dimensional scale, the catalyst usually presents significant surface effect and size effect. Due to the low-dimensional structure, catalyst has a large specific surface area, which can greatly increase the number of atoms on the surface, so that the catalyst has a higher surface activity. However, the powdery low-dimensional catalyst particles are easily agglomerated, thereby affecting their dispersion, separation, and recycling. As a catalyst carrier, the E-spun fibers can be a good way to overcome the above shortcomings owing to its small fiber diameter, good flexibility, and ease of operation. At present, E-spun fibers are commonly used in fuel cell electrode catalytic carriers, photocatalyst, biosensor catalysts, and gas sensor catalytic carriers.

Fuel Cell Electrode Catalyst Carrier

In recent years, excessive exploitation of fossil fuels has caused energy crisis, environmental pollution, and other issues. Among many solutions, the development of new energy is one of the effective ways to solve the above problems. A fuel cell is an efficient power plant that converts the chemical energy of a fuel into electrical energy by electrochemical reaction without direct combustion. The principle of a fuel cell is the same of a general battery. The single cell is composed of positive and negative electrodes and the electrolyte. The difference is that the active material of the battery is stored inside the battery, thus limiting the battery capacity. The positive and negative of the fuel cell itself does not contain the active substance, just a catalytic conversion element. Therefore, the fuel cell is truly a chemical energy into the energy conversion machine. When the battery is in operation, the fuel and oxidant are supplied externally and react. In principle, as long as the reactants continue to enter, the reaction products continue to be excluded, the fuel cell can continue to generate electricity. The fuel can be hydrogen, methane, carbon monoxide, methanol, fuel, coal, etc. Fuel cell research is becoming a hot spot because of its high efficiency and energy conversion efficiency. However, the fuel cell also needs suitable electrode catalytic material to further reduce the activation energy of electrode reaction and improve the reaction rate. Therefore, the preparation of high performance and low-cost electrode catalytic materials has become the key to fuel cell development. Zhi et al. fabricated lanthanum strontium cobalt ferrite (LSCF) nanofibers by the E-spinning method (Zhi et al. 2012). The as-prepared LSCF nanofibers were used as the cathode of an intermediate-temperature solid oxide fuel cell (SOFC) with yttria-stabilized zirconia (YSZ) electrolyte. Due to the advantages of 3D nanofiber network cathode, such as high porosity, high percolation, continuous pathway for charge transport, good thermal stability at the operating temperature, and excellent scaffold for infiltration, the fuel cell exhibits a power density of 0.90 W cm^{-2} at 1.9 A cm^{-2} at 750ºC. Zhang and Pintauro fabricated nanofiber mat electrode in which Nafion was used as the ionomer component of the catalyst ink and poly(acrylic acid) (PAA) was chosen as the carrier polymer (Zhang and Pintauro 2011). The nanofiber-structured electrode was tested as the cathode in proton exchange membrane H_2/air and H_2/O_2 fuel cell membrane electrode assembly. E-spun nanofiber mats which have an average diameter of 470 nm were composed of Pt/C catalyst powder with Nafion and PAA binders. The as-prepared fuel cell achieved very high fuel cell power densities at a low Pt loading. The catalyst mass activity was exceptionally high at 0.23 Amg_{Pt}^{-1}. Due to a more uniform distribution of Pt/C particles and Nafion binder in a nanofiber membrane, which can provide a high electrochemically active catalyst surface area and better proton and oxygen transport to catalyst sites during fuel cell operation, the E-spun electrodes present excellent performance.

Photocatalyst

With the development of social development, environmental pollution has attracted more and more attention. Water pollution, indoor and outdoor air pollution, soil pollution has seriously affected people's health and normal life. Photocatalyst is a kind of material which can promote the chemical reaction of the material. Photocatalysis converts natural light energy into a chemical reaction. Most of the semiconductor photocatalysts are n-type semiconductor materials which have a unique band structure different from that of metal or insulating material. There is a band gap between valence band (VB) and conduction band (CB). Since the light absorption threshold of the semiconductor and the bandgap have the relation K = 1240/Eg (eV), the absorption wavelength thresholds of the commonly used wide bandgap semiconductors are mostly in the ultraviolet region. When the incident photon energy is higher than the bandgap of semiconductor, the semiconductor absorbs the photon and photogenerate electron (e^-) on the CB and hole (h^+) on the VB. The photogenerated electrons and holes may migrate to the semiconductor's surface. At this time, the dissolved oxygen in the reaction system trapped electrons adsorbed on the semiconductor's surface form a superoxide anion ($\bullet O_2^-$). The photogenerated combined with the hydroxide ions adsorbed on the surface of the catalyst form the hydroxyl radicals ($\bullet OH^-$). The superoxide anion ($\bullet O_2^-$) and hydroxyl radicals ($\bullet OH^-$) with a strong oxidation can degrade the vast majority of organic matter and even some of the inorganic can also be completely decomposed. Benefit from the photocatalytic performance of semiconductor, the water, or airborne organic pollutants can be completely degraded into carbon dioxide, water, and inorganic acids at room temperature. Compared to conventional photocatalytic materials, E-spun nanofibers have a large specific surface area and can be more fully contacted with the reaction components. On the one hand, the spinning solution containing the inorganic precursor solution can be E-spun into nanofibers, followed by sintering the fibers to obtain fibrous photocatalyst. On the other hand, the inorganic photocatalytic material can be directly supported on the organic high molecular material fiber flexible photocatalytic composites. So far, the photocatalyst obtained from these methods has been widely used in the treatment of waste water and waste gas.

Arun et al. synthesized coral and rice shape TiO_2 from acid etching TiO_2–ZnO composites fabricated by E-spinning followed by a sintering process (Arun et al. 2014). The rice-like TiO_2 were ~ 500 nm and ~ 250 nm in length and breadth, respectively. The coral and rice shape TiO_2 present good photocatalytic performance in the photodegradation of methyl orange dye and phenol under UV light. The coral and rice shape TiO_2 present 92% and 75% photocatalytic efficiency in photodegradation of methyl orange after UV irradiation for 50 minutes. The coral and rice shape TiO_2 present 72% and 48% photocatalytic efficiency in photodegradation of phenol after UV irradiation for 100 minutes. Zhang et al. synthesized ZnO-SnO_2 nanofibers by E-spinning solution containing the inorganic precursor solution followed by a sintering process (Zhang et al. 2010). The diameters of the ZnO-SnO_2 nanofibers were in the range from 100 to 150 nm. The average size of ZnO and SnO_2 grains were 42 and 13 nm, respectively. The ZnO-SnO_2 nanofibers present excellent photocatalytic performance in the photodegradation of Rh B dye under UV light, due to the formation of a ZnO-SnO_2 heterojunction and its high specific surface area. The RhB dyes were completely photodegradated in 30 minutes.

In addition to the preparation of inorganic nanofiber photocatalysts, a series of new catalysts with photocatalytic particles supported by E-spun fibers have been extensively studied. Tan et al. fabricated various morphology rutile TiO_2 stuctures on amorphous tetrabutyl titanate@PVDF nanowires (Tan et al. 2012). Before the hydrothermal treatment, the amorphous tetrabutyl titanate was mainly distributed in the core of tetrabutyl titanate@PVDF nanowires. After hydrothermal treatment, some leaf-like rutile TiO_2 stuctures appeared on the surface of tetrabutyl titanate@ PVDF nanowires. As the increase of the amount of tetrabutyl titanate added to the hydrothermal reaction solution, the morphology of the rutile TiO_2 changed from dendrite-like and leaf-like to flower-like shape. Photomethanation researches with the as-prepared sample proceeded until the 24th hour of UV irradiation. The methane yield exceeded 35 μmol g^{-1} h^{-1} under UV light illumination.

In addition to TiO_2 and ZnO, there are many photocatalysts prepared by E-spinning methods are widely studied, such as Cr-$SrTiO_3$, Pt-ZnO, $Bi_4Ti_3O_{12}$, TiO_2-Graphene, SnS_2-TiO_2, MoS_2/CdS-TiO_2, CdS/ZnO, etc. (Hou et al. 2015, Hou et al. 2013, Qin et al. 2017, Yang et al. 2013, Yu et al. 2013, Nair et al. 2012, Zhang et al. 2013).

Enzyme Catalyst Carrier

Enzyme refers to a kind of protein produced by the organism with high efficiency and specific catalytic function. Enzyme catalysts and viable cell catalysts can be called biocatalysts. In the body of a creature, the enzyme is involved in catalyzing almost all of the material transformation process. Therefore, the enzyme and life activities are closely related. *In vitro*, enzymes as a catalyst for industrial production have a high conversion efficiency in this respect—especially under mild conditions (room temperature, atmospheric pressure, neutral), it is extremely effective. The catalytic efficiency is 10^9–10^{12} times as compared to other non-biological catalysts. Enzyme catalyst selectivity (also known as specificity) is extremely high, meaning one enzyme usually only catalyzes one or a class of reactions, and can only catalyze the transformation of one or a class of reactants (also known as substrates), including stereoselective structural selectivity. In addition, compared to the living cell catalyst, the enzyme catalyst also has no side effects, is non-toxic, harmless, and separation process easy handling, as well as other advantages.

Due to the aforementioned advantages, the E-spun nanofibers can be used as a catalyst support carrier. The enzyme is loaded on the surface of the nanofibers, and the shape of the catalyst, size, and mechanical light can be easy control. So that it can meet the operational requirements of industrial reactions, it is easy to control the enzymatic reaction. At the same time, since the E-spun fibers have a large specific surface area, the catalytic efficiency of the active mass per unit mass is effectively improved. And the amount of enzyme loading and the degree of binding to the carrier can be adjusted by changing the nature of the spinning process and the properties of the polymer, thereby improving the operability and utilization of the enzyme catalyst. El-Aassar reported that β-glutamate and α-galactosidase could be immobilized on amino-functionalized poly(acrylonitrile-methyl methacrylate) (PAN-co-MMA) nanofibers (El-Aassar 2013), with the fiber of electrostatic as the supporting template. As a result, it was found that the immobilized enzyme had better heat resistance, pH resistance, and better stability than the free enzyme. This method opens up a new way to prepare E-spun fibers with biocatalytic activity that can effectively control enzyme activity, enzyme loading, and enzyme stability.

Laccase can catalyze the oxidation of phenols, chlorophenols, aromatic compounds, and other easily oxidizable compounds, so it can be applied to some polluting industries, such as textile printing and dyeing, paper making, so that pollutants can undergo biological degradation or transformation. However, if the free enzyme is treating contaminants directly, the enzyme is easily deactivated in the aqueous solution, and therefore limits its large-scale use, so it needs to be immobilized on the carrier to improve its application performance. Xu et al. used PAN as a template effectively removal of 2,4,6-trichlorophenol (TCP) from water by immobilizing the enzyme on the PAN nanofibrous membranes through ethanol/HCl method of amidination reaction (Xu et al. 2013). The immobilized laccase still exhibited 72% of the free enzyme activity and kept 60% of its initial activity after 10 operation cycles. The immobilized enzyme activity still remained at 92% catalytic activity compared to the free enzyme, and the free enzyme remained at only 92% catalytic activity after the same time (18 days). It is proved that the immobilized enzyme after a period of time can still maintain good stability activity, as shown in Fig. 9.17.

Lipase is widely found in animals and plants and microorganisms. Lipase is one of the important industrial enzyme preparations, and can catalyze the decomposition of lipids, ester exchange, ester synthesis, and other reactions, and is widely used in oil processing, food, medicine, cosmetics, and other industries. Different sources of lipase have different catalytic properties and catalytic activity. Huang et al. have prepared a highly efficient lipase catalyst (Huang et al. 2011). The cellulose nanofibers were prepared by E-spinning, and the Candida rugosa lipase was then immobilized on the fiber membrane. Studies have shown that lipase immobilization on fibers with a larger specific surface area has a low diffusion resistance and can effectively prevent lipase agglomeration, thereby improving its catalytic efficiency.

After the chemical treatment of fibers with ethylenediamine (EDA) and glutaraldehyde (GA) the urease (EC 3.5.1.5) was then covalently immobilized on dispersed microfibrous PAN mats to effectively catalyze the urea which was reported by Daneshfar et al. (Daneshfar et al. 2015). Then a new type of catalytic material

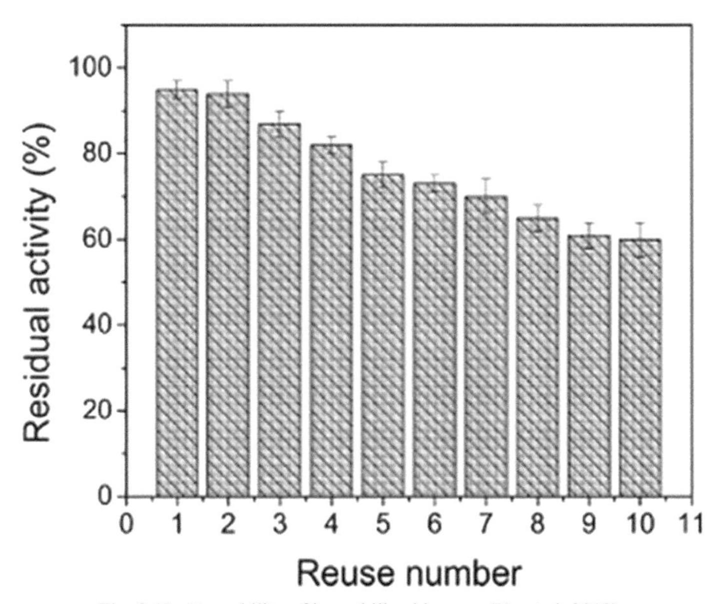

Fig. 9.17: Reusability of immobilized laccase (Xu et al. 2013).

was obtained. The amount of loaded urease reached 157 lg/mg mat, still exhibiting 54% catalytic activity of the free urease. The sensitivity of the immobilized urease to PH is also reduced, especially under acidic conditions. Compared to free urease, acid and alkali resistance has improved. After 15 cycles of experience, the activity of the immobilized enzyme was 70% less than the initial activity. New catalyst which is predicted may be applied in dialysate liquid regeneration system in artificial kidney machines.

Precious Metal Catalyst Carrier

Precious metals mainly involved metal, silver, and platinum group metals (rhodium, palladium, iridium, platinum), and other metal elements. As all metals are transition metal elements, with an empty *d* band with orbit, thus its reaction molecules are electrophilic, nucleophilic, and redox capacity. It is widely used in hydrogenation, dehydrogenation, oxidation, reduction, isomerization, synthesis, aromatization, and other reactions, and plays a catalytic role. Therefore, precious metal catalyst in the petroleum, pharmaceutical, environmental protection, chemical, and new energy field has a significant application value. In addition, in order to increase the specific surface area of the catalyst and to ensure that it has sufficient thermal stability and mechanical strength, activated carbon, porous ceramics and SiO_2 are used as carriers for noble metal catalysts in the industry. In recent years, the researchers prepared nanofibers fabricated by E-spinning technology with good conductivity, chemical stability, high specific surface area, and stable pore structure of as a noble metal catalyst carrier. Zhang et al. successfully prepared a new type of catalytic material for the uniform silver particles loading on the fibers combined with electrospinning and sol-gel technology using PVP as template (Zhang et al. 2011). The catalyst has a good catalytic effect on 4-nitrophenol. Since highly dispersed silver nanoparticles are exposed to the inner and outer surfaces of the E-spun SNT, it is allowed to interact effectively with the reactants and catalytic reaction. The material preparation process is shown in Fig. 9.18. In addition to E-spun fiber as a carrier can prevent the agglomeration of nano-silver particles, but also with the E-spun fiber composite, resulting in synergistic effect, to further enhance the catalytic performance of precious metal catalyst.

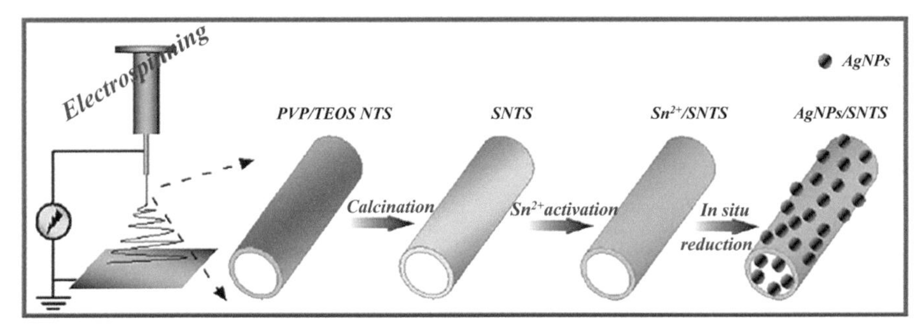

Fig. 9.18: Schematic diagram of the fabrication of AgNPs/SNTs tubular nanocomposites (Zhang et al. 2011).

Table 9.5: Catalytic materials synthesized by E-spinning method.

Catalyst type	Catalyst	E-spun polymer	Degration materials	References
Fuel cell catalytic	Lanthanum strontium cobalt ferrite (LSCF)	PAN	–	Zhi et al. 2012
	Pt/C	PAA	–	Zhang and Pintauro 2011
	Pt/NbO_2 @ Pt/Ti_4O_7	PVP	–	Senevirathne et al. 2012
	Pt/Nb-TiO_2	PEEK	–	Chevallier et al. 2012
Photocatalytic	TiO_2–ZnO	PVAc	methyl orange	Arun et al. 2014
	ZnO-SnO_2	PVP	RB	Zhang et al. 2010
	TiO_2	PVDF	–	Tan et al. 2012
	Cr-$SrTiO_3$	PAN	RhB	Hou et al. 2015
	Pt-ZnO	PEG	acid orange II	Yu et al. 2013
	$Bi_4Ti_3O_{12}$	PVP	RhB	Hou et al. 2013
	TiO_2-Graphene	PVAc	Methyl orange	Nair et al. 2012
	SnS_2-TiO_2	PVP	RhB, MO, 4-NP	Zhang et al. 2013
	MoS_2/CdS-TiO_2	PVP	lactic acid aqueous solution	Qin et al. 2017
	CdS/ZnO	PVP	Na_2S/Na_2SO_3 aqueous solution	Yang et al. 2013
	ZnO	PVP	Rhodamine B	Zhu et al. 2012
Enzyme catalyst	β-galactosidase	PMMA/PAN	glutaraldehyde	El-Aassar 2013
	Laccase	PAN	2,4,6-Trichlorophenol	Xu et al. 2013
	Lipase	cellulose nanofiber	–	Huang et al. 2011
	Urease	PAN	urea	Daneshfar et al. 2015
Precious metal catalyst carrier	Ag	PVP	4-nitrophenol	Zhang et al. 2011

Energy Conversion and Storage

Sensitized Solar Cells

Solar energy is a pollution-free and inexhaustible clean energy and has been a hot spot in the field of new energy. Dye-sensitised solar cells (DSSCs) due to their low cost, light weight, and high power conversion efficiency have attracted a lot of attention (Grätzel 2003). The working principle of the DSSC is that when the dye absorbs photons, excitions (bound electron-hole pairs) can be generated and released as free electrons and holes. The free electrons are injected into the adjacent semiconductor and transported for collection at the conductive substrate. Then, these electrons flow through the external circuit to the counter electrode and form a current (Thavasi et al. 2008).

Due to the large specific surface area, high porosity, and good mechanical properties, E-spun nanofibers have been introduced as the photoanode of DSSCs (Kumar et al. 2014). The high specific surface area and good interconnectivity of the E-spun nanofibers lead to the enhanced absorption of photosensitizing dye, better charge conduction, and reduced charge-carrier recombination. Furthermore, the high porosity of nanofibers is beneficial to the penetration of the viscous polymer gel electrolyte.

Based on these advantages, several recent works reported E-spun photoanode materials based DSSCs showed enhanced performance. Li (Li et al. 2012), fabricated highly transparent nanocrystalline TiO_2 films by adding an appropriate amount of tri-ethanolamine (TEOA) to the precursor. The photoluminescence (PL) spectroscopy test show that the films prepared via transmutation from E-spun nanofibers possess rich bulk oxygen vacancies, which result in large open-circuit voltage (V_{oc}) and fill factor (*FF*) improvements in DSSCs, and thus a large improvement of energy conversion efficiency. In Mali's work (Mali et al. 2014), kesterite Cu_2ZnSnS_4 (CZTS) nanofibers were obtained by E-spinning process using polyvinylpyrrolidone (PVP) and cellulose acetate (CA) solvent separately. CZTS nanofibers synthesized from PVP are a single crystalline, while CA assisted CZTS nanofibers are polycrystalline in nature. Then the CZTS nanofibers were served as counter electrodes for DSSCs. DSSC based on PVP-CZTS and CA-CZTS counter shows 3.10% and 3.90%, respectively. The present study showed that CZTS will be a potential alternative material in counter electrodes in DSSCs application.

Piezoelectric Nanogenerators

Nanogenerators (NGs) which can harvest energy from the surrounding local energy in our daily life into electricity have attracted great attention, especially, flexible NGs due to their potential application in self-powered wearable devices have been a hot topic. E-spun nanofibers showing high flexibility have made some progress in the field of flexible NGs. Several works have reported E-spun nanofiber based NGs by using piezoelectric materials.

PVDF, a piezoelectric polymer commonly used for E-spinning, has been used as an active material in NGs. PVDF is a semi-crystalline polymer with four distinct crystal phases: α, β, γ, and δ phase. Among all these phases, the β phase has the most favourable piezoelectric property. The key to achieve excellent piezoelectricity in PVDF is the formation of high β crystal phase content and well oriented molecular dipoles within the structure. In the E-spinning process, the applied high voltage, high stretching ratio of the polymer solution's jet, and high evaporation rate for the solvent also can effectively improve the formation of the desired phase (β-phase) in the fiber structure for power generation application (Gheibi et al. 2014).

In 2010, Chang et al. reported a NG fabricated by directly written PVDF onto flexible plastic substrate using near field electrospinning (see Fig. 9.19a), and conductive silver paste was used on either side as electrodes, as shown in Fig. 9.19b (Chang et al. 2010). The NG can effectively convert mechanical stretching energy

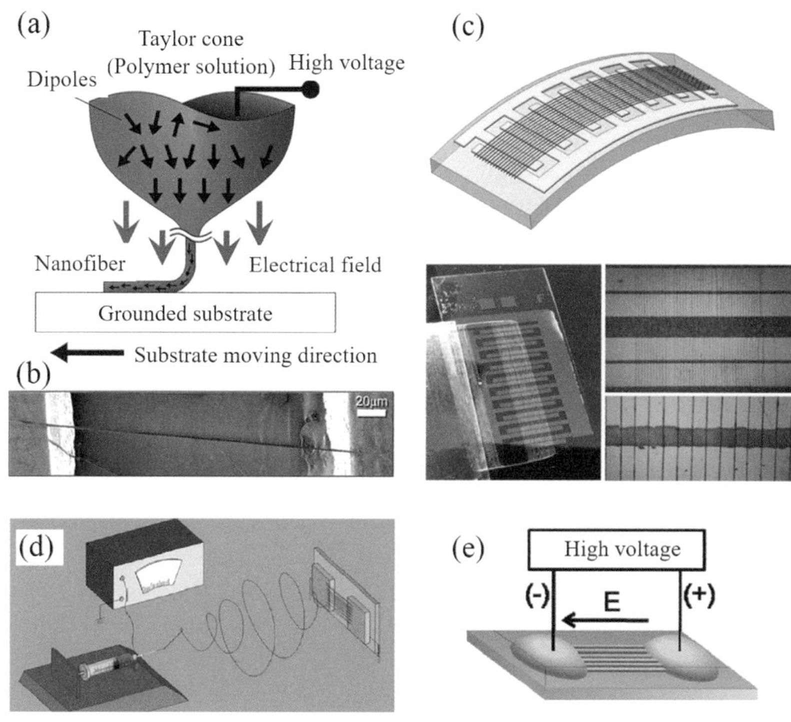

Fig. 9.19: (a) Schematic diagram of the near-field E-spinning process that creating and placing piezoelectric nanogenerators onto a substrate. During the formation of PVDF nanofiber the PVDF polymer solution experiences mechanical stretching and *in situ* electrical poling (Chang et al. 2010) Copyright 2010, American Chemical Society. (b) SEM image of the E-spun PVDF nanofiber on two contact electrodes (Chang et al. 2010) Copyright 2010, American Chemical Society. (c) Fabrication results showing an arrayed PVDF fiber structures and a comb-shape electrode on top a flexible substrate. And the optical images of the NG (Chang and Lin 2011) Copyright 2011, IEEE. (d) Schematic of the E-spinning process to obtain aligned PVDF nanofibers through two grounded copper pieces collector. (e) Schematic of the NG that these PVDF nanofibers were fixed with silver paste (Hansen et al. 2010) Copyright 2010, American Chemical Society.

into electrical energy. In order to increase the total electrical outputs, a sample method was increased the number of the PVDF fibers. Experimentally, a total of 50 parallel PVDF fibers were E-spun using the near field E-spinning process with a designed pattern on top of comb-shape gold electrodes fabricated on a flexible substrate, as shown in Fig. 9.19c (Chang and Lin 2011). The device was designed to amplify the current output using the parallel connection to collect possible current generations from each individual nanofiber, and the monitored peak current was about 30 nA. The parallel fibers can also be fabricated using conventional E-spinning. Hansen et al. have demonstrated an energy harvesting system combining a PVDF nanogenerator and a biofuel cell (Hansen et al. 2010). The E-spun fibers were collected onto two grounded copper pieces with a 2 cm gap and the fibers were electrostatically aligned across the electrode gap, as shown in Fig. 9.19d. Then the nanofibers that were fixed with silver paste (see Fig. 9.19e).

The active sites of above-mentioned well-aligned PVDF nanofiber nanogenerators were very limited. Fang et al. reported that randomly oriented E-spun PVDF nanofiber membrane can be directly used to prepare a piezoelectric NG by sandwiching the PVDF membrane between two metal foils (Fang et al. 2011). Under impacting at a high frequency, the PVDF membrane was able to generate up to 7 V. The energy output was able to be stored in a capacitor and then light up an LED. Subsequently, several works were focused on increasing content of the β phase to improve the piezoelectric output. Dhakras et al. reported that the addition of the nickel chloride was found to enhance the β phase by about 30%. The voltage generated for the PVDF-nickel chloride E-spun nanofiber membrane was a factor of 3 higher than that for pure PVDF nanofiber membrane (Dhakras et al. 2012). In Yu's work, the addition of multi-walled carbon nanotubes in the PVDF fibers promote the formation of the β phase (Yu et al. 2013). Fang's work prepared PVDF nanofiber membrane by using needleless E-spinning technology. It found that, with increasing the applied voltage in the E-spinning process, a higher β phase was formed in the resulting PVDF nanofiber membrane, leading to enhanced piezoelectric output of the device (Fang et al. 2013).

In addition to PVDF, there are some inorganic materials used in the preparation of piezoelectric NG by E-spinning. Wu et al. reported a flexible and wearable NG made by using a lead zirconate titanate textile in which nanowires are parallel to each other (Wu et al. 2012), as shown in Fig. 9.20a. The output voltage and output current reached 6 V and 45 nA, respectively, which are large enough to power a

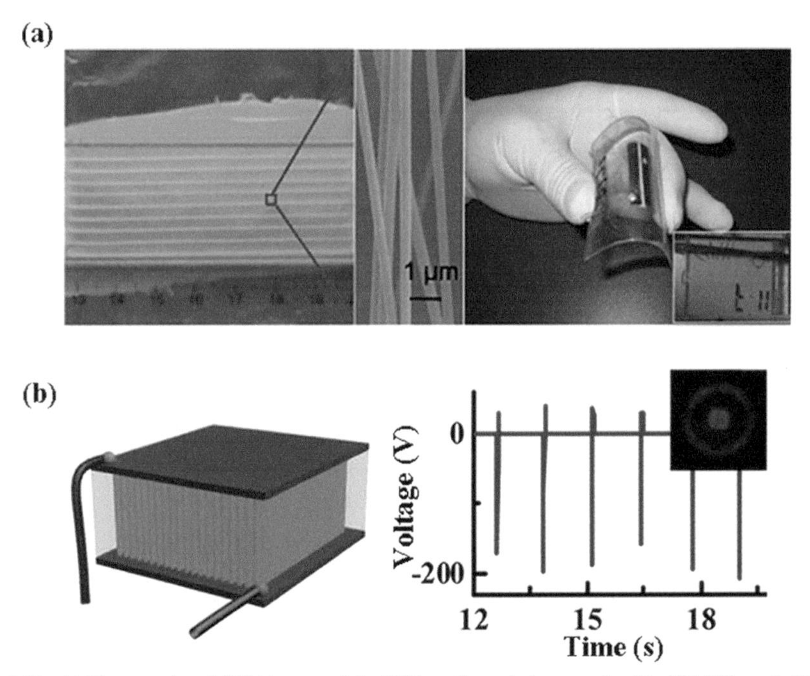

Fig. 9.20: (a) Photograph and SEM images of the PZT textile, and photograph of the NG (Wu et al. 2012) Copyright 2012, American Chemical Society. (b) Schematic of the NG, and the output voltage which could directly light a LED (Gu et al. 2012) Copyright 2013, American Chemical Society.

liquid crystal display and a UV sensor. Gu et al. developed a new type of integrated NG on the basis of a vertically aligned ultralong $Pb(Zr_{0.52}Ti_{0.48})O_3$ (PZT) nanowire array fabricated using E-spinning nanofibers (Gu et al. 2012), as shown in Fig. 9.20b. The NG can generate an ultrahigh output voltage of 209 V and current density of 23.5 μA cm^{-2}. The output electricity can be directly power a commercial LED without the energy storage process.

Triboelectric Nanogenerator

Triboelectrification is a most common phenomenon in our daily life, but it has been ignored as an energy source for electricity. Recently, reliance on the conjunction of triboelectrification and electrostatic induction, triboelectric nanogenerator (TENG) has been invented which has been demonstrated as an effective way to harvest mechanical energy into electricity. The work mechanism of the TENG is schematically shown in Fig. 9.21 (Zhu et al. 2012). Generally, The TENG consists of two polymer films with conductive films on their back sides. The two polymer films have different electron-attracting abilities. When the two films are driven to contact, friction happens on the surfaces, and as a result, equal amount but opposite signs of charges generate at the two surfaces. Thus, the triboelectric potential is formed. As the two films contact and separate repeatedly, the alternative potential can result in output of alternating current in the external circuit.

E-spun NFMs are suitable as the triboelectric due to its nano-level rough surface, good connectivity, fine flexibility, and so on (Li et al. 2017). Recently, TENGs consist of E-spun NFM have made important progress. Zheng et al. reported a TENG that is developed based on E-spun PVDF and nylon nanowires (Zheng et al. 2014), as shown in Fig. 9.22a. The output open-circuit voltage and short-circuit current density reached up to 1163 V and 11.5 mA cm^2, respectively, and the peak power density was 26.6 W m^2. The NG could directly power a direct-current motor without an energy storage system. In Sun's work (Sun et al. 2017), a flexible TENG based on E-spun PAN and carbon paper was reported, as shown in Fig. 9.22b. Furthermore, an all E-spun paper based supercapacitors (EP-SCs) was combined with the TENG, that formed an ultralight and flexible self-charging power system. As an effective

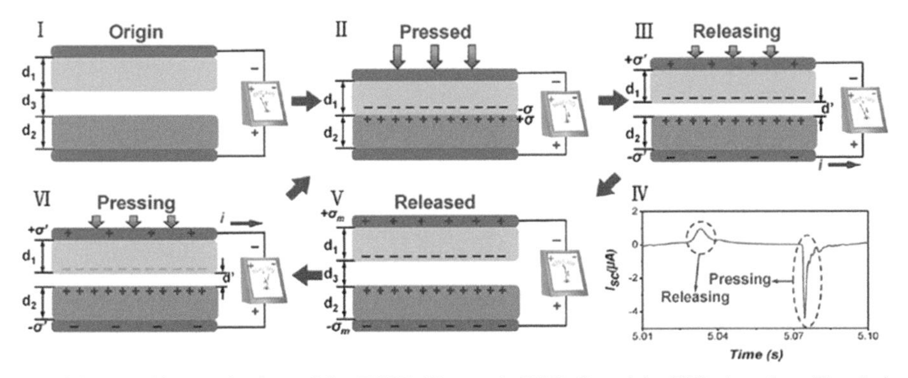

Fig. 9.21: Working mechanism of the TENG (Zhu et al. 2012) Copyright 2012, American Chemical Society.

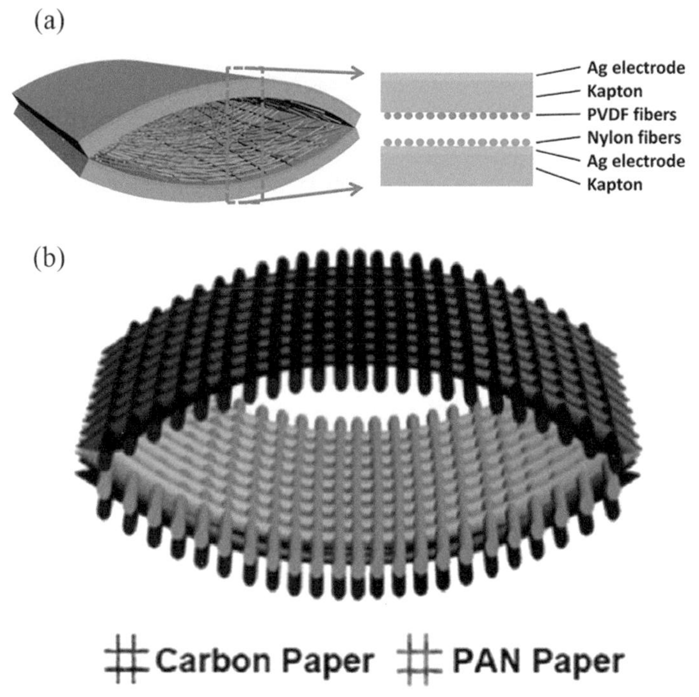

Fig. 9.22: (a) Schematic of the TENG showing that the triboelectric layers consist of E-spun PVDF and nylon nanowires (Zheng et al. 2014) Copyright 2014, The Royal Society of Chemistry. (b) Schematic illustration of the E-spun PAN paper and carbon paper based TENG (Sun et al. 2017) Copyright 2017, Elsevier.

power provider, the self-charging system was able to power an electronic watch and calculator.

E-spun Nanofibers for Batteries

The current commercial lithium-ion battery is composed of four parts. They are cathode, anode, electrolyte, and separator. E-spun nanofibers can be used as anodes or separators, and in some rare case as cathodes. Transition metal oxides, such as lithium cobalt oxide, lithium manganese oxide, lithium iron phosphate, and so on are used as cathode, where discharge occurs along with the reduction reaction. Carbon materials, such as graphite, coke, and asphalt mesophase, carbon microspheres are used as anode, where discharge occurs with the oxidation reaction. Oxides are can also be used as anode, inseted onto the carbon scaffold. Some typical anodes are displayed in Table 9.6 and Table 9.7. Lithium-ion battery electrolyte is pure ion conductor, usually dissolved in lithium salt (such as $LiPF_6$) organic carbonate solution. Porous polymer is placed as a separator between the cathode and anode, to provide electronic isolation for the cathode and anode, because direct contact of the cathode and anode lead to short-circuit of the battery.

The anode materials mainly include carbon-based materials (graphitized carbon materials and amorphous carbon materials), alloys, oxides, silicon-based materials.

Table 9.6: Electrochemical performance of some E-spun nanomaterials.

Materials	**Structure**	**Performance**	**Rates**	**References**
SnO_2	nanofiber	446 mA h g^{-1} after 50 cycles	100 mA g^{-1}	Yang et al. 2010
CuO	nanofiber	425 mA h g^{-1} after 100 cycles	100 mA g^{-1}	Kumar et al. 2012
V_2O_5	nanofiber	228 mA h g^{-1} after 50 cycles	35 mA g^{-1}	Yan et al. 2011
TiO_2	nanofiber	188 mA h g^{-1} after 100 cycles	40 mA g^{-1}	Lee et al. 2013
Fe_2O_3	nanorod	1095 mA h g^{-1} after 50 cycles	50 mA g^{-1}	Cherian et al. 2012
NiO	porous nanofiber	638 mA h g^{-1} after 50 cycles	40 mA g^{-1}	Wang et al. 2012
SnO_2	porous nanotube	808 mA h g^{-1} after 50 cycles	180 mA g^{-1}	Li et al. 2012
$ZnFe_2O_4$	nanofiber	733 mA h g^{-1} after 30 cycles	60 mA g^{-1}	Pei et al. 2011
NiO-ZnO	porous nanofiber	949 mA h g^{-1} after 120 cycles	200 mA g^{-1}	Qiao et al. 2013
$ZnCo_2O_4$	porous nanotube	1454 mA h g^{-1} after 30 cycles	100 mA g^{-1}	Luo et al. 2012

Table 9.7: Electrochemical performance of some E-spun 1D composites with carbon nanofibers (CNFs).

Material	**Performance**	**Rates**	**References**
Co-Sn alloy/CNFs	560 mA h g^{-1} after 80 cycles	161 mA g^{-1}	Jiang and Liu 2013
ultra-uniform SnO_x/CNFs	608 mA h g^{-1} after 200 cycles	500 mA g^{-1}	Zhou et al. 2014
porous TiO_2/CNFs	680 mA h g^{-1} after 250 cycles	100 mA g^{-1}	Li et al. 2014
porous Co_3O_4/CNFs	534 mA h g^{-1} after 20 cycles	100 mA g^{-1}	Zhang et al. 2011
MoO_2/CNFs	762.7 mA h g^{-1} after 100 cycles	50 mA g^{-1}	Luo et al. 2011
Fe_3O_4/CNFs	1000 mA h g^{-1} after 80 cycles	200 mA g^{-1}	Wang et al. 2008
Si/CNFs core-shell fibers	> 1250 mA h g^{-1} after 300 cycles	2750 mA g^{-1}	Hwang et al. 2012
Si/CNFs/hollow graphitized carbon	1601 mA h g^{-1} after 50 cycles	100 mA g^{-1}	Kong et al. 2013
Si/TiO_2/CNFs	720 mA h g^{-1} after 55 cycles	48 mA g^{-1}	Wu et al. 2014

The lithium ion battery anode materials must have the following basic requirements: (1) The off/imbedding redox potential of the lithium ion should be as low as possible, close to the potential of lithium metal, thus ensuring the battery output voltage. (2) The reversibility of the off/imbedding process should be as large as possible. Reduction of capacity in the first cycle should be small. Coulomb efficiency should be high. (3) During the whole off/imbedding process, the cell volume change of the negative electrode material should be small, and it should have high structural stability, chemical stability, and thermal stability. This helps to maintain the stability of the electrode and keep the circulation capacity. (4) The diffusion coefficient of the off/imbedding material should be relatively large, so as to improve the charge and discharge efficiency of the battery and the rate of lithium ion insertion/detachment, which is beneficial to the rapid charge and discharge. (5) The electronic conductivity and ionic conductivity of the anode material should be as large as possible, which can reduce the polarization, increasing the rate of charge and discharge. (6) The host material has a good surface structure, which can form a good solid-electrolyte (SEI) film at the interface with the liquid electrolyte. (7) The inserted compound has good chemical stability over the entire voltage range and does not react with the electrolyte after the formation of the SEI film. (8) From a practical point of view, the anode material should be resource-rich, cheap, and easy-preparating. The anode material should also be stable in the air, non-toxic, and environment friendly.

Cho et al. prepared a novel nanomaterial denoted as "bubble-nanorodstructured Fe_2O_3-C composite nanofibers" (Cho et al. 2015). They used the H_2/Ar atmospheres and air sintering for two consecutive steps to convert the precursor fibers into Fe_2O_3-C composite fibers. The Kirkendell effect plays a key role in this transformation process. They also obtained pure Fe_2O_3 hollow fibers by simply sintering in the air. As a lithium ion battery electrode material, bubble-nanorodstructured Fe_2O_3-C composite nanofibers showed superior electrochemical properties comparatively. The hollow nanospheres accommodate the volume change that occurs during cycling, keeping a discharge capacity of 812 mA h g^{-1} after 300 cycles at a current density of 1.0 A g^{-1}.

Separator is the safeguard of lithium-ion battery. It should prevent direct contact of cathode and anode so as to effectively avoid short-circuit phenomenon, while allowing lithium ions to move rapidly between the cathode and anode to achieve the purpose of charge and discharge. The heat stability of the separator and its ability to prevent being damaged at certain circumstances like impact are drawing more and more attention for safety problems of the lithium ion battery. Cathode material provides lithium in the process of lithium-ion charge and discharge. Lithium-ion battery cathode materials are generally lithium embedded compounds.

Polyimide (PI) is a high-performance engineering polymer, because of its excellent high temperature performance, chemical stability, and mechanical properties. This makes it an excellent lithium-ion battery separator. Wang et al. fabricated PI nanofiber membrane by E-spinning (Wang et al. 2014). They found that the PI E-spun membrane has good thermal stability at 500°C, while the Celgard membrane thermally shrinks at 150°C and melts at 167°C. In addition, batteries with PI separator show high specific capacity, low impedance, and excellent rate performance. The obtained PI E-spinning membrane was subjected to hot pressing treatment as a lithium ion battery separator with excellent thermal stability, and its breaking strength was 30 MPa (Jiang and Liu 2013).

E-spun Nanofibers for Supercapacitors

Supercapacitor, also known as electrochemical capacitor (EC), Farah capacitors. It contains two types, as electrostatic double-layer capacitors and electrochemical pseudocapacitor, both of which worked by polarizing the electrolyte to store energy. It is an electrochemical element, the energy storage process is reversible, and one can repeatedly charge and discharge hundreds of thousands of times (Zhang et al. 2016, Thavasi et al. 2008, Xie et al. 2013). The supercapacitor can be considered as two non-reactive porous electrode plates suspended in the electrolyte, energized on the plate, the positive plate attracts negative ions in the electrolyte, and the negative plate attracts positive ions, actually forming two capacitive stores layer, separated from the positive ions in the vicinity of the negative plate, negative ions in the vicinity of the positive plate, is a kind of battery and capacitor between the new special components.

In order to reduce the size and cost of supercapacitors, high-energy-density dielectrics are highly desirable. Nanocomposites with sandwich structure were prepared by hot-pressing method. The PVDF nanocomposite layer filled with graphene oxide nanosheets coated with TiO_2 nanoparticles or $Ba_{0.6}Sr_{0.4}TiO_3$ nanofibers was cast from solution and assembled into nanostructures with a sandwich structure with a reverse topology composite. The geometrical energy density of about 14.6 J cm^{-3} is achieved in nanocomposites (Shen et al. 2015). $Ba_{0.3}Sr_{0.7}TiO_3$/P(VDF-TrFE) was prepared by solution casting using Bi_2O_3 doped $Ba_{0.3}Sr_{0.7}TiO_3$ fiber prepared by E-spinning and modified by dopamine as a dielectric filler. The dielectric constant and the breakdown strength of the flexible composites are improved, especially as the saturation polarization of the composites is higher and the residual polarization is lower, and accordingly the discharge energy density calculated from the D-E loop is greatly enhanced than pure P(VDF-TrFE) films (Hu et al. 2013). $BaTiO_3$@TiO_2 nanofibers, in which $BaTiO_3$ nanoparticles were embedded in TiO_2 nanofibers by improved E-spinning, were used to enhance the interfacial polarization of PVDF-based nanocomposites. The composites produce ultra-high energy $\approx$ 20 J cm^{-3} (Zhang et al. 2015). These $BaTiO_3$@TiO_2 nanofibers are fused with highly flexible nanocomposite films by low cost solution casting process and poly(vinylidene fluoride-hexafluoropropylene) (P(VDF-HFP)) matrix. Due to these advantageous features, it is converted to a giant energy density of $\approx$ 31.2 J cm^{-3}. This is by far the highest energy density obtained in the polymer nanocomposite dielectric. More importantly, even ~ 800 kV mm^{-1} can achieve ~ 78% high discharge efficiency (Zhang et al. 2016b).

Summary and Outlook

The E-spun nanofibers have a decent performance in many fields. This chapter only summarizes some promising applications of E-spun nanostructures. However, there are still distances toward some practical applications requesting further exploration and optimization. For instance, in the field of filtration and separation, the practical application of water treatment shows that the improvement of fiber strength is of vital importance. In the field of sensors, the stability and sensitivity of the device are the urgent requirements of practical application. The applications in the biomedical field have a strict requirement, and the relevant materials need to be non-toxic and harmless to the human body. In the field of catalysis, the researchers are still in the pursuit of higher catalytic efficiency, lower costs, and more reliable recovery. In the field of energy, the novel E-spun structures/materials are likely to breed the embryonic form of the future power supply. Some new applications are also emerging, such as food-related field, agriculture, cosmetics, military, aviation, and so on. Applications in these fields also have a broad prospect.

Acknowledgements

This work was supported by the National Natural Science Foundation of China (51673103 and 51373082), the Taishan Scholars Program of Shandong Province,

China (ts20120528), the Key Research and Development Plan of Shandong Province, China (2016GGX102011), and the Postdoctoral Scientific Research Foundation of Qingdao.

References

Abdelgawad, A. M., Hudson, S. M. and Rojas, O. J. 2014. Antimicrobial wound dressing nanofiber mats from multicomponent (chitosan/silver-NPs/polyvinyl alcohol) systems. Carbohydrate Polymers 100(100): 166.

Abdouss, M., Shoushtari, A. M., Simakani, A. M., Akbari, S. and Haji, A. 2014. Citric acid-modified acrylic micro and nanofibers for removal of heavy metal ions from aqueous media. Desalination & Water Treatment 52(37-39): 7133–7142.

Ahmad, M., Pan, C., Luo, Z. and Zhu, J. 2010. A single ZnO nanofiber-based highly sensitive amperometric glucose biosensor. J. Phys. Chem. C 114(20): 9308–9313.

Albuquerque, M. T. P., Ryan, S. J., Münchow, E. A., Kamocka, M. M., Gregory, R. L., Valera, M. C. et al. 2015. Antimicrobial effects of novel triple antibiotic Pasteâ mimic scaffolds on actinomyces naeslundii biofilm. Journal of Endodontics 41(8): 1337.

An, J., Zhang, H., Zhang, J., Zhao, Y. and Yuan, X. 2009. Preparation and antibacterial activity of electrospun chitosan/poly(ethylene oxide) membranes containing silver nanoparticles. Colloid & Polymer Science 287(12): 1425–1434.

Arun, T. A., Madhavan, A. A., Chacko, D. K., Anjusree, G. S., Deepak, T. G., Thomas, S. et al. 2014. A facile approach for high surface area electrospun TiO_2 nanostructures for photovoltaic and photocatalytic applications. Dalton Transactions 43(12): 4830–4837.

Augustine, R., Dominic, E. A., Reju, I., Kaimal, B., Kalarikkal, N. and Thomas, S. 2014. Investigation of angiogenesis and its mechanism using zinc oxide nanoparticle-loaded electrospun tissue engineering scaffolds. Rsc Advances 4(93): 51528–51536.

Aussawasathien, D., Sahasithiwat, S. and Menbangpung, L. 2008. Electrospun camphorsulfonic acid doped poly(o-toluidine)–polystyrene composite fibers: Chemical vapor sensing. Synthetic Metals 158(7): 259–263.

Bagchi, S. and Ghanshyam, C. 2017. Understanding the gas sensing properties of polypyrrole coated tin oxide nanofiber mats. Journal of Physics D Applied Physics 50(10): 105302.

Bai, M. Y. and Tsai, J. C. 2014. Preparation of electrospun EDTA/PVDF blend nonwoven mats and their use in removing heavy metal ions from electropolishing electrolyte. Fibers & Polymers 15(11): 2265–2271.

Bao, J., Yang, B., Sun, Y., Zu, Y. and Deng, Y. 2013. A berberine-loaded electrospun poly-(epsilon-caprolactone) nanofibrous membrane with hemostatic potential and antimicrobial property for wound dressing. Journal of Biomedical Nanotechnology 9(7): 1173.

Barani, H. 2014. Antibacterial continuous nanofibrous hybrid yarn through *in situ* synthesis of silver nanoparticles: Preparation and characterization. Mater. Sci. Eng. C Mater. Biol. Appl. 43(43): 50–57.

Bie, Y. Q., Liao, Z. M., Zhang, H. Z., Li, G. R., Ye, Y., Zhou, Y. B. et al. 2011. Self-powered, ultrafast, visible-blind UV detection and optical logical operation based on ZnO/GaN nanoscale p-n junctions. Advanced Materials 23(5): 649–653.

Bishop-Haynes, A. and Gouma, P. 2007. Electrospun polyaniline composites for NO_2 detection. Advanced Manufacturing Processes 22(6): 764–767.

Bjorge, D., Daels, N., Vrieze, S. D., Dejans, P., Camp, T. V., Audenaert, W. et al. 2009. Performance assessment of electrospun nanofibers for filter applications. Desalination 249(3): 942–948.

Byun, J., Patel, H. A. and Yavuz, C. T. 2014. Magnetic $BaFe_{12}O_{19}$ nanofiber filter for effective separation of Fe_3O_4 nanoparticles and removal of arsenic. Journal of Nanoparticle Research 16(12): 1–12.

Çalamak, S., Erdoğdu, C., Özalp, M. and Ulubayram, K. 2014. Silk fibroin based antibacterial bionanotextiles as wound dressing materials. Materials Science & Engineering C 43: 11–20.

Calamak, S., Aksoy, E. A., Ertas, N., Erdogdu, C., Sagıroglu, M. and Ulubayram, K. 2015. Ag/silk fibroin nanofibers: Effect of fibroin morphology on Ag^+ release and antibacterial activity. European Polymer Journal 67: 99–112.

Cash, K. J. and Clark, H. A. 2010. Nanosensors and nanomaterials for monitoring glucose in diabetes. Trends in Molecular Medicine 16(12): 584–593.

Celebioglu, A., Umu, O. C., Tekinay, T. and Uyar, T. 2014. Antibacterial electrospun nanofibers from triclosan/cyclodextrin inclusion complexes. Colloids & Surfaces B Biointerfaces 116(2): 612.

Chae, S. K., Park, H., Yoon, J., Lee, C. H., Ahn, D. J. and Kim, J. M. 2007. Polydiacetylene supramolecules in electrospun microfibers: Fabrication, micropatterning, and sensor applications. Advanced Materials 19(4): 521–524.

Chahal, S., Hussain, F. S. J., Kumar, A., Rasad, M. S. B. A. and Yusoff, M. M. 2016. Fabrication, characterization and *in vitro* biocompatibility of electrospun hydroxyethyl cellulose/poly (vinyl) alcohol nanofibrous composite biomaterial for bone tissue engineering. Chemical Engineering Science 144: 17–29.

Chang, C., Tran, V. H., Wang, J., Fuh, Y. -K. and Lin, L. 2010. Direct-write piezoelectric polymeric nanogenerator with high energy conversion efficiency. Nano Letters 10(2): 726–731.

Chang, J. and Lin, L. 2011. Large array electrospun PVDF nanogenerators on a flexible substrate. Solid-State Sensors, Actuators and Microsystems Conference (TRANSDUCERS), 2011 16th International (pp. 747–750): IEEE.

Chao, Y. K., Lee, C. H., Liu, K. S., Wang, Y. C., Wang, C. W. and Liu, S. J. 2015. Sustained release of bactericidal concentrations of penicillin in the pleural space via an antibiotic-eluting pigtail catheter coated with electrospun nanofibers: results from *in vivo* and *in vitro* studies. International Journal of Nanomedicine 10: 3329.

Chen, H., Huang, J., Yu, J., Liu, S. and Gu, P. 2011. Electrospun chitosan-graft-poly(ε-caprolactone)/poly(ε-caprolactone) cationic nanofibrous mats as potential scaffolds for skin tissue engineering. International Journal of Biological Macromolecules 48(1): 13–19.

Chen, L., Bromberg, L., Hatton, T. A. and Rutledge, G. C. 2008. Electrospun cellulose acetate fibers containing chlorhexidine as a bactericide. Polymer 49(5): 1266–1275.

Chen, Y., Liu, S., Hou, Z., Ma, P., Yang, D., Li, C. et al. 2015. Multifunctional electrospinning composite fibers for orthotopic cancer treatment *in vivo*. Nano Research 8(6): 1917–1931.

Cherian, C. T., Sundaramurthy, J., Ragupathy, P., Kumar, P. S., Thavasi, V., Mhaisalkar, S. G. et al. 2012. Electrospun α-Fe_2O_3 nanorods as stable, high capacity anode material for Li-ion batteries. Journal of Materials Chemistry 22(24): 12198–12204.

Chevallier, L., Bauer, A., Cavaliere, S., Hui, R., Rozière, J. and Jones, D. J. 2012. Mesoporous nanostructured Nb-doped titanium dioxide microsphere catalyst supports for PEM fuel cell electrodes. Acs Appl. Mater. Interfaces 4(3): 1752–1759.

Cho, J. S., Hong, Y. J. and Kang, Y. C. 2015. Design and synthesis of bubble-nanorod-structured Fe_2O_3-carbon nanofibers as advanced anode material for Li-Ion batteries. Acs Nano 9(4): 4026–4035.

Choi, S. H., Ankonina, G., Youn, D. Y., Oh, S. G., Hong, J. M., Rothschild, A. et al. 2009. Hollow ZnO nanofibers fabricated using electrospun polymer templates and their electronic transport properties. Acs Nano 3(9): 2623.

Cohn, C. 2016. Lipid-mediated protein functionalization of electrospun polycaprolactone fibers. Express Polymer Letters 10(5): 430–437.

Courjean, O. and Mano, N. 2011. Recombinant glucose oxidase from Penicillium amagasakiense for efficient bioelectrochemical applications in physiological conditions. Journal of Biotechnology 151(1): 122–129.

Daneshfar, A., Matsuura, T., Emadzadeh, D., Pahlevani, Z. and Ismail, A. F. 2015. Urease-carrying electrospun polyacrylonitrile mat for urea hydrolysis. Reactive & Functional Polymers 87: 37–45.

de Valence, S., Tille, J. -C., Mugnai, D., Mrowczynski, W., Gurny, R., Möller, M. et al. 2012. Long term performance of polycaprolactone vascular grafts in a rat abdominal aorta replacement model. Biomaterials 33(1): 38–47.

Deng, C., Yong, P., Lei, S., Liu, Y. N. and Zhou, F. 2012. On-line removal of redox-active interferents by a porous electrode before amperometric blood glucose determination. Analytica Chimica Acta 719(719): 52–56.

Dhakras, D., Borkar, V., Ogale, S. and Jog, J. 2012. Enhanced piezoresponse of electrospun PVDF mats with a touch of nickel chloride hexahydrate salt. Nanoscale 4(3): 752–756.

Dhand, C., Venkatesh, M., Barathi, V. A., Harini, S., Bairagi, S., Goh, T. L. E. et al. 2017. Bio-inspired crosslinking and matrix-drug interactions for advanced wound dressings with long-term antimicrobial activity. Biomaterials 138: 153.

Ding, B., Kim, J., Miyazaki, Y. and Shiratori, S. 2004. Electrospun nanofibrous membranes coated quartz crystal microbalance as gas sensor for NH3 detection. Sensors and Actuators B: Chemical 101(3): 373–380.

Ding, B., Yamazaki, M. and Shiratori, S. 2005. Electrospun fibrous polyacrylic acid membrane-based gas sensors. Sensors and Actuators B: Chemical 106(1): 477–483.

Ding, B., Wang, M., Yu, J. and Sun, G. 2009. Gas sensors based on electrospun nanofibers. Sensors 9(3): 1609–1624.

Ding, Y., Wang, Y., Li, B. and Lei, Y. 2010. Electrospun hemoglobin microbelts based biosensor for sensitive detection of hydrogen peroxide and nitrite. Biosensors & Bioelectronics 25(9): 2009.

Ding, Y., Wang, Y., Su, L., Bellagamba, M., Zhang, H. and Lei, Y. 2010. Electrospun Co_3O_4 nanofibers for sensitive and selective glucose detection. Biosensors and Bioelectronics 26(2): 542–548.

Ding, Y., Wang, Y., Su, L., Zhang, H. and Lei, Y. 2010. Preparation and characterization of NiO–Ag nanofibers, NiO nanofibers, and porous Ag: towards the development of a highly sensitive and selective non-enzymatic glucose sensor. Journal of Materials Chemistry 20(44): 9918–9926.

Dinis, T. M., Elia, R., Vidal, G., Dermigny, Q., Denoeud, C., Kaplan, D. L. et al. 2015. 3D multi-channel bi-functionalized silk electrospun conduits for peripheral nerve regeneration. Journal of the Mechanical Behavior of Biomedical Materials 41: 43–55.

Dirote, E. V. 2006. Nanotechnology at the Leading Edge. Nova Publishers.

Dong, H., Fey, E., Anna Gandelman, A. and Jones, W. E. 2011. Synthesis and assembly of metal nanoparticles on electrospun poly(4-vinylpyridine) fibers and poly(4-vinylpyridine) composite fibers. Chemistry of Materials 18(8): 2008–2011.

Dong, R. H., Qin, C. C., Qiu, X., Yan, X., Yu, M., Cui, L. et al. 2015. *In situ* precision electrospinning as an effective delivery technique for cyanoacrylate medical glue with high efficiency and low toxicity. Nanoscale 7(46): 19468.

Dong, R. H., Jia, Y. X., Qin, C. C., Zhan, L., Yan, X., Cui, L. et al. 2016. *In situ* deposition of a personalized nanofibrous dressing via a handy electrospinning device for skin wound care. Nanoscale 8(6): 3482–3488.

Duman, M., Saber, R. and Pişkin, E. 2003. A new approach for immobilization of oligonucleotides onto piezoelectric quartz crystal for preparation of a nucleic acid sensor for following hybridization. Biosensors & Bioelectronics 18(11): 1355–1363.

El-Aassar, M. R. 2013. Functionalized electrospun nanofibers from poly (AN-co-MMA) for enzyme immobilization. Journal of Molecular Catalysis B Enzymatic 85-86(1): 140–148.

Fang, J., Wang, X. and Lin, T. 2011. Electrical power generator from randomly oriented electrospun poly (vinylidene fluoride) nanofibre membranes. Journal of Materials Chemistry 21(30): 11088–11091.

Fang, J., Niu, H., Wang, H., Wang, X. and Lin, T. 2013. Enhanced mechanical energy harvesting using needleless electrospun poly (vinylidene fluoride) nanofibre webs. Energy & Environmental Science 6(7): 2196–2202.

Fouda, M. M., Elaassar, M. R. and Aldeyab, S. S. 2013. Antimicrobial activity of carboxymethyl chitosan/polyethylene oxide nanofibers embedded silver nanoparticles. Carbohydr. Polym. 92(2): 1012–1017.

Gao, Y., Li, X., Gong, J., Fan, B., Su, Z. and Qu, L. 2008. Polyaniline nanotubes prepared using fiber mats membrane as the template and their gas-response behavior. Journal of Physical Chemistry C 112(22): 8215–8222.

Ge, J., Zhang, J., Wang, F., Li, Z., Yu, J. and Ding, B. 2017. Superhydrophilic and underwater superoleophobic nanofibrous membrane with hierarchical structured skin for effective oil-in-water emulsion separation. Journal of Materials Chemistry A 5(2): 497–502.

Gheibi, A., Latifi, M., Merati, A. A. and Bagherzadeh, R. 2014. Piezoelectric electrospun nanofibrous materials for self-powering wearable electronic textiles applications. Journal of Polymer Research 21(7): 469.

Gliścińska, E., Gutarowska, B., Brycki, B. and Krucińska, I. 2013. Electrospun polyacrylonitrile nanofibers modified by quaternary ammonium salts. Journal of Applied Polymer Science 128(1): 767–775.

Gomes, M. T. S. R., Nogueira, P. S. T. and Oliveira, J. A. B. P. 2000. Quantification of CO_2, SO_2, NH_3, and H_2S with a single coated piezoelectric quartz crystal. Sensors & Actuators B Chemical 68(1-3): 218–222.

Gopal, R., Kaur, S., Ma, Z., Chan, C., Ramakrishna, S. and Matsuura, T. 2006. Electrospun nanofibrous filtration membrane. Journal of Membrane Science 281(1): 581–586.

Grätzel, M. 2003. Dye-sensitized solar cells. Journal of Photochemistry and Photobiology C: Photochemistry Reviews 4(2): 145–153.

Gu, L., Cui, N., Cheng, L., Xu, Q., Bai, S., Yuan, M. et al. 2012. Flexible fiber nanogenerator with 209 V output voltage directly powers a light-emitting diode. Nano Letters 13(1): 91–94.

Guascito, M. R., Chirizzi, D., Malitesta, C., Siciliano, M., Siciliano, T. and Tepore, A. 2012. Amperometric non-enzymatic bimetallic glucose sensor based on platinum tellurium microtubes modified electrode. Electrochemistry Communications 22(8): 45–48.

Guo, H., Qiao, T., Jiang, S., Li, T., Song, P., Zhang, B. et al. 2016. Aligned poly (glycolide-lactide) fiber membranes with conducting polypyrrole. Polymers for Advanced Technologies 28(4): 484–490.

Guo, W., Wang, J., Wang, C., He, J. Q., He, X. W. and Cheng, J. P. 2002. Design, synthesis, and enantiomeric recognition of dicyclodipeptide-bearing calix[4]arenes: a promising family for chiral gas sensor coatings. Tetrahedron Letters 43(32): 5665–5667.

Hang, A. T., Tae, B. and Park, J. S. 2010. Non-woven mats of poly(vinyl alcohol)/chitosan blends containing silver nanoparticles: Fabrication and characterization. Carbohydrate Polymers 82(2): 472–479.

Hang, T. A., Lan, N. P., Vu, T. H. T. and Park, J. S. 2012. Fabrication of an antibacterial non-woven mat of a poly(lactic acid)/chitosan blend by electrospinning. Macromolecular Research 20(1): 51–58.

Hansen, B. J., Liu, Y., Yang, R. and Wang, Z. L. 2010. Hybrid nanogenerator for concurrently harvesting biomechanical and biochemical energy. Acs Nano 4(7): 3647–3652.

He, X., Arsat, R., Sadek, A. Z., Wlodarski, W., Kalantar-zadeh, K. and Li, J. 2010. Electrospun PVP fibers and gas sensing properties of PVP/36° YX $LiTaO_3$ SAW device. Sensors and Actuators B: Chemical 145(2): 674–679.

Heunis, T., Bshena, O., Klumperman, B. and Dicks, L. 2011. Release of bacteriocins from nanofibers prepared with combinations of poly(d,l-lactide) (PDLLA) and poly(Ethylene Oxide) (PEO). International Journal of Molecular Sciences 12(4): 2158–2173.

Hou, D., Luo, W., Huang, Y., Yu, J. C. and Hu, X. 2013. Synthesis of porous $Bi_4Ti_3O_{12}$ nanofibers by electrospinning and their enhanced visible-light-driven photocatalytic properties. Nanoscale 5(5): 2028.

Hou, D., Hu, X., Ho, W., Hu, P. and Huang, Y. 2015. Facile fabrication of porous Cr-doped $SrTiO_3$ nanotubes by electrospinning and their enhanced visible-light-driven photocatalytic properties. Journal of Materials Chemistry A 3(7): 3935–3943.

Hu, J., Kai, D., Ye, H., Tian, L., Ding, X., Ramakrishna, S. et al. 2017. Electrospinning of poly (glycerol sebacate)-based nanofibers for nerve tissue engineering. Materials Science and Engineering: C 70: 1089–1094.

Hu, P., Song, Y., Liu, H., Shen, Y., Lin, Y. and Nan, C. W. 2013. Largely enhanced energy density in flexible P(VDF-TrFE) nanocomposites by surface-modified electrospun $BaSrTiO_3$ fibers. Journal of Materials Chemistry A 1(5): 1688–1693.

Huang, R., Long, Y. Z., Tang, C. C. and Zhang, H. D. 2014. Fabrication of nano-branched coaxial polyaniline/polyvinylidene fluoride fibers via electrospinning for strain sensor. Advanced Materials Research 853: 79–82.

Huang, X. J., Chen, P. C., Huang, F., Ou, Y., Chen, M. R. and Xu, Z. K. 2011. Immobilization of Candida rugosa lipase on electrospun cellulose nanofiber membrane. Journal of Molecular Catalysis B Enzymatic 70(3-4): 95–100.

Huang, Y., Hu, D., Wen, S., Shen, M., Zhu, M. and Shi, X. 2014. Selective removal of mercury ions using thymine-grafted electrospun polymer nanofibers. New Journal of Chemistry 38(4): 1533–1539.

Hwang, S. and Jeong, S. 2011. Electrospun nano composites of poly(vinyl pyrrolidone)/nano-silver for antibacterial materials. Journal of Nanoscience & Nanotechnology 11(1): 610.

Hwang, T. H., Yong, M. L., Kong, B. S., Seo, J. S. and Choi, J. W. 2012. Electrospun core–shell fibers for robust silicon nanoparticle-based lithium ion battery anodes. Nano Letters 12(2): 802.

Ildoo Kim, Avner Rothschild, B. H. L., Young Kim, D., Seong Mu Jo. and Harry L. Tuller. 2006. Ultrasensitive chemiresistors based on electrospun TiO_2 nanofibers. Nano Letters 6(9): 2009–2013.

Iqbal, S., Rashid, M. H., Arbab, A. S. and Khan, M. 2017. Encapsulation of anticancer drugs (5-Fluorouracil and Paclitaxel) into Polycaprolactone (PCL) nanofibers and *in vitro* testing for sustained and targeted therapy. Journal of Biomedical Nanotechnology 13(4): 355.

Ji, S., Li, Y. and Yang, M. 2008. Gas sensing properties of a composite composed of electrospun poly(methyl methacrylate) nanofibers and *in situ* polymerized polyaniline. Sensors & Actuators B Chemical 133(2): 644–649.

Jia, X. S., Tang, C. C., Yan, X., Yu, G. F., Li, J. T., Zhang, H. D. et al. 2016. Flexible polyaniline/polymethyl methacrylate composite fibers via electrospinning and *in situ* polymerization for ammonia gas sensing and strain sensing. Journal of Nanomaterials (2016-12-21), 2016: 32.

Jiang, K., Long, Y. Z., Chen, Z. J., Liu, S. L., Huang, Y. Y., Jiang, X. et al. 2014. Airflow-directed *in situ* electrospinning of a medical glue of cyanoacrylate for rapid hemostasis in liver resection. Nanoscale 6(14): 7792–7798.

Jiang, W. and Liu, Z. 2013. A high temperature operating nanofibrous polyimide separator in Li-ion battery. Solid State Ionics 232(2013): 44–48.

Jin, L., Feng, Z. -Q., Zhu, M. -L., Wang, T., Leach, M. K. and Jiang, Q. 2012. A novel fluffy conductive polypyrrole nano-layer coated PLLA fibrous scaffold for nerve tissue engineering. Journal of Biomedical Nanotechnology 8(5): 779–785.

Kai, D., Tan, M. J., Prabhakaran, M. P., Chan, B. Q. Y., Liow, S. S., Ramakrishna, S. et al. 2016. Biocompatible electrically conductive nanofibers from inorganic-organic shape memory polymers. Colloids and Surfaces B: Biointerfaces 148: 557–565.

Kalinov, K., Ignatova, M., Maximova, V., Rashkov, I. and Manolova, N. 2014. Modification of electrospun poly(ε-caprolactone) mats by formation of a polyelectrolyte complex between poly(acrylic acid) and quaternized chitosan for tuning of their antibacterial properties. European Polymer Journal 50(1): 18–29.

Kang, M. S., Kim, J. H., Singh, R. K., Jang, J. H. and Kim, H. W. 2015. Therapeutic-designed electrospun bone scaffolds: mesoporous bioactive nanocarriers in hollow fiber composites to sequentially deliver dual growth factors. Acta Biomaterialia 16: 103.

Kaur, S., Ma, Z., Gopal, R., Singh, G., Ramakrishna, S. and Matsuura, T. 2007. Plasma-induced graft copolymerization of poly(methacrylic acid) on electrospun poly(vinylidene fluoride) nanofiber membrane. Langmuir 23(26): 13085–13092.

Kayaci, F., Umu, O. C., Tekinay, T. and Uyar, T. 2013. Antibacterial electrospun poly(lactic acid) (PLA) nanofibrous webs incorporating triclosan/cyclodextrin inclusion complexes. Journal of Agricultural & Food Chemistry 61(16): 3901–3908.

Kessick, R. and Tepper, G. 2006. Electrospun polymer composite fiber arrays for the detection and identification of volatile organic compounds. Sensors & Actuators B Chemical 117(1): 205–210.

Khampieng, T., Wnek, G. E. and Supaphol, P. 2014. Electrospun DOXY-h loaded-poly(acrylic acid) nanofiber mats: *in vitro* drug release and antibacterial properties investigation. J. Biomater. Sci. Polym. Ed. 25(12): 1292–1305.

Kijeńska, E., Prabhakaran, M. P., Swieszkowski, W., Kurzydlowski, K. J. and Ramakrishna, S. 2012. Electrospun bio-composite P(LLA-CL)/collagen I/collagen III scaffolds for nerve tissue engineering. Journal of Biomedical Materials Research Part B: Applied Biomaterials 100(4): 1093–1102.

Kim, B. S., Kimura, N., Kim, H. K., Watanabe, K. and Kim, I. S. 2011. Thermal insulation, antibacterial and mold properties of breathable nanofiber-laminated wallpapers. Journal of Nanoscience & Nanotechnology 11(6): 4929.

Kim, J. H., Abideen, Z. U., Zheng, Y. and Sang, S. K. 2016. Improvement of toluene-sensing performance of SnO_2 nanofibers by Pt functionalization. Sensors 16(11): 1857.

Kim, J. M., Chang, S. M., Suda, Y. and Muramatsu, H. 1999. Stability study of carbon graphite covered quartz crystal. Sensors & Actuators A Physical 72(2): 140–147.

Kim, K., Luu, Y. K., Chang, C., Fang, D., Hsiao, B. S., Chu, B. et al. 2004. Incorporation and controlled release of a hydrophilic antibiotic using poly(lactide-co-glycolide)-based electrospun nanofibrous scaffolds. Journal of Controlled Release Official Journal of the Controlled Release Society 98(1): 47.

Kim, S. S. and Lee, J. 2014. Antibacterial activity of polyacrylonitrile-chitosan electrospun nanofibers. Carbohydrate Polymers 102(3): 231–237.

Kong, J., Yee, W. A., Wei, Y., Yang, L., Ang, J. M., Phua, S. L. et al. 2013. Silicon nanoparticles encapsulated in hollow graphitized carbon nanofibers for lithium ion battery anodes. Nanoscale 5(7): 2967–2973.

Kumar, P. S., Sahay, R., Aravindan, V., Sundaramurthy, J., Chui Ling, W., Thavasi, V. et al. 2012. Free-standing electrospun carbon nanofibres—a high performance anode material for lithium-ion batteries. Journal of Physics D Applied Physics 45(26): 493–500.

Kumar, P. S., Sundaramurthy, J., Sundarrajan, S., Babu, V. J., Singh, G., Allakhverdiev, S. I. et al. 2014. Hierarchical electrospun nanofibers for energy harvesting, production and environmental remediation. Energy & Environmental Science 7(10): 3192–3222.

Lala, N., Thavasi, V. and Ramakrishna, S. 2009. Preparation of surface adsorbed and impregnated multi-walled carbon nanotube/nylon-6 nanofiber composites and investigation of their gas sensing ability. Sensors 9(1): 86.

Landau, O., Rothschild, A. and Zussman, E. 2008. Processing-microstructure-properties correlation of ultrasensitive gas sensors produced by electrospinning. Chemistry of Materials 21(1): 9–11.

Lee, J. H., Park, J. H., El-Fiqi, A., Kim, J. H., Yun, Y. R., Jang, J. H. et al. 2014. Biointerface control of electrospun fiber scaffolds for bone regeneration: engineered protein link to mineralized surface. Acta Biomaterialia 10(6): 2750–2761.

Lee, K. and Lee, S. 2012. Multifunctionality of poly(vinyl alcohol) nanofiber webs containing titanium dioxide. Journal of Applied Polymer Science 124(5): 4038–4046.

Lee, S., Ha, J., Choi, J., Song, T., Lee, J. W. and Paik, U. 2013. 3D cross-linked nanoweb architecture of binder-free TiO_2 electrodes for lithium ion batteries. Acs Applied Materials & Interfaces 5(22): 11525.

Leung, W. F., Hung, C. H. and Yuen, P. T. 2010. Effect of face velocity, nanofiber packing density and thickness on filtration performance of filters with nanofibers coated on a substrate. Separation & Purification Technology 71(1): 30–37.

Li, J. 2013. Needleless electro-spun nanofibers used for filtration of small particles. Express Polymer Letters 7(8): 683–689.

Li, M., Liu, W., Sun, J., Xianyu, Y., Wang, J., Zhang, W. et al. 2013. Culturing primary human osteoblasts on electrospun poly (lactic-co-glycolic acid) and poly (lactic-co-glycolic acid)/nanohydroxyapatite scaffolds for bone tissue engineering. Acs Applied Materials & Interfaces 5(13): 5921–5926.

Li, M., Wang, S., Jiang, J., Sun, J., Li, Y., Huang, D. et al. 2015. Surface modification of nano-silica on the ligament advanced reinforcement system for accelerated bone formation: primary human osteoblasts testing *in vitro* and animal testing *in vivo*. Nanoscale 7(17): 8071–8075.

Li, P., Li, Y., Ying, B. and Yang, M. 2009. Electrospun nanofibers of polymer composite as a promising humidity sensitive material. Sensors & Actuators B Chemical 141(2): 390–395.

Li, X., Gao, C., Wang, J., Lu, B., Chen, W., Song, J. et al. 2012. TiO_2 films with rich bulk oxygen vacancies prepared by electrospinning for dye-sensitized solar cells. Journal of Power Sources 214: 244–250.

Li, X., Chen, Y., Zhou, L., Mai, Y. W. and Huang, H. 2014. Exceptional electrochemical performance of porous TiO_2–carbon nanofibers for lithium ion battery anodes. Journal of Materials Chemistry A 2(11): 3875–3880.

Li, Y., Li, P., Yang, M., Lei, S., Chen, Y. and Guo, X. 2010. A surface acoustic wave humidity sensor based on electrosprayed silicon-containing polyelectrolyte. Sensors & Actuators B Chemical 145(1): 516–520.

Li, Z., Zhang, H., Zheng, W., Wang, W., Huang, H., Wang, C. et al. 2008. Highly sensitive and stable humidity nanosensors based on LiCl doped TiO_2 electrospun nanofibers. Journal of the American Chemical Society 130(15): 5036–5037.

Li, Z., Shen, J., Abdalla, I., Yu, J. and Ding, B. 2017. Nanofibrous membrane constructed wearable triboelectric nanogenerator for high performance biomechanical energy harvesting. Nano Energy 36: 341–348.

Lin, D. P., Long, Y. Z., He, H. W., Huang, Y. Y., Han, W. P., Yu, G. F. et al. 2014. Twisted microropes for stretchable devices based on electrospun conducting polymer fibers doped with ionic liquid. Journal of Materials Chemistry C 2(42): 8962–8966.

Lin, H. B. and Shih, J. S. 2003. Fullerene C60-cryptand coated surface acoustic wave quartz crystal sensor for organic vapors. Sensors & Actuators B Chemical 92(3): 243–254.

Lin, J., Ding, B., Yang, J., Yu, J. and Sun, G. 2012. Subtle regulation of the micro- and nanostructures of electrospun polystyrene fibers and their application in oil absorption. Nanoscale 4(1): 176–182.

Lin, J., Shang, Y., Ding, B., Yang, J., Yu, J. and Al-Deyab, S. S. 2012. Nanoporous polystyrene fibers for oil spill cleanup. Marine Pollution Bulletin 64(2): 347–352.

Lin, J., Tian, F., Shang, Y., Wang, F., Ding, B., Yu, J. et al. 2013. Co-axial electrospun polystyrene/polyurethane fibres for oil collection from water surface. Nanoscale 5(7): 2745–2755.

Liu, B., Zhang, S., Wang, X., Yu, J. and Ding, B. 2015. Efficient and reusable polyamide-56 nanofiber/nets membrane with bimodal structures for air filtration. Journal of Colloid & Interface Science 457: 203.

Liu, H., Kameoka, J., Czaplewski, D. A. and Craighead, H. G. 2004. Polymeric nanowire chemical sensor. Nano Letters 4(4): 671–675.

Liu, M., Liu, R. and Chen, W. 2013. Graphene wrapped Cu_2O nanocubes: non-enzymatic electrochemical sensors for the detection of glucose and hydrogen peroxide with enhanced stability. Biosensors & Bioelectronics 45(45C): 206.

Liu, M., Duan, X. -P., Li, Y. -M., Yang, D. -P. and Long, Y. -Z. 2017a. Electrospun nanofibers for wound healing. Materials Science and Engineering: C 76: 1413–1423.

Liu, M., Duan, X. P., Li, Y. M., Yang, D. P. and Long, Y. Z. 2017b. Electrospun nanofibers for wound healing. Materials Science & Engineering C Materials for Biological Applications 76: 1413.

Liu, X., Lin, T., Gao, Y., Xu, Z., Huang, C., Yao, G. et al. 2012. Antimicrobial electrospun nanofibers of cellulose acetate and polyester urethane composite for wound dressing†. J. Biomed. Mater. Res. B Appl. Biomater. 100B(6): 1556–1565.

Liu, Y., Teng, H., Hou, H. and You, T. 2009. Nonenzymatic glucose sensor based on renewable electrospun Ni nanoparticle-loaded carbon nanofiber paste electrode. Biosensors & Bioelectronics 24(11): 3329–3334.

Liu, Y., Park, M., Ding, B., Kim, J., El-Newehy, M., Al-Deyab, S. S. et al. 2015. Facile electrospun Polyacrylonitrile/poly(acrylic acid) nanofibrous membranes for high efficiency particulate air filtration. Fibers & Polymers 16(3): 629–633.

Liu, Y. J., Zhang, H. D., Yan, X., Zhao, A. J., Zhang, Z. G., Si, W. Y. et al. 2016. Effect of Ce doping on the optoelectronic and sensing properties of electrospun ZnO nanofibers. Rsc Advances 6(89): 85727–85734.

Luo, W., Hu, X., Sun, Y. and Huang, Y. 2011. Electrospinning of carbon-coated MoO_2 nanofibers with enhanced lithium-storage properties. Physical Chemistry Chemical Physics PCCP 13(37): 16735.

Luo, W., Hu, X., Sun, Y. and Huang, Y. 2012. Electrospun porous $ZnCo_2O_4$ nanotubes as a high-performance anode material for lithium-ion batteries. Journal of Materials Chemistry 22(18): 8916–8921.

Luoh, R. and Hahn, H. T. 2006. Electrospun nanocomposite fiber mats as gas sensors. Composites Science & Technology 66(14): 2436–2441.

Ma, Z., Kotaki, M. and Ramakrishna, S. 2005. Electrospun cellulose nanofiber as affinity membrane. Journal of Membrane Science 265(1): 115–123.

Ma, Z., Masaya, K. and Ramakrishna, S. 2006. Immobilization of Cibacron blue F3GA on electrospun polysulphone ultra-fine fiber surfaces towards developing an affinity membrane for albumin adsorption. Journal of Membrane Science 282(1-2): 237–244.

Mahapatra, A., Garg, N., Nayak, B. P., Mishra, B. G. and Hota, G. 2012. Studies on the synthesis of electrospun PAN-Ag composite nanofibers for antibacterial application. Journal of Applied Polymer Science 124(2): 1178–1185.

Mali, S. S., Patil, P. S. and Hong, C. K. 2014. Low-cost electrospun highly crystalline kesterite Cu2ZnSnS4 nanofiber counter electrodes for efficient dye-sensitized solar cells. Acs Applied Materials & Interfaces 6(3): 1688–1696.

Manesh, K. M., Gopalan, A. I., Lee, K. P., Santhosh, P., Song, K. D. and Lee, D. D. 2007. Fabrication of functional nanofibrous ammonia sensor. IEEE Transactions on Nanotechnology 6(5): 513–518.

Manesh, K. M., Santhosh, P., Gopalan, A. and Lee, K. P. 2007. Electrospun poly(vinylidene fluoride)/poly(aminophenylboronic acid) composite nanofibrous membrane as a novel glucose sensor. Analytical Biochemistry 360(2): 189–195.

Mannucci, P. M., Harari, S., Martinelli, I. and Franchini, M. 2015. Effects on health of air pollution: a narrative review. Internal & Emergency Medicine 10(6): 657.

Mayorga-Martinez, C. C., Guix, M., Madrid, R. E. and Merkoçi, A. 2012. Bimetallic nanowires as electrocatalysts for nonenzymatic real-time impedancimetric detection of glucose. Chemical Communications 48(11): 1686.

Mei, Y., Yao, C., Fan, K. and Li, X. 2012. Surface modification of polyacrylonitrile nanofibrous membranes with superior antibacterial and easy-cleaning properties through hydrophilic flexible spacers. Journal of Membrane Science 417-418(1): 20–27.

Min, W. L., An, S., Latthe, S. S., Lee, C., Hong, S. and Yoon, S. S. 2013. Electrospun polystyrene nanofiber membrane with superhydrophobicity and superoleophilicity for selective separation of water and low viscous oil. ACS Applied Materials & Interfaces 5(21): 10597–10604.

Nguyen, T. H., Lee, K. H. and Lee, B. T. 2010. Fabrication of Ag nanoparticles dispersed in PVA nanowire mats by microwave irradiation and electro-spinning. Materials Science & Engineering C 30(7): 944–950.

Nguyen, T. T. T., Chung, O. H. and Park, J. S. 2011. Coaxial electrospun poly(lactic acid)/chitosan (core/shell) composite nanofibers and their antibacterial activity. Carbohydrate Polymers 86(4): 1799–1806.

Nicosia, A., Gieparda, W., Foksowicz-Flaczyk, J., Walentowska, J., Wesołek, D., Vazquez, B. et al. 2015. Air filtration and antimicrobial capabilities of electrospun PLA/PHB containing ionic liquid. Separation & Purification Technology 154(10): 154–160.

Pant, H. R., Pandeya, D. R., Nam, K. T., Baek, W. I., Hong, S. T. and Kim, H. Y. 2011. Photocatalytic and antibacterial properties of a TiO_2/nylon-6 electrospun nanocomposite mat containing silver nanoparticles. Journal of Hazardous Materials 189(1-2): 465.

Park, J. Y., Choi, S. W., Lee, J. W., Lee, C. and Sang, S. K. 2009. Synthesis and gas sensing properties of TiO_2–ZnO core-shell nanofibers. Journal of the American Ceramic Society 92(11): 2551–2554.

Pei, F. T., Sharma, Y., Pramana, S. S. and Srinivasan, M. 2011. Nanoweb anodes composed of one-dimensional, high aspect ratio, size tunable electrospun $ZnFe_2O_4$ nanofibers for lithium ion batteries. Journal of Materials Chemistry 21(38): 14999–15008.

Pinto, N. J., Ramos, I., Rojas, R., Wang, P. C. and Jr, A. T. J. 2008. Electric response of isolated electrospun polyaniline nanofibers to vapors of aliphatic alcohols. Sensors & Actuators B Chemical 129(2): 621–627.

Prabhakaran, M. P., Vatankhah, E. and Ramakrishna, S. 2013. Electrospun aligned PHBV/ collagen nanofibers as substrates for nerve tissue engineering. Biotechnology and Bioengineering 110(10): 2775–2784.

Qi, Q., Zhang, T., Liu, L., Zheng, X. and Lu, G. 2009. Improved NH_3, C_2H_5OH, and CH_3COCH_3 sensing properties of SnO_2 nanofibers by adding block copolymer P123. Sensors & Actuators B Chemical 141(1): 174–178.

Qi, R., Guo, R., Zheng, F., Liu, H., Yu, J. and Shi, X. 2013. Controlled release and antibacterial activity of antibiotic-loaded electrospun halloysite/poly(lactic-co-glycolic acid) composite nanofibers. Colloids & Surfaces B Biointerfaces 110(10): 148–155.

Qiao, L., Wang, X., Qiao, L., Sun, X., Li, X., Zheng, Y. et al. 2013. Single electrospun porous NiO-ZnO hybrid nanofibers as anode materials for advanced lithium-ion batteries. Nanoscale 5(7): 3037–3042.

Qin, N., Xiong, J., Liang, R., Liu, Y., Zhang, S., Li, Y. et al. 2017. Highly efficient photocatalytic H_2 evolution over MoS_2/CdS-TiO_2 nanofibers prepared by an electrospinning mediated photodeposition method. Applied Catalysis B Environmental 202: 374–380.

Ren, G., Xu, X., Liu, Q., Cheng, J., Yuan, X., Wu, L. et al. 2006. Electrospun poly(vinyl alcohol)/glucose oxidase biocomposite membranes for biosensor applications. Reactive & Functional Polymers 66(12): 1559–1564.

Ren, X., Kocer, H. B., Worley, S. D., Broughton, R. M. and Huang, S. T. 2013. Biocidal nanofibers via electrospinning. Journal of Applied Polymer Science 127(4): 3192–3197.

Ru, C., Wang, F., Pang, M., Sun, L., Chen, R. and Sun, Y. 2015. Suspended, shrinkage-free, electrospun PLGA nanofibrous scaffold for skin tissue engineering. Acs Applied Materials & Interfaces 7(20): 10872–10877.

Rui, Z., Xiang, L., Sun, B., Ying, Z., Zhang, D., Tang, Z. et al. 2014. Electrospun chitosan/sericin composite nanofibers with antibacterial property as potential wound dressings. International Journal of Biological Macromolecules 68(7): 92.

Sahner, K., Gouma, P. and Moos, R. 2007. Electrodeposited and sol-gel precipitated p-type $SrTi_{1-x}Fe_xO_{3-\delta}$ semiconductors for gas sensing. Sensors 7(9): 1871–1886.
Saraf, A., Baggett, L. S., Raphael, R. M., Kasper, F. K. and Mikos, A. G. 2010. Regulated non-viral gene delivery from coaxial electrospun fiber mesh scaffolds. Journal of Controlled Release Official Journal of the Controlled Release Society 143(1): 95.
Sarhan, W. A. and Azzazy, H. M. 2015. High concentration honey chitosan electrospun nanofibers: biocompatibility and antibacterial effects. Carbohydrate Polymers 122: 135.
Sarioglu, O. F., Keskin, N. O. S., Celebioglu, A., Tekinay, T. and Uyar, T. 2017. Bacteria encapsulated electrospun nanofibrous webs for remediation of methylene blue dye in water. Colloids Surf B Biointerfaces 152: 245–251.
Sasaki, I., Tsuchiya, H., Nishioka, M., Sadakata, M. and Okubo, T. 2002. Gas sensing with zeolite-coated quartz crystal microbalances—principal component analysis approach. Sensors & Actuators B Chemical 86(1): 26–33.
Sawicka, K., Gouma, P. and Simon, S. 2005. Electrospun biocomposite nanofibers for urea biosensing. Sensors & Actuators B Chemical 108(1): 585–588.
Senevirathne, K., Hui, R., Campbell, S., Ye, S. and Zhang, J. 2012. Electrocatalytic activity and durability of Pt/NbO_2 and Pt/Ti_4O_7 nanofibers for PEM fuel cell oxygen reduction reaction. Electrochimica Acta 59: 538–547.
Shang, Y., Si, Y., Raza, A., Yang, L., Mao, X., Ding, B. et al. 2012. An *in situ* polymerization approach for the synthesis of superhydrophobic and superoleophilic nanofibrous membranes for oil-water separation. Nanoscale 4(24): 7847–7854.
Sharma, D. K., Li, F. and Wu, Y. N. 2014. Electrospinning of nafion and polyvinyl alcohol into nanofiber membranes: A facile approach to fabricate functional adsorbent for heavy metals. Colloids & Surfaces A Physicochemical & Engineering Aspects 457(1): 236–243.
Shen, Y., Hu, Y., Chen, W., Wang, J., Guan, Y., Du, J. et al. 2015. Modulation of topological structure induces ultrahigh energy density of graphene/$Ba_{0.6}Sr_{0.4}TiO_3$ nanofiber/polymer nanocomposites. Nano Energy 18: 176–186.
Shi, W., Lu, W. and Jiang, L. 2009. The fabrication of photosensitive self-assembly Au nanoparticles embedded in silica nanofibers by electrospinning. J. Colloid Interface Sci. 340(2): 291–297.
Shinde, S. S. and Rajpure, K. Y. 2011. Fast response ultraviolet Ga-doped ZnO based photoconductive detector. Materials Research Bulletin 46(10): 1734–1737.
Si, Y., Mao, X., Zheng, H., Yu, J. and Ding, B. 2014. Silica nanofibrous membranes with ultra-softness and enhanced tensile strength for thermal insulation. RSC Advances 5(8): 6027–6032.
Singh, A., Poshtiban, S. and Evoy, S. 2013. Recent Advances in bacteriophage based biosensors for food-borne pathogen detection. Sensors 13(2): 1763.
Sohrabi, A., Shaibani, P. M., Etayash, H., Kaur, K. and Thundat, T. 2013. Sustained drug release and antibacterial activity of ampicillin incorporated poly(methyl methacrylate)–nylon6 core/shell nanofibers. Polymer 54(11): 2699–2705.
Son, W. K., Youk, J. H. and Park, W. H. 2006. Antimicrobial cellulose acetate nanofibers containing silver nanoparticles. Carbohydrate Polymers 65(4): 430–434.
Song, X., Qi, Q., Zhang, T. and Wang, C. 2009. A humidity sensor based on KCl-doped SnO_2 nanofibers. Sensors & Actuators B Chemical 138(1): 368–373.
Song, X., Wang, Z., Liu, Y., Wang, C. and Li, L. 2009. A highly sensitive ethanol sensor based on mesoporous ZnO–SnO_2 nanofibers. Nanotechnology 20(7): 75501.
Souzandeh, H., Wang, Y. and Zhong, W. H. K. 2016. "Green" nano-filters: Fine nanofibers of natural protein for high efficiency filtration of particulate pollutants and toxic gases. RSC Advances 6(107): 105948–105956.
Spasova, M., Manolova, N., Paneva, D. and Rashkov, I. 2004. Preparation of chitosan-containing nanofibres by electrospinning of chitosan/poly(ethylene oxide) blend solutions. e-Polymers 4(1): 624–635.
Stefani, I. and Cooper-White, J. 2016. Development of an in-process UV-crosslinked, electrospun PCL/aPLA-co-TMC composite polymer for tubular tissue engineering applications. Acta Biomaterialia 36: 231–240.

Sun, B., Long, Y. Z., Liu, S. L., Huang, Y. Y., Ma, J., Zhang, H. D. et al. 2013. Fabrication of curled conducting polymer microfibrous arrays via a novel electrospinning method for stretchable strain sensors. Nanoscale 5(15): 7041.

Sun, N., Wen, Z., Zhao, F., Yang, Y., Shao, H., Zhou, C. et al. 2017. All flexible electrospun papers based self-charging power system. Nano Energy 38: 210–217.

Sun, Y., Buck, H. and Mallouk, T. E. 2001. Combinatorial discovery of alloy electrocatalysts for amperometric glucose sensors. Analytical Chemistry 73(7): 1599–1604.

Syritski, V., Reut, J., Öpik, A. and Idla, K. 1999. Environmental QCM sensors coated with polypyrrole. Synthetic Metals 102(1): 1326–1327.

Tabatabaeefar, A., Keshtkar, A. R. and Moosavian, M. A. 2015. Preparation and characterization of a novel electrospun ammonium molybdophosphate/polyacrylonitrile nanofiber adsorbent for cesium removal. Journal of Radioanalytical & Nuclear Chemistry 305(2): 1–12.

Tan, J. Z. Y., Zeng, J., Kong, D., Bian, J. and Zhang, X. 2012. Growth of crystallized titania from the cores of amorphous tetrabutyl titanate@PVDF nanowires. Journal of Materials Chemistry 22(35): 18603–18608.

Tang, H., Yan, F., Tai, Q. and Chan, H. L. 2010. The improvement of glucose bioelectrocatalytic properties of platinum electrodes modified with electrospun TiO_2 nanofibers. Biosensors & Bioelectronics 25(7): 1646–1651.

Thakur, R. A., Florek, C. A., Kohn, J. and Michniak, B. B. 2008. Electrospun nanofibrous polymeric scaffold with targeted drug release profiles for potential application as wound dressing. International Journal of Pharmaceutics 364(1): 87–93.

Thavasi, V., Singh, G. and Ramakrishna, S. 2008. Electrospun nanofibers in energy and environmental applications. Energy & Environmental Science 1(2): 205–221.

Thunberg, J., Kalogeropoulos, T., Kuzmenko, V., Hägg, D., Johannesson, S., Westman, G. et al. 2015. *In situ* synthesis of conductive polypyrrole on electrospun cellulose nanofibers: scaffold for neural tissue engineering. Cellulose 22(3): 1459–1467.

Toncheva, A., Paneva, D., Maximova, V., Manolova, N. and Rashkov, I. 2012. Antibacterial fluoroquinolone antibiotic-containing fibrous materials from poly(L-lactide-co-D,L-lactide) prepared by electrospinning. European Journal of Pharmaceutical Sciences 47(4): 642–651.

Torres-Giner, S., Martinez-Abad, A., Gimeno-Alcañiz, J. V., Ocio, M. J. and Lagaron, J. M. 2012. Controlled delivery of gentamicin antibiotic from bioactive electrospun polylactide-based ultrathin fibers. Advanced Engineering Materials 14(4): B112–B122.

Valle, L. J. D., Camps, R., Díaz, A., Franco, L., Rodríguez-Galán, A. and Puiggalí, J. 2011. Electrospinning of polylactide and polycaprolactone mixtures for preparation of materials with tunable drug release properties. Journal of Polymer Research 18(6): 1903–1917.

Vitchuli, N., Shi, Q., Nowak, J., Kay, K., Caldwell, J. M., Breidt, F. et al. 2011. Multifunctional ZnO/Nylon 6 nanofiber mats by an electrospinning–electrospraying hybrid process for use in protective applications. Science & Technology of Advanced Materials 12(5): 055004.

Wan, H., Na, W., Yang, J., Si, Y., Chen, K., Ding, B. et al. 2014. Hierarchically structured polysulfone/titania fibrous membranes with enhanced air filtration performance. J. Colloid Interface Sci. 417(3): 18–26.

Wang, G., Ji, Y., Huang, X., Yang, X., Gouma, P. I. and Dudley, M. 2006. Fabrication and characterization of polycrystalline WO_3 nanofibers and their application for ammonia sensing. Journal of Physical Chemistry B 110(47): 23777.

Wang, B., Cheng, J. L., Wu, Y. P., Wang, D. and He, D. N. 2012. Porous NiO fibers prepared by electrospinning as high performance anode materials for lithium ion batteries. Electrochemistry Communications 23(8): 5–8.

Wang, C., Ma, C., Wu, Z., Liang, H., Yan, P., Song, J. et al. 2015. Enhanced bioavailability and anticancer effect of curcumin-loaded electrospun nanofiber: *in vitro* and *in vivo* study. Nanoscale Research Letters 10(1): 439.

Wang, C., Wu, S., Jian, M., Xie, J., Xu, L., Yang, X. et al. 2016. Silk nanofibers as high efficient and lightweight air filter. Nano Research 9(9): 1–8.

Wang, G., Lu, X., Zhai, T., Ling, Y., Wang, H., Tong, Y. et al. 2012a. Free-standing nickel oxide nanoflake arrays: synthesis and application for highly sensitive non-enzymatic glucose sensors. Nanoscale 4(10): 3123–3127.

Wang, L., Yu, Y., Chen, P. C., Zhang, D. W. and Chen, C. H. 2008. Electrospinning synthesis of C/Fe_3O_4 composite nanofibers and their application for high performance lithium-ion batteries. Journal of Power Sources 183(2): 717–723.

Wang, N., Raza, A., Si, Y., Yu, J., Sun, G. and Ding, B. 2013. Tortuously structured polyvinyl chloride/polyurethane fibrous membranes for high-efficiency fine particulate filtration. Journal of Colloid & Interface Science 398(19): 240–246.

Wang, N., Si, Y., Wang, N., Sun, G., El-Newehy, M., Al-Deyab, S. S. et al. 2014. Multilevel structured polyacrylonitrile/silica nanofibrous membranes for high-performance air filtration. Separation & Purification Technology 126(15): 44–51.

Wang, N., Zhu, Z., Sheng, J., Aldeyab, S. S., Yu, J. and Ding, B. 2014. Superamphiphobic nanofibrous membranes for effective filtration of fine particles. Journal of Colloid & Interface Science 428: 41.

Wang, N., Yang, Y., Aldeyab, S. S., Elnewehy, M., Yu, J. and Ding, B. 2015. Ultra-light 3D nanofibre-nets binary structured nylon 6–polyacrylonitrile membranes for efficient filtration of fine particulate matter. Journal of Materials Chemistry A 3(47): 23946–23954.

Wang, Q., Song, W. L., Wang, L., Song, Y., Shi, Q. and Fan, L. Z. 2014. Electrospun polyimide-based fiber membranes as polymer electrolytes for lithium-ion batteries. Electrochimica Acta 132(3): 538–544.

Wang, S., Zheng, F., Huang, Y., Fang, Y., Shen, M., Zhu, M. et al. 2012b. Encapsulation of amoxicillin within laponite-doped poly(lactic-co-glycolic acid) nanofibers: preparation, characterization, and antibacterial activity. Acs Applied Materials & Interfaces 4(11): 6393.

Wang, S., Zhao, X., Yin, X., Yu, J. and Ding, B. 2016. Electret Polyvinylidene fluoride nanofibers hybridized by polytetrafluoroethylene nanoparticles for high-efficiency air filtration. ACS Applied Materials & Interfaces 8(36): 23985.

Wang, W., Li, Z., Zheng, W., Yang, J., Zhang, H. and Wang, C. 2009. Electrospun palladium (IV)-doped copper oxide composite nanofibers for non-enzymatic glucose sensors. Electrochemistry Communications 11(9): 1811–1814.

Wang, X., Fang, D., Yoon, K., Hsiao, B. S. and Chu, B. 2006. High performance ultrafiltration composite membranes based on poly(vinyl alcohol) hydrogel coating on crosslinked nanofibrous poly(vinyl alcohol) scaffold. Journal of Membrane Science 278(1-2): 261–268.

Wang, X., Ding, B., Yu, J., Wang, M. and Pan, F. 2009. A highly sensitive humidity sensor based on a nanofibrous membrane coated quartz crystal microbalance. Nanotechnology 21(5): 055502.

Wang, X., Drew, C., Lee, S. H., Senecal, K. J., Kumar, J. and Samuelson, L. A. 2002. Electrospun nanofibrous membranes for highly sensitive optical sensors. Nano Letters 2(11): 1273–1275.

Wang, Y., Ramos, I. and Santiago-Aviles, J. J. 2007. Detection of moisture and methanol gas using a single electrospun tin oxide nanofiber. IEEE Sensors Journal 7(9): 1347–1348.

Wang, Y., Jia, W., Strout, T., Schempf, A., Zhang, H., Li, B. et al. 2009. Ammonia gas sensor using polypyrrole-coated TiO_2/ZnO nanofibers. Electroanalysis 21(12): 1432–1438.

Wang, Y., Li, W., Xia, Y., Jiao, X. and Chen, D. 2014. Electrospun flexible self-standing g-alumina fibrous membranes and their potential as high-efficiency fine particulate filtration media. Journal of Materials Chemistry A 2(36): 15124–15131.

Wang, Z., Li, Z., Li, L., Xu, X., Zhang, H., Wang, W. et al. 2010. A novel alcohol detector based on ZrO_2-doped SnO_2 electrospun nanofibers. Journal of the American Ceramic Society 93(3): 634–637.

Wei, Z., Zhao, H., Zhang, J., Deng, L., Wu, S., He, J. et al. 2014. Poly(vinyl alcohol) electrospun nanofibrous membrane modified with spirolactam-rhodamine derivatives for visible detection and removal of metal ions. Rsc Advances 4(93): 51381–51388.

Weiss, D., Skrybeck, D., Misslitz, H., Nardini, D., Kern, A., Kreger, K. et al. 2016. Tailoring supramolecular nanofibers for air filtration applications. Acs Applied Materials & Interfaces 8(23): 14885.

Wu, H., Sun, Y., Lin, D., Zhang, R., Zhang, C. and Pan, W. 2009. GaN nanofibers based on electrospinning: Facile synthesis, controlled assembly, precise doping, and application as high performance UV photodetector. Advanced Materials 21(2): 227–231.

Wu, J., Wang, N., Wang, L., Dong, H., Zhao, Y. and Jiang, L. 2012. Electrospun porous structure fibrous film with high oil adsorption capacity. Acs Applied Materials & Interfaces 4(6): 3207.

Wu, Q., Tran, T., Lu, W. and Wu, J. 2014. Electrospun silicon/carbon/titanium oxide composite nanofibers for lithium ion batteries. Journal of Power Sources 258(21): 39–45.

Wu, W., Bai, S., Yuan, M., Qin, Y., Wang, Z. L. and Jing, T. 2012. Lead zirconate titanate nanowire textile nanogenerator for wearable energy-harvesting and self-powered devices. Acs Nano 6(7): 6231–6235.
Wu, W. Y., Ting, J. M. and Huang, P. J. 2009. Electrospun ZnO nanowires as gas sensors for ethanol detection. Nanoscale Research Letters 4(6): 513.
Xia, Q., Liu, Z., Wang, C., Zhang, Z., Xu, S. and Han, C. C. 2015. A biodegradable trilayered barrier membrane composed of sponge and electrospun layers: Hemostasis and antiadhesion. Biomacromolecules 16(9): 3083–3092.
Xiaoqi, L., Na, W., Gang, F., Jianyong, Y., Jing, G., Gang, S. et al. 2015. Electreted polyetherimide-silica fibrous membranes for enhanced filtration of fine particles. Journal of Colloid & Interface Science 439: 12.
Xie, Z., Paras, C. B., Weng, H., Punnakitikashem, P., Su, L. C., Vu, K. et al. 2013. Dual growth factor releasing multi-functional nanofibers for wound healing. Acta Biomaterialia 9(12): 9351.
Xu, R., Chi, C., Li, F. and Zhang, B. 2013. Laccase-polyacrylonitrile nanofibrous membrane: highly immobilized, stable, reusable, and efficacious for 2,4,6-trichlorophenol removal. Acs Appl. Mater. Interfaces 5(23): 12554–12560.
Yüksel, E. and Karakeçili, A. 2014. Antibacterial activity on electrospun poly(lactide-co-glycolide) based membranes via Magainin II grafting. Materials Science & Engineering C 45: 510–518.
Yan, L., Si, S., Chen, Y., Yuan, T., Fan, H., Yao, Y. et al. 2011. Electrospun *in situ* hybrid polyurethane/nano-TiO_2 as wound dressings. Fibers & Polymers 12(2): 207–213.
Yan, L. C., Gupta, N., Pramana, S. S., Aravindan, V., Wee, G. and Srinivasan, M. 2011. Morphology, structure and electrochemical properties of single phase electrospun vanadium pentoxide nanofibers for lithium ion batteries. Journal of Power Sources 196(15): 6465–6472.
Yang, A., Tao, X., Wang, R., Lee, S. and Surya, C. 2007. Room temperature gas sensing properties of SnO_2/multiwall-carbon-nanotube composite nanofibers. Applied Physics Letters 91(13): 151.
Yang, C. F. 2012. Aerosol filtration application using fibrous media an industrial perspective. Chinese Journal of Chemical Engineering 20(1): 1–9.
Yang, G., Yan, W., Zhang, Q., Shen, S. and Ding, S. 2013. One-dimensional CdS/ZnO core/shell nanofibers via single-spinneret electrospinning: tunable morphology and efficient photocatalytic hydrogen production. Nanoscale 5(24): 12432–12439.
Yang, J., Zhang, W. D. and Gunasekaran, S. 2010. An amperometric non-enzymatic glucose sensor by electrodepositing copper nanocubes onto vertically well-aligned multi-walled carbon nanotube arrays. Biosensors & Bioelectronics 26(1): 279–284.
Yang, M., Xie, T., Peng, L., Zhao, Y. and Wang, D. 2007. Fabrication and photoelectric oxygen sensing characteristics of electrospun Co doped ZnO nanofibres. Applied Physics A Materials Science & Processing 89(2): 427–430.
Yang, X., Salles, V., Kaneti, Y. V., Liu, M., Maillard, M., Journet, C. et al. 2015. Fabrication of highly sensitive gas sensor based on Au functionalized WO3 composite nanofibers by electrospinning. Sensors & Actuators B Chemical 220(1): 1112–1119.
Yang, Z., Du, G., Feng, C., Li, S., Chen, Z., Zhang, P. et al. 2010. Synthesis of uniform polycrystalline tin dioxide nanofibers and electrochemical application in lithium-ion batteries. Electrochimica Acta 55(19): 5485–5491.
Ying, W., Jia, W., Timothy, S., Yu, D. and Yu, L. 2009. Preparation, characterization and sensitive gas sensing of conductive core-sheath TiO_2-PEDOT nanocables. Sensors 9(9): 6752–6763.
Yoon, K., Hsiao, B. S. and Chu, B. 2009. High flux nanofiltration membranes based on interfacially polymerized polyamide barrier layer on polyacrylonitrile nanofibrous scaffolds. Journal of Membrane Science 326(2): 484–492.
Yu, C., Yang, K., Xie, Y., Fan, Q., Yu, J. C., Shu, Q. et al. 2013. Novel hollow Pt-ZnO nanocomposite microspheres with hierarchical structure and enhanced photocatalytic activity and stability. Nanoscale 5(5): 2142–2151.
Yu, G. F., Yan, X., Yu, M., Jia, M. Y., Pan, W., He, X. X. et al. 2016. Patterned, highly stretchable and conductive nanofibrous PANI/PVDF strain sensors based on electrospinning and *in situ* polymerization. Nanoscale 8(5): 2944–2950.

Yu, H., Huang, T., Lu, M., Mao, M., Zhang, Q. and Wang, H. 2013. Enhanced power output of an electrospun PVDF/MWCNTs-based nanogenerator by tuning its conductivity. Nanotechnology 24(40): 405401.

Zahedi, P., Karami, Z., Rezaeian, I., Jafari, S. H., Mahdaviani, P., Abdolghaffari, A. H. et al. 2012. Preparation and performance evaluation of tetracycline hydrochloride loaded wound dressing mats based on electrospun nanofibrous poly(lactic acid)/poly(ϵ-caprolactone) blends. Journal of Applied Polymer Science 124(5): 4174–4183.

Zhang, B., Kang, F., Tarascon, J. M. and Kim, J. K. 2016. Recent advances in electrospun carbon nanofibers and their application in electrochemical energy storage. Progress in Materials Science 76: 319–380.

Zhang, B., Zhang, Z. G., Yan, X., Wang, X. X., Zhao, H., Guo, J. et al. 2017. Chitosan nanostructures by *in situ* electrospinning for high-efficiency PM2.5 capture. Nanoscale 9(12): 4154–4161.

Zhang, H., Li, Z., Liu, L., Wang, C., Wei, Y. and Macdiarmid, A. G. 2009. Mg^{2+}/Na^{+}-doped rutile TiO_2 nanofiber mats for high-speed and anti-fogged humidity sensors. Talanta 79(3): 953–958.

Zhang, H. D., Yan, X., Zhang, Z. H., Yu, G. F., Han, W. P., Zhang, J. C. et al. 2016a. Electrospun PEDOT:PSS/PVP nanofibers for CO gas sensing with quartz crystal microbalance technique. International Journal of Polymer Science (2016-4-27), 2016: 1–6.

Zhang, J., Dai, C., Su, X. and O'Shea, S. J. 2002. Determination of liquid density with a low frequency mechanical sensor based on quartz tuning fork. Sensors & Actuators B Chemical 84(2-3): 123–128.

Zhang, J., Wang, X., Liu, T., Liu, S. and Jing, X. 2014. Antitumor activity of electrospun polylactide nanofibers loaded with 5-fluorouracil and oxaliplatin against colorectal cancer. Drug Delivery 23(3): 794.

Zhang, K., Jinglei, W., Huang, C. and Mo, X. 2013. Fabrication of silk fibroin/P (LLA-CL) aligned nanofibrous scaffolds for nerve tissue engineering. Macromolecular Materials and Engineering 298(5): 565–574.

Zhang, P., Guo, Z. P., Huang, Y., Jia, D. and Liu, H. K. 2011. Synthesis of Co_3O_4/carbon composite nanowires and their electrochemical properties. Journal of Power Sources 196(16): 6987–6991.

Zhang, S., Liu, H., Yin, X., Li, Z., Yu, J. and Ding, B. 2017. Tailoring mechanically robust poly(m-phenylene isophthalamide) nanofiber/nets for ultrathin high-efficiency air filter. Scientific Reports 7: 40550.

Zhang, S., Liu, H., Zuo, F., Yin, X., Yu, J. and Ding, B. 2017. Air filtration: A controlled design of ripple-like polyamide-6 nanofiber/nets membrane for high-efficiency air filter (Small 10/2017). Small 13(10).

Zhang, W. and Pintauro, P. N. 2011. High-performance nanofiber fuel cell electrodes. Chemsuschem 4(12): 1753.

Zhang, X., Shen, Y., Zhang, Q., Gu, L., Hu, Y., Du, J. et al. 2015. Ultrahigh energy density of polymer nanocomposites containing $BaTiO_3$@TiO_2 nanofibers by atomic-scale interface engineering. Advanced Materials 27(5): 819.

Zhang, X., Shen, Y., Xu, B., Zhang, Q., Gu, L., Jiang, J. et al. 2016b. Giant energy density and improved discharge efficiency of solution-processed polymer nanocomposites for dielectric energy storage. Advanced Materials 28(10): 2055–2061.

Zhang, Y., He, X., Li, J., Miao, Z. and Huang, F. 2008. Fabrication and ethanol-sensing properties of micro gas sensor based on electrospun SnO_2 nanofibers. Sensors & Actuators B Chemical 132(1): 67–73.

Zhang, Z., Li, X., Wang, C., Wei, L., Liu, Y. and Shao, C. 2009. ZnO hollow nanofibers: Fabrication from facile single capillary electrospinning and applications in gas sensors. Journal of Physical Chemistry C 113(45): 19397–19403.

Zhang, Z., Shao, C., Li, X., Zhang, L., Xue, H., Wang, C. et al. 2010. Electrospun nanofibers of $ZnO-SnO_2$ heterojunction with high photocatalytic activity. J. Phys. Chem. C 114(17): 7920–7925.

Zhang, Z., Shao, C., Sun, Y., Mu, J., Zhang, M., Zhang, P. et al. 2011. Tubular nanocomposite catalysts based on size-controlled and highly dispersed silver nanoparticles assembled on electrospun silica nanotubes for catalytic reduction of 4-nitrophenol. Journal of Materials Chemistry 22(4): 1387–1395.

Zhang, Z., Shao, C., Li, X., Sun, Y., Zhang, M., Mu, J. et al. 2013. Hierarchical assembly of ultrathin hexagonal SnS_2 nanosheets onto electrospun TiO_2 nanofibers: enhanced photocatalytic activity based on photoinduced interfacial charge transfer. Nanoscale 5(2): 606.

Zhao, J., Wei, L., Peng, C., Su, Y., Yang, Z., Zhang, L. et al. 2013. A non-enzymatic glucose sensor based on the composite of cubic Cu nanoparticles and arc-synthesized multi-walled carbon nanotubes. Biosensors & Bioelectronics 47(17): 86–91.

Zhao, X., Li, Y., Hua, T., Jiang, P., Yin, X., Yu, J. et al. 2017. Low-resistance dual-purpose air filter releasing negative ions and effectively capturing PM2.5. ACS Applied Materials & Interfaces 9(13): 12054.

Zheng, F., Wang, S., Shen, M., Zhu, M. and Shi, X. 2013. Antitumor efficacy of doxorubicin-loaded electrospun nano-hydroxyapatite–poly(lactic-co-glycolic acid) composite nanofibers. Polymer Chemistry 4(4): 933–941.

Zheng, W., Li, Z., Zhang, H., Wang, W., Wang, Y. and Wang, C. 2009. Electrospinning route for α-Fe_2O_3 ceramic nanofibers and their gas sensing properties. Materials Research Bulletin 44(6): 1432–1436.

Zheng, W., Lu, X., Wang, W., Li, Z., Zhang, H., Wang, Z. et al. 2009. Assembly of Pt nanoparticles on electrospun In_2O_3 nanofibers for H_2S detection. J. Colloid Interface Sci. 338(2): 366–370.

Zheng, Y., Cheng, L., Yuan, M., Wang, Z., Zhang, L., Qin, Y. et al. 2014. An electrospun nanowire-based triboelectric nanogenerator and its application in a fully self-powered UV detector. Nanoscale 6(14): 7842–7846.

Zhi, M., Lee, S., Miller, N., Menzler, N. H. and Wu, N. 2012. An intermediate-temperature solid oxide fuel cell with electrospun nanofiber cathode. Energy & Environmental Science 5(5): 7066–7071.

Zhou, C., Lin, X., Jian, S., Xing, R., Xu, S., Liu, D. et al. 2014. Ultrasensitive non-enzymatic glucose sensor based on three-dimensional network of ZnO-CuO hierarchical nanocomposites by electrospinning. Scientific Reports 4: 7382.

Zhou, X., Dai, Z., Liu, S., Bao, J. and Guo, Y. G. 2014. Ultra-uniform SnOx/carbon nanohybrids toward advanced lithium-ion battery anodes. Advanced Materials 26(23): 3943–3949.

Zhu, C., Lu, B., Su, Q., Xie, E. and Lan, W. 2012. A simple method for the preparation of hollow ZnO nanospheres for use as a high performance photocatalyst. Nanoscale 4(10): 3060.

Zhu, G., Pan, C., Guo, W., Chen, C. -Y., Zhou, Y., Yu, R. et al. 2012. Triboelectric-generator-driven pulse electrodeposition for micropatterning. Nano Letters 12(9): 4960–4965.

Zhu, H., Qiu, S., Jiang, W., Wu, D. and Zhang, C. 2011. Evaluation of electrospun polyvinyl chloride/polystyrene fibers as sorbent materials for oil spill cleanup. Environmental Science & Technology 45(10): 4527–4531.

Zhu, M., Han, J., Wang, F., Shao, W., Xiong, R., Zhang, Q. et al. 2017. Electrospun nanofibers membranes for effective air filtration. Macromolecular Materials & Engineering 302: 1600353.

Zhu, P., Nair, A. S., Shengjie, P., Shengyuan, Y. and Ramakrishna, S. 2012. Facile fabrication of TiO_2-graphene composite with enhanced photovoltaic and photocatalytic properties by electrospinning. Acs Appl. Mater. Interfaces 4(2): 581–585.

Chapter 10

Mass Production and Issues in Electrospinning Technology

Yun-Ze Long,[1,*] *Xiao-Xiong Wang,*[1] *Miao Yu*[1,2] and *Xu Yan*[3]

Introduction

Electrospinning (E-spinning) is a process capable of fabricating nonwoven webs and well-aligned arrays of continuous nanofibers with controlled morphology, size, and structure from polymer solutions or melts in high-voltage electrostatic field. Till date, many advances have been made including the spinning mechanism, new E-spinning devices, fabrication of various nanofiber assemblies, coaxial E-spinning, as well as combination of sol-gel and E-spinning. These electrospun (E-spun) nanofibers have considerable potential applications in filtration, nanofiber reinforcement, catalysis, textiles, biomedical applications tissue engineering, wound healing (Dong et al. 2016, Dong et al. 2015, Liu et al. 2017), transport and release of drugs, energy materials, sensors, and so on (He et al. 2017, Sun et al. 2014, Zhang et al. 2017). Moreover, it has been one of the main approaches to producing nanofiber webs in textiles, especially in the nonwoven industry. E-spun nanofibers were first commercialized for filter applications, as one part of the nonwoven industry. Therefore, mass production of the E-spun nanofibers is highly desired and a lot of researches have been devoted for this field.

In recent years, single jet E-spinning in laboratory-scale has attracted much attention. However, nanofibers produced by this single jet E-spinning set-up are obtained in low yield, which has limited the practical and industrial applications of E-spinning. The essential parameter determining whether or not high efficiency can

[1] Collaborative Innovation Center for Nanomaterials & Devices, College of Physics, Qingdao University, Qingdao 266071, China.
[2] Qingdao Junada Technology Co. Ltd., Qingdao International Academician Park, Qingdao 266199, China.
[3] Industrial Research Institute of Nonwovens & Technical Textiles, Qingdao University, Qingdao 266071, China.
* Corresponding author: yunze.long@163.com or yunze.long@qdu.edu.cn

be achieved in the E-spinning process is flow rate. However, viewed in the concept of E-spinning, flow rate is largely determined by the strength of the electrostatic field limited by the electric breakdown strength of the spinning atmosphere. In addition, the diameters of E-spun fibers usually increase with flow rate. Finally, E-spun fibers have diameters of a few tens to hundreds of nanometers and are two or three orders of magnitude thinner than conventionally spun polymer fibers. To produce the same amount by weight of conventional fibers, the jet velocity for E-spun fibers increases by four or six orders of magnitude, which nearly amounts to the order of the speed of sound in air. Therefore, it is not feasible to substantially increase the throughput of the single-jet E-spinning process where a balance must be made between applied voltage and flow rate. Recently, several techniques have been proposed and developed to enhance the E-spinning throughput that can be roughly classified into single-needle E-spinning, multi-needle E-spinning, and needleless E-spinning methods. Here we discuss the recent advances in the scale-up of nanofiber production by E-spinning, with primary focus on multi-needle and needleless E-spinning.

Multi-jets from Single-needle E-spinning

One interesting E-spinning method was first carried out using a needle with a grooved tip, from the branches of which multi-jets were formed when a high voltage was applied. The spray from a noncrystalline polymer solution forms multiple branches easily. However, this multi spray is unstable and droplets can still form because the drawing force on the multi cone-jet becomes weak amongst several branches. It was previously described that this problem could be solved by decreasing the nozzle diameter and shortening the distance between the nozzles to the target. A thick nozzle diameter and spinning to a wide area will be efficient for multi spray, however, the fiber diameter is larger than from a single spray. That is, E-spinning can carried out with branches at the tip of the nozzle. These cones are shown in Fig. 10.1 (Yamashita et al. 2007a). More recently, the formation of multiple jets from multiple Taylor cones was also observed in an E-spinning process using curved collectors with significant curvature, as shown in Fig. 10.2 (Vaseashta 2007). Another alternative approach to scaling up the single-needle E-spinning process is to split the polymer jet

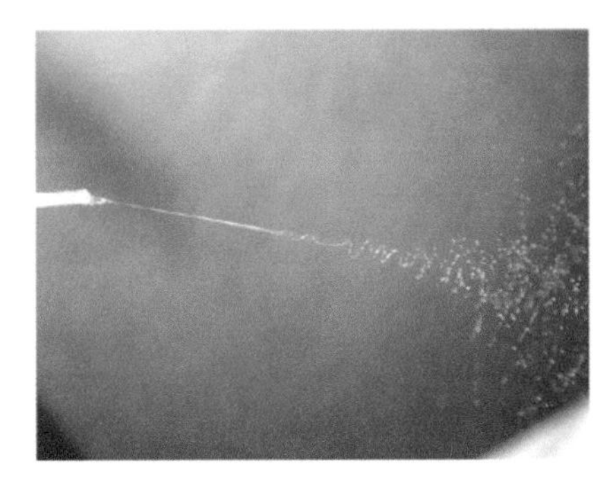

(a) Single cone spun

(b) Multi-cone spun

Fig. 10.1: Photograph of single and multi cone E-spinning (Yamashita et al. 2007a).

into two separate sub-filaments during its flight to the fiber collector. For example, it has been demonstrated that jet splitting could be controllably achieved by applying a sufficiently large tangential stress to the cross-section of the jet, as shown in Fig. 10.3 (Paruchuri and Brenner 2007).

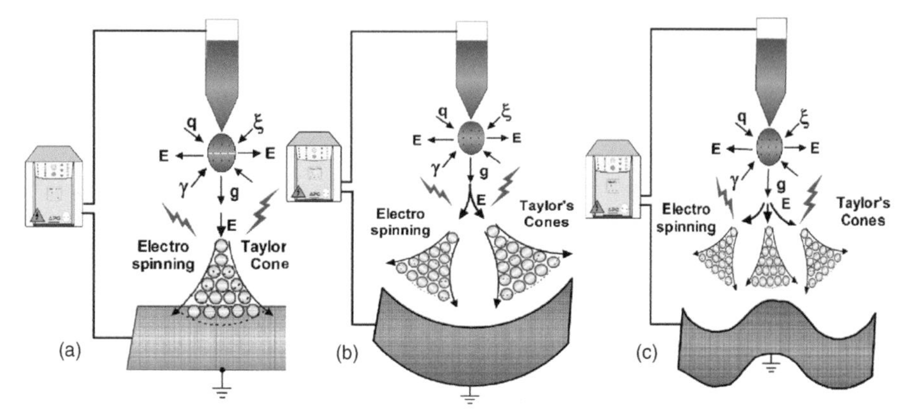

Fig. 10.2: Conceptual framework showing possible mechanism for (a) one, (b) two, and (c) three Taylor cone formation (Vaseashta 2007).

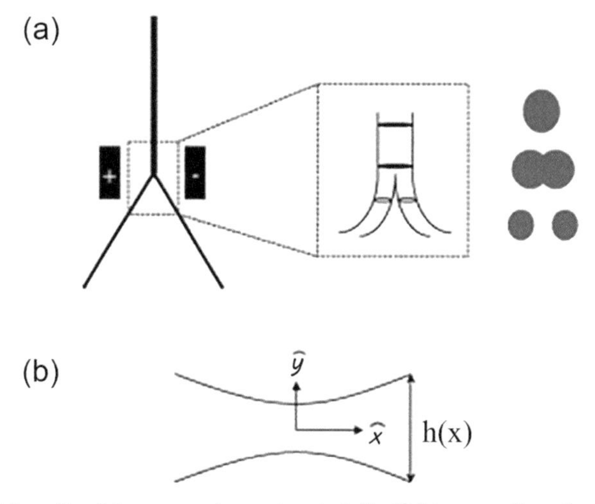

Fig. 10.3: (a) Schematic of the proposed experiment. A liquid jet passes through a forcing element (e.g., an electric capacitor) that stretches out the cross section of the jet. The stretching splits the jets into two sub-filaments. Near the splitting event the cross section of the jet stretches from a circle to break into two separate pieces. (b) The dynamics of the splitting is described by deriving equations for evolution of the thickness $h(x; t)$ of the cross section (Paruchuri and Brenner 2007).

Multi-jets from Multi-needle E-spinning

Multi-jet E-spinning based on multi-spinneret components can be arranged in two types: one-dimensional (1D) linear arrays (Fig. 10.4) and two-dimensional (2D) arrays (Fig. 10.5) (Bowman et al. 2002, Theron et al. 2005, Tomaszewksi and

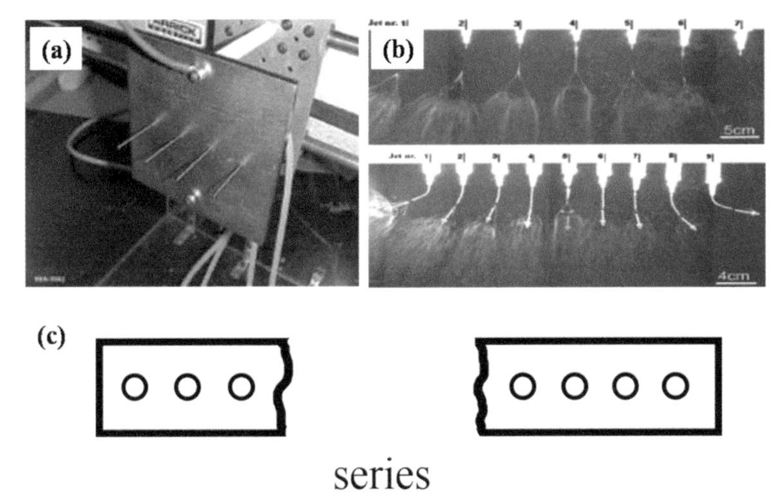

Fig. 10.4: Schematic of (a) four-needle E-spinning set-up with linear arrangement (Bowman et al. 2002), (b) seven- and nine-needle with linear array (Theron et al. 2005), (c) 26-needle set-up with linear array (Tomaszewksi and Szadkowski 2005).

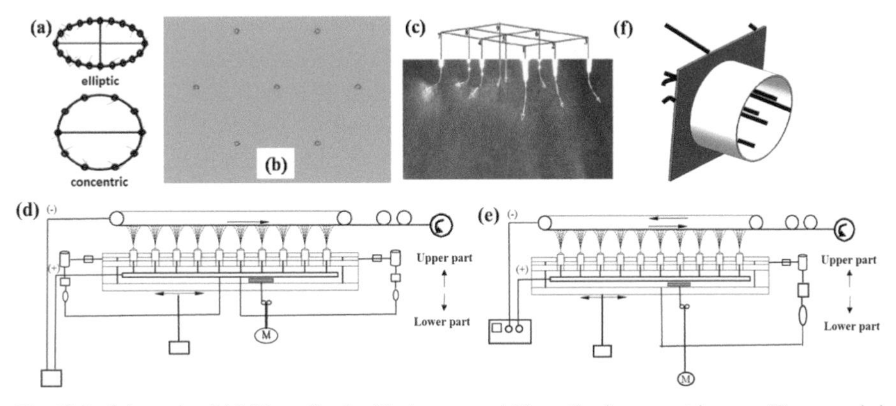

Fig. 10.5: Schematic of (a) 26 needles in elliptic array and 10 needles in concentric array (Tomaszewksi and Szadkowski 2005), (b) concentric array (Tomaszewksi and Szadkowski 2005), (c) nine needles in 3 × 3 matrix array (Theron et al. 2005), (d) conjugated multi-needle E-spinning set-up (Kim and Park 2008), (e) bottom-up multi-needle E-spinning set-up (Kim 2005), (f) (Yamashita et al. 2008) five needles with auxiliary cylindrical electrode (Kim et al. 2006).

Szadkowski 2005, Yamashita et al. 2007b, Yang et al. 2006). These approaches can be useful for increasing the overall set-up throughput and the thickness of resulting mats and can also be applied for large area deposition and for mixing fibers made of different materials.

Only a few works have been published that deal with the effects of needle configuration on flow rate and electric field distribution (Burger et al. 2006, Yamashita et al. 2008). The smallest needle spacing is defined as the distance where droplets suspended on the tip of neighboring needles do not bond together at the initial stage of multi-needle E-spinning (Yamashita et al. 2008). The needle spacing in multineedle E-spinning set-up depends not only on the nozzle gauge but also on the properties of the solutions to be E-spun. For instance, the smallest spacing in the case of 18-gauge needles was found to be 3.5 mm, where poly(vinyl alcohol) (PVA) or polyacrylonitrile (PAN) solution droplets did not bond. Several examples of multiple needles in a linear array as the simplest arrangement to scale up the E-spinning process have been reported in the last few years. For example, linear multi-needle E-spinning set-ups with four needles have been designed to produce nanofiber webs at reasonable scales (Fig. 10.4a) (Bowman et al. 2002). However, it was revealed that the nanofibers produced were unevenly deposited on fibrous substrates because of the distorted electric field in the linear multi-needle E-spinning set-up. Burger et al. also exploited a prototype multi-needle E-spinning set-up with a linear needle array to investigate the possibility of scale-up (Burger et al. 2006). Spinnerets with seven and nine needles each, equidistantly arranged in a straight line were employed to observe the behavior of jets in multi-needle E-spinning (Fig. 10.4b) (Theron et al. 2005). Experimental and simulated results showed that the behavior of border jets along the linear array was different from that of central jets, such as bending direction and envelope cone. However, every jet in both linear configurations was subject to the characteristic bending instability similar to that in single-jet E-spinning. In a study of a 26-needle spinning head with a linear arrangement (Fig. 10.4c), it was observed that the jets were generated only from a few border needles on either side, while the needles closer to the center were quite inactive (Tomaszewksi and Szadkowski 2005). This phenomenon is most probably caused by electric field shielding near the central needles. The use of an auxiliary electrode, usually known as the extractor, may be essential in linear multi-needle E-spinning as in high-compactness multi-jet electrospraying (e-spraying) (Bocanegra et al. 2005). Therefore, it remains unclear whether linear multi-needle set-ups are feasible for increasing the throughput rate in the E-spinning process.

Most investigations on multi-needle E-spinning have been focused on processes with 2D needle arrays. For instance, multi-needle spinning heads with elliptic and circular arrangements were designed to improve the process stability and also the production efficiency (Fig. 10.5a) (Tomaszewksi and Szadkowski 2005). It was demonstrated that the set-up with the circular needle array achieved higher production efficiency for PVA nanofibers (1 *vs* 0.4 mg min^{-1} for each needle) and was also a more stable process. Yang et al. developed a seven-needle E-spinning set-up, arranged as a regular hexagon with one needle located in the center (Fig. 10.5b) (Yang et al. 2006). They found that at a 10 mm spacing the middle jet behaved as in a normal single-jet E-spinning process but the jets from peripheral needles bended outwards due to the resultant electric forces. Theron et al. reported a much more complex multi-needle design with a 3 × 3 needle matrix (Fig. 10.5c) (Theron et al. 2005). The production rate achieved in this nine-needle E-spinning process ranged from 22.5 mL cm^{-2} min^{-1} to 22.5 L cm^{-2} min^{-1}. A phenomenon similar to that reported by Yang et al. was observed for the jets in this 2D needle array (Yang et al. 2006). Kim patented a bottom-up multi-needle E-spinning device, which was expected to enhance the productivity, but also overcome droplet formation by dropping solution on nanofiber webs (Fig. 10.5e) (Kim 2005). A more advanced multi-needle process, known as conjugated E-spinning, was patented by Kim and Park, which was able to mass produce composite nanofiber webs from several kinds of polymer solutions (Fig. 10.5d) (Kim and Park 2008). More recently, an industrial multi-needle E-spinning set-up with 1,000 needles was reported (Fig. 10.5f) (Yamashita et al. 2008). Here, it is noted that the use of an auxiliary cylindrical electrode (Kim et al. 2006) could reduce bending instability, and thus show a more stable spinning process and a higher productivity than that without auxiliary electrode.

Needleless E-spinning

The traditional "needle-type" E-spinning method is, in principle, subject to problems related to polymer clogging at the spinneret nozzle, which may limit the achievable throughput of continuous production processes. Such clogging events can occur because of the fineness of the needle and become more frequent for high solution concentrations or when spinning composite blends embedding nanoparticles. Although multi-needle E-spinning can improve the fiber fabrication efficiency obviously, this way is not efficient enough in practical use because of the needles blocking and repulsion among the jets (Niu et al. 2012). Namely, needle E-spinning method meets some difficulties in fabricating of nanofibers on industrial scale.

In this respect, "needleless" E-spinning (also named free-surface E-spinning) techniques possess greater up-scaling potentiality. These processes are based on the formation of an E-spinning jet from the free surface of a liquid as long as the electrostatic force exceeds the critical value of the E-spinning solution, without using needles or nozzles. Table 10.1 summarizes the comparison of multi-needle E-spinning and needleless E-spinning methods.

In 1979, the first needleless E-spinning set-up was proposed by Simm et al. which used an annular electrode as spinneret (Simm et al. 1979). Then Lukas et al. investigated the self-organization of charged jets initiated from the free liquid surface in the E-spinning process (Lukas et al. 2008). They found that the critical electric field intensity E_c for E-spinning nanofibers could be calculated as Eq. 10.1.

$$E_c = \sqrt[4]{4\gamma\rho g / \varepsilon^2} \tag{10.1}$$

Equation 10.1, where γ is the surface tension of the solution, ρ is the liquid mass density, g is the gravity acceleration, and ε is the permittivity.

Free-surface E-spinning may, in principle, produce micrometer-sized droplets resulting in defects deposited on the collector target, and significant variation of fiber

Table 10.1: Comparison of multi-needle E-spinning and needleless E-spinning.

	Multi-needle E-spinning	**Needleless E-spinning**
Advantages	• Easy to set up • Relatively low voltage • Uniform fiber diameter	➢ Easy maintenance ➢ No clogged orifices ➢ High volume output
Disadvantages	• Low output • Clogging of spinneret • Electric field disturbance • Heterogeneous feed rate	➢ Relatively high voltage ➢ Difficult to maintain consistent solution concentration and viscosity

diameter and limited configurability of the fabricated fiber assembly (e.g., general lack of fiber alignment, impracticability of core-shell fiber structures, etc.). Specific free-surface E-spinning apparatus has to be designed to overcome some of these issues and to meet industrial requirements for functional nanofibers in the different application fields.

Upward E-spinning from Stationary Spinnerets

In order to reduce solution droplet from the electrodes onto the collector, upward needleless E-spinning (the fiber generators are right below the fiber collectors) is often used. Table 10.2 shows the comparison of different modifications of stationary spinnerets for upward E-spinning. In 2007, a new bottom-up gas-jet E-spinning process for mass production was presented by Liu and He., which was also called "bubble E-spinning" (Liu and He 2007). In this method, bubbles produced by compressed air or nitrogen are blown into the solution. As the bubbles burst on the surface, multiple temporary jets are created and E-spinning is initiated (as shown in Fig. 10.6a). This technique has a higher fiber throughput than needle E-spinning, however, strip-like and sphere-like morphologies occurred easily because of the bubble broken (He et al. 2012). Till date, polymers such as polyethyleneoxide (PEO) (Liu and He 2007), PVA (Yang et al. 2009), polylacticacid (PLA) (Yang et al. 2010), and PAN (He et al. 2008) have been E-spun into fibers successfully by this bubble E-spinning. Recently, an innovative single bubble E-spinning was developed, which is based on maintenance of the bubble during E-spinning (as shown in Fig. 10.6b). In the single bubble E-spinning, the large surface area of the spinning bubble encourages multiple jets initiation from its surface, which differs from surface disruption by bubbles. In other words, in single bubble E-spinning, the concept is based on maintenance of the bubble not to burst during E-spinning. Similarly, Higham et al. reported a new approach of needleless E-spinning utilizing a sample spanning high gas volume fraction foam (as shown in Fig. 10.6c) (Higham et al.

Table 10.2: Modifications of stationary spinnerets for upward E-spinning.

Spinneret design	Features	Advantages	Disadvantages
Bubble spinneret	PVA fiber with minimal diameter 46.8 nm	High production rate with low cost; simple process; no clogging	Solvent easy to evaporate; solvent recovery problem
Cleft	PVA output: 6–15 ml h^{-1}; high voltage: 0–50 kV	No clogging	Precise design of spinneret; complex test for some materials
Magnetic liquid	PEO output: 12 times of the typical E-spinning process	Higher throughput	Heterogeneous fiber diameter distribution
Stepped pyramid	PVA output: 4 g h^{-1}; high voltage: 55 kV to 70 kV; diameter distribution: 142–205 nm	High production with narrower diameter distribution	Solvent easy to evaporate
Stationary wire	PA6 output: 0.45 g min^{-1}; electrode length: 1.0 m	Higher production rate	Inconsistent feed rate; higher spinning voltage; wash of supply pipe

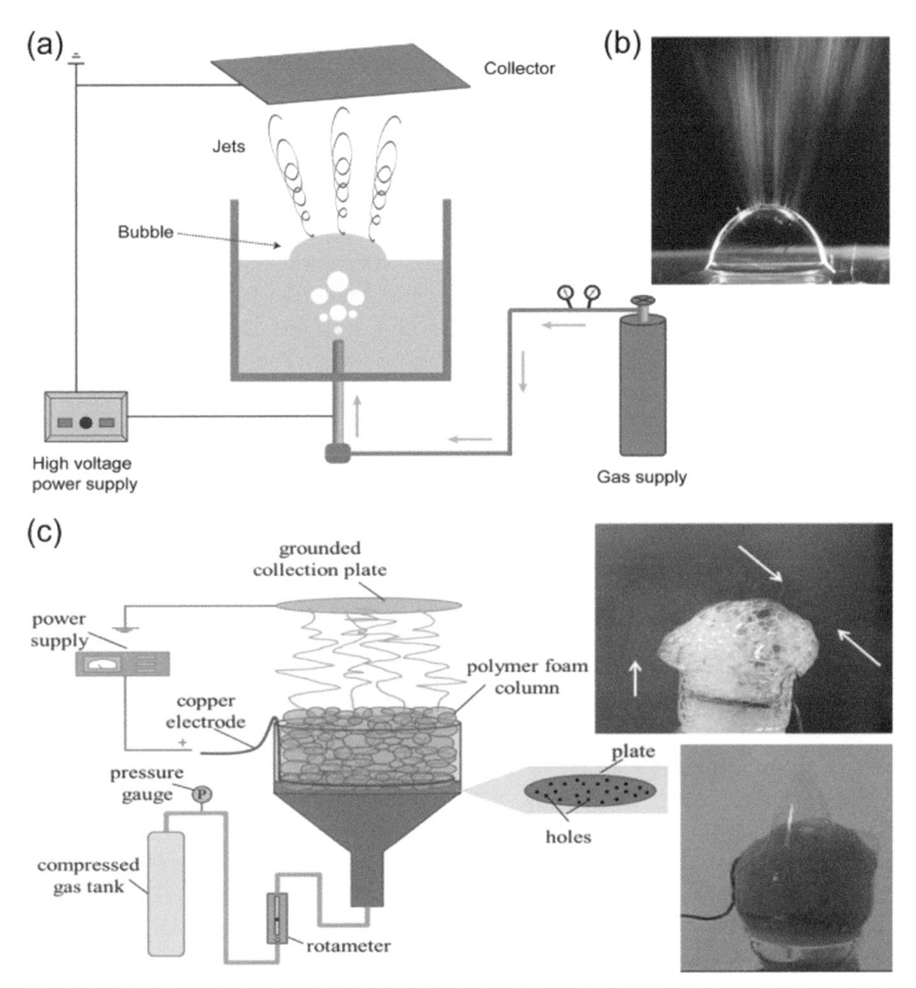

Fig. 10.6: (a) Schematic drawing of the bubble E-spinning (Liu and He 2007); (b) ejected jet from bubble E-spinning captured by a digital camera (Stellenbosch Nanofiber Company, SNC); (c) schematic of the foam E-spinning apparatus. Reproduced with permission (Higham et al. 2014). Copyright 2014, John Wiley & Sons Inc.

2014). Compressed gas through a porous surface was injected into polymer solutions to form charged multiple jets. The key design, processing, and solution parameters for producing uniform fibers were also identified. Above all, with more established needleless high-throughput E-spinning processes, it remains to see whether bubble E-spinning can present some new features.

Yarin and Zussman proposed a two-layer-fluid needleless approach (Yarin and Zussman 2004), in which the lower fluid layer was a ferromagnetic suspension and the upper layer was the polymer solution. During the E-spinning process, when a magnetic field was employed to the device, constant vertical spikes of magnetic suspension were shaped perturbing the interlayer interface. When high voltage was applied to the system, fibers were E-spun from numerous cones located at the free surface of the upper polymer solution layer. This approach leads to a 12-fold enhancement of the production rate of the typical E-spinning process, as well as eliminates clogging problem.

Jiang et al. presented a novel approach to needleless E-spinning which used a stepped pyramid-shaped fiber generator as spinneret (Jiang et al. 2013). During the E-spinning, multiple jets distributed in three dimensions were observed to form on the edges of the stepped pyramid-shaped spinneret simultaneously (as shown in Fig. 10.7). Experiments showed that under a given concentration and working distance, the fiber productivity greatly increased corresponding to the applied voltage. The highest PVA fiber productivity could reach to 5.7 g h^{-1}. Lukas et al. reported a novel E-spinning set-up which used linear clefts as spinneret (Fig. 10.8a) and developed a 1D electrohydrodynamic theory to explain the E-spinning process of conductive liquids from an open plane surface (as shown in Fig. 10.8b) (Lukas et al. 2008).

Though many designs have potential in scaling up the E-spun fibers, very few can make practical or commercial application. Elmarco Co. introduced production lines with stationary wire spinneret in 2010. During the E-spinning process, the high voltage is applied on the stationary wires and polymer solution is loaded on the surface of the wires by a reciprocating movement polymer solution container. Then, numerous jets are generated from the wires. For example, the new design

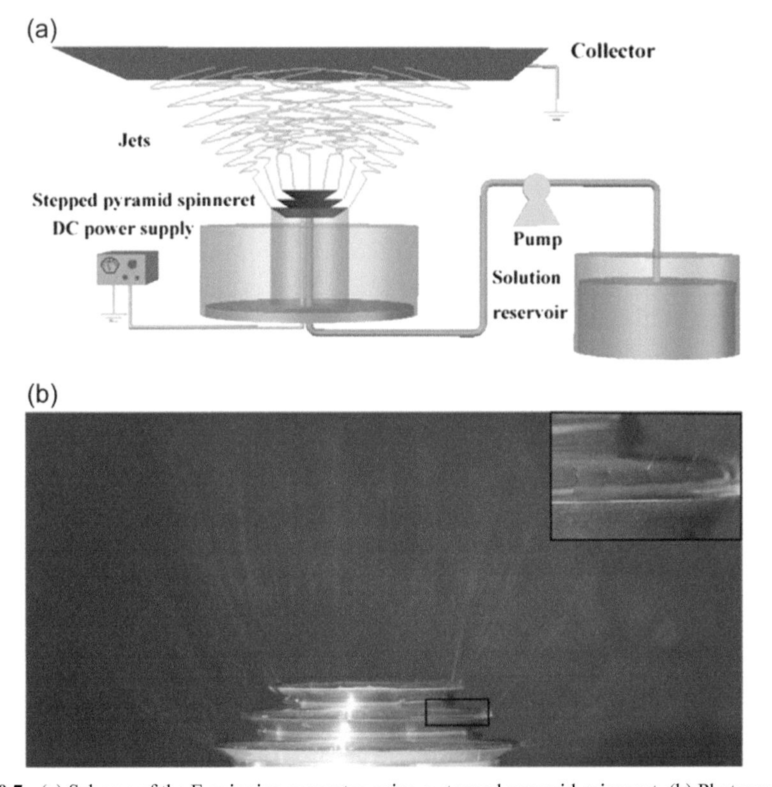

Fig. 10.7: (a) Scheme of the E-spinning apparatus using a stepped pyramid spinneret. (b) Photograph of multi-jets in E-spinning process. Inset: a magnified image of the jets generated sites. Reproduced with permission (Jiang et al. 2013). Copyright 2014, Elsevier Ltd.

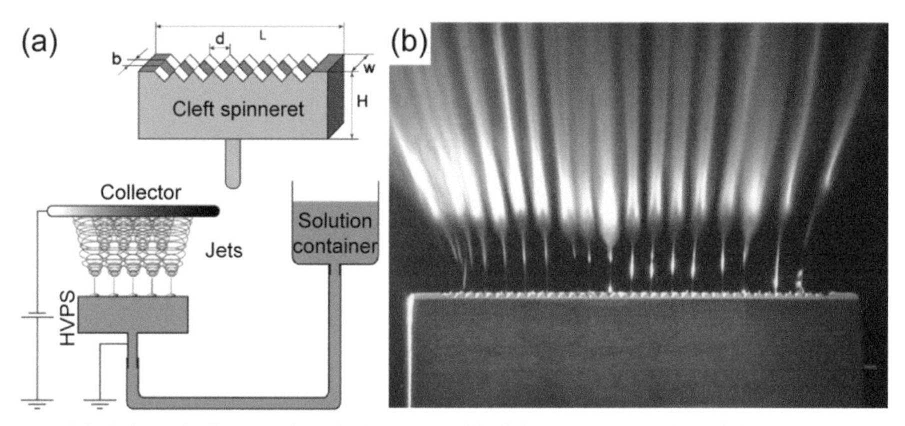

Fig. 10.8: Schematic diagram of E-spinning set-up with cleft spinneret (a), polymeric jets emitting from linear clefts (b). Reproduced with permission (Lukas et al. 2008). Copyright 2008, AIP Publishing LLC.

of Elmarco's commercialized Nanospider™ has a better performance in producing nanofibers in industrial-level than its first generation of Nanospider™ which uses rotary cylinder as spinneret.

Downward E-spinning from Stationary Spinnerets

Not all the stationary spinnerets are suitable for the upward E-spinning. The spinnerets like flat electrode with holes (Zheng et al. 2013, Zhou et al. 2009a, Zhou et al. 2009b, Zhou et al. 2010), tube with linear holes (Varabhas et al. 2008), and conical wire coil (Wang et al. 2009), always take the form of downward E-spinning with the fiber collector right below the spinnerets. Table 10.3 summarizes the modifications of stationary spinnerets for downward E-spinning.

E-spinning from flat electrode with holes was first reported by Zhou et al. in 2009 (Zhou et al. 2009a). Till now, spinnerets with one (Zhou et al. 2009a, 2010), three (Zhou et al. 2009b), four (Zhou et al. 2009a), and seven (Zheng et al. 2013) holes have been studied. Figure 10.9a shows the schematic of multi hole experimental set-up. During the E-spinning, only one jet ejects from the hole which is similar to needle E-spinning. However, the significance of multi hole E-spinning study is that the electric field can keep uniform when the number of holes increased, which is different from the complex electric field of the multi-needle E-spinning (Zhou et al. 2010). For example, Zheng et al. made a comparison between 7-hole spinneret and 7-needle spinneret (Zheng et al. 2013). Experimental results showed that 7-hole E-spinning system obtains lower jets divergence degree and smaller jets whipping amplitude than those in the seven-needle system (Fig. 10.9c–d), which results in more concentrated and smaller fiber mats on the collector. The main reason is that the 7-hole E-spinning system produced a much uniform electric field intensity distribution than 7-needle system.

In addition, Varabhas et al. reported a device for transmitting multiple jets from a hollow porous cylindrical tube (Varabhas et al. 2008). The tube is oriented with its axis horizontal and holes drilled partway into the wall are aligned along the bottom. When an air pressure of 1–2 kPa is supplied, due to the lower flow resistance, polymer

Table 10.3: Modifications of stationary spinnerets for downward E-spinning.

Spinneret design	Features	Advantages	Disadvantages
Tube with linear holes	13-cm-long tube with 20 holes: 0.3–0.5 g h^{-1} of PVP fibers	Simple operation compared with multiple E-spinning	Production rate limited by number & distribution of holes
Flat electrode with holes	PEO output: 28.0 ml h^{-1} of four-hole spinneret	Simple design and construction compared with multi-channel spinnerets	Interference between holes
Conical wire coil	PVA output: up to 30.6 ml h^{-1}	High productivity; smaller average fiber diameter	Unstable electric field intensity; low quality fibers

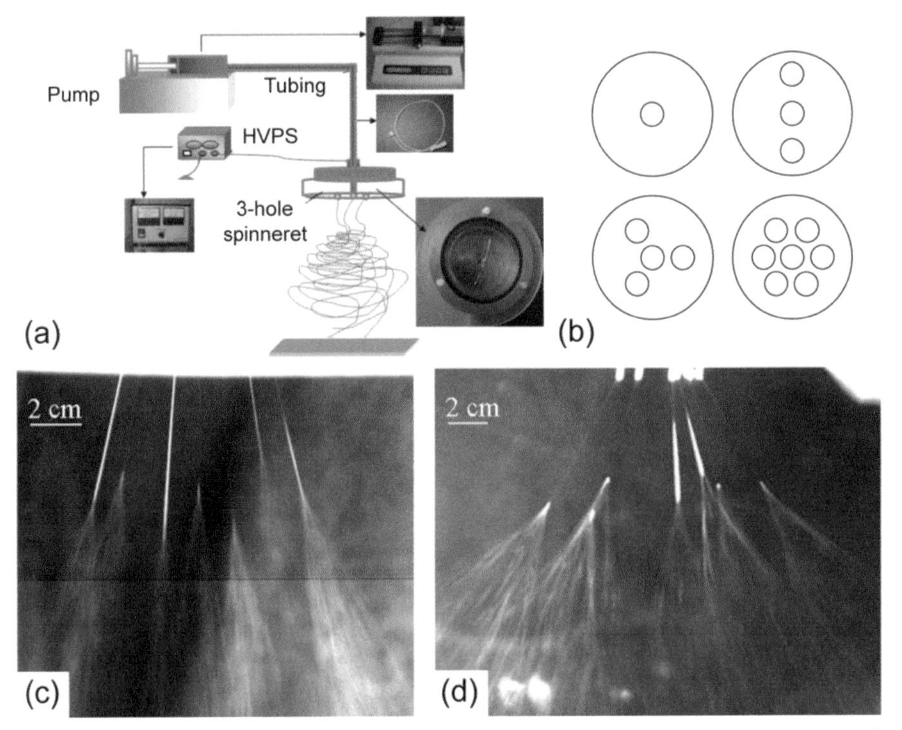

Fig. 10.9: (a) Schematic of multi-hole experimental set-up. Reproduced with permission (Zhou et al. 2009b); (b) spinnerets with 1, 3, 4, and 7 holes, respectively; photographs taken at long exposure times (33.33 ms) of the E-spinning processes of (c) 7-needle system and (d) 7-hole system. Reproduced with permission (Zheng et al. 2013). Copyright 2013, John Wiley & Sons Inc.

solution will flow out of the linear holes. Then jets are produced from the holes after applying high voltage. This approach generates smooth fibers at large rates greater than the single needle E-spinning. It was reported that 13-cm-long tube with 20 holes could produce 0.3–0.5 g h^{-1} of nanofibers. Like E-spinning from flat electrode with holes (Zheng et al. 2013, Zhou et al. 2009a, Zhou et al. 2009b), as mentioned above, the fibrous mats on the collector were unevenly distributed due to strong charge-repulsion between the jets.

Wang et al. also reported a needleless E-spinning by using a conical metal wire-coil as spinneret (Wang et al. 2009). It included a cone-shaped nozzle made from the copper wire coil and the wire was connected to a high-voltage power supply. In spinning process, PVA solution was filled into the wire cone and high voltage was employed to the wire coil, then charged solution moved down and covered the outer surface of the wire and mass of jets were produced on the conical wire surface. No "corona discharge" was caused even at a high voltage of up to 70 kV. Compared to conventional needle E-spinning, this approach generated finer fibers on great larger scale, and the fiber processing has much less relationship with the applied voltage.

Sideward E-spinning from Stationary Spinnerets

Besides the upward and downward E-spinning, some kinds of spinnerets, such as flat electrode edge (Thoppey et al. 2010), porous tube (Dosunmu et al. 2006), and bowl edge (Thoppey et al. 2011), are much more appropriate to put the collector sideward, which is called "sideward E-spinning" in this article. Table 10.4 summarizes the comparison of different modifications of stationary spinnerets for sideward E-spinning. E-spinning utilizing edge-plate configuration (Fig. 10.10a) functions in a remarkably similar manner to traditional needle E-spinning (Thoppey et al. 2010). However, this method is much easily-implemented, without the possibility of clogging and has high scale-up potential. During E-spinning, polymer solution flow down to the edge along the plate surface under gravity. Subsequently, E-spinning jet will be generated near the plate edge, as shown in Fig. 10.10b.

Unlike hollow porous polytetrafluoroethylene-ethylene (PTFE) tube drilled with linear holes (Varabhas et al. 2008), porous polyethylene tube was used as a spinneret by Dosunmu et al. with a surrounded circle collector, as shown in Fig. 10.11a (Dosunmu et al. 2006). The polymer filled in the porous walled cylindrical tube was pushed through the pores to form drops on the outer surface of the tube. When the solution was charged, jets issued from the drops and formed many E-spun fibers. The length weighted fiber diameters have a similar mean diameter to those from a single jet, but are broader in distribution. However, the mass production rate from the porous tube is 250 times greater than that of a single jet. Holopaninen et al. developed a needless twisted wire E-spinning set-up, which achieved high production rates (as shown in Fig. 10.11b) (Holopainen et al. 2014). The polymer solution was E-spun from the surface of a twisted wire set to a high voltage and collected on a cylindrical

Table 10.4: Modifications of stationary spinnerets for sideward E-spinning.

Spinneret design	Features	Advantages	Disadvantages
Edge	PEO output: 0.27 g h^{-1}	Without the possibility of clogging; 5x higher throughput	Hard to transfer to industry
Porous tube	PA6 output: 5 g min^{-1} m^{-1}	Easy to achieve industry	Clogging of small holes; wider fiber diameter distribution
Bowl	PEO output: 0.6 g h^{-1} for single-batch bowl	40 times production rate increase; simple geometry	Difficult to continuously solution feed

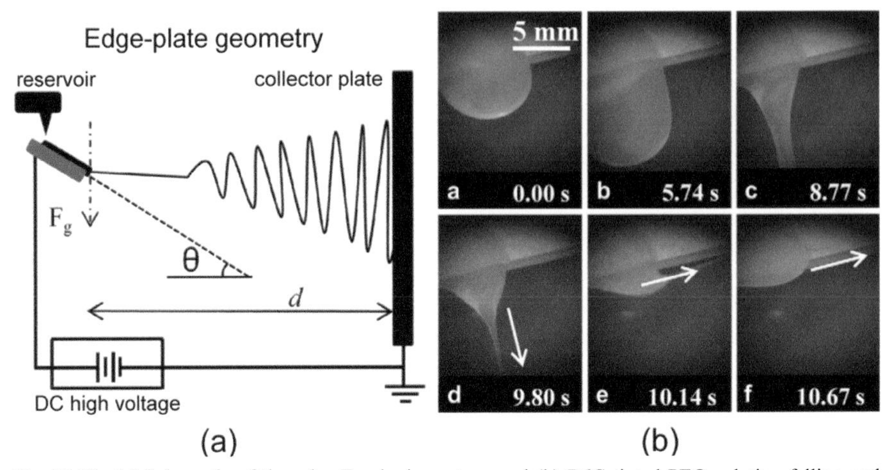

Fig. 10.10: (a) Schematic of the edge E-spinning set-up and (b) R6G-tinted PEO solution falling and ejecting from the source plate. Reproduced with permission (Thoppey et al. 2010). Copyright 2010, Elsevier Ltd.

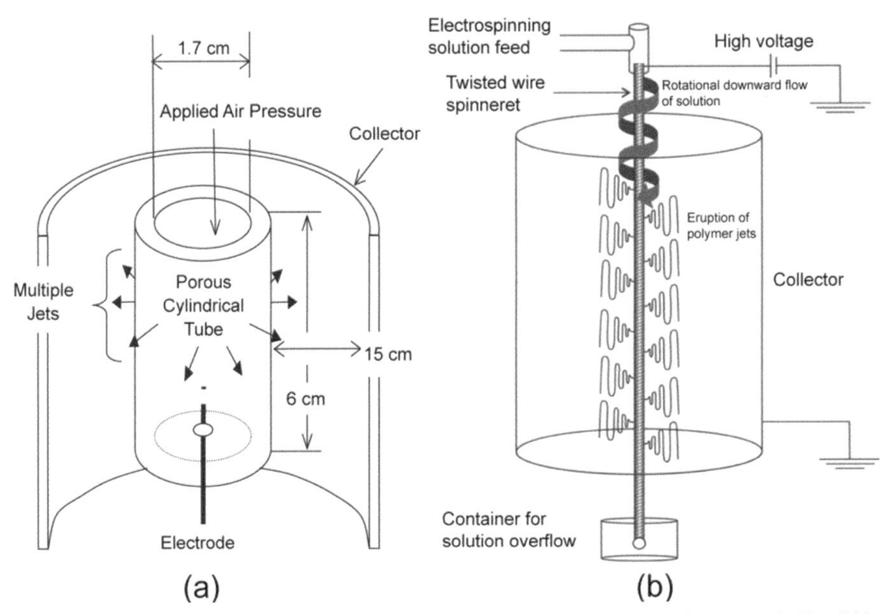

Fig. 10.11: (a) Cutaway view showing the cylindrical porous tube with its axis oriented vertically within a coaxial cylindrical collector. Reproduced with permission (Dosunmu et al. 2006). Copyright 2006, IOP Publishing; (b) schematic diagram of needleless twisted wire E-spinning set-up showing the main components in the system. Reproduced with permission (Holopainen et al. 2014). Copyright 2014, IOP Publishing.

collector around the wire. Multiple Taylor cones were simultaneously self-formed in the downward flowing solution. However, the major limiting factors of this method are fast drying of the polymer solution on the wire and cleaning of wire electrode.

At last, edge E-spinning from a bowl-shaped spinneret is also a kind of sideward E-spinning (Thoppey et al. 2011). As shown in Fig. 10.12a, under the high voltage

Fig. 10.12: Schematic of the bowl E-spinning apparatus as viewed from the top looking down (a), digital SLR camera image of 37 stabilized jets E-spinning radially from the bowl edge (b), SEM images of PEO ultrathin fibers (c). Reproduced with permission (Thoppey et al. 2011). Copyright 2011, IOP Publishing.

initiation, the jets spontaneously form directly on the fluid surface and rearrange along the circumference of the bowl to provide approximately equal spacing between spinning sites (Fig. 10.12b). Different from the edge-plate geometry (Thoppey et al. 2010), jets form and spinning occurs directly from the fluid surface within the bowl, rather than utilizing falling or elongated droplets; moreover, the bowl itself serves as the source of the polymer solution instead of gravity-assisted fluid streams. Nanofibers produced from bowl E-spinning (Fig. 10.12c) are identical in quality to those fabricated by conventional needle E-spinning with a demonstrated ~40 times increase in the production rate. Further investigations of this technique were also carried out by Thoppey et al., using PEO as a model polymer (Thoppey et al. 2012). Maximum of 40 jets from a bowl were obtained at the same time and the total number of jets tends to decrease with the increase of the viscosity of the solution.

Needleless E-spinning from Rotary Spinnerets

Besides the stationary spinnerets E-spinning, E-spinning with rotary spinnerets has also attracted much attention. Each of the spinnerets is connected to a high voltage power supply and driven by a motor. Polymer solution is usually loaded on the spinneret surface at the far side from the collector and E-spun at the near side. These needleless rotary spinnerets are featured as simplicity of design and high throughput of E-spun nanofibers. Table 10.5 shows the comparison of different modifications of rotary spinnerets for needleless E-spinning.

Mostly, the rotary spinnerets are partially immersed into the polymer solution, and the nanofibers are E-spun upward, which effectively prevents the polymer liquid from dropping onto the fiber collector. In 2005, the design of E-spinning from a rotatory cylinder was patented by Jirsak et al. (Jirsak et al. 2009), which aimed to create an industrially applicable device and reach a high spinning capacity. In this design, a cylinder is used as fiber generator and the polymer solution is delivered from the lower part of the cylinder to the upper part by the surface. Numerous solution jets will be generated from the upper part of the spinneret that is closer to the collector when high voltage is supplied (Fig. 10.13b). This method has a PVA fiber production rate of 108 g h^{-1}, which is much higher than that of the conventional single jet

Table 10.5: Modifications of rotary spinnerets for needleless E-spinning.

Spinneret design	Features	Advantages	Disadvantages
Cylinder/disk/ball	High voltage: 30–120 kV; self-optimized distance between Taylor cones	High throughput; smaller fiber diameter; economical operation and easy maintenance	Very high voltage needed; problem of solvent evaporation and solution concentration
Spiral wire coil/ rack	High voltage: 25 ~ 60 kV	Active process of jets generation; high quality of fiber; no limit of polymer category	Solvent evaporation; unstable solution concentration
Splashing/cone	Output: 10–14 g h^{-1}	High throughput; uniform fiber diameter	Problem of consistent solution supply
Rotary beaded chain	PVP output: 40 g h^{-1}; voltage: 30 kV	High production rate; lower applied voltage	Uneven fiber diameter

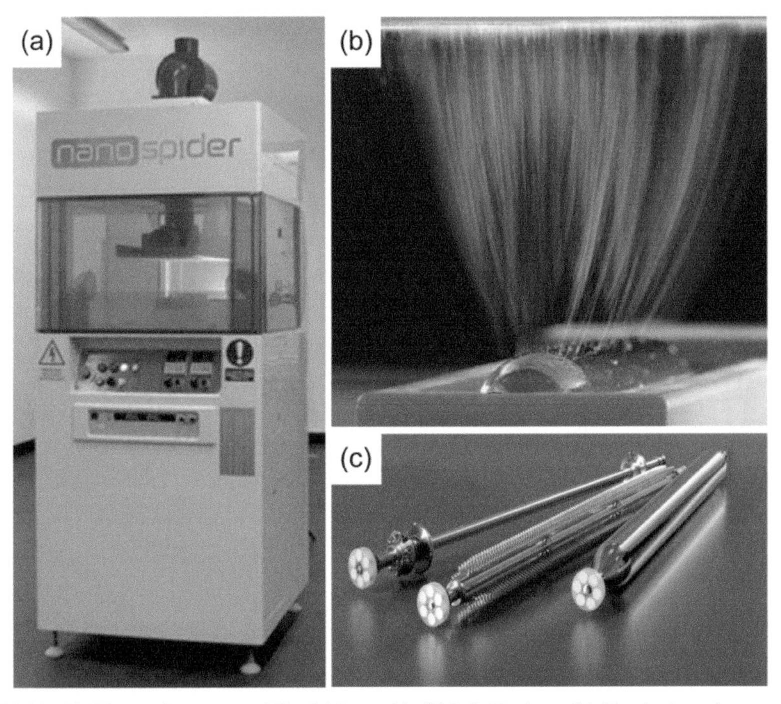

Fig. 10.13: (a) Elmarco's commercialized Nanospider™ Lab Product. (b) E-spinning photograph of the rotary cylinder spinneret. Reproduced with permission (Jirsak et al. 2009). (c) Several types of the spinnerets.

E-spinning with 0.1 ~ 1 g h^{-1}. Beside PVA (Jirsak et al. 2009), PVA composites with carbon nanotubes (CNTs) (Kostakova et al. 2009), polyamic acid (Jirsak et al. 2010), and PEO (Wang et al. 2013) have also been E-spun into fibers by this method. This set-up has been commercialized by Elmarco Co. with the brand name "Nanospide™" (Fig. 10.13a), and three kinds of industrial spinnerets are provided (Fig. 10.13c):

cylinder spinneret (Jirsak et al. 2009, Wang et al. 2013), rotary needles spinneret (Soukup et al. 2012), and rotary wires spinneret (Forward and Rutledge 2012).

Similar to the design of cylinder spinneret, rotating cylinder, ball, and disk spinnerets have been analyzed by Niu et al. (Niu et al. 2009). The rim radius of cylinder spinneret can reduce the discrepancy of electric field intensity and influence the fiber productivity. Thinner disk spinnerets increased the electric field intensity, leading to finer nanofibers and higher throughput. Ball spinnerets generated evenly distributed electric field, but failed to E-spun nanofibers when the diameters were below 60 mm.

In 2012, a needleless E-spinning technique using a rotating spiral wire coil as a spinneret (Fig. 10.14) was reported by Wang et al. to prepare PVA and PAN ultrathin fibers (Wang et al. 2012). Compared to the conventional needle E-spinning, this spiral coil E-spinning produced finer fibers with a narrower diameter distribution and had a PAN fiber production rate of 9.42 g h^{-1} which was much higher than 0.21 g h^{-1} of single jet E-spinning. Han et al. studied the interactive impact of spiral coil E-spinning process parameters (polymer solution concentration, spinning distance, applied electric voltage, and wire diameter) on E-spun fiber morphology (Han et al. 2014). Niu et al. compared the E-spinning process, electric field intensity, fiber diameter distribution, and fiber productivity of cylinder, disk, and spiral coil, respectively (Han et al. 2014).

Liu et al. presented a novel technique using needle-disk as spinneret to enhance fiber throughput and maintain high quality nanofibers (as shown in Fig. 10.15a, b) (Liu et al. 2016). The needle-disk spinneret makes it easier to initiate charged jets and

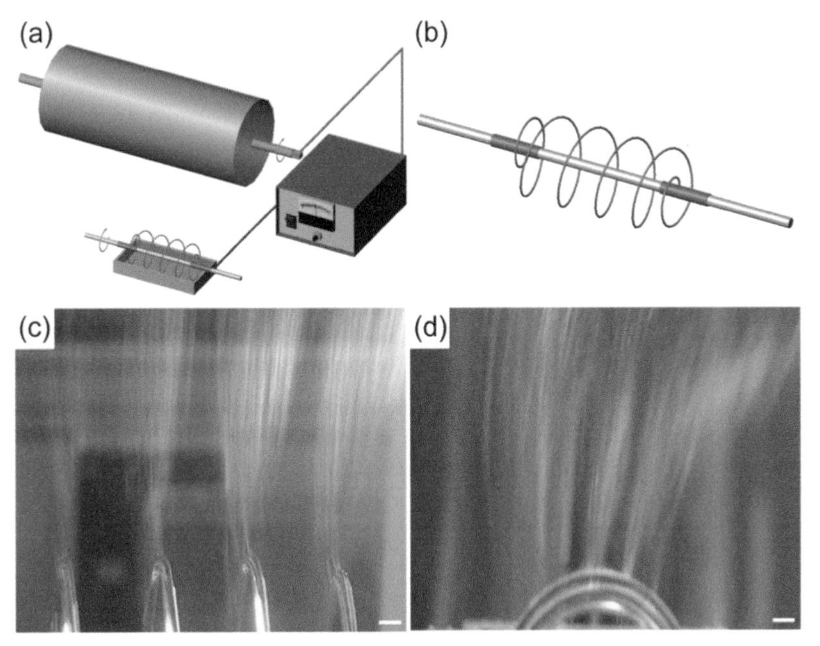

Fig. 10.14: (a) Schematics of spiral coil E-spinning set-up; (b) magnified view of the coil; photos of spiral coil spinning processes; (c) front view and (d) side view. Scale bar = 1 cm. Reproduced with permission (Wang et al. 2012). Copyright 2012, Association for Computing Machinery.

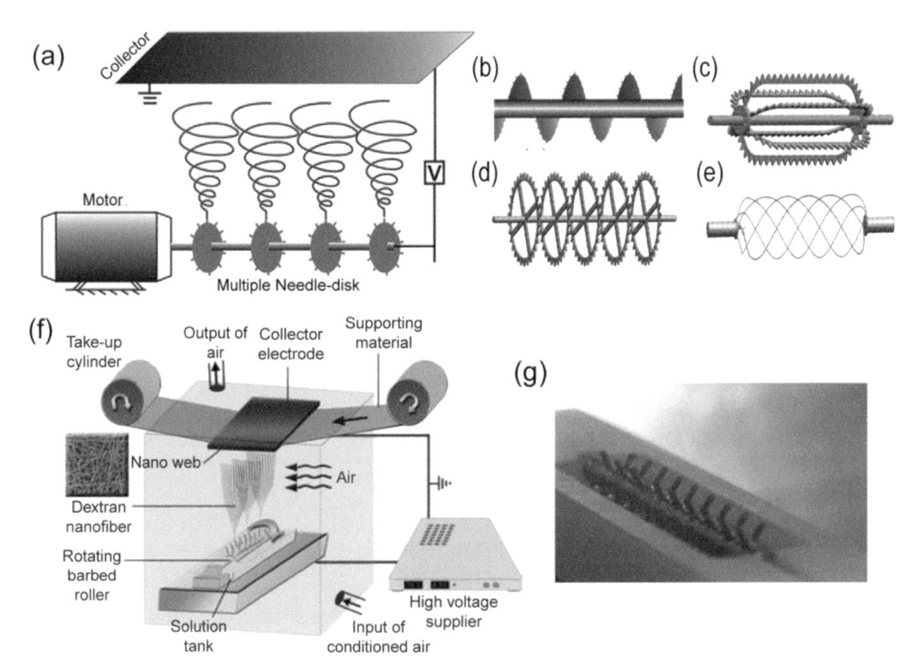

Fig. 10.15: Schematic of needle-disk E-spinning apparatus (a). Reproduced with permission (Liu et al. 2016). Copyright 2016, Elsevier Ltd. needle-helix slice spinneret. (b); arched rack spinneret (c); circular rack spinneret (d); continuous helix rack spinneret (e); schematic diagram of the barbed roller E-spinning method (f). Reproduced with permission (Liu 2015); barbed roller (g).

shows stable spinning process. During the E-spinning, electric field concentrated on the tip of every needle and same parameters of every needle make the spinneret process more controllable. Figure 10.15c–e shows the other promising types of spinnerets for rotary electrode E-spinning, which utilizes rack spinneret to efficiently generate spinning jets (Liu 2015). Similarly, a barbed roller (a total of 54 barbs) was used for transferring charges to the surface to improve the fiber spinning process (as shown in Fig. 10.15f–g) (Cengizçallıoğlu 2014). As another modified roller E-spinning, this system has the throughput of 0.6726 g min^{-1} m^{-1} and the average diameter is about 162 nm. However, the maintenance of consistent solution concentration and viscosity is worth consideration due to the open solution reservoir.

Although the E-spinning set-ups with rotary E-spinning spinnerets partially immersed into the polymer solution have simpler structures, the flow rate of polymer solution cannot be controlled precisely. Moreover, the polymer concentration will increase gradually because of the open design of the solution container, and cannot protect the solvent of polymer solution from evaporation. Splashing needleless E-spinning was first reported by Tang et al. (Tang et al. 2010) in 2010. This set-up introduced a novel design in the solution supplying system. This mode of solution supplying was classified as upper solution supplying, which was different from Nanospider's (Jirsak et al. 2009) inferior or Varabhas's (Varabhas et al. 2008) inner solution supplying modes in previous designs. The production rate of this device is 24–45 times larger than that of the single needle E-spinning system. Lu et al.

reported a needleless E-spinning apparatus, using an electriferous rotating cone as the spinneret (Lu et al. 2010). The polymer solution continuously fed on the surface of the cone through a tube. When high voltage was supplied, numerous jets generated from the rim of the cone. The PVP fiber production throughput of this approach was about 10 g min^{-1} with a diameter of 220–320 nm, which is 950 times faster than that of the single-needle E-spinning system (Liu et al. 2014). These two methods inspire the researchers to feed the polymer solution on the rotary spinnerets by a pump instead of partially immersed into the polymer solution. Thus the researchers can protect the polymer solution effectively and control the flow rate precisely under the premise of high productivity.

In addition, Long et al. (Liu et al. 2014, Long et al. 2012) reported an interesting design of efficient needleless E-spinning, in which circular rotary bead-wire driven by DC motor was used as needleless electrode (as shown in Fig. 10.16). When applied high voltage, numerous solution jets are generated from the bead-wire electrode through wrapping the electrode with solution from the solution brush. This method not only efficiently enhances the production rate of E-spinning, but also avoids the heterogeneous electric field distribution and solution clogging.

Based on the above, the needleless E-spinning has shown great potentials in large-scale production of nanofibers for both industrial production and laboratory research. Nevertheless, there has been significantly less industrial transposition

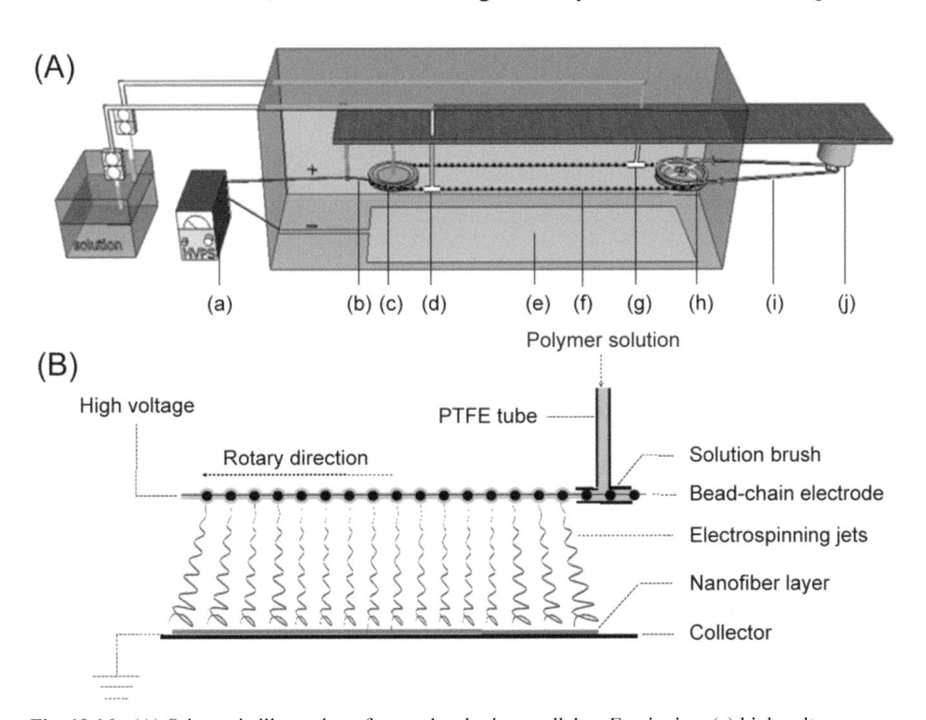

Fig. 10.16: (A) Schematic illustration of rotary bead-wire needleless E-spinning: (a) high voltage power supply (HVPS), (b) electric brush, (c) single groove pulley, (d) left solution brush, (e) grounded collector, (f) bead-wire electrode, (g) right solution brush, (h) double groove pulley, (i) belt, and (j) DC motor; (B) a schematic illustration of E-spinning process. Reproduced with permission (Liu et al. 2014). Copyright 2014, Informa UK Limited.

of this technique compared to melt blowing and melt spinning. On one hand, the inherent challenge of achieving sufficient economies of scale has been a key factor in the commodity fiber industry. On the other hand, for example, even the most commercialized Nanospider™ has drawbacks, such as unstable jets and electric field intensity around the spinnerets, which leads to the limitation of this technology in industry. To our knowledge, the formation of multiple jets from needleless E-spinning has been demonstrated that the waves of electrically conductive liquid self-organize in mesoscopic scale and finally form jets when the applied electric field intensity is above a critical value (Lukas et al. 2008). Therefore, the formation and maintenance of charged jets in needleless E-spinning will be highly influenced by the geometry of the spinneret. Furthermore, like multi-nozzles E-spinning, the negative impact of jet-to-jet interactions increases the compression in the E-spinning jet cone. Therefore, further studies are still needed for needleless E-spinning, although this method is very promising.

Mass Production of Ultrathin Fibers by Melt E-spinning

The above discussed large-scale production techniques are mainly based on solution E-spinning. Melt E-spinning, as an ecofriendly spinning technique which produces fibers from polymer melt, has drawn much attention in recent years (Zhang et al. 2016). Moreover, melt E-spun fibers provide opportunities without any residual solvent in many areas such as tissue engineering, wound dressings, filtration, and textiles. Mass production of ultra-thin fibers via melt E-spinning is very important to fulfill their potential applications.

Several research groups have proposed different technique designs to improve fiber throughput (Erisken et al. 2008, Fang et al. 2012, He et al. 2004, Komárek and Martinová 2010, Li et al. 2014a, Lyons et al. 2004, Shimada et al. 2010, Yang and Li 2014). For instance, Shimada et al. used a customized linear laser source to heat a membrane, by which a line of Taylor cones was produced and then a row of jets was generated, as shown in Fig. 10.17a (Shimada et al. 2010). Compared to point laser source, this design could increase the fiber throughput, but the yield was still low. Komarek and Martinova suggested a rod style spinning head and a slot-shaped spinning head (Fig. 10.17b) (Komárek and Martinová 2010). This slot-shaped spinning head may combine a screw continuous extrusion device to increase the melt distribution at the slot and the number of E-spun fibers. As shown in Fig. 10.17c, Fang et al. reported a disk as a melt E-spinning device for mass production (Fang et al. 2012). However, this device was only suitable for polymer melts with very low viscosity. Particularly, a multi-needle melt E-spinning apparatus was produced by ITA Aachen of Germany (Fig. 10.17d) (Hacker et al. 2009, Hutmacher and Dalton 2011). However, the weaknesses for this device are of not very high efficiency, high cost, and complex hot runner. In addition, He et al. proposed a vibration melt E-spinning device which applied high frequency shear force field to the fluid to reduce the fluid flow viscosity (He et al. 2004). Lyons et al. (Lyons et al. 2004) and Erisken et al. (Erisken et al. 2008) tried using single-screw and twin-screw to melt e-spin polymer microfibers. However, complex designing was involved in these set-ups to avoid electrical interference between heating system and high voltage spinning system.

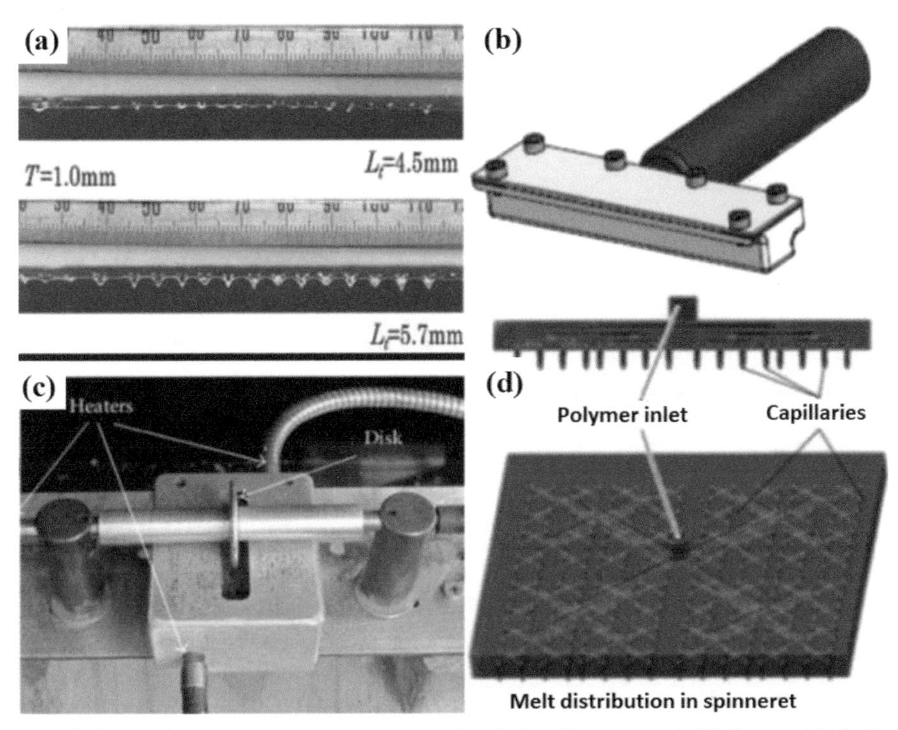

Fig. 10.17: (a) Picture of linear laser melt E-spinning device (Shimada et al. 2010), copyright 2010, with permission from Wiley Periodicals, Inc; (b) Schematic illustration of slot-shaped melt E-spinning device (Komárek and Martinová 2010); (c) Picture of needleless melt E-spinning set-up (Fang et al. 2012), copyright 2012, with permission from the authors; (d) Schematic diagram of a multi-needle melt E-spinning apparatus (Hutmacher and Dalton 2011), copyright 2011, with permission from Wiley-VCH Verlag GmbH & Co. KGa.

Recently, Yang's research group proposed a needleless melt E-spinning device using an umbellate spinneret for mass production (Li et al. 2014a, Li et al. 2014b, Li et al. 2014c, Yang and Li 2014, Yong et al. 2012). As shown in Fig. 10.18, the device was mainly consisted of melt inlet, melt runner, and differential nozzle. And the differential nozzle was an umbrella-like and cone-shaped nozzle. The use of umbrella spinneret avoided fluid plugging and high maintenance cost caused by simple combination of multiple needles. An umbrella spinneret with bottom rim diameter of 10 mm could yield almost 30 jets. The productivity of umbrella spinneret was improved efficiently, contrasted with conventional melt E-spinning set-up. However, the influence or interference of different umbrella spinnerets could not be avoided, and the stability of jetting would be affected at the same time. In addition, different polymers such as polypropylene (PP), polylacticacid (PLA), and polycaprolactone (PCL) were successfully melted and E-spun into fibers by this apparatus. The emergence of melt E-spinning method with umbrella spinneret provides powerful driving force for the study of mass production by melt E-spinning, which will provide opportunities in many applications, such as water filtration and marine oil-spill cleanup (Li et al. 2014a, Li et al. 2014b, Li et al. 2014c). Although melt E-spinning is considered a safe, cost effective, and environmental friendly

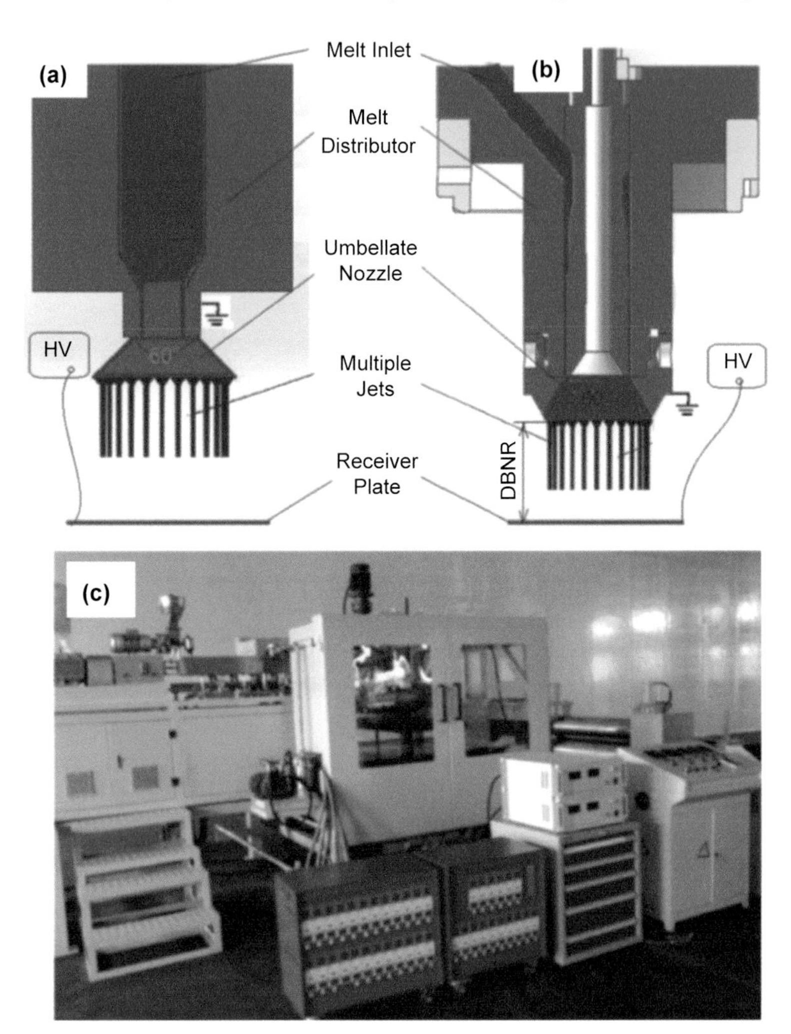

Fig. 10.18: Structure diagram of the two melt differential nozzels (a–b) and picture of the pilot line prototype (c) of the needleless polymer melt differential E-spinning apparatus (Yang and Li 2014), copyright 2014, with permission from IOP Publishing Ltd.

technique, one of the drawbacks or challenges of this technique is that the average diameter of melt E-spun fibers is too large, ranging from several hundred nanometers to dozens of microns (Qin et al. 2015, Yan et al. 2017, Yan et al. 2016). In order to bring out the nano effects of melt E-spun fibers, it is very important to decrease their average diameter to less than 200 nm, and even less than 100 nm.

Summary and Outlook

In summary, the examples illustrated in this chapter have shown that E-spinning can be a promising technology for the mass production of continuous polymeric nanofibers. Efforts to increase nanofiber throughput from polymer solutions or melts are focused on multi-jet processes from multi nozzles, or 1D, 2D, and even 3D surfaces (electrodes). Although exciting advances have been made in large-scale production of ultra-thin fibers via solution E-spinning and melt E-spinning, compared to conventional fiber production technologies, further researches are still needed to improve nanofiber throughput and decrease nanofiber diameter down to tens of nanometers. With this ability in hand, the challenge is to incorporate the tailor-made nanofibers into woven, knitted, and braided fabrics using weaving, knitting, and braiding processes, respectively. It is expected that nanofibers will find applications and show enhanced performance in fields where traditional fibers are being used, but also in some new fields, such as tissue engineering and reinforced composites.

Acknowledgements

This work was supported by the National Natural Science Foundation of China (51673103 and 51373082), the Taishan Scholars Program of Shandong Province, China (ts20120528), the Key Research and Development Plan of Shandong Province, China (2016GGX102011), and the Postdoctoral Scientific Research Foundation of Qingdao.

References

Bocanegra, R., Galán, D., Márquez, M., Loscertales, I. G. and Barrero, A. 2005. Multiple electrosprays emitted from an array of holes. Journal of Aerosol Science 36(12): 1387–1399.

Bowman, J., Taylor, M., Sharma, V., Lynch, A. and Chadha, S. 2002. Multispinneret methodologies for high throughput electrospun nanofiber. MRS Proceedings 752.

Burger, C., Hsiao, B. S. and Chu, B. 2006. Nanofibrous materials and their applications. Annual Review of Materials Research 36(1): 333–368.

Cengizçallıoğlu, F. 2014. Dextran nanofiber production by needleless electrospinning process. E-Polymers 14(1): 5–13.

Dong, R. H., Qin, C. C., Qiu, X., Yan, X., Yu, M., Cui, L. et al. 2015. *In situ* precision electrospinning as an effective delivery technique for cyanoacrylate medical glue with high efficiency and low toxicity. Nanoscale 7(46): 19468.

Dong, R. H., Jia, Y. X., Qin, C. C., Zhan, L., Yan, X., Cui, L. et al. 2016. *In situ* deposition of a personalized nanofibrous dressing via a handy electrospinning device for skin wound care. Nanoscale 8(6): 3482–3488.

Dosunmu, O. O., Chase, G. G., Kataphinan, W. and Reneker, D. H. 2006. Electrospinning of polymer nanofibres from multiple jets on a porous tubular surface. Nanotechnology 17(4): 1123.

Erisken, C., Kalyon, D. M. and Wang, H. 2008. A hybrid twin screw extrusion/electrospinning method to process nanoparticle-incorporated electrospun nanofibres. Nanotechnology 19(16): 165302.

Fang, J., Sutton, D., Sutton, D., Wang, X. and Lin, T. 2012. Needleless melt-electrospinning of polypropylene nanofibres. Journal of Nanomaterials 2012(3): 16.

Forward, K. M. and Rutledge, G. C. 2012. Free surface electrospinning from a wire electrode. Chemical Engineering Journal 183(3): 492–503.

Hacker, C., Jungbecker, P. A., Seide, G. H., Gries, T., Hassounah, I., Thomas, H. et al. 2009. Electrospinning of polymer melt: steps towards an upscaled multi-jet process, in: 80 Years department of textiles: Proceedings international conference Latest advances in high tech textiles and textile-based materials, Universiteit Gent (Hrsg.), 23–25 September 2009, Het Pand, Ghent, Belgium, S. 71–76.

Han, W., Nurwaha, D., Li, C. and Wang, X. 2014. Free surface electrospun fibers: The combined effect of processing parameters. Polymer Engineering & Science 54(1): 189–197.

He, J. H., Wan, Y. Q. and Yu, J. Y. 2004. Application of vibration technology to polymer electrospinning. International Journal of Nonlinear Sciences & Numerical Simulation 5(3): 253–262.

He, J. H., Liu, Y., Xu, L., Yu, J. Y. and Sun, G. 2008. BioMimic fabrication of electrospun nanofibers with high-throughput. Chaos Solitons & Fractals 37(3): 643–651.

He, J. H., Kong, H. Y., Yang, R. R. and Hao, D. 2012. Review on fiber morphology obtained by bubble electrospinning and blown bubble spinning. Thermal Science 16(5): 1263–1279.

He, X. X., Zheng, J., Yu, G. F., You, M. H., Yu, M., Ning, X. et al. 2017. Near-field electrospinning: progress and applications. Journal of Physical Chemistry C 121(16).

Higham, A. K., Tang, C., Lry, A. M., Pridgeon, M. C., Lee, E. M., Andrady, A. L. et al. 2014. Foam electrospinning: a multiple jet, needle-less process for nanofiber production. Aiche Journal 60(4): 1355–1364.

Holopainen, J., Penttinen, T., Santala, E. and Ritala, M. 2014. Needleless electrospinning with twisted wire spinneret. Nanotechnology 26(2): 025301.

Hutmacher, D. W. and Dalton, P. D. 2011. Melt electrospinning. Chemistry—An Asian Journal 6(1): 44–56.

Jiang, G., Zhang, S. and Qin, X. 2013. High throughput of quality nanofibers via one stepped pyramid-shaped spinneret. Materials Letters 106(106): 56–58.

Jirsak, O., Sanetrnik, F., Lukas, D., Kotek, V., Martinova, L. and Chaloupek, J. 2009. Method of nanofibres production from a polymer solution using electrostatic spinning and a device for carrying out the method. US Patent, 7585437.

Jirsak, O., Sysel, P., Sanetrnik, F., Hruza, J. and Chaloupek, J. 2010. Polyamic acid nanofibers produced by needleless electrospinning. Journal of Nanomaterials 2010(1): 49.

Kim, G. H., Cho, Y. S. and Wan, D. K. 2006. Stability analysis for multi-jets electrospinning process modified with a cylindrical electrode. European Polymer Journal 42(9): 2031–2038.

Kim, H. Y. 2005. A bottom-up electrospinning devices, and nanofibers prepared by using the same, KR Patent, 2005, WO2005073441.

Kim, H. Y. and Park, J. C. 2008. Conjugate electrospinning devices, conjugate nonwoven and filament comprising nanofibers prepared by using the same, KR Patent, 2007, WO2007035011.

Komárek, M. and Martinová, L. 2010. Design and evaluation of melt-electrospinning electrodes. Antennas and Propagation Society International Symposium, 1997. IEEE., 1997 Digest (pp. 1708–1711 vol. 1703).

Kostakova, E., Meszaros, L. and Gregr, J. 2009. Composite nanofibers produced by modified needleless electrospinning. Materials Letters 63(28): 2419–2422.

Li, H. Y., Bubakir, M. M., Xia, T., Zhong, X. F., Ding, Y. M. and Yang, W. M. 2014a. Mass production of ultra-fine fibre by melt electrospinning method using umbellate spinneret. Materials Research Innovations 18(sup4): S4-921-S924-925.

Li, H., Wu, W., Bubakir, M. M., Chen, H., Zhong, X., Liu, Z. et al. 2014b. Polypropylene fibers fabricated via a needleless melt-electrospinning device for marine oil-spill cleanup. Journal of Applied Polymer Science 131(7): 2540–2540.

Li, X., Zhang, Y., Li, H., Chen, H., Ding, Y. and Yang, W. 2014c. Effect of oriented fiber membrane fabricated via needleless melt electrospinning on water filtration efficiency. Desalination 344(344): 266–273.

Liu, M., Duan, X. P., Li, Y. M., Yang, D. P. and Long, Y. Z. 2017. Electrospun nanofibers for wound healing. Materials Science & Engineering C Materials for Biological Applications 76: 1413.

Liu, S. L., Huang, Y. Y., Zhang, H. D., Sun, B., Zhang, J. C. and Long, Y. Z. 2014. Needleless electrospinning for large scale production of ultrathin polymer fibres. Materials Research Innovations 18(sup4): S4-833-S834-837.

Liu, Y. and He, J. H. 2007. Bubble electrospinning for mass production of nanofibers. International Journal of Nonlinear Sciences & Numerical Simulation 8(3): 393–396.

Liu, Y. B. 2015. A saw-toothed needleless electrospinning device.

Liu, Z., Chen, R. and He, J. 2016. Active generation of multiple jets for producing nanofibres with high quality and high throughput. Materials & Design 94: 496–501.

Long, Y. Z., Liu, S. L., Sun, B., Zhang, H. D. and Cao, G. Q. 2012. An electrospinning apparatus for large-scale fabrication of micro-/nanoscale fibers.

Lu, B., Wang, Y., Liu, Y., Duan, H., Zhou, J., Zhang, Z. et al. 2010. Superhigh-throughput needleless electrospinning using a rotary cone as spinneret. Small 6(15): 1612–1616.

Lukas, D., Sarkar, A. and Pokorny, P. 2008. Self-organization of jets in electrospinning from free liquid surface: A generalized approach. Journal of Applied Physics 103(8): 084309-084309-084307.

Lyons, J., Li, C. and Ko, F. 2004. Melt-electrospinning part I: processing parameters and geometric properties. Polymer 45(22): 7597–7603.

Niu, H., Lin, T. and Wang, X. 2009. Needleless electrospinning. I. A comparison of cylinder and disk nozzles. Journal of Applied Polymer Science 114(6): 3524–3530.

Niu, H., Wang, X. and Lin, T. 2012. Upward needless electrospinning of nanofibres. Journal of Engineered Fabrics & Fibers 7(3): 17–22.

Paruchuri, S. and Brenner, M. P. 2007. Splitting of a liquid jet. Physical Review Letters 98(13).

Qin, C. C., Duan, X. P., Wang, L., Zhang, L. H., Yu, M., Dong, R. H. et al. 2015. Melt electrospinning of poly(lactic acid) and polycaprolactone microfibers by using a hand-operated Wimshurst generator. Nanoscale 7(40): 16611–16615.

Shimada, N., Tsutsumi, H., Nakane, K., Ogihara, T. and Ogata, N. 2010. Poly(ethylene-co-vinyl alcohol) and Nylon 6/12 nanofibers produced by melt electrospinning system equipped with a line-like laser beam melting device. Journal of Applied Polymer Science 116(5): 2998–3004.

Simm, W., Gosling, C., Bonart, R. and Von Falkai, B. 1979. Fibre fleece of electrostatically spun fibres and methods of making same. US.

Soukup, K., Petráš, D., Topka, P., Slobodian, P. and Šolcová, O. 2012. Preparation and characterization of electrospun poly(p-phenylene oxide) membranes. Molecular & General Genetics Mgg 193(1): 165–171.

Sun, B., Long, Y. Z., Zhang, H. D., Li, M. M., Duvail, J. L., Jiang, X. Y. et al. 2014. Advances in three-dimensional nanofibrous macrostructures via electrospinning. Progress in Polymer Science 39(5): 862–890.

Tang, S., Zeng, Y. and Wang, X. 2010. Splashing needleless electrospinning of nanofibers. Polymer Engineering & Science 50(11): 2252–2257.

Theron, S. A., Yarin, A. L., Zussman, E. and Kroll, E. 2005. Multiple jets in electrospinning: experiment and modeling. Polymer 46(9): 2889–2899.

Thoppey, N. M., Bochinski, J. R., Clarke, L. I. and Gorga, R. E. 2010. Unconfined fluid electrospun into high quality nanofibers from a plate edge. Polymer 51(21): 4928–4936.

Thoppey, N. M., Bochinski, J. R., Clarke, L. I. and Gorga, R. E. 2011. Edge electrospinning for high throughput production of quality nanofibers. Nanotechnology 22(34): 345301.

Thoppey, N. M., Gorga, R. E., Bochinski, J. R. and Clarke, L. I. 2012. Effect of solution parameters on spontaneous jet formation and throughput in edge electrospinning from a fluid-filled bowl. Macromolecules 45(16): 6527–6537.

Tomaszewksi, W. and Szadkowski, M. 2005. Polymeric nanofibers via flat spinneret electrospinning. Fibers Textiles Eastern Eur. 13: 22.

Varabhas, J. S., Chase, G. G. and Reneker, D. H. 2008. Electrospun nanofibers from a porous hollow tube. Polymer 49(19): 4226–4229.

Vaseashta, A. 2007. Controlled formation of multiple Taylor cones in electrospinning process. Applied Physics Letters 90(9): 093115-093115-093113.

Wang, X., Niu, H., Lin, T. and Wang, X. 2009. Needleless electrospinning of nanofibers with a conical wire coil. Polymer Engineering & Science 49(8): 1582–1586.

Wang, X., Niu, H., Wang, X. and Lin, T. 2012. Needleless electrospinning of uniform nanofibers using spiral coil spinnerets. Journal of Nanomaterials 2012(10): 3.

Wang, X., Hu, X., Qiu, X., Huang, X., Wu, D. and Sun, D. 2013. An improved tip-less electrospinning with strip-distributed solution delivery for massive production of uniform polymer nanofibers. Materials Letters 99(20): 21–23.

Yamashita, Y., Ko, F., Tanaka, A. and Miyake, H.. 2007a. Characteristics of elastomeric nanofiber membranes produced by electrospinning. Journal of Textile Engineering 53(4): 137–142.

Yamashita, Y., Miyake, H. and Higashiyama, A. 2007b. Practical use of nanofiber made by electrospinning process [C]. Proceedings of the 9th Asian Textile Conference, Taiwan, China.

Yamashita, Y., Ko, F., Miyake, H. and Higashiyama, A. 2008. Establishment of nanofiber preparation technique by electrospinning. Sen-ito Kogyo 64(1): 24–28.

Yan, X., Yu, M., Zhang, L. H., Jia, X. S., Li, J. T., Duan, X. P. et al. 2016. A portable electrospinning apparatus based on a small solar cell and a hand generator: design, performance and application. Nanoscale 8(1): 209.

Yan, X., Duan, X. P., Yu, S. X., Li, Y. M., Lv, X., Li, J. T. et al. 2017. Portable melt electrospinning apparatus without an extra electricity supply. Rsc Advances 7(53): 33132–33136.

Yang, R., He, J., Xu, L. and Yu, J. 2009. Bubble-electrospinning for fabricating nanofibers. Polymer 50(24): 5846–5850.

Yang, R. R., He, J. H., Yu, J. Y. and Xu, L. 2010. Bubble-electrospinning for fabrication of nanofibers with diameter of about 20 nm. International Journal of Nonlinear Sciences & Numerical Simulation 11(Supplement): 163–164.

Yang, W. and Li, H. 2014. Principle and equipment of polymer melt differential electrospinning preparing ultrafine fiber. IOP Conf. Series: Materials Science and Engineering 64: 012013–012017.

Yang, Y., Jia, Z., Li,.Q., Hou, L., Gao, H. and Wang, L. 2006. 8th International Conference on Properties and Applications of Dielectric Materials, Bali, 2006, p. 940.

Yarin, A. L. and Zussman, E. 2004. Upward needleless electrospinning of multiple nanofibers. Polymer 45(9): 2977–2980.

Yong, L., Zhao, F., Chi, Z. and Zhang, J. 2012. Solvent-free preparation of polylactic acid fibers by melt electrospinning using umbrella-like spray head and alleviation of problematic thermal degradation. Journal of the Serbian Chemical Society 77(8): 1071–1082.

Zhang, B., Yan, X., He, H. W., Yu, M., Ning, X. and Long, Y. Z. 2017. Solvent-free electrospinning: opportunities and challenges. Polymer Chemistry 8(1): 333–352.

Zhang, L. H., Duan, X. P., Yan, X., Yu, M., Ning, X., Zhao, Y. et al. 2016. Recent advances in melt electrospinning. RSC Advances 6(58): 53400–53414.

Zheng, Y., Liu, X. and Zeng, Y. 2013. Electrospun nanofibers from a multihole spinneret with uniform electric field. Journal of Applied Polymer Science 130(5): 3221–3228.

Zhou, F. L., Gong, R. H. and Porat, I. 2009a. Polymeric nanofibers via flat spinneret electrospinning. Polymer Engineering & Science 49(12): 2475–2481.

Zhou, F. L., Gong, R. H. and Porat, I. 2009b. Three-jet electrospinning using a flat spinneret. Journal of Materials Science 44(20): 5501–5508.

Zhou, F. L., Gong, R. H. and Porat, I. 2010. Needle and needleless electrospinning for nanofibers. Journal of Applied Polymer Science 115(5): 2591–2598.

Index